Nuclear Facilities

Related titles

Nuclear fuel cycle science and engineering
(ISBN 978-0-85709-073-7)

Nuclear decommissioning
(ISBN 978-0-85709-115-4)

Nuclear energy: An introduction to the concepts, systems, and applications of nuclear processes, 7e
(ISBN 978-0-12416-654-7)

Woodhead Publishing Series in Energy:
Number 112

Nuclear Facilities

A Designer's Guide

Bill Collum

ELSEVIER

AMSTERDAM • BOSTON • CAMBRIDGE • HEIDELBERG
LONDON • NEW YORK • OXFORD • PARIS • SAN DIEGO
SAN FRANCISCO • SINGAPORE • SYDNEY • TOKYO
Woodhead Publishing is an imprint of Elsevier

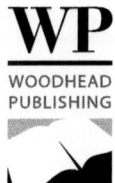

WP
WOODHEAD
PUBLISHING

Woodhead Publishing is an imprint of Elsevier
The Officers' Mess Business Centre, Royston Road, Duxford, CB22 4QH, United Kingdom
50 Hampshire Street, 5th Floor, Cambridge, MA 02139, United States
The Boulevard, Langford Lane, Kidlington, OX5 1GB, United Kingdom

Notices
Knowledge and best practice in this field are constantly changing. As new research and experience broaden our understanding, changes in research methods, professional practices, or medical treatment may become necessary.

Practitioners and researchers must always rely on their own experience and knowledge in evaluating and using any information, methods, compounds, or experiments described herein. In using such information or methods they should be mindful of their own safety and the safety of others, including parties for whom they have a professional responsibility.

To the fullest extent of the law, neither the Publisher nor the authors, contributors, or editors, assume any liability for any injury and/or damage to persons or property as a matter of products liability, negligence or otherwise, or from any use or operation of any methods, products, instructions, or ideas contained in the material herein.

Library of Congress Cataloging-in-Publication Data
A catalog record for this book is available from the Library of Congress

British Library Cataloguing-in-Publication Data
A catalogue record for this book is available from the British Library

ISBN: 978-0-08-101938-2 (print)
ISBN: 978-0-08-101939-9 (online)

For information on all Woodhead publications
visit our website at https://www.elsevier.com/

Working together
to grow libraries in
developing countries

www.elsevier.com • www.bookaid.org

Publisher: Joe Hayton
Acquisition Editor: Cari Owen
Editorial Project Manager: Charlotte Kent
Production Project Manager: Jason Mitchell
Cover Designer: Mark Rogers

Typeset by SPi Global, India

Dedication

This book is dedicated to my wife Maggie, who kept the
coffee flowing and uncomplainingly shared our home with this book
while patiently waiting for me to finish it.
It only took 6 years!

Chapters

1   Nuclear fuel cycle   1

2   Radiation   45

3   Radiological zoning   61

4   Radiological changerooms   77

5   Structural   105

6   Process engineering   139

7   Mechanical engineering   185

8   Ventilation   231

9   Cranes   275

10   Electrical   313

11   Radiometric instruments   349

12   Project planning   369

13   Waste management   395

14   Safety   405

15   Decommissioning planning   455

16   Future-proofing   505

17   Design development   533

Contents

List of figures   xxv
Author biography   xxxv
Acknowledgments   xxxvii
Author's note   xxxix
Woodhead Publishing Series in Energy   xli
Introduction   xlvii

1 Nuclear fuel cycle   **1**
    1.1   Uranium mining and purification   3
    1.2   The atom   5
    1.3   Enrichment   7
    1.4   Fuel fabrication   9
    1.5   Nuclear reactors   11
          Heat exchanger   13
    1.6   Nuclear reaction   16
          1.6.1   Neutron capture   16
          1.6.2   Nuclear fission   17
          1.6.3   Half-life   18
    1.7   Control rods   18
    1.8   Burnable poison   19
    1.9   Neutron activation   19
    1.10 Decay chain   22
    1.11 Plutonium creation   22
    1.12 MOX fuel   23
    1.13 Fast breeder fuel   25
    1.14 Spent fuel removal   26
          1.14.1 Charge floors   26
          1.14.2 Cooling ponds   28
    1.15 Spent fuel routing   29
    1.16 Reprocessing   32
    1.17 High level waste   35
          1.17.1 HLW conditioning and packaging   35
          1.17.2 HLW interim storage   35

1.18	Intermediate level waste	**36**
	1.18.1 ILW conditioning and packaging	**36**
	1.18.2 ILW interim storage	**39**
1.19	Permanent disposal	**41**
	1.19.1 HLW and ILW	**41**
	1.19.2 LLW	**42**

2 Radiation — **45**

2.1	Radiation and contamination	**45**
2.2	Electromagnetic spectrum	**46**
2.3	Nonionizing radiation	**47**
2.4	Ionizing radiation	**48**
	2.4.1 Alpha	**48**
	2.4.2 Beta	**49**
	2.4.3 Gamma	**50**
2.5	Exposure to radiation	**50**
	2.5.1 Background radiation	**50**
	2.5.2 Dose equivalent	**52**
	2.5.3 Nuclear workers	**53**
2.6	Protection from radiation	**54**
	2.6.1 Shielding	**54**
	2.6.2 Distance	**55**
2.7	Criticality	**56**
2.8	Personal dose measurement	**57**
	2.8.1 Thermoluminescent dosimeter	**57**
	2.8.2 Electronic personal dosimeter	**59**

3 Radiological zoning — **61**

3.1	Zoning rationale	**61**
3.2	Naming conventions	**62**
3.3	Radiation zones	**63**
	3.3.1 R1 (white)	**64**
	3.3.2 R2 (green)	**64**
	3.3.3 R3 (amber)	**64**
	3.3.4 R4 (red)	**65**
3.4	Contamination zones	**65**
	3.4.1 Sievert versus Becquerel	**66**
	3.4.2 C1 (white)	**66**
	3.4.3 C2 (green)	**67**
	3.4.4 C3 (amber)	**68**
	3.4.5 C4 (red)	**68**
3.5	Guiding principles	**69**
	3.5.1 Designations need not match	**69**
	3.5.2 Apply to normal operations	**69**
	3.5.3 One step at a time	**69**

3.6 Dual classifications   71
     3.6.1 Start sequence   71
     3.6.2 Monitoring—pass   72
     3.6.3 Monitoring—fail   72
3.7 Surface contamination   73
     3.7.1 Indoors   73
     3.7.2 Outdoors   74
3.8 Depicting zones   75

4 Radiological changerooms   **77**

4.1 Generic types   78
4.2 Changerooms   78
     4.2.1 Locker rooms and basics   78
     4.2.2 Access to C2 areas   79
            4.2.2.1 Monitoring and coverall areas   79
            4.2.2.2 Boot barrier—inward   82
     4.2.3 Egress from C2 areas   84
            4.2.3.1 Airflow   84
            4.2.3.2 Coveralls   85
            4.2.3.3 Boot barrier—return   85
            4.2.3.4 Monitoring   86
                     Hand monitoring   87
                     Frisking station   88
                     Installed personnel monitor   89
     4.2.4 Health physics   92
     4.2.5 Duplicating equipment   94
     4.2.6 Restrooms   94
     4.2.7 Access to outdoor radiation controlled areas   94
4.3 Sub changerooms   95
     4.3.1 Generic types   95
     4.3.2 Sub changeroom—C2/C3   96
            4.3.2.1 Clothing and protective equipment   96
            4.3.2.2 Personal monitoring devices   98
            4.3.2.3 Disposable items   98
            4.3.2.4 Containment   98
            4.3.2.5 Monitoring procedure   98
     4.3.3 Sub changeroom—C3/C4   99
            4.3.3.1 Trans-zone protocol   100
            4.3.3.2 Clothing and protective equipment   100
            4.3.3.3 Monitoring procedure   101
     4.3.4 Open sub changeroom—C2/C3 (S)   102
     4.3.5 Equipment transfer lobby   103

5 Structural   **105**

5.1 Steelwork structures   106
     5.1.1 Structural grid   106

	5.1.1.1	Grid arrangements	**107**
	5.1.1.2	Load paths	**108**
	5.1.1.3	Grid adjustments	**109**
5.1.2	Floor construction		**110**
5.1.3	Bracing		**111**
	5.1.3.1	Bracing types	**112**
	5.1.3.2	Bracing arrangements	**112**
5.1.4	Portal frame		**114**
5.2	Concrete structures		**115**
5.2.1	Precast concrete		**115**
5.2.2	In situ concrete		**117**
5.2.3	Wall penetrations		**119**
	5.2.3.1	Concrete banding	**120**
5.2.4	Phased release of structural information		**121**
5.3	Combined concrete and steel structures		**121**
5.3.1	Differential movement		**122**
5.4	Seismic analysis		**123**
5.4.1	Phased release of wall penetration information		**123**
5.4.2	Seismic evaluation		**124**
5.4.3	Response spectra		**124**
5.5	Extreme environmental events		**125**
5.5.1	Seismic		**126**
	5.5.1.1	Design basis earthquake	**126**
	5.5.1.2	Operating basis earthquake	**127**
	5.5.1.3	Seismic categories	**127**
	5.5.1.4	Displacement	**128**
	5.5.1.5	Subterranean construction	**129**
	5.5.1.6	Steelwork	**129**
5.5.2	Wind		**129**
	5.5.2.1	Building movement	**129**
	5.5.2.2	Design basis—1 in 10,000-year wind	**130**
	5.5.2.3	Protection from airborne debris	**130**
	5.5.2.4	Suction forces	**132**
	5.5.2.5	Wind simulation modeling	**133**
5.5.3	Snow		**134**
	5.5.3.1	Design basis	**134**
	5.5.3.2	Shifting snow	**134**
5.5.4	Temperature		**135**
	5.5.4.1	Design basis	**136**
	5.5.4.2	Embrittlement	**136**
5.5.5	Rain		**136**
	5.5.5.1	Flood planning	**136**
	5.5.5.2	Design basis	**137**
	5.5.5.3	Rainfall management	**137**

6   Process engineering   **139**

6.1   Closed cells   139
6.2   Mass balance   143
6.3   Feedstock analysis   146
    6.3.1   Conditions for acceptance (CFA)   146
    6.3.2   Stratification   147
    6.3.3   Organics   148
    6.3.4   Orphan streams   149
6.4   End product   149
6.5   Transfer devices   150
    6.5.1   Mechanical pumping   150
        6.5.1.1   Progressive cavity pump   150
        6.5.1.2   Centrifugal pump   150
            Suction head   151
            Double block and bleed   152
            Shielding bulge   153
        6.5.1.3   Remotely maintainable pump   154
    6.5.2   Fluidic pumping   156
        6.5.2.1   Reverse flow diverter (RFD)   156
        6.5.2.2   Fluidic pumping system   157
            Barometric head   158
            Pumping sequence   159
            RFD arrangements   159
            Propellant   160
    6.5.3   Steam ejectors   161
    6.5.4   Gravity flow systems   161
    6.5.5   Breakpot   161
    6.5.6   Transfer device selection   164
6.6   Services distribution   166
    6.6.1   Service risers   167
    6.6.2   Access   168
    6.6.3   Services hierarchy   168
    6.6.4   Joggle boxes   169
6.7   Agitation systems   170
    6.7.1   Mechanical agitation   170
        6.7.1.1   Pump-round   171
    6.7.2   Fluidic agitation   172
        6.7.2.1   Pulse jet mixer   172
    6.7.3   Air sparge   173
6.8   Overflows   174
6.9   Volume reduction   175
    6.9.1   Evaporation   175
6.10 Solids removal   177
    6.10.1   Settling   177

6.10.2	Centrifuge	**178**
6.10.3	Cross-flow filtration	**179**
	6.10.3.1 Ultrafiltration	**179**
6.11	Ion exchange (IX)	**179**
6.12	Off gas treatment	**182**
	6.12.1 Packed column	**183**

7 Mechanical engineering — **185**

7.1	Mechanical handling caves	**186**
7.2	Shielding windows	**187**
	7.2.1 Composition	**187**
	7.2.2 Construction	**188**
	7.2.3 Refraction	**189**
	7.2.4 Containment	**190**
	7.2.5 Tint and degradation	**191**
	7.2.6 Change-out	**191**
7.3	Manipulators	**192**
	7.3.1 Master-slave manipulator	**193**
	7.3.1.1 Components	**193**
	7.3.1.2 Gearing	**194**
	7.3.1.3 Human factors	**195**
	7.3.1.4 Reach	**195**
	7.3.1.5 Change-out	**196**
	7.3.2 Electrical and hydraulic manipulators	**197**
	7.3.3 Power manipulator	**199**
	7.3.3.1 Deployment	**199**
	7.3.3.2 Decontamination and maintenance	**201**
	7.3.3.3 End Effectors	**202**
7.4	Shield doors	**203**
	7.4.1 Construction	**203**
	7.4.2 Personnel access doors	**204**
	7.4.3 Pressure relief vent	**204**
	7.4.4 Maintenance access	**205**
	7.4.5 Vertical shield doors	**205**
	7.4.6 Combination vertical and horizontal shield doors	**206**
	7.4.7 Vertical shield door recovery	**207**
	7.4.8 Pivoting shield doors	**208**
	7.4.9 Concrete filled shield doors	**209**
7.5	Bogies	**209**
7.6	Decontamination	**210**
	7.6.1 Swabbing	**211**
	7.6.2 Water	**212**
	7.6.3 Submersion	**212**
	7.6.4 Dry pellets	**213**
	7.6.5 Carbon dioxide	**213**

7.7	Cave arrangements		**214**
	7.7.1	Shared crane rails	**214**
		7.7.1.1 Combined crane and power manipulator on a two girder bridge	**214**
		7.7.1.2 Combined crane and power manipulator on a four girder bridge	**215**
		7.7.1.3 Independent crane and power manipulator	**215**
		7.7.1.4 Single decontamination and maintenance bays	**215**
		7.7.1.5 Duplicate decontamination and maintenance bays	**216**
	7.7.2	Twin crane rails	**217**
	7.7.3	Cave arrangement selection	**218**
7.8	Flasks		**219**
	7.8.1	Top loading flasks	**219**
		7.8.1.1 Cuboidal and cylindrical flasks	**219**
		7.8.1.2 Impact testing	**221**
		7.8.1.3 Lifting above assigned drop height	**223**
		7.8.1.4 Fire testing	**223**
		7.8.1.5 Lid removal	**223**
	7.8.2	Bottom loading flasks	**226**
		7.8.2.1 Permanently installed gamma gates	**227**
		7.8.2.2 Mobile gamma gates	**228**
		7.8.2.3 Nappy	**228**
	7.8.3	Flask design durations	**229**

8	**Ventilation**	**231**
8.1	Role of nuclear ventilation system	**231**
8.2	Integration with radiological zoning	**233**
8.3	Cascade philosophy	**234**
	8.3.1 Reliability	**234**
	8.3.2 Air pressure	**235**
	8.3.3 Velocity	**236**
8.4	Engineered gaps	**238**
8.5	Maintaining containment at truck bays	**240**
8.6	Maintaining containment on building perimeter	**241**
8.7	Filtration	**243**
	8.7.1 HEPA filters	**244**
	8.7.2 HEPA filtration—Beta-gamma plants	**245**
	8.7.3 HEPA filtration—Alpha plants	**246**
	8.7.4 Mobile filtration unit	**247**
	8.7.5 Air in-bleed filters	**248**
	8.7.6 Manual versus remote filter change	**250**
	8.7.7 Filter life	**253**
	8.7.8 Determining when to change filters	**253**
	8.7.9 Filter disposal	**254**
	8.7.10 Push-through filters	**255**

8.8   Air conditioning   256
8.9   Heat recovery   257
8.10   Solar heat gain   258
8.11   The ventilation sequence   259
        8.11.1   Generic arrangement of ventilation plant rooms   259
        8.11.2   Ductwork distribution network   262
8.12   Air handling units   264
        8.12.1   AHU location   265
8.13   Air quality   266
8.14   Vessel ventilation   267
8.15   Gloveboxes   268
        8.15.1   Purpose   269
        8.15.2   Ventilation   269
        8.15.3   Glove ports   271
        8.15.4   Shielding   272
        8.15.5   Maintenance   273

9 Cranes   **275**

9.1   Conventional cranes and high integrity nuclear cranes   275
        9.1.1   Crane components   275
        9.1.2   Design standards   278
        9.1.3   Operation   278
        9.1.4   Lifting accessories   280
                  9.1.4.1   Spreader frame   282
                  9.1.4.2   Lifting beam   283
        9.1.5   Pintle   284
        9.1.6   Performance   284
        9.1.7   Hook approach   287
        9.1.8   Safe working load   289
        9.1.9   Annual examination   290
        9.1.10   Load limiting devices   290
        9.1.11   Protection against dropped loads   291
        9.1.12   Zoning   293
        9.1.13   Seismic performance   294
        9.1.14   Polar crane   296
9.2   In-cave cranes   297
        9.2.1   Cartesian crane   297
        9.2.2   Polar jib crane   298
        9.2.3   Modularization   299
        9.2.4   Written scheme of examination   299
        9.2.5   Recovery   300
        9.2.6   Power supply   306
                  9.2.6.1   Catenary   306
                  9.2.6.2   Cable reeling   307

		9.2.6.3	Busbars	**309**
		9.2.6.4	Drag chain	**310**
	9.2.7	Guidance systems		**311**

10 Electrical   **313**

10.1	Electricity supply			**313**
	10.1.1	Site ring main		**313**
	10.1.2	Transformer		**314**
	10.1.3	Distribution board		**316**
	10.1.4	Cable color coding		**317**
	10.1.5	Backup electricity		**318**
		10.1.5.1	Non-firm power supply	**318**
		10.1.5.2	Firm power supply	**318**
		10.1.5.3	Guaranteed interruptible power supply	**320**
		10.1.5.4	Guaranteed uninterruptible power supply (UPS)	**324**
	10.1.6	Cable handling		**326**
	10.1.7	Fire resistant cable		**326**
10.2	Control systems			**327**
	10.2.1	Control hierarchy		**327**
	10.2.2	Basic control		**328**
	10.2.3	Sequence control		**328**
	10.2.4	Motor control center (MCC)		**329**
		10.2.4.1	Direct on line (DOL)	**329**
		10.2.4.2	Soft start	**330**
		10.2.4.3	Variable frequency drive (VFD)	**330**
	10.2.5	Automation		**330**
	10.2.6	Remote operation		**332**
	10.2.7	Supervisory control and data acquisition (SCADA)		**332**
	10.2.8	Hardwiring		**334**
10.3	Instrumentation			**335**
	10.3.1	Intelligent instruments		**335**
	10.3.2	Level measurement		**335**
		10.3.2.1	Bubbler system	**336**
		10.3.2.2	Vibrating forks	**337**
		10.3.2.3	Radar systems	**338**
		10.3.2.4	Conductive sensors	**339**
		10.3.2.5	Capacitance sensors	**340**
		10.3.2.6	Ultrasonic	**340**
	10.3.3	Temperature measurement		**341**
		10.3.3.1	Resistance temperature detector (RTD)	**342**
		10.3.3.2	Thermocouple	**342**
	10.3.4	Pressure measurement		**342**
		10.3.4.1	Bourdon gauge	**344**
		10.3.4.2	Isolated diaphragm gauge	**344**

		10.3.5	Flow measurement	346
			10.3.5.1 Orifice meter	346
			10.3.5.2 Pitot tube	347

11   Radiometric instruments   **349**

11.1   Monitoring requirements   349
11.2   Detection technologies   350
       11.2.1   Gas-filled detectors   350
                11.2.1.1   Detection process   351
                11.2.1.2   Proportional counter   352
                11.2.1.3   Gas flow proportional counter   352
                11.2.1.4   Geiger-Müller (G-M) tube   352
                11.2.1.5   Gas types   353
       11.2.2   Scintillation detectors   353
                11.2.2.1   Scintillator materials   353
                11.2.2.2   Scintillation process   354
                11.2.2.3   Photomultiplier tube   354
                11.2.2.4   Measurement   355
       11.2.3   Semiconductor detectors   355
                11.2.3.1   Conductors   355
                11.2.3.2   Semiconductors   356
                11.2.3.3   Transistors   357
                11.2.3.4   Semiconductor detection process   357
                11.2.3.5   Germanium detectors   359
11.3   Technology selection   360
       11.3.1   Cost   361
       11.3.2   Resolution   361
       11.3.3   Efficiency   362
11.4   Instruments   362
       11.4.1   Health physics   363
       11.4.2   Environmental   363
       11.4.3   Assay   365
       11.4.4   Instrument technologies   366
11.5   Safeguards   367

12   Project planning   **369**

12.1   Client specification   369
12.2   Project controls   370
       12.2.1   Project control network   370
                12.2.1.1   Work breakdown structure (WBS)   372
                12.2.1.2   Cost breakdown structure (CBS)   374
                12.2.1.3   Organization breakdown structure (OBS)   374
                12.2.1.4   Integrated breakdown structures   375

		12.2.2	Project execution plan	375
		12.2.3	Risk management	376
			12.2.3.1 Risk register	376
	12.3	Project management		377
	12.4	Programming		377
		12.4.1	Nuclear factors	377
			12.4.1.1 Division of costs	378
			12.4.1.2 Regulatory constraints	379
		12.4.2	Programming logic	379
		12.4.3	Activity links	380
		12.4.4	Critical path	381
		12.4.5	Project phasing	381
		12.4.6	Hold points	382
	12.5	Project phase activities		383
		12.5.1	Project inception	383
			12.5.1.1 Investment justification	384
			12.5.1.2 Technology identification	384
			12.5.1.3 Research and development (R&D)	384
		12.5.2	Optioneering	385
			12.5.2.1 Preferred option selection	385
			12.5.2.2 Design philosophy documents	387
			12.5.2.3 Mass and energy balance	387
			12.5.2.4 Layout development	387
		12.5.3	Option refinement	387
			12.5.3.1 Refine preferred option	388
			12.5.3.2 Review criteria	388
		12.5.4	Detail design	388
			12.5.4.1 Maintaining design intent	389
			12.5.4.2 Specification, design, and manufacturing	389
		12.5.5	Construction	390
			12.5.5.1 Construction programming	391
			12.5.5.2 Phased release of information	391
			12.5.5.3 Equipment installation	392
		12.5.6	Commissioning	392
			12.5.6.1 Phased commissioning	393
	12.6	Feedback		394
13	**Waste management**			**395**
	13.1	Waste management hierarchy—conventional		395
		13.1.1	Prevent	396
		13.1.2	Minimize	396
		13.1.3	Reuse	396
		13.1.4	Recycle	397

13.1.5   Replace   397
13.1.6   Energy recovery   397
13.1.7   Disposal   398
13.2 Waste management hierarchy—radiological   398
Primary and secondary wastes   398
13.2.1   Very low level waste (VLLW)   399
13.2.2   Low level waste (LLW)   399
13.2.3   Intermediate level waste (ILW)   399
13.2.4   High level waste (HLW)   400
13.3 Evolution of a waste management strategy   400
13.3.1   Waste management—project inception   400
13.3.1.1   Waste management strategy   400
13.3.1.2   On-site waste treatment options   401
13.3.1.3   National waste management strategy   401
13.3.2   Waste management—optioneering   401
13.3.2.1   Cross-site integration   402
13.3.2.2   Waste minimization   402
13.3.3   Waste management—option refinement   402
13.3.3.1   Hierarchy fine-tuning   402
13.3.4   Waste management—detail design   403
13.3.4.1   Trials   403
13.3.4.2   Embed waste management strategy   403
13.3.5   Waste management—construction and installation   403
13.3.5.1   Waste management plan   403
13.3.6   Waste management—commissioning   404
13.3.6.1   Refine waste management plans   404

14 Safety   405
14.1 Occupational safety   405
14.1.1   Asphyxiation   406
14.1.2   Confined spaces   406
14.1.3   Noise   407
14.2 Nuclear safety   409
14.2.1   Site license conditions   410
14.3 Safety case   413
14.3.1   Safety committees   413
14.3.1.1   Nuclear safety committee   414
14.3.1.2   Site safety committee   414
14.3.2   Safety case categories   414
14.3.2.1   Category 1   415
14.3.2.2   Category 2   415
14.3.2.3   Category 3   416
14.3.2.4   Category 4   416
14.3.3   Evolution of a safety case   416

14.4   Hazard analysis studies   418
       14.4.1   Hazard appraisal   418
       14.4.2   Hazard and operability studies   419
                14.4.2.1   HAZOP 0   420
                14.4.2.2   HAZOP 1   420
                14.4.2.3   HAZOP 1.5   420
                14.4.2.4   HAZOP 2   421
       14.4.3   Hazard analysis (HAZAN)   421
                14.4.3.1   Fault schedule   422
       14.4.4   Engineering schedule   422
                14.4.4.1   Safety functional requirements   425
                14.4.4.2   Structures, systems, and components   425
                14.4.4.3   Safety performance requirements   425
                14.4.4.4   Design justification reports   425
       14.4.5   Operating rules   426
14.5   Risk   426
       14.5.1   Hazard control hierarchy   427
                14.5.1.1   Eliminate   427
                14.5.1.2   Reduce   428
                14.5.1.3   Isolate   428
                14.5.1.4   Manage   429
                14.5.1.5   Personal protective equipment   429
       14.5.2   Fault sequence   429
                14.5.2.1   Unmitigated hazards   430
                14.5.2.2   Mitigated hazards   430
       14.5.3   Beyond design basis   432
       14.5.4   Probabilistic risk assessment   433
       14.5.5   Risk classification   433
14.6   As low as reasonably practicable   436
14.7   Safety reports   438
       14.7.1   Preliminary safety report   439
       14.7.2   Preconstruction safety report   439
       14.7.3   Preinactive commissioning safety report   441
       14.7.4   Preactive commissioning safety report   442
       14.7.5   Operations safety report (OSR)   444
14.8   Periodic reviews   444
       14.8.1   Short-term reviews   445
       14.8.2   Long-term reviews   445
14.9   Safety case integration   446

15   Decommissioning planning   455
15.1   Site license conditions   455
15.2   Planning factors   456
       15.2.1   Hazard management   457
                15.2.1.1   Safety   457
                15.2.1.2   Environmental   458

	15.2.2	Learning from experience	**459**
	15.2.3	Financial	**460**
		15.2.3.1 Cost	**460**
		15.2.3.2 Funding	**461**
	15.2.4	Existing facility capability	**461**
		15.2.4.1 Refurbish	**462**
		15.2.4.2 Remodel and reuse	**464**
	15.2.5	Site zoning plan	**465**
	15.2.6	Socioeconomic	**467**
	15.2.7	Resources	**470**
	15.2.8	Program	**471**
15.3	Operations phase		**472**
	15.3.1	Post operational clean out	**473**
	15.3.2	Operations—Secondary phases	**475**
15.4	Decommissioning phases		**476**
	15.4.1	Primary phases	**476**
		15.4.1.1 Surveillance and maintenance	**476**
		15.4.1.2 Decontamination and dismantling	**477**
		VLL materials	**478**
		LL materials	**478**
		IL materials	**480**
		15.4.1.3 Care and maintenance	**482**
		15.4.1.4 Demolition	**483**
	15.4.2	Secondary phases	**484**
		15.4.2.1 Advance D&D	**485**
		15.4.2.2 Remediation	**486**
	15.4.3	Tertiary phases	**486**
		15.4.3.1 Finalize POCO	**487**
		15.4.3.2 Advance demolition	**488**
15.5	Decommissioning strategy		**489**
	15.5.1	Immediate dismantling	**490**
	15.5.2	Safe enclosure	**492**
	15.5.3	Entombment	**495**
	15.5.4	Strategy selection	**498**
15.6	Decommissioning plan		**499**
16	**Future-proofing**		**505**
16.1	Design for decommissioning (DfD)		**505**
	16.1.1	Self-perpetuating cycle	**506**
16.2	Design to remodel and reuse (DRR)		**509**
	16.2.1	Reuse—self-decommissioning	**511**
	16.2.2	Reuse—specific purpose	**512**
	16.2.3	Reuse—non-specific purpose	**513**
16.3	Integration with decommissioning strategy		**514**

16.4 Principles of DfD and DRR 515
16.4.1 Do not compromise primary operations 515
16.4.2 Simplify transition to decommissioning or reuse 516
16.4.3 Maximize use of original P&E 516
16.5 Funding future-proofing 517
16.5.1 Decommissioning cost estimate 519
16.6 Future-proofing enablers 519
16.6.1 Radiological zoning and ventilation 520
16.6.1.1 Contamination 520
16.6.1.2 Radiation 521
16.6.2 Utilities and active pipework 522
16.6.3 Remote waste handling 523
16.6.4 Fire compartmentation 525
16.6.5 Segregation 526
16.6.6 Space 528
16.7 Responsibilities 530
16.7.1 Client role 530
16.7.2 Design team role 531

17 Design development 533
17.1 Layout preparations 533
17.1.1 Building flow diagram (BFD) 534
17.1.1.1 BFD process 535
17.1.2 Constraints 538
17.1.3 Assumptions 539
17.1.4 BFD validation 539
17.1.5 BFD iterations 540
17.2 Layout development 541
17.3 Vehicle bay 541
17.3.1 Layout considerations—rail only 541
17.3.2 Layout considerations—road only 544
17.3.3 Layout considerations—rail and road 544
17.3.3.1 Buffering 544
17.3.3.2 Vehicle characteristics 545
17.3.3.3 Platforms 545
17.3.3.4 Means of escape 546
17.3.3.5 Consumables inspection 546
17.3.3.6 Local office 546
17.3.3.7 Engine fumes 546
17.3.3.8 Fire compartmentation 547
17.3.3.9 Wind turbulence 547
17.3.3.10 Floor level 547
17.3.3.11 Changeroom access 547
17.3.3.12 Crane hook approach 548
17.3.3.13 Flask lifting 549
17.3.3.14 Future-proofing 550

17.4   Ventilation stack   **551**
        17.4.1   Layout considerations   **552**
                 17.4.1.1   Flue arrangements   **552**
                 17.4.1.2   Aerial dispersal   **553**
                 17.4.1.3   Prevailing winds   **554**
                 17.4.1.4   Stack location   **555**
                            Freestanding stack   **555**
                            Through-roof stack   **556**
                 17.4.1.5   Strakes   **557**
                 17.4.1.6   C2 stub stack   **558**
                 17.4.1.7   Link bridge   **558**
                 17.4.1.8   Stack monitoring   **559**
                 17.4.1.9   Future-proofing   **561**

Index   **563**

List of figures

Chapter 1—Nuclear fuel cycle

Fig. 1.1   Nuclear reactor dome. © *EDF Energy*   **1**

Fig. 1.2   Nuclear fuel cycle. © *Bill Collum*   **2**

Fig. 1.3   Opencast uranium mine. *John Carnemolla/Shutterstock.com*   **3**

Fig. 1.4   Yellowcake. *Istock*   **4**

Fig. 1.5   Magnox fuel rod. © *Nuclear Decommissioning Authority*   **4**

Fig. 1.6   Atom constituents. © *Bill Collum*   **5**

Fig. 1.7   The atom. *general-fmv/Shutterstock.com*   **6**

Fig. 1.8   Composition of natural uranium. © *Bill Collum*   **7**

Fig. 1.9   Zippe-type gas centrifuge. *Fastfission*   **8**

Fig. 1.10 Centrifuge cascade. © *URENCO*   **9**

Fig. 1.11 Nuclear fuel pellets. *Science Photo Library*   **10**

Fig. 1.12 AGR fuel assembly. © *Nuclear Decommissioning Authority*   **10**

Fig. 1.13 PWR fuel assembly. © *Westinghouse*   **11**

Fig. 1.14 Advanced gas-cooled reactor (AGR). *Science Photo Library*   **12**

Fig. 1.15 Pressurized water reactor (PWR). *Science Photo Library*   **12**

Fig. 1.16 Heat exchanger. © *Bill Collum*   **13**

Fig. 1.17 Boiling water reactor (BWR). *Science Photo Library*   **14**

Fig. 1.18 Turbine hall. © *EDF Energy*   **14**

Fig. 1.19 Reactor charge floor. *Science Photo Library*   **15**

Fig. 1.20 PWR fuel loading. *Science Photo Library*   **15**

Fig. 1.21 Neutron capture. © *Bill Collum*   **17**

Fig. 1.22 Nuclear fission. *BlueRingMedia/Shutterstock.com*   **17**

Fig. 1.23 Control rods. *Designua/Shutterstock.com*   **18**

Fig. 1.24 Example chemical element variations. © *Bill Collum*   **20**

Fig. 1.25 Activation of cobalt-59. © *Bill Collum*   **21**

Fig. 1.26 Activation of cobalt-59 and subsequent decay. © *Bill Collum*   **21**

Fig. 1.27 Formation of plutonium in a reactor. © *Bill Collum*   **22**

Fig. 1.28 Distribution of the approximately 1% plutonium in irradiated fuel. © *Bill Collum*   **23**

Fig. 1.29 Gloveboxes. *Science Photo Library*   **24**

Fig. 1.30 Fast breeder reactor (FBR). *Science Photo Library*   **25**

Fig. 1.31 Reactor refueling machine. © *EDF Energy*   **27**

Fig. 1.32 Fuel cooling pond. © *Nuclear Decommissioning Authority*   **28**

Fig. 1.33 Initial fuel cooling periods. © *Bill Collum*   **28**

Fig. 1.34 Fuel transport flasks. © *Nuclear Decommissioning Authority*   **29**

Fig. 1.35 Spent fuel routing options. © *Bill Collum*   **30**

Fig. 1.36 Dry fuel storage cask. © *EDF Energy*   **31**
Fig. 1.37 Copper disposal canister. © *Posiva Oy*   **31**
Fig. 1.38 Reprocessing. © *Bill Collum*   **33**
Fig. 1.39 De-cladding Magnox fuel rod. © *Nuclear Decommissioning Authority*   **33**
Fig. 1.40 Constituents of spent fuel. © *Bill Collum*   **34**
Fig. 1.41 Vitrified radioactive waste. © *Nuclear Decommissioning Authority*   **35**
Fig. 1.42 Vitrified waste store. © *Nuclear Decommissioning Authority*   **36**
Fig. 1.43 Five hundred liter drum. © *Nuclear Decommissioning Authority*   **37**
Fig. 1.44 Three cubic meter box. © *Nuclear Decommissioning Authority*   **37**
Fig. 1.45 Magnox swarf encapsulated in a 500 L drum. © *Nuclear Decommissioning Authority*   **38**
Fig. 1.46 In-drum mixing. © *Nuclear Decommissioning Authority*   **38**
Fig. 1.47 Shielded overpacks. © *Nuclear Decommissioning Authority*   **39**
Fig. 1.48 Modular ILW storage vault. © *Nuclear Decommissioning Authority*   **40**
Fig. 1.49 ILW drum being lowered into storage tube. © *Nuclear Decommissioning Authority*   **40**
Fig. 1.50 Deep geological disposal. © *Nuclear Decommissioning Authority*   **41**
Fig. 1.51 LLW disposal drum. © *Nuclear Decommissioning Authority*   **42**
Fig. 1.52 LLW storage vault. © *Nuclear Decommissioning Authority*   **43**

Chapter 2—Radiation
Fig. 2.1 Confining light rays. *Visual3Dfocus/Shutterstock.com*   **45**
Fig. 2.2 Transporting radioactive material. © *Nuclear Decommissioning Authority*   **46**
Fig. 2.3 Electromagnetic spectrum. *Peter Hermes Furian/Shutterstock.com*   **47**
Fig. 2.4 Shielding against ionizing radiation. © *Bill Collum*   **48**
Fig. 2.5 Gloveboxes. *Science Photo Library*   **49**
Fig. 2.6 Quantifying exposure to radiation. © *Bill Collum*   **51**
Fig. 2.7 Radiation exposure in the United Kingdom. © *Bill Collum. Data from Public Health England*   **51**
Fig. 2.8 Radiation exposure limits. © *Bill Collum*   **53**
Fig. 2.9 Indicative shielding equivalents. © *Bill Collum*   **54**
Fig. 2.10 Spherical dispersion. *Science Photo Library*   **55**
Fig. 2.11 Reduction in radiation energy over distance. © *Bill Collum*   **56**
Fig. 2.12 Thermoluminescent dosimeter (TLD). *Science Photo Library*   **58**
Fig. 2.13 Fingertip TLD. *Science Photo Library*   **59**
Fig. 2.14 Electronic personal dosimeter (EPD). *Science Photo Library*   **59**

Chapter 3—Radiological zoning
Fig. 3.1 Radiation zones. © *Bill Collum*   **63**
Fig. 3.2 Contamination zones. © *Bill Collum*   **67**
Fig. 3.3 Containment zone interfaces. © *Bill Collum*   **70**
Fig. 3.4 Start sequence. © *Bill Collum*   **71**
Fig. 3.5 Monitoring—pass. © *Bill Collum*   **72**
Fig. 3.6 Monitoring—fail. © *Bill Collum*   **72**
Fig. 3.7 Dual radiological classification. © *Bill Collum*   **75**

Chapter 4—Radiological changerooms

Fig. 4.1 Nuclear facility with administration building attached. © *Nuclear Decommissioning Authority* 77

Fig. 4.2 Nonradiological access routes. © *Bill Collum* 79

Fig. 4.3 Nuclear worker "basics." © *Nuclear Decommissioning Authority* 80

Fig. 4.4 Security turnstile. © *Nuclear Decommissioning Authority* 81

Fig. 4.5 Boot barrier. © *Nuclear Decommissioning Authority* 81

Fig. 4.6 Boot barrier procedure—step one. © *Nuclear Decommissioning Authority* 82

Fig. 4.7 Boot barrier procedure—step two. © *Nuclear Decommissioning Authority* 83

Fig. 4.8 Boot barrier procedure—step three. © *Nuclear Decommissioning Authority* 83

Fig. 4.9 Airflow direction. © *Nuclear Decommissioning Authority* 85

Fig. 4.10 Monitoring area. © *Nuclear Decommissioning Authority* 86

Fig. 4.11 Monitoring sequence. © *Bill Collum* 87

Fig. 4.12 Hand monitoring. © *Nuclear Decommissioning Authority* 88

Fig. 4.13 Frisking station. © *CANBERRA* 89

Fig. 4.14 Installed personnel monitor (IPM). © *CANBERRA* 90

Fig. 4.15 IPM in use—step one. © *Nuclear Decommissioning Authority* 91

Fig. 4.16 IPM in use—step two. © *Nuclear Decommissioning Authority* 92

Fig. 4.17 Health physics office. © *Bill Collum* 93

Fig. 4.18 C2/C3 sub changeroom. © *Bill Collum* 96

Fig. 4.19 Disposable boot covers. © *Nuclear Decommissioning Authority* 97

Fig. 4.20 Full face respirator. © *Nuclear Decommissioning Authority* 97

Fig. 4.21 C3/C4 sub changeroom. © *Bill Collum* 100

Fig. 4.22 Air-fed PVC suit being fitted. © *Nuclear Decommissioning Authority* 101

Fig. 4.23 Health physicist conducting monitoring. © *Nuclear Decommissioning Authority* 102

Chapter 5—Structural

Fig. 5.1 Nuclear facilities. © *Nuclear Decommissioning Authority* 105

Fig. 5.2 Steelwork grid. © *Nuclear Decommissioning Authority* 106

Fig. 5.3 Steel stanchion. © *Bill Collum* 107

Fig. 5.4 Structural grid. © *Bill Collum* 108

Fig. 5.5 Steelwork structure. *James Steidl/Shutterstock.com* 109

Fig. 5.6 Steel column off grid. © *Bill Collum* 110

Fig. 5.7 Steel beams supporting floor construction. © *Nuclear Decommissioning Authority* 111

Fig. 5.8 Unbraced cube. © *Bill Collum* 111

Fig. 5.9 Braced cube. © *Bill Collum* 112

Fig. 5.10 Bracing types. © *Bill Collum* 113

Fig. 5.11 Typical bracing arrangement. © *Nuclear Decommissioning Authority* 113

Fig. 5.12 Portal frame. © *Nuclear Decommissioning Authority* 114

Fig. 5.13 Overhead traveling crane. © *Nuclear Decommissioning Authority* 115

Fig. 5.14 Precast concrete wall panel. *Istock* 116

Fig. 5.15 Precast concrete planks. *wrangler/Shutterstock.com* 116

Fig. 5.16 In situ concrete pouring. © *Nuclear Decommissioning Authority*   **118**
Fig. 5.17 Shear wall reinforcing bars. © *Nuclear Decommissioning Authority*   **118**
Fig. 5.18 Concrete banding. © *Bill Collum*   **120**
Fig. 5.19 Concrete structure within a steel framed building. © *Nuclear Decommissioning Authority*   **122**
Fig. 5.20 Sliding bearing. © *Bill Collum*   **123**
Fig. 5.21 Phased concrete penetration information. © *Bill Collum*   **123**
Fig. 5.22 Seismic categories. © *Bill Collum*   **128**
Fig. 5.23 Aftermath of a hurricane. *John Huntington/Shutterstock.com*   **131**
Fig. 5.24 Cladding panel. © *Nuclear Decommissioning Authority*   **132**
Fig. 5.25 Wind tunnel test. *Science Photo Library*   **133**
Fig. 5.26 Shifting snow. *Istock*   **135**
Fig. 5.27 Extreme rainfall. *Paco Espinoza/Shutterstock.com*   **137**
Fig. 5.28 Box gutter. © *Nuclear Decommissioning Authority*   **137**

Chapter 6—Process engineering
Fig. 6.1   Closed cell processing facility. © *Nuclear Decommissioning Authority*   **139**
Fig. 6.2   Internals of a maintenance-free processing vessel. © *Nuclear Decommissioning Authority*   **140**
Fig. 6.3   Closed cell during commissioning. © *Nuclear Decommissioning Authority*   **141**
Fig. 6.4   Closed cell computer model. © *Nuclear Decommissioning Authority*   **142**
Fig. 6.5   Mass balance of a fruit cake. © *Bill Collum*   **144**
Fig. 6.6   Stratification. © *Bill Collum*   **147**
Fig. 6.7   Centrifugal pump. *Korotkevich/Shutterstock.com*   **151**
Fig. 6.8   Self-priming pump. © *Bill Collum*   **152**
Fig. 6.9   Double block and bleed. © *Bill Collum*   **153**
Fig. 6.10 Remotely maintainable pump. © *Bill Collum*   **155**
Fig. 6.11 Reverse flow diverter. © *Elsevier*   **156**
Fig. 6.12 Fluidic pumping system—RFD internal. © *Bill Collum*   **157**
Fig. 6.13 Pipetting. © *Bill Collum*   **158**
Fig. 6.14 Fluidic pumping system—RFD external. © *Bill Collum*   **160**
Fig. 6.15 Siphoning. © *Bill Collum*   **162**
Fig. 6.16 Breakpot. © *Bill Collum*   **163**
Fig. 6.17 Services. © *Bill Collum*   **166**
Fig. 6.18 Services distribution. *imantsu/Shutterstock.com*   **167**
Fig. 6.19 Services hierarchy. © *Bill Collum*   **168**
Fig. 6.20 Joggle box. © *Nuclear Decommissioning Authority*   **169**
Fig. 6.21 Pump-round arrangement. © *Bill Collum*   **172**
Fig. 6.22 Pulse jet mixer. © *Bill Collum*   **173**
Fig. 6.23 Air sparge ring. © *Bill Collum*   **174**
Fig. 6.24 Evaporation process. © *Bill Collum*   **176**
Fig. 6.25 Settling tank—continuous flow. © *Bill Collum*   **177**
Fig. 6.26 Settling tank—batch process. © *Bill Collum*   **178**
Fig. 6.27 Ion exchange column. © *Bill Collum*   **180**
Fig. 6.28 Packed column. © *Bill Collum*   **182**

Chapter 7—Mechanical engineering

Fig. 7.1   Drum handling grab. © *Nuclear Decommissioning Authority*   185
Fig. 7.2   Remotely operated mechanical equipment. © *Nuclear Decommissioning Authority*   186
Fig. 7.3   Mechanical handling cave. © *Nuclear Decommissioning Authority*   187
Fig. 7.4   Shielding window. © *Nuclear Decommissioning Authority*   188
Fig. 7.5   Shielding window—out-cave view. © *Nuclear Decommissioning Authority*   189
Fig. 7.6   Refraction. *Science Photo Library*   190
Fig. 7.7   Shielding window—in-cave view. © *Nuclear Decommissioning Authority* 190
Fig. 7.8   MSM components. © *Wälischmiller*   193
Fig. 7.9   MSM drive wires and pulleys. © *Wälischmiller*   194
Fig. 7.10 MSM—cold arm controls. © *Nuclear Decommissioning Authority*   195
Fig. 7.11 MSMs—in-cave view. © *Nuclear Decommissioning Authority*   196
Fig. 7.12 Cave face corridor. © *Nuclear Decommissioning Authority*   197
Fig. 7.13 Electrical manipulator—rail mounted. © *Wälischmiller*   198
Fig. 7.14 Electrical manipulator deployed underwater. © *Wälischmiller*   198
Fig. 7.15 Hydraulic manipulator. © *Nuclear Decommissioning Authority*   199
Fig. 7.16 Power manipulator. © *Wälischmiller*   200
Fig. 7.17 Telescopic mast. © *Wälischmiller*   201
Fig. 7.18 Power manipulator with cutting disc. © *Wälischmiller*   202
Fig. 7.19 Laminated shield door. © *Bill Collum*   203
Fig. 7.20 PA door with integral frame. © *Nuclear Decommissioning Authority* 204
Fig. 7.21 Combination vertical and horizontal shield doors. © *Nuclear Decommissioning Authority*   206
Fig. 7.22 Interface between vertical and horizontal shield doors. © *Nuclear Decommissioning Authority*   207
Fig. 7.23 Vertical shield door recovery. © *Nuclear Decommissioning Authority* 208
Fig. 7.24 Remotely operated bogie with raise and lower capability. © *Nuclear Decommissioning Authority*   210
Fig. 7.25 Waste drum after dry pellet cleaning process. © *Nuclear Decommissioning Authority*   213
Fig. 7.26 Combined crane and power manipulator on a four girder bridge. © *Nuclear Decommissioning Authority*   215
Fig. 7.27 50 tonne cuboidal flask on transport bogie. © *Nuclear Decommissioning Authority*   219
Fig. 7.28 Cylindrical flask. © *Nuclear Decommissioning Authority*   220
Fig. 7.29 Cylindrical flask with shock absorbers fitted. *Science Photo Library* 221
Fig. 7.30 Flask crash test. *Science Photo Library*   222
Fig. 7.31 Flask after crash test. © *EDF Energy*   222
Fig. 7.32 Flask unloading underwater. *Science Photo Library*   224
Fig. 7.33 Lid lifting and flask unloading sequence. © *Nuclear Decommissioning Authority*   224
Fig. 7.34 Pintle arrangement. © *Nuclear Decommissioning Authority*   225

Fig. 7.35 Bottom loading flask. © *Nuclear Decommissioning Authority*   **226**
Fig. 7.36 Permanently installed gamma gate. © *Nuclear Decommissioning Authority* **227**
Fig. 7.37 Mobile gamma gates. *Science Photo Library*   **228**
Fig. 7.38 Flask nappy. © *Nuclear Decommissioning Authority*   **229**

Chapter 8—Ventilation
Fig. 8.1   Extract velocity contours. © *Bill Collum*   **232**
Fig. 8.2   Cascade philosophy—direction of airflow. © *Bill Collum*   **234**
Fig. 8.3   Three times 50% ventilation system. © *Bill Collum*   **235**
Fig. 8.4   Two times 100% ventilation system. © *Bill Collum*   **235**
Fig. 8.5   Differential pressure between radiological zones. © *Bill Collum* **236**
Fig. 8.6   Airspeed at interface between radiological zones. © *Bill Collum* **237**
Fig. 8.7   Airflow across C2/C3 sub changeroom. © *Bill Collum*   **238**
Fig. 8.8   Airflow through open shield door. © *Bill Collum*   **239**
Fig. 8.9   Adjustable airspeed fin. © *Bill Collum*   **240**
Fig. 8.10 Interlocked doors. © *Bill Collum*   **241**
Fig. 8.11 Wind suction. © *Bill Collum*   **242**
Fig. 8.12 Containment of a C4 zone. © *Bill Collum*   **243**
Fig. 8.13 HEPA filter types. © *M C Air Filtration*   **244**
Fig. 8.14 HEPA filter bank. © *M C Air Filtration*   **245**
Fig. 8.15 HEPA filtration—beta-gamma plants. © *Bill Collum*   **246**
Fig. 8.16 HEPA filtration—alpha plants. © *Bill Collum*   **247**
Fig. 8.17 Mobile filtration unit. © *M C Air Filtration*   **248**
Fig. 8.18 Air in-bleed filter. © *Bill Collum*   **249**
Fig. 8.19 Manual filter change (testing). © *M C Air Filtration*   **251**
Fig. 8.20 Remote filter change. © *Nuclear Decommissioning Authority*   **252**
Fig. 8.21 Shielding between filter banks. © *Bill Collum*   **252**
Fig. 8.22 Push-through filter partly inserted into housing. © *M C Air Filtration* **255**
Fig. 8.23 Generic arrangement of ventilation plant rooms—beta-gamma plant.
              © *Bill Collum*   **260**
Fig. 8.24 Ductwork distribution network—quartering. © *Bill Collum*   **263**
Fig. 8.25 Air handling unit. © *Bill Collum*   **264**
Fig. 8.26 Typical airflow by volume. © *Bill Collum*   **267**
Fig. 8.27 Glovebox suite. © *Nuclear Decommissioning Authority*   **269**
Fig. 8.28 Glovebox ventilation. © *Nuclear Decommissioning Authority*   **270**
Fig. 8.29 Push-through filter. © *Nuclear Decommissioning Authority*   **271**
Fig. 8.30 Glovebox housing automated processes. © *Nuclear
              Decommissioning Authority*   **272**

Chapter 9—Cranes
Fig. 9.1   Single girder EOT crane. *Istock*   **276**
Fig. 9.2   Twin girder EOT crane. *SasinTipchai/Shutterstock.com*   **277**
Fig. 9.3   Rope drum. *Pnor Tkk/Shutterstock.com*   **277**
Fig. 9.4   Wireless crane operation. *Istock*   **279**
Fig. 9.5   Pendant crane operation. © *Nuclear Decommissioning Authority*   **279**

Fig. 9.6 On-board crane operation. *moomsabuy/Shutterstock.com* **280**
Fig. 9.7 Lifting accessory—mechanical grab. © *Nuclear Decommissioning Authority* **281**
Fig. 9.8 Lifting feature—crane hook. *Jamesbin/Shutterstock.com* **281**
Fig. 9.9 Spreader frame. *Zelfit/Shutterstock.com* **282**
Fig. 9.10 Twist-lock. © *Nuclear Decommissioning Authority* **282**
Fig. 9.11 Four drum stillage. © *Nuclear Decommissioning Authority* **283**
Fig. 9.12 Lifting beam. © *Nuclear Decommissioning Authority* **284**
Fig. 9.13 Pintle. © *Nuclear Decommissioning Authority* **285**
Fig. 9.14 Shielded transport flask. © *Nuclear Decommissioning Authority* **286**
Fig. 9.15 Hook approach. © *Nuclear Decommissioning Authority* **288**
Fig. 9.16 Polar crane. *Getty Images* **296**
Fig. 9.17 Polar jib crane handling vitrified waste container. © *Nuclear Decommissioning Authority* **298**
Fig. 9.18 Scissor jack. *pryzmat/Shutterstock.com* **302**
Fig. 9.19 Catenary cable. *PoohFotoz/Shutterstock.com* **306**
Fig. 9.20 Cable reeling system. © *Metreel Ltd* **307**
Fig. 9.21 Busbar power supply. © *Metreel Ltd* **309**
Fig. 9.22 Drag chain. © *Metreel Ltd* **310**
Fig. 9.23 Rack and pinion. *PhotoProRo/Shutterstock.com* **311**

Chapter 10—Electrical
Fig. 10.1 132 kV–11 kV Substation. © *Nuclear Decommissioning Authority* **314**
Fig. 10.2 Ring main. © *Bill Collum* **315**
Fig. 10.3 Transformer. © *Bill Collum* **315**
Fig. 10.4 Distribution board. © *Bill Collum* **317**
Fig. 10.5 Non-firm power supply. © *Bill Collum* **319**
Fig. 10.6 Firm power supply. © *Bill Collum* **319**
Fig. 10.7 Guaranteed interruptible power supply. © *Bill Collum* **321**
Fig. 10.8 11 kV Generator. *Rehan Qureshi/Shutterstock.com* **321**
Fig. 10.9 Mobile generator. *Radovan1/Shutterstock.com* **322**
Fig. 10.10 Wattage units. © *Bill Collum* **322**
Fig. 10.11 EPD panels. © *Nuclear Decommissioning Authority* **323**
Fig. 10.12 Guaranteed uninterruptible power supply (UPS). © *Bill Collum* **324**
Fig. 10.13 Battery room. *cpaulfell/Shutterstock.com* **325**
Fig. 10.14 Cable installation. *Dmitry Kalinovsky/Shutterstock.com* **326**
Fig. 10.15 Cable trays. *Misterdone Image/Shutterstock.com* **327**
Fig. 10.16 Basic control. © *Bill Collum* **328**
Fig. 10.17 Sequence control. © *Bill Collum* **329**
Fig. 10.18 Automation panel. © *Bill Collum* **331**
Fig. 10.19 Automated sequence. © *Bill Collum* **331**
Fig. 10.20 Remote operation. *branislavpudar/Shutterstock.com* **332**
Fig. 10.21 SCADA system. © *Bill Collum* **333**
Fig. 10.22 Central control room. © *EDF Energy* **334**
Fig. 10.23 Bubbler system. © *Bill Collum* **336**
Fig. 10.24 Vibrating fork. © *Bill Collum* **337**

Fig. 10.25 Radar sensors. © *Bill Collum*   338
Fig. 10.26 Conductive sensors. © *Bill Collum*   339
Fig. 10.27 Capacitance sensors. © *Bill Collum*   340
Fig. 10.28 Ultrasonic. © *Bill Collum*   341
Fig. 10.29 Resistance temperature detectors. © *Elsevier*   343
Fig. 10.30 Bourdon tube. © *Bill Collum*   344
Fig. 10.31 Isolated diaphragm gauge. © *Bill Collum*   345
Fig. 10.32 Orifice meter. © *Bill Collum*   346
Fig. 10.33 Pitot tube. © *Bill Collum*   347

Chapter 11—Radiometric instruments
Fig. 11.1   Gas-filled detector. © *Bill Collum*   351
Fig. 11.2   Scintillation detector. © *Bill Collum*   355
Fig. 11.3   Transistor. © *Bill Collum*   357
Fig. 11.4   Semiconductor detector—passive state. © *Bill Collum*   358
Fig. 11.5   Semiconductor detector—active state. © *Bill Collum*   359
Fig. 11.6   Technology capabilities. © *Bill Collum*   360
Fig. 11.7   Resolution comparison. © *Bill Collum*   361
Fig. 11.8   Health physics instruments. © *CANBERRA*   363
Fig. 11.9   Environmental monitoring instruments. © *CANBERRA*   364
Fig. 11.10 Through-wall gamma monitor. © *Nuclear Decommissioning Authority* 364
Fig. 11.11 Assay equipment. © *CANBERRA*   365
Fig. 11.12 Radiometric instrument technologies. © *Bill Collum*   367

Chapter 12—Project planning
Fig. 12.1   Project control network. © *Bill Collum*   371
Fig. 12.2   Work breakdown structure. © *Bill Collum*   373
Fig. 12.3   Integrated breakdown structures. © *Bill Collum*   375
Fig. 12.4   Nuclear versus conventional expenditure. © *Bill Collum*   378
Fig. 12.5   Gantt chart. *Paul Barnwell/Shutterstock.com*   380
Fig. 12.6   Primary project phases. © *Bill Collum*   381
Fig. 12.7   Project hold points. © *Bill Collum*   382
Fig. 12.8   Escalating cost of change over time. © *Bill Collum*   386
Fig. 12.9   Evolution of mechanical equipment design. © *Bill Collum*   389

Chapter 13—Waste management
Fig. 13.1   Conventional waste management hierarchy. © *Bill Collum*   395
Fig. 13.2   Radiological waste management hierarchy. © *Bill Collum*   398

Chapter 14—Safety
Fig. 14.1   Confined space. © *Nuclear Decommissioning Authority*   407
Fig. 14.2   Summary of site license conditions. © *Bill Collum. Based on data from*
                the Office for Nuclear Regulation (ONR)   411
Fig. 14.3   Safety case categories. © *Bill Collum*   415
Fig. 14.4   Primary components of a safety case. © *Bill Collum*   417
Fig. 14.5   Example key words. © *Bill Collum*   419

Fig. 14.6 Fault schedule. © *Bill Collum*   **423**

Fig. 14.7 Satisfying a safety functional requirement. © *Bill Collum*   **424**

Fig. 14.8 Risk. © *Bill Collum*   **426**

Fig. 14.9 Hazard control hierarchy. © *Bill Collum*   **427**

Fig. 14.10 Fault sequence—unmitigated. © *Bill Collum*   **431**

Fig. 14.11 Fault sequence—mitigated. © *Bill Collum*   **432**

Fig. 14.12 Risk classification. © *Bill Collum*   **434**

Fig. 14.13 Mitigation measures and reliability. © *Bill Collum*   **435**

Fig. 14.14 Application of ALARP principles. © *Bill Collum*   **437**

Fig. 14.15 Safety reports—common themes. © *Bill Collum*   **438**

Fig. 14.16 Sealed radioactive source. *Science Photo Library*   **443**

Fig. 14.17 Safety case development—project inception. © *Bill Collum*   **447**

Fig. 14.18 Safety case development—optioneering. © *Bill Collum*   **448**

Fig. 14.19 Safety case development—option refinement. © *Bill Collum*   **449**

Fig. 14.20 Safety case development—detail design. © *Bill Collum*   **450**

Fig. 14.21 Safety case development—construction and installation. © *Bill Collum*   **451**

Fig. 14.22 Safety case development—inactive and active commissioning. © *Bill Collum*   **452**

Fig. 14.23 Safety case development—operations. © *Bill Collum*   **453**

Chapter 15—Decommissioning planning

Fig. 15.1 Site license condition 35. © *Bill Collum. Based on data from the Office for Nuclear Regulation (ONR)*   **455**

Fig. 15.2 Planning factors. © *Bill Collum*   **456**

Fig. 15.3 Refurbishment. © *Nuclear Decommissioning Authority*   **463**

Fig. 15.4 Shielded pipebridge. © *Nuclear Decommissioning Authority*   **467**

Fig. 15.5 Nuclear site in a costal location. © *Nuclear Decommissioning Authority*   **468**

Fig. 15.6 Fuel cooling pond. © *Nuclear Decommissioning Authority*   **473**

Fig. 15.7 Operations—primary phases. © *Bill Collum*   **473**

Fig. 15.8 POCO—preparing a cave for shutdown. © *Nuclear Decommissioning Authority*   **474**

Fig. 15.9 Operations—primary and secondary phases. © *Bill Collum*   **475**

Fig. 15.10 Decommissioning—primary phases. © *Bill Collum*   **477**

Fig. 15.11 Drummed LLW. © *Nuclear Decommissioning Authority*   **479**

Fig. 15.12 Low level waste store. *Science Photo Library*   **479**

Fig. 15.13 Scabbling concrete. © *Nuclear Decommissioning Authority*   **480**

Fig. 15.14 Manual D&D operations. © *Nuclear Decommissioning Authority*   **480**

Fig. 15.15 Remote dismantling. © *Nuclear Decommissioning Authority*   **481**

Fig. 15.16 Demolition of redundant nuclear plant. © *Nuclear Decommissioning Authority*   **483**

Fig. 15.17 In-ground sampling. © *Nuclear Decommissioning Authority*   **484**

Fig. 15.18 Decommissioning—primary and secondary phases. © *Bill Collum*   **485**

Fig. 15.19 Decommissioning—primary, secondary, and tertiary phases. © *Bill Collum*   **487**

Fig. 15.20 Advance demolition. © *Nuclear Decommissioning Authority*   **488**

Fig. 15.21 Operations and decommissioning phases. © *Bill Collum*   **489**

Fig. 15.22 Immediate dismantling. © *Bill Collum*   **490**
Fig. 15.23 Glovebox dismantling. © *Nuclear Decommissioning Authority*   **491**
Fig. 15.24 Safe enclosure. © *Bill Collum*   **493**
Fig. 15.25 Entombment. © *Bill Collum*   **495**
Fig. 15.26 Dry fuel storage cask. © *EDF Energy*   **496**
Fig. 15.27 Entombment concept. © *Woodhead Publishing*   **497**
Fig. 15.28 Decommissioning strategies. © *Bill Collum*   **498**
Fig. 15.29 Decommissioning plan. © *Bill Collum*   **501**
Fig. 15.30 Decommissioning planning. © *Bill Collum*   **503**

Chapter 16—Future-proofing
Fig. 16.1   In situ dismantling—manual. © *Nuclear Decommissioning Authority*   **506**
Fig. 16.2   In situ dismantling—remote. © *Nuclear Decommissioning Authority*   **507**
Fig. 16.3   Self-perpetuating cycle. © *Bill Collum*   **508**
Fig. 16.4   Incorporation of DfD and DRR in design process. © *Bill Collum*   **510**
Fig. 16.5   Maximize ability to self-decommission. © *Bill Collum*   **512**
Fig. 16.6   Integration with decommissioning strategy. © *Bill Collum*   **515**
Fig. 16.7   Rebar. *Kokliang/Shutterstock.com*   **521**
Fig. 16.8   ILW stored in concrete overpacks. © *Nuclear Decommissioning Authority* **522**
Fig. 16.9   Fire sealing. © *Nuclear Decommissioning Authority*   **526**
Fig. 16.10 Movement joint. © *Nuclear Decommissioning Authority*   **528**

Chapter 17—Design development
Fig. 17.1   Precursors to layout development. © *Bill Collum*   **534**
Fig. 17.2   Building flow diagram. © *Bill Collum*   **536**
Fig. 17.3   On-site rail track geometry. © *Bill Collum*   **542**
Fig. 17.4   Rail track influence on facility location. © *Bill Collum*   **543**
Fig. 17.5   Influence of crane geometry on a vehicle bay. © *Bill Collum*   **549**
Fig. 17.6   Tambour door. © *Nuclear Decommissioning Authority*   **550**
Fig. 17.7   Air discharge flues. © *Bill Collum*   **552**
Fig. 17.8   Effective stack height. © *Bill Collum*   **554**
Fig. 17.9   Freestanding concrete stack. © *Nuclear Decommissioning Authority* **555**
Fig. 17.10 Through-roof stack. © *Bill Collum*   **556**
Fig. 17.11 Strakes. © *Nuclear Decommissioning Authority*   **557**
Fig. 17.12 Stack monitoring ductwork. © *Bill Collum*   **560**

Author biography

Bill Collum began his nuclear career over 30 years ago with the former British Nuclear Fuels Ltd. (BNFL). He established the Layout Centre of Excellence, and developed the processes and procedures used by the company to layout new nuclear facilities, starting from the point of a *blank sheet* and through to a frozen design. At the same time, he developed processes used to embed design for future decommissioning into proposals for new nuclear facilities.

Bill has taken a lead role in coordinating the front-end design development for both new-build and decommissioning projects, for multiple clients across the United Kingdom and around the world, most notably in the United States and Europe. To operate at this level, he has amassed considerable multidisciplinary knowledge, along with skills to incorporate this knowledge into all types of nuclear project.

More latterly, Bill operates as an independent nuclear consultant, providing expert advice and conducting reviews on behalf of his clients. It is this wealth of experience that he brings to these pages.

E-mail: enquiries@collumnuclear.com

Acknowledgments

Before setting out to write this book, I contemplated how the whole exercise might unfold. I figured that for some of the chapters, I could get on with writing a first draft. Whilst for others I could develop an outline of the story I wanted to tell, but would effectively need to interview a subject matter expert before committing pen to paper. In all cases, I determined that peer review would be essential. In addition, I felt its pages should be as well illustrated as possible, which was not going to be easy. In other words, a book of this type covers far too much ground to be written in isolation, so I was going to need some help. Happily, those I approached were invariably enthusiastic about my endeavor and, most importantly, willing to lend their support.

Among subject matter experts, I am particularly indebted to the following:

Phil Rees, who, before he retired, patiently allowed me to quiz him relentlessly on the nuclear fuel cycle; Jonathan Dobson, for dissecting the intricacies of a nuclear safety case; Mark Davies, who revealed the underling complexities of a whole host of mechanical equipment; Derek Calland, whose breadth of knowledge across the whole gamut of electrical engineering is quite remarkable; and Richard Hunter, who had the unenviable task of explaining the mysteries of radiometric instruments to me.

In addition, I am grateful to several experts who raised my existing understanding of their particular discipline to a whole new level, in particular, Sarah Warner-Jones for sharing her encyclopedic knowledge of all things radiometric with me; Dr. Ray Doig, who added to my grasp of nuclear ventilation and demonstrated why he is so highly regarded within the industry. Dai Davies kindly scrutinized my discourse on nuclear cranes and provided essential details for me to weave into the text. Mike Prescott expounded ways in which the project planning exercise is even more intertwined than I had previously appreciated. Chris Bolton conducted a penetrating analysis of my chapter on structural engineering and provided feedback which hit the spot perfectly. Dr. Jim Honeywill waded through my thoughts on the various aspects of decommissioning and made insightful comments which were much appreciated. Dr. Paul McMorn added his perspective to the nuclear fuel cycle and Dave Knight shared his insights on the complexities of distributing electricity on an industrial scale. As the book neared completion, the Nuclear Institute kindly conducted a selective review across several chapters. For this I must thank Tim Chittenden, John Robertson, Kevin Allars, and John Warden.

At the time that I was writing this book, I was employed by Cavendish Nuclear, so early on in the process I informed the executive about what I was up to in my spare time. To help me along in this personal endeavor, I was generously granted as much flexibility as possible in my working pattern, along with access to printing and photocopying services for countless draft manuscripts. In addition, several of the subject matter experts that I consulted were drawn from across the company. I am indebted to Cavendish Nuclear for their encouragement and for the latitude accorded to me while working on this book. For this, I wish to record my sincere thanks.

When describing nuclear subject matter, it helps tremendously to have access to bespoke technical illustrations which can be embedded within the text, including images from 3D computer models. I am therefore extremely grateful to Sellafield Ltd. who, through the auspices of the Nuclear Decommissioning Authority (NDA), prepared many of the book's more technical illustrations. John O'Brien kindly arranged for members of his engineering team to prepare the illustrations and I must say they could not possibly have been more helpful, or indeed skillful. For this I must express particular thanks to Paul Whitworth, Chris Hamlett, Dave Jones, Roy Pitt, and Dave Mason.

Other technical illustrations were provided by Steve Coffey and Glenn Graham, both of whom I have worked closely with for more years than any of us would care to remember. The remaining charts and diagrams I managed to prepare myself, a feat which I must say kept me quiet for a while.

Nothing brings a page to life in quite the same way as a photograph, so my deepest gratitude goes to those who provided the many examples on these pages. A full list of copyright owners appears in the List of figures, but I must make special mention of several companies who kindly supplied photographs of their products: CANBERRA, for a selection from their range of radiometric instruments; Wälischmiller, for manipulators, including MSMs, electrical manipulators, and power manipulators; M C Air Filtration, for HEPA filters, including mobile filtration; Metreel Ltd., for systems delivering electrical power to cranes and URENCO for the centrifuge cascade.

In addition, Juliette Sanders of EDF Energy somehow managed to track down several photographs associated with nuclear reactors and fuel transport, all of which fitted perfectly with the relevant text, and in one case even arranging for a photograph to be specially taken.

I must give very special thanks to Sue Lawson of the NDA who, over a period of several months, sourced over 80 of the book's photographs. When you consider that finding just one or two appropriate photographs can be terribly time consuming, it highlights just what a mammoth exercise this has been. In every case, Sue was fastidious in obtaining "just the right" image, often providing several alternatives for me to choose from. Where suitable images were unavailable, arrangements would be made and I would find myself gazing at yet another image that I had almost given up on. Amassing so many wholly appropriate photographs is a fabulous accomplishment and one for which I am extremely grateful.

Finally, I must thank the NDA's Dr. Adrian Simper. His encouragement to me personally and his wider influence have played a major role in bringing this book to fruition.

Author's note

For purposes of clarity, I must state upfront that our discussion here will concentrate exclusively on *civil* nuclear facilities and exclude examination of any matters that are security related.

Woodhead Publishing Series in Energy

1 **Generating power at high efficiency: Combined cycle technology for sustainable energy production**
Eric Jeffs

2 **Advanced separation techniques for nuclear fuel reprocessing and radioactive waste treatment**
Edited by Kenneth L. Nash and Gregg J. Lumetta

3 **Bioalcohol production: Biochemical conversion of lignocellulosic biomass**
Edited by Keith W. Waldron

4 **Understanding and mitigating ageing in nuclear power plants: Materials and operational aspects of plant life management (PLiM)**
Edited by Philip G. Tipping

5 **Advanced power plant materials, design and technology**
Edited by Dermot Roddy

6 **Stand-alone and hybrid wind energy systems: Technology, energy storage and applications**
Edited by John K. Kaldellis

7 **Biodiesel science and technology: From soil to oil**
Jan C. J. Bart, Natale Palmeri and Stefano Cavallaro

8 **Developments and innovation in carbon dioxide (CO_2) capture and storage technology Volume 1: Carbon dioxide (CO_2) capture, transport and industrial applications**
Edited by M. Mercedes Maroto-Valer

9 **Geological repository systems for safe disposal of spent nuclear fuels and radioactive waste**
Edited by Joonhong Ahn and Michael J. Apted

10 **Wind energy systems: Optimising design and construction for safe and reliable operation**
Edited by John D. Sørensen and Jens N. Sørensen

11 **Solid oxide fuel cell technology: Principles, performance and operations**
Kevin Huang and John Bannister Goodenough

12 **Handbook of advanced radioactive waste conditioning technologies**
Edited by Michael I. Ojovan

13 **Membranes for clean and renewable power applications**
Edited by Annarosa Gugliuzza and Angelo Basile

14 **Materials for energy efficiency and thermal comfort in buildings**
Edited by Matthew R. Hall

15 **Handbook of biofuels production: Processes and technologies**
Edited by Rafael Luque, Juan Campelo and James Clark

16 **Developments and innovation in carbon dioxide (CO_2) capture and storage technology Volume 2: Carbon dioxide (CO_2) storage and utilisation**
   Edited by M. Mercedes Maroto-Valer

17 **Oxy-fuel combustion for power generation and carbon dioxide (CO_2) capture**
   Edited by Ligang Zheng

18 **Small and micro combined heat and power (CHP) systems: Advanced design, performance, materials and applications**
   Edited by Robert Beith

19 **Advances in clean hydrocarbon fuel processing: Science and technology**
   Edited by M. Rashid Khan

20 **Modern gas turbine systems: High efficiency, low emission, fuel flexible power generation**
   Edited by Peter Jansohn

21 **Concentrating solar power technology: Principles, developments and applications**
   Edited by Keith Lovegrove and Wes Stein

22 **Nuclear corrosion science and engineering**
   Edited by Damien Féron

23 **Power plant life management and performance improvement**
   Edited by John E. Oakey

24 **Electrical drives for direct drive renewable energy systems**
   Edited by Markus Mueller and Henk Polinder

25 **Advanced membrane science and technology for sustainable energy and environmental applications**
   Edited by Angelo Basile and Suzana Pereira Nunes

26 **Irradiation embrittlement of reactor pressure vessels (RPVs) in nuclear power plants**
   Edited by Naoki Soneda

27 **High temperature superconductors (HTS) for energy applications**
   Edited by Ziad Melhem

28 **Infrastructure and methodologies for the justification of nuclear power programmes**
   Edited by Agustín Alonso

29 **Waste to energy conversion technology**
   Edited by Naomi B. Klinghoffer and Marco J. Castaldi

30 **Polymer electrolyte membrane and direct methanol fuel cell technology Volume 1: Fundamentals and performance of low temperature fuel cells**
   Edited by Christoph Hartnig and Christina Roth

31 **Polymer electrolyte membrane and direct methanol fuel cell technology Volume 2: *In situ* characterization techniques for low temperature fuel cells**
   Edited by Christoph Hartnig and Christina Roth

32 **Combined cycle systems for near-zero emission power generation**
   Edited by Ashok D. Rao

33 **Modern earth buildings: Materials, engineering, construction and applications**
   Edited by Matthew R. Hall, Rick Lindsay and Meror Krayenhoff

34 **Metropolitan sustainability: Understanding and improving the urban environment**
   Edited by Frank Zeman

35 **Functional materials for sustainable energy applications**
   Edited by John A. Kilner, Stephen J. Skinner, Stuart J. C. Irvine and Peter P. Edwards

36 **Nuclear decommissioning: Planning, execution and international experience**
   Edited by Michele Laraia

37 **Nuclear fuel cycle science and engineering**
   Edited by Ian Crossland
38 **Electricity transmission, distribution and storage systems**
   Edited by Ziad Melhem
39 **Advances in biodiesel production: Processes and technologies**
   Edited by Rafael Luque and Juan A. Melero
40 **Biomass combustion science, technology and engineering**
   Edited by Lasse Rosendahl
41 **Ultra-supercritical coal power plants: Materials, technologies and optimisation**
   Edited by Dongke Zhang
42 **Radionuclide behaviour in the natural environment: Science, implications
   and lessons for the nuclear industry**
   Edited by Christophe Poinssot and Horst Geckeis
43 **Calcium and chemical looping technology for power generation and carbon dioxide
   (CO_2) capture: Solid oxygen- and CO_2-carriers**
   Paul Fennell and E. J. Anthony
44 **Materials' ageing and degradation in light water reactors: Mechanisms,
   and management**
   Edited by K. L. Murty
45 **Structural alloys for power plants: Operational challenges and high-temperature
   materials**
   Edited by Amir Shirzadi and Susan Jackson
46 **Biolubricants: Science and technology**
   Jan C. J. Bart, Emanuele Gucciardi and Stefano Cavallaro
47 **Advances in wind turbine blade design and materials**
   Edited by Povl Brøndsted and Rogier P. L. Nijssen
48 **Radioactive waste management and contaminated site clean-up: Processes,
   technologies and international experience**
   Edited by William E. Lee, Michael I. Ojovan, Carol M. Jantzen
49 **Probabilistic safety assessment for optimum nuclear power plant life management
   (PLiM): Theory and application of reliability analysis methods for major power plant
   components**
   Gennadij V. Arkadov, Alexander F. Getman and Andrei N. Rodionov
50 **The coal handbook: Towards cleaner production Volume 1: Coal production**
   Edited by Dave Osborne
51 **The coal handbook: Towards cleaner production Volume 2: Coal utilisation**
   Edited by Dave Osborne
52 **The biogas handbook: Science, production and applications**
   Edited by Arthur Wellinger, Jerry Murphy and David Baxter
53 **Advances in biorefineries: Biomass and waste supply chain exploitation**
   Edited by Keith Waldron
54 **Geological storage of carbon dioxide (CO_2): Geoscience, technologies, environmental
   aspects and legal frameworks**
   Edited by Jon Gluyas and Simon Mathias
55 **Handbook of membrane reactors Volume 1: Fundamental materials science,
   design and optimisation**
   Edited by Angelo Basile
56 **Handbook of membrane reactors Volume 2: Reactor types
   and industrial applications**
   Edited by Angelo Basile

57 **Alternative fuels and advanced vehicle technologies for improved environmental performance: Towards zero carbon transportation**
    Edited by Richard Folkson
58 **Handbook of microalgal bioprocess engineering**
    Christopher Lan and Bei Wang
59 **Fluidized bed technologies for near-zero emission combustion and gasification**
    Edited by Fabrizio Scala
60 **Managing nuclear projects: A comprehensive management resource**
    Edited by Jas Devgun
61 **Handbook of Process Integration (PI): Minimisation of energy and water use, waste and emissions**
    Edited by Jiří J. Klemeš
62 **Coal power plant materials and life assessment**
    Edited by Ahmed Shibli
63 **Advances in hydrogen production, storage and distribution**
    Edited by Ahmed Basile and Adolfo Iulianelli
64 **Handbook of small modular nuclear reactors**
    Edited by Mario D. Carelli and Dan T. Ingersoll
65 **Superconductors in the power grid: Materials and applications**
    Edited by Christopher Rey
66 **Advances in thermal energy storage systems: Methods and applications**
    Edited by Luisa F. Cabeza
67 **Advances in batteries for medium and large-scale energy storage**
    Edited by Chris Menictas, Maria Skyllas-Kazacos and Tuti Mariana Lim
68 **Palladium membrane technology for hydrogen production, carbon capture and other applications**
    Edited by Aggelos Doukelis, Kyriakos Panopoulos, Antonios Koumanakos and Emmanouil Kakaras
69 **Gasification for synthetic fuel production: Fundamentals, processes and applications**
    Edited by Rafael Luque and James G. Speight
70 **Renewable heating and cooling: Technologies and applications**
    Edited by Gerhard Stryi-Hipp
71 **Environmental remediation and restoration of contaminated nuclear and NORM sites**
    Edited by Leo van Velzen
72 **Eco-friendly innovation in electricity networks**
    Edited by Jean-Luc Bessede
73 **The 2011 Fukushima nuclear power plant accident: How and why it happened**
    Yotaro Hatamura, Seiji Abe, Masao Fuchigami and Naoto Kasahara.
    Translated by Kenji Iino
74 **Lignocellulose biorefinery engineering: Principles and applications**
    Hongzhang Chen
75 **Advances in membrane technologies for water treatment: Materials, processes and applications**
    Edited by Angelo Basile, Alfredo Cassano and Navin Rastogi
76 **Membrane reactors for energy applications and basic chemical production**
    Edited by Angelo Basile, Luisa Di Paola, Faisal Hai and Vincenzo Piemonte

77 **Pervaporation, vapour permeation and membrane distillation: Principles and applications**
Edited by Angelo Basile, Alberto Figoli and Mohamed Khayet

78 **Safe and secure transport and storage of radioactive materials**
Edited by Ken Sorenson

79 **Reprocessing and recycling of spent nuclear fuel**
Edited by Robin Taylor

80 **Advances in battery technologies for electric vehicles**
Edited by Bruno Scrosati, Jürgen Garche and Werner Tillmetz

81 **Rechargeable lithium batteries: From fundamentals to applications**
Edited by Alejandro A. Franco

82 **Calcium and chemical looping technology for power generation and carbon dioxide (CO_2) capture**
Edited by Paul Fennell and Ben Anthony

83 **Compendium of Hydrogen Energy Volume 1: Hydrogen Production and Purificiation**
Edited by Velu Subramani, Angelo Basile and T. Nejat Veziroglu

84 **Compendium of Hydrogen Energy Volume 2: Hydrogen Storage, Transmission, Transportation and Infrastructure**
Edited by Ram Gupta, Angelo Basile and T. Nejat Veziroglu

85 **Compendium of Hydrogen Energy Volume 3: Hydrogen Energy Conversion**
Edited by Frano Barbir, Angelo Basile and T. Nejat Veziroglu

86 **Compendium of Hydrogen Energy Volume 4: Hydrogen Use, Safety and the Hydrogen Economy**
Edited by Michael Ball, Angelo Basile and T. Nejat Veziroglu

87 **Advanced district heating and cooling (DHC) systems**
Edited by Robin Wiltshire

88 **Microbial Electrochemical and Fuel Cells: Fundamentals and Applications**
Edited by Keith Scott and Eileen Hao Yu

89 **Renewable Heating and Cooling: Technologies and Applications**
Edited by Gerhard Stryi-Hipp

90 **Small Modular Reactors: Nuclear Power Fad or Future?**
Edited by Daniel T. Ingersoll

91 **Fuel Flexible Energy Generation: Solid, Liquid and Gaseous Fuels**
Edited by John Oakey

92 **Offshore Wind Farms: Technologies, Design and Operation**
Edited by Chong Ng & Li Ran

93 **Uranium for Nuclear Power: Resources, Mining and Transformation to Fuel**
Edited by Ian Hore-Lacy

94 **Biomass Supply Chains for Bioenergy and Biorefining**
Edited by Jens Bo Holm-Nielsen and Ehiaze Augustine Ehimen

95 **Sustainable Energy from Salinity Gradients**
Edited by Andrea Cipollina and Giorgio Micale

96 **Membrane Technologies for Biorefining**
Edited by Alberto Figoli, Alfredo Cassano and Angelo Basile

97 **Geothermal Power Generation: Developments and Innovation**
Edited by Ronald DiPippo

 98 **Handbook of Biofuels' Production: Processes and Technologies (Second Edition)**
     Edited by Rafael Luque, Carol Sze Ki Lin, Karen Wilson and James Clark
 99 **Magnetic Fusion Energy: From Experiments to Power Plants**
     Edited by George H. Neilson
100 **Advances in Ground-Source Heat Pump Systems**
     Edited by Simon Rees
101 **Absorption-Based Post-Combustion Capture of Carbon Dioxide**
     Edited by Paul Feron
102 **Advances in Solar Heating and Cooling**
     Edited by Ruzhu Wang and Tianshu Ge
103 **Handbook of Generation IV Nuclear Power Reactors**
     Edited by Igor Pioro
104 **Materials for Ultra-Supercritical and Advanced Ultra-Supercritical Power Plants**
     Edited by Augusto Di Gianfrancesco
105 **The Performance of Photovoltaic Systems: Modelling, Measurement and Assessment**
     Edited by Nicola Pearsall
106 **Structural Materials for Generation IV Nuclear Reactors**
     Edited by Pascal Yvon
107 **Organic Rankine Cycle (ORC) Power Systems: Technologies and Applications**
     Edited by Ennio Macchi and Marco Astolfi
108 **Advances in Steam Turbines for Modern Power Plants**
     Edited by Tadashi Tanuma
109 **The Performance of Concentrated Solar Power Systems: Modelling, Measurement
     and Assessment**
     Edited by Peter Heller
110 **Advances in Concentrating Solar Thermal Research and Technology**
     Edited by Manuel Blanco and Lourdes Ramirez Santigosa
111 **Integrated Gasification Combined Cycle (IGCC) Technologies**
     Edited by Ting Wang and Gary Stiegel
112 **Nuclear Facilities: A Designer's Guide**
     Bill Collum

Introduction

My first impression on joining the nuclear industry was that I barely understood what everyone around me was talking about. Experienced personnel were using words and acronyms that I did not recognize, tackling problems I could just occasionally get the gist of and discussing concepts that were science fiction to my ears. Of course I did my best to join in, but to be honest, those early days were quite bewildering. Jump forward a few decades and I got to thinking maybe, just maybe, I could write a book that explains how all this nuclear stuff works. So what happened in between?

Well, I found myself in the fortunate position of rubbing shoulders with engineers and other specialists who were awfully good at what they did, many of them national or even international experts. Even more fortuitous, when I asked probing questions about their particular field of expertise—which I did continually—they were invariably generous with their time and enthusiastic to share as much knowledge as I could possibly absorb. As a consequence, I spent many a happy hour deep in conversation, scribbling feverishly and feeding my enduring fascination with all things nuclear. Added to that, I had the good fortune to work on a wide variety of projects, both in the United Kingdom and internationally, from new-build facilities to decommissioning those that had seen out their days, and with budgets ranging from a few tens of thousands of pounds to vast multibillion pound complexes.

The area that has always interested me most is the front-end stage, whether it be planning a new building or figuring out how to decommission one which already exists. In the case of new build, it begins with what is literally a blank sheet and ends with a frozen scheme which is ready for detail design. For decommissioning, the challenge is compounded by having to grapple with radiological issues which constrain how a building can safely be taken apart. As challenges go nuclear front-end is right up there, so I was keen to get involved.

I soon realized that, more than any other design phase, the period from a project's inception to reaching a frozen design is all about compromise—after all, with so many interested parties to satisfy, there has to be a high degree of give and take. Furthermore, I determined that the main skill required to coordinate this stage is arbitration, and to do that well would require the widest possible knowledge about what all of the various disciplines were grappling with every day, otherwise it would be impossible to become an effective *referee*. I resolved to listen up, gather as much multidisciplinary knowledge as I possibly could, and strive to earn a seat at the front-end table.

It was a long road, very long indeed, but finally I reached the point where I knew enough to look across the various design phases, see a clear picture (well, clear-ish), and eventually play a strategic role in pulling the whole exercise together. Along the way I often lamented the fact that hardly any of the information I craved was actually written down, hence my motivation for writing this book.

Whatever your interest in *nuclear*, I do hope you will enjoy reading the book and, for those of you who work in the industry, I hope it will make some contribution to the accumulation of knowledge necessary for your existing role and others you may perform in the future.

Nuclear fuel cycle

1

If there is any image that can justifiably claim to represent the nuclear industry, then it must be that of a dome-topped reactor building such as that shown in Fig. 1.1. On the face of it their trademark profile seems to epitomize all things nuclear, but in truth it is just one building, one step in a long process known as the *nuclear fuel cycle*.

Fig. 1.1 Nuclear reactor dome.
© EDF Energy.

To appreciate the breadth of facilities involved and the role each of them plays, we need to start by examining the entire cycle, then with the scene set in later chapters we can delve into a closer look at how such facilities function and what it takes to put them together. To help us keep track, Fig. 1.2 captures this entire chapter in diagrammatic form and, surprisingly enough for such a hi-tech industry, we shall see that it all begins with some very rough and tumble engineering.

It is quite improbable, isn't it, to think that rock which is mined in a conventional way and begins its journey on trucks the size of a house, will one day find itself in a nuclear reactor giving off fierce heat and generating the electricity that powers our lives, but unlikely as it may seem, there it is. So how does the transformation take place?

Nuclear Facilities. http://dx.doi.org/10.1016/B978-0-08-101938-2.00001-5

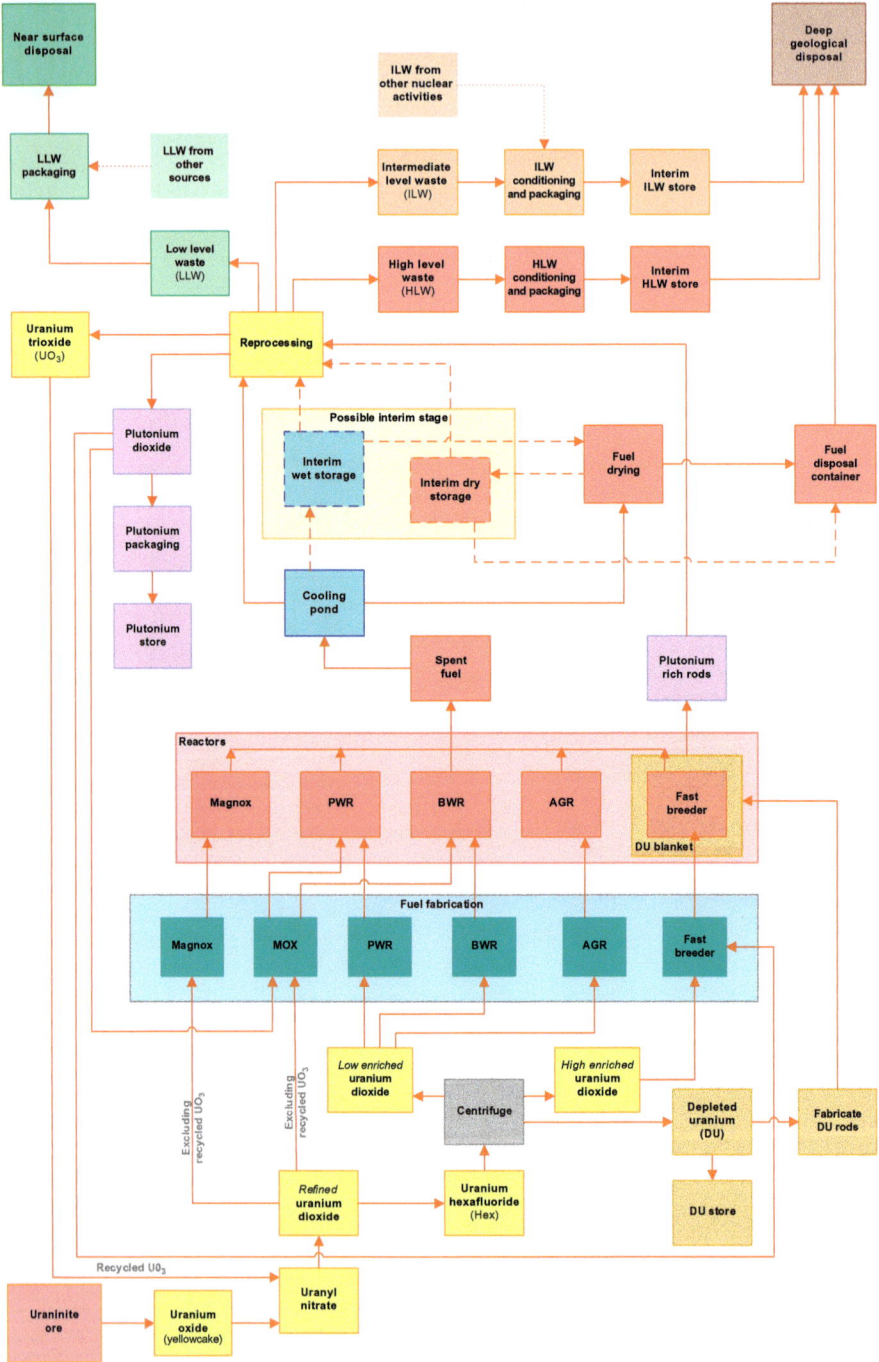

Fig. 1.2 Nuclear fuel cycle.
© Bill Collum.

1.1   Uranium mining and purification

Just as some rock bears gold, silver, copper, and so on, others contain uranium; in fact, it can be found pretty much all over the globe, particularly in granite but also to some extent in all soils and even in seawater. Ordinarily the amounts present in rock are measured in just a handful of parts per million, so mining is not a realistic proposition. However, there are areas of the world, such as Australia, Canada and Namibia, along with several others, where uranium exists in sufficient quantities to make its extraction economically viable.

Fig. 1.3 Opencast uranium mine.
Shutterstock.com.

The depth at which suitable rock exists varies, to the extent that while some is deep underground, most uranium is recovered from in-situ leaching or extracted from opencast mines (Fig. 1.3). Whatever its origin, once mined the uraninite ore, or *pitchblende*, is crushed and milled, and conventional metal extraction techniques used to liberate the uranium bound up within it. The resulting product is uranium oxide, also known as uranium ore concentrate but more commonly referred to as yellowcake.

During these early stages of the nuclear fuel cycle, there is no risk from penetrating radiation, although there is some radiological hazard due to uranium dust and the presence of radon gas. Such hazards are discussed in the next chapter, so I will just say here that dust masks and a mine's ventilation systems are employed to protect the workforce. Other hazards arise as a result of regular mining operations, but these are of a more conventional nature.

Once produced, the yellowcake (Fig. 1.4) is loaded into drums and transported to a uranium processing and fuel fabrication facility, where the first process is to remove any remaining contaminants by dissolving the yellowcake in nitric acid and turning it into a purified liquid solution called uranyl nitrate. The uranyl nitrate then undergoes a chemical process which converts it into uranium dioxide (UO_2). However, as we shall see later, there are several variations of UO_2, a bit like the way coffee is coffee but comes in different strengths, so at this stage we can refer to it as *refined* uranium dioxide.

Fig. 1.4 Yellowcake.
Istock.

Now that we have a refined uranium product, the next step depends on the particular fuel being fabricated. Magnox fuel, for example, which powered many of the UK's older reactors, was made from purified natural uranium, so required no additional treatment other than to convert the solution back into the uranium metal from which Magnox fuel was manufactured. Incidentally, if you ever get the opportunity to pick up a Magnox fuel rod (Fig. 1.5) brace yourself. Uranium is more than one and a half times the weight of lead, so these fuel rods, which are about a meter in length, take a bit of lifting.

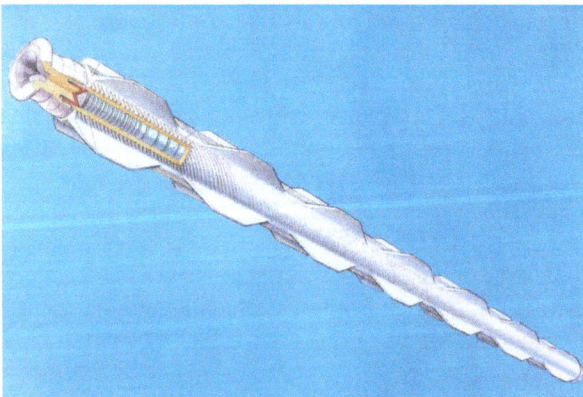

Fig. 1.5 Magnox fuel rod.
© Nuclear Decommissioning Authority.

Unlike Magnox, uranium for modern fuels such as those used in an advanced gas-cooled reactor (AGR), pressurized water reactor (PWR), and boiling water reactor (BWR), needs some additional processing before it is ready to be manufactured. However, before we get to that I need to explain something of the composition of purified natural uranium then we can look at how it is adjusted to produce a more efficient fuel.

1.2   The atom

From here on there will be some reference to *elements*, *isotopes* and parts of the atom so we need to take a brief tour of what the different terms mean (Fig. 1.6).

Atom	Collection of protons, neutrons, and electrons
Element	Pure substance formed by just one type of atom, which typically exists in a "family" of several variations
Isotope	A member of the family of atoms which form a given element
Proton	Within the family of atoms which form a given element, the number is always the same Carries a positive charge
Neutron	Within the family of atoms which form a given element, the number can vary Does not carry a charge
Electron	Within the family of atoms that form a given element, always equals the number of protons Orbits the nucleus and carries a negative charge
Nucleus	Core of an atom within which protons and neutrons reside

Fig. 1.6 Atom constituents.
© Bill Collum.

If you were to take a small block of sandstone, place it on a table and have a good look at it, it would quite clearly be one object, a solid lump so to speak. But then if you scraped it hard enough with a tough steel chisel small particles, or grains of sand, would come off and settle beside it. Moreover, if you happened to have a powerful microscope handy and were to analyze the individual grains, you would find that they were all different: their color, shape, size, and weight would vary a little. Well, at an atomic level something very similar is going on.

Fig. 1.7 The atom.
Shutterstock.com.

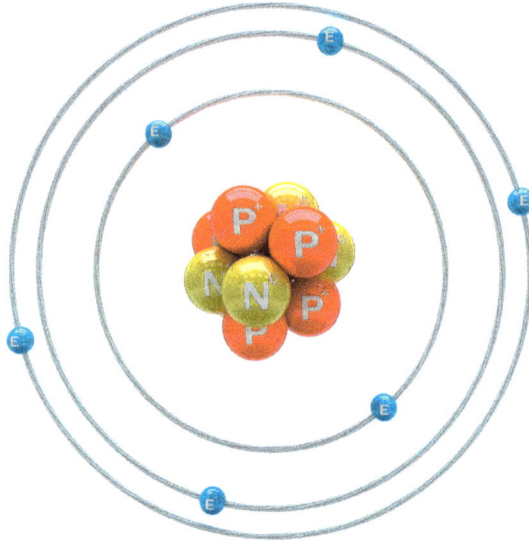

Everything around us, whether it be solid, liquid, or gas, is actually made up from billions of infinitely microscopic components that we know as atoms (Fig. 1.7) and just as with our grains of sandstone the atoms that make up solid objects are, in most cases, not all identical. Furthermore, all pure substances, such as copper, silver, uranium, plutonium, and many more are known as *chemical elements*, with the vast majority comprising a group of atoms referred to as *isotopes*.

Happily, when it comes to differences at an atomic level, we find more order than is revealed in grains of sand; copper, for example, is limited to just 6 different variations of atom (or isotope), silver 11, carbon 5 variations, and so on. To complete this particular theme, with a bit more science, we need to be aware that within their nucleus, atoms contain protons and neutrons which vary in number from one type of atom to another. However, for any given element the number of protons in each of its isotopes is always identical, so their differences arise from the number of neutrons within them. What's more, if you scrutinize atoms within any given element closely enough, you will find the number of positively charged protons and negatively charged electrons is always equal.

A full list of the more than 100 chemical elements can be found in the *periodic table*. If you examine it carefully, you will find that a small number of elements such as Dubnium (Db-258) and Hassium (Hs-265) are indeed formed from just a single atom type, rather than a family of isotopes, which is equivalent to grains of sandstone all being identical. Ordinarily, though, and certainly for our discussion here, we can think of elements as comprising multiple isotopes.

We shall cover all of this in more detail over the coming pages, so for the moment it is just important to note that, at an atomic level, all of the elements we shall be discussing are made up from several isotopes, or atom types, each with their own distinct characteristics.

1.3 Enrichment

Uranium isotopes	Percentage
U-238	99.27
U-235	0.72
U-234	0.01

Fig. 1.8 Composition of natural uranium.
© Bill Collum.

Fig. 1.8 shows how natural uranium consists of three different isotopes with over 99% being U-238. Interestingly though it is the U-235, which although it accounts for less than 1%, is the fissile component that actually makes a nuclear reactor work. For our purposes here uranium's minute proportion of U-234 does not need to be considered, so we can concentrate on the U-235 and U-238.

While reactors such as the Magnox type do operate with refined natural uranium, others perform more effectively by using uranium fuel that has a higher fissile content. This *enriched* uranium has an increased concentration of U-235, typically (for civil reactors) up to around 4% or 5%, which enables it to generate heat much more efficiently and therefore use less fuel to deliver the same, or even improved performance from a reactor.

Enrichment begins by using a chemical process to convert the refined UO_2 into a compound by the name of uranium hexafluoride, more commonly known as *hex*. The great benefit of hex for this particular process is that it exists as a gas at around 57°C, a relatively low temperature which in this context is easily achieved. It is crucial factor because once you have a gas it becomes possible to carry out molecular separation, and from there to increase the relative quantity of uranium's U-235 isotope.

Enrichment can be performed by either a centrifuge or diffusion process, but for modern plants gas centrifuge is the norm. Actually centrifuge technology has quite widespread applications, ranging from the medical field where it is used to separate plasma from red blood cells, to those huge machines that spin fighter pilots so they can have a taste of what awaits them when they get airborne. As for the centrifuges we are discussing here, there is much about the detail of their workings that is closely guarded; happily, though, there is no restriction on discussing the basic principles, which are widely disseminated.

Enrichment centrifuges create a vortex into which the hex gas is introduced. Once inside, it behaves in exactly the same way as any substance within a vortex; heavier particles are always thrown out towards the perimeter, while lighter particles swirl around at its core. However, the molecular weights of U-235 and U-238 are very similar with just 0.0086% difference between the heavier U-238 and its lighter companion. This makes it very difficult to separate the two, even within equipment which for enrichment processes spins at around the speed of sound. Fig. 1.9 gives an impression of the process within a Zippe-type centrifuge. With this particular design the base is heated, which creates convection currents that improve their efficiency.

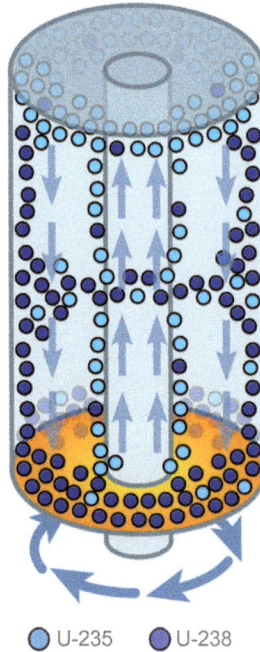

Fig. 1.9 Zippe-type gas centrifuge.
Fastfission.

With such a close correlation between their weights, a single centrifuge can only achieve minimal separation of the two isotopes. The gas is therefore sent on from one centrifuge to another in what is known as a cascade (Fig. 1.10) each enriching the hex U-235 content a little more than the last until it reaches its required concentration. For mainstream reactors such as PWR and BWR the hex is brought to a level which is termed *low enriched*, while that destined for the less commonplace fast breeder reactor (FBR) is raised to a *high enriched* level. All that remains then, is to use a chemical process to convert the enriched hex back into the refined UO_2 powder I mentioned earlier, from which fuel can be produced.

To complete this particular part of the picture we need to stay for a moment with what is happening inside the centrifuges. Clearly their primary purpose is to create enriched hex U-235, for later fuel production, but this still leaves the remaining hex to deal with. As we have seen, the heavier U-238 is forced out towards the perimeter of a centrifuge, from where it is drawn off in the same way as enriched

Fig. 1.10 Centrifuge cascade.
© URENCO.

hex. The product, known as *tails*, that is produced via this route is depleted uranium (DU), so called because it has been substantially stripped of its U-235 and now contains around just 0.2% of this isotope.

Because of its high density, DU is occasionally used in medical and industrial equipment to provide shielding from radioactive sources, which, bearing in mind that the other product from a centrifuge will 1 day be emitting fierce radiation from within a reactor, always strikes me as quite an elegant symmetry.

1.4 Fuel fabrication

Now that we have UO_2 we can begin the actual fuel fabrication process. The powder is compressed to form small pellets which are heated, or *sintered* to be precise, and then machined to very tight tolerances. Exact dimensions vary from one reactor type to another but nuclear fuel pellets (Fig. 1.11) are generally around 1 cm in diameter and a little more than that in length.

Having been machined and thoroughly checked, pellets are clad within one of the long thin rods, or *pins* that characterize modern fuels. And while AGR fuel cladding is fabricated from stainless steel, PWR and BWR fuel cladding is formed from the much more exotic sounding zirconium alloy.

Once rods are fabricated they are inserted into fuel assemblies, with their exact configuration depending on the reactor type for which the fuel is destined. AGR fuel is bundled in a circular assembly housed within a graphite sleeve (Fig. 1.12) while

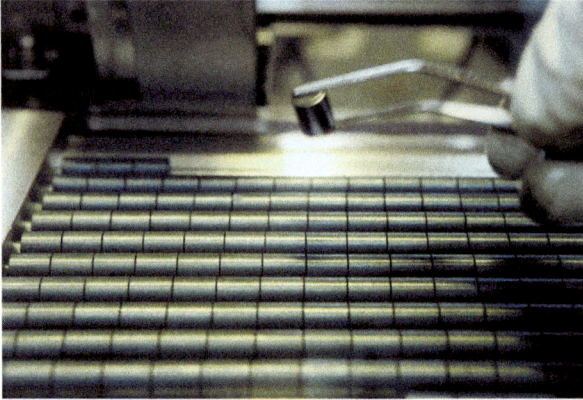

Fig. 1.11 Nuclear fuel pellets.
Science Photo Library.

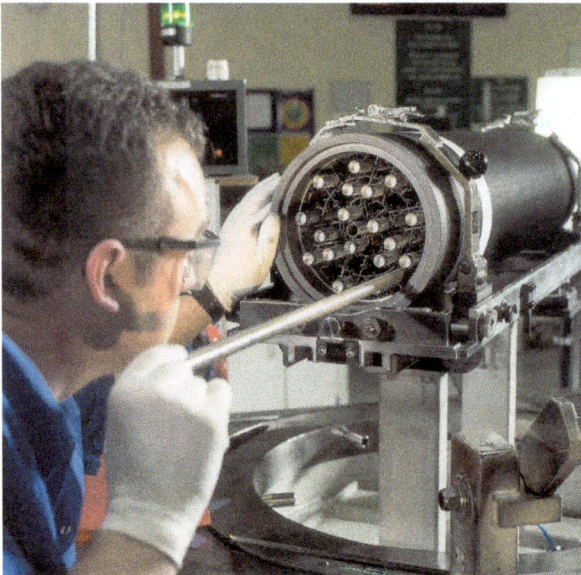

Fig. 1.12 AGR fuel assembly.
© Nuclear Decommissioning Authority.

BWR and PWR (Fig. 1.13) fuel assemblies are arranged in square grids which, like the rods, are formed mainly from zirconium alloy.

The fuel assemblies are now ready to be transported to their various reactors and by the way are still perfectly safe to approach. In fact they will not emit

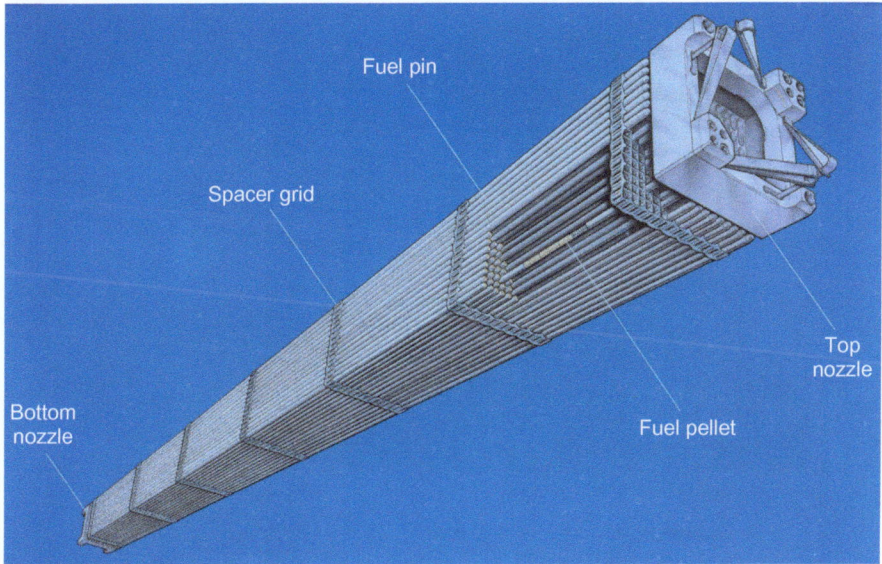

Fig. 1.13 PWR fuel assembly.
© Westinghouse.

penetrating radiation until the point of entering an operating nuclear reactor, so at this stage it is simply their sheer weight that rules out any possibility of manually handling them.

1.5 Nuclear reactors

As for reactors themselves they could almost be described as boilers, granted on a grand scale and very sophisticated to say the least, but when you get right down to it their job is to provide heat. It might seem like a lot of trouble to boil water, but then they are very good at it. As we have seen, modern civil thermal reactors come in three primary guises, PWR, BWR, and AGR, so we shall concentrate on those. And while Magnox reactors are no longer being built I will, for completeness, make some reference to them as well.

If you were to look into the core of an operating AGR (Fig. 1.14) or a Magnox reactor, you would find a pressurized gas, normally carbon dioxide (CO_2) circulating among their fuel assemblies, while, as their names suggest, PWR and BWR have water occupying the same space. In all cases the ferocious heat radiating from fuel rods is transferred to the gas or water which comes into contact with them.

All that remains then is to get that heat out of a reactor so that it can be put to work in driving steam turbines. To achieve this, AGR and Magnox reactors utilize a heat exchanger, while PWRs (Fig. 1.15) employ a steam generator which operates on similar principles.

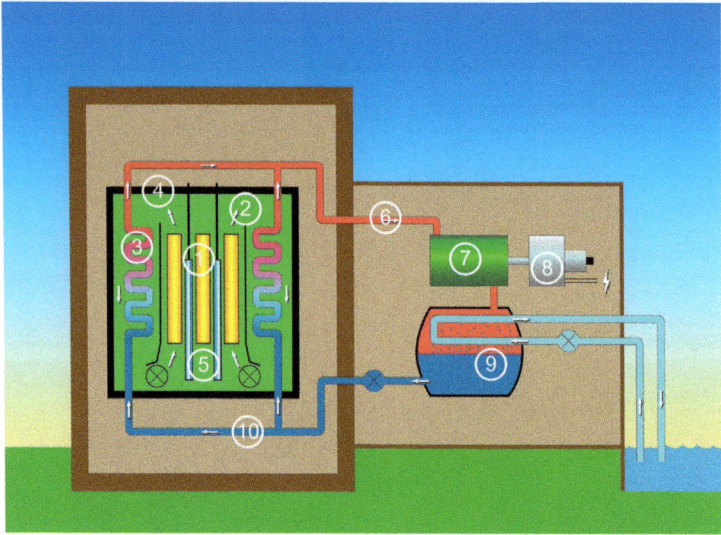

1 Fuel	2 Gas	3 Heat exchanger
4 Charge floor	5 Control rods	6 Steam
7 Turbine	8 Electricity generator	9 Condenser
10 Water		

Fig. 1.14 Advanced gas-cooled reactor (AGR).
Science Photo Library.

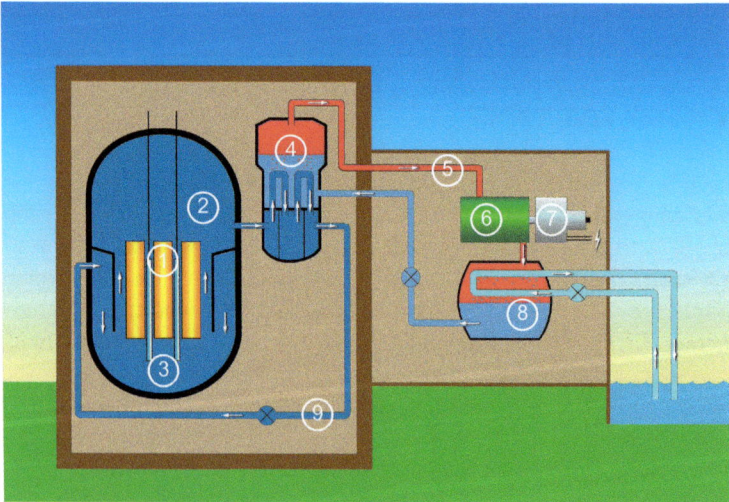

1 Fuel	2 Pressurised water	3 Control rods
4 Steam generator	5 Steam	6 Turbine
7 Electricity generator	8 Condenser	9 Water

Fig. 1.15 Pressurized water reactor (PWR).
Science Photo Library.

Heat exchanger

If you have ever left a garden hose stretched out in bright sunlight and then turned on a tap to start the water flowing, you will have experienced something of how a heat exchanger operates.

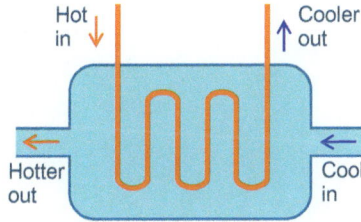

Fig. 1.16 Heat exchanger.
© Bill Collum.

Sunlight beaming down will heat an exposed garden hose and much of that heat is transferred to the water inside it, so much so that when a tap is turned on warm water will flow for a little while until fresher cold water pushes it out. Heat exchangers (Fig. 1.16) operate on much the same principle, where the medium heated within a reactor's core transfers its heat to water flowing through pipework on a different circuit.

The efficiency of a heat exchanger depends in large measure on the transfer medium, with water being much more effective than gas. However, for maximum efficiency we need look no further than the FBRs we shall discuss later. They use liquid metals such as sodium or lead-bismuth eutectic which are fantastically efficient at carrying the heat absorbed from reactor fuel rods.

BWRs (Fig. 1.17) operate on a different principle which enables them to omit a heat exchanger from their simpler heat transfer circuit. Instead, steam produced by a reactor is routed to its turbines (Fig. 1.18) after which it is condensed out before being sent back to the reactor to be superheated again.

Apart from their heat transfer medium, the other major difference in reactor design is the way fuel is loaded into and removed from them. AGR and Magnox reactors conduct transfers via a *charge face*, typically in the form of a *charge floor* (Fig. 1.19) directly above a reactor's core, while fuel loading for PWR and BWR takes place entirely underwater.

In practice, then, PWR and BWR, which also operate under pressure, must be shut down while fuel transfers take place, whereas lower pressure, albeit less efficient, AGR and Magnox reactors can continue operating during their equivalent transfer sequence. As a result PWR and BWR fuels are transferred in a major campaign, while operators of AGR and Magnox reactors can conduct refueling operations at a more piecemeal, steadier pace. Fig. 1.20 shows a PWR pressure vessel with its heavily shielded lid removed and floodlights illuminating the area which is about to receive a new fuel assembly.

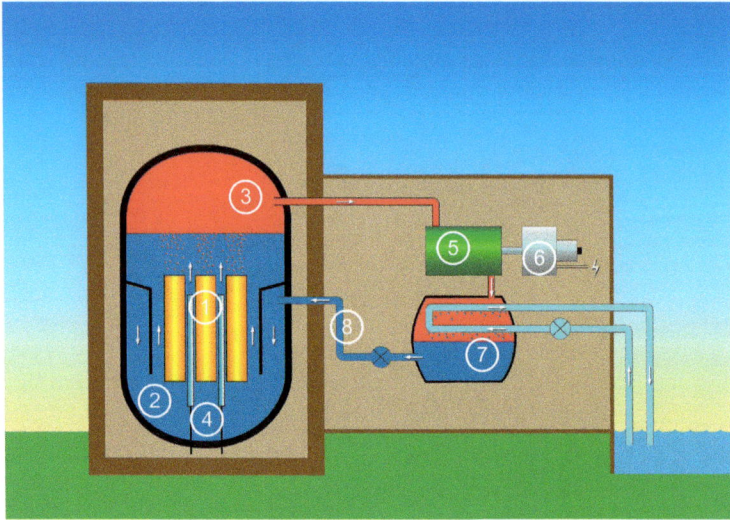

1 Fuel   2 Boiling water   3 Steam

4 Control rods   5 Turbine   6 Electricity generator

7 Condenser   8 Water

Fig. 1.17 Boiling water reactor (BWR).
Science Photo Library.

Fig. 1.18 Turbine hall.
© EDF Energy.

Fig. 1.19 Reactor charge floor.
Science Photo Library.

Fig. 1.20 PWR fuel loading.
Science Photo Library.

Before we leave the subject of reactor types, it is worth noting that by no means are all of them housed beneath a dome topped containment building. Many reside within cylindrical, or *can* shaped buildings, while others occupy structures with a more conventional boxlike profile. It all comes down to the specifics of each case and, in no small part, the preference of design teams.

1.6   Nuclear reaction

Now that we have seen something of the different types of reactor, we can move on to examine the nuclear reaction that takes place inside them. Personally I still find it quite remarkable to think that a fuel assembly which you could safely hug, if you had a mind to, just before it enters a reactor, will soon afterwards reach temperatures measured in several hundred degrees centigrade and be so radioactive that even momentary exposure would be lethal. So what happens?

Natural uranium, bound up in rock for example, spontaneously emits neutrons, primarily from the U-238 isotope which makes up over 99% of its bulk. In its natural setting the neutrons disperse with no significant consequences. So unless they are being observed by someone with very sophisticated equipment and a keen interest in such things, no one notices and life goes on. However, if you put enough pure uranium together—and there can be over 100 tonnes in many civil reactors—then the number of neutrons emitted and colliding with U-235 atoms around them increases significantly. And once this happens we are on our way to creating the conditions necessary to power a nuclear reactor. Better still, if uranium is enriched by increasing its proportion of U-235 and it is arranged it in just the right configuration, then the chances of those collisions occurring multiplies even further.

Left to their own devices, neutrons travel so fast that they will glance off the U-235 atoms like a flat stone skipping across a lake, and once again nothing of any consequence will happen. What we need to get a reaction going is to slow the neutrons down, but not too much, otherwise they will have insufficient energy to do what is required of them. So getting their speed right is a fine balance.

The medium used to surround fuel rods and slow hurtling neutrons is called a *moderator*; Magnox and AGR reactors utilize graphite, while, as their names suggest, PWR and BWR use water for the same purpose. In fact, water in the primary circuits that we discussed earlier doubles up to provide this moderator function. Actually, to be absolutely precise, it is particular atoms within a moderator that enable it to fulfill its role: thus, carbon atoms within graphite and hydrogen within water.

1.6.1   Neutron capture

Once they have been slowed to just the right speed, a process known as *neutron capture* occurs, where neutrons leaving the nucleus of their host U-238 atom are absorbed by U-235 atoms with which they collide. It is somewhat akin to a baseball being enveloped by a catcher's glove, which is why *capture* is such an appropriate term.

Fig. 1.21 Neutron capture. © Bill Collum.

Having another neutron on-board (Fig. 1.21) briefly transforms a U-235 nucleus into U-236 which is unstable, to the extent that the atoms are torn apart in a violent reaction that we know as nuclear *fission*. And this is the pivotal moment from which so much follows. Incidentally when it comes to capturing neutrons different atoms have varying levels of success; their adeptness is quantifiable, however, and is expressed on a scale known as the *barn*.

1.6.2 Nuclear fission

As an atom begins to tear itself apart a sequence of events is set in motion, some of which are over the instant they begin, while others live on for millions of years. In fact at that frozen moment of time, there can hardly be a nanosecond on Earth with the potential for such abiding consequences.

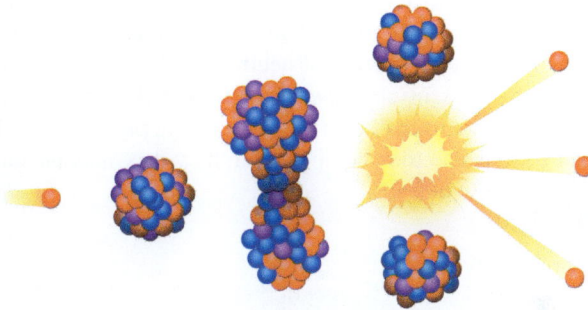

Fig. 1.22 Nuclear fission. Shutterstock.com.

When a U-235 atom splits asunder it results in the creation of other atoms, such as caesium-137, strontium-90, iodine-131 and many more, and with that the heat generation process begins. More importantly it releases more neutrons, which in turn collide with more U-235 atoms. And so it goes, creating the cycle known as a *chain reaction*. So nuclear fission (Fig. 1.22) is the process whereby atoms are split apart, while a chain reaction occurs in the controlled conditions that exist within in a nuclear reactor, where fission can be sustained pretty much indefinitely. And an out of control chain reaction can generate sufficient heat to melt metal fuel cladding, so appropriately enough is known as *meltdown*.

As if all of the energy bursting from rapidly dividing atoms wasn't generating enough heat, once it begins, additional reactions join the fray to elevate temperatures even further. It comes about because one of the consequences of a rapidly unfurling and violent chain reaction is heat emanating from the decay of newly created radioactive atoms.

1.6.3 Half-life

The lifespan of an isotope is defined in terms of how long it takes for half of its atoms to decay, a duration referred to as its *half-life*. For some isotopes such as iodine-131 their half-life is measured in just a few days. Others, for example caesium-137 need just over 30 years for the same process to occur, while at the far extreme, isotopes such as uranium-238 have a half-life that spans millions of years. And because radiation is a by-product of decay the intensity of radiation, and heat, they emit also follows a similar pattern.

As a result of this decay process there are two immediate consequences. Firstly, the radiation bursting out from short lived isotopes dictates that reactors need considerable shielding around them to protect those in the vicinity. And secondly, when a reactor is shutdown it does not gradually cool in the same way as say a furnace or the irons we use to press our clothes. In fact decaying short half-life atoms continue to self-generate heat, to the extent that a reactor needs its cooling systems to operate for some months after a shutdown takes place.

1.7 Control rods

Bearing in mind that a chain reaction is caused by neutrons clashing with U-235 atoms, then to maintain a reactor's equilibrium, or to be more precise its *neutron flux*, we need some way halting a proportion of neutrons while they are in mid-flight. All reactors deploy control rods or shutdown rods to absorb surplus neutrons but, to varying degrees, other systems may also be employed in parallel. We shall look first at the role control of rods and then come back to how they may be supplemented.

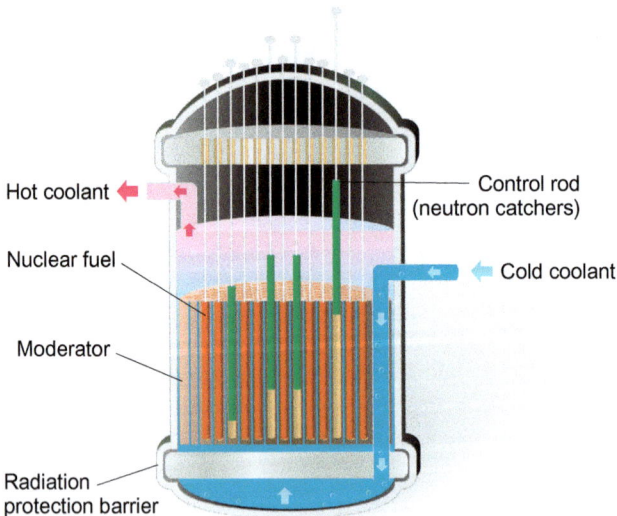

Fig. 1.23 Control rods.
Shutterstock.com.

Control rods are manufactured from materials such as cadmium, hafnium, silver and boron, which can absorb neutrons without any drama. Primarily because the isotopes produced are not fissionable, so halt propagation of a chain reaction. Normally the rods are deployed vertically into a reactor's core, either from above (Fig. 1.23) or below, and when fully extended will halt the fission process entirely. For reactors relying exclusively on control rods to maintain their neutron flux; if they were withdrawn while a reactor was loaded with fresh fuel, an uncontrolled chain reaction would quickly ensue and white-hot heat would not be too far away.

In these cases, then, control rods are effectively a reactor's temperature control mechanism, such that deploying them at just the right point holds a core at its optimum performance level. As the months go by, though, the fissile content of fuel decreases. To counter this reduction, control rods are withdrawn incrementally so as to sustain neutron levels and maintain a reactor's efficiency.

In an emergency, a chain reaction can be halted more quickly by deploying control rods, or *shutdown rods* into a reactor's core. If needed, they can be linked to safety systems which deploy them automatically if a potential problem is detected. When triggered this process is referred to as *tripping a reactor*, or alternatively a *scram*. Apart from an emergency shutdown situation, some reactors must halt their fission process while they are being refueled. Where this is the case, a similar shutdown mechanism is utilized.

1.8 Burnable poison

One of the other methods of maintaining neutron flux levels is to seed a neutron absorber, or *poison*, within a reactor's bioshield. With this passive system, surplus neutrons are absorbed by a compound such as gadolinium oxide (Gd_2O_3) also referred to as *gadolinia*, which is either incorporated within fuel pins during the fabrication process, or loaded into separate pins. This approach is particularly useful when fresh fuel is loaded into a reactor, as it can act as a damper to its initial high reactivity.

To capitalize on its neutron absorbing characteristics, gadolinium oxide is spread around a reactor in a predetermined pattern, one that enables it to soak up surplus neutrons at a rate which matches that of the surrounding fuel's gradually diminishing performance. In other words, even though their properties are gradually being altered as a result of the fission process, the Gd_2O_3 and fuel continue to cancel each other out at around the same rate. All that is needed then, is fine adjustments from a reactor's control rods to keep its neutron flux balanced at just the right level.

For completeness, I should add that similar effects can be achieved by boronating the water in a reactor's primary cooling circuit, although clearly this cannot be targeted in the same way as seeding neutron absorbers in specific locations.

1.9 Neutron activation

In the next chapter we shall look in some detail at the subject of radiation. However, there is a particular radiation issue which is primarily associated with reactors so for completeness I need to mention it here.

Ordinarily, if a radioactive source is placed close to an object that is not active in any way, say stainless steel which is a material that abounds in nuclear facilities; then once the radioactive source is moved away, it would be perfectly safe to approach the metal and handle it normally. In other words the stainless steel would not be affected in any way by radiation emanating from an object nearby, even if it was there indefinitely. However, neutrons streaming from the core of a nuclear reactor result in some atoms within the same stainless steel item becoming radioactive, due to a phenomenon known as *neutron activation*.

Before we look at how it works, I need to set the scene with a few words about the number we often see attached to the right side of an element's chemical symbol. It signifies its atomic mass, but more relevant for our discussion here these numbers also denote the sum of an element's protons and neutrons, which also defines what particular isotope it is.

Element	Symbol and atomic mass	Number of protons/ electrons	Number of neutrons
Uranium	U-238	92	146
Uranium	U-235	92	143
Uranium	U-234	92	142
Plutonium	Pu-239	94	145
Plutonium	Pu-240	94	146
Plutonium	Pu-241	94	147
Plutonium	Pu-242	94	148
Cobalt	Co-59	27	32
Cobalt	Co-60	27	33

Fig. 1.24 Example chemical element variations.
© Bill Collum.

As we saw earlier, uranium exists in several forms each with a different atomic mass, and interestingly exactly the same phenomenon applies to many other elements, such as the plutonium and cobalt shown in Fig. 1.24 It may seem like a fairly inconsequential difference but when the number of neutrons changes it actually alters the fundamental characteristics of an element, such as U-235 being fissionable but not U-238.

If we go back now to our item of stainless steel in a reactor and were able to examine its atomic structure, we would find that like all stainless steel it contains cobalt-59 which in itself is not radioactive. Being in a reactor environment, though, cobalt-59 atoms are prone to capturing the hurtling neutrons with which they are being bombarded and therefore transforming themselves to become cobalt-60 (Fig. 1.25).

However, unlike the U-235 atoms that we looked at earlier, cobalt-60 is not torn apart by the experience of capturing another neutron. It does though become unstable and in so doing changes its characteristics to become highly radioactive, or *activated*.

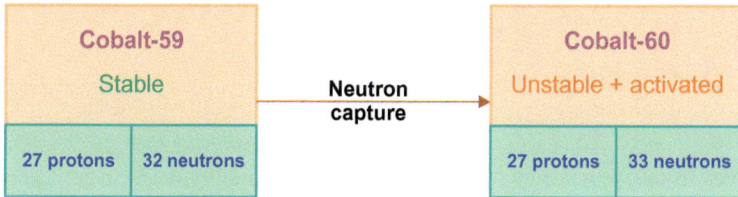

Fig. 1.25 Activation of cobalt-59.
© Bill Collum.

The same activation process is also visited on other elements within stainless steel, such as nickel and manganese, but to a lesser extent. And in any event the majority of these isotopes have a comparatively short life and the radiation they emit is contained within a reactor's bioshield. Normally then when activation is discussed in the nuclear industry, particularly in a decommissioning context, it is cobalt-60 which is considered to be the main culprit for turning benign stainless steel into a metal that is very radioactive indeed.

It might seem that as far as cobalt goes the story would end there and that it would simply decay through its half-life cycle, which in the case of cobalt-60 is 5.3 years. It turns out, though, there is a twist in the tale. Where an atom absorbs a neutron and becomes unstable, it immediately begins a process of wrestling within itself to try and become stable again. In the case of cobalt-60 this results in one of its neutrons being transformed into a proton. So rather than having 27 protons and 33 neutrons it ends up with 28 and 32, respectively, a feat which you have to admit is pretty impressive. Now, hold that thought for a moment.

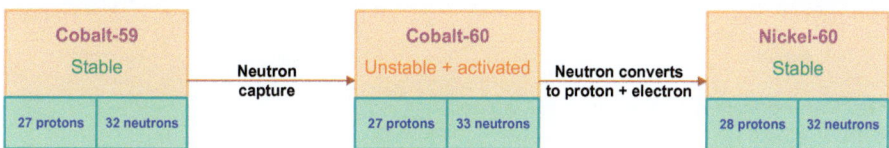

Fig. 1.26 Activation of cobalt-59 and subsequent decay.
© Bill Collum.

This is a good time to mention that the number of protons in an element is the factor that determines what material it is; so uranium always has 92, zinc 30, iron 26, and so on. The upshot is that once cobalt-59 transforms through cobalt-60 and eventually finds itself with another proton, it cannot stay as cobalt any longer. In fact, in a process known as nuclear transmutation, of one of its neutrons transforms into a proton (along with an electron) changing as it does into nickel-60 (Fig. 1.26) which is indeed made up of 28 protons and 32 neutrons. And crucially, in a nuclear context, nickel-60 is a stable isotope.

As a result, during its decay cycle the once activated stainless steel gradually decreases in radioactivity, transforming its cobalt-60 into nickel-60 until it comes benign again. Having said that, the irradiation level is so intense that activated steel needs to decay for around 30 years before it is safe to handle.

1.10   Decay chain

It might seem reasonable to assume that the phenomenon of isotopes transforming themselves from one material to another could only possibly occur in the uncommon confines of a nuclear reactor. But actually it has been happening in the natural environment since the beginning of time and in this context is known as the *decay chain*. It affects quite a wide range of elements but the one we are most interested here is uranium, which in its natural form (as we saw earlier) comprises over 99% U-238.

Left to its own devices U-238 will undergo radioactive decay, transitioning as it does through thorium, protactinium, radon, polonium and other wonderfully sounding elements, until after billions of years it eventually settles down into its final form as lead-206, which is a stable isotope. Interestingly, though, when the same U-238 spends time in a nuclear reactor some of it can take quite a different path, primarily because instead of losing neutrons it can gain another one.

1.11   Plutonium creation

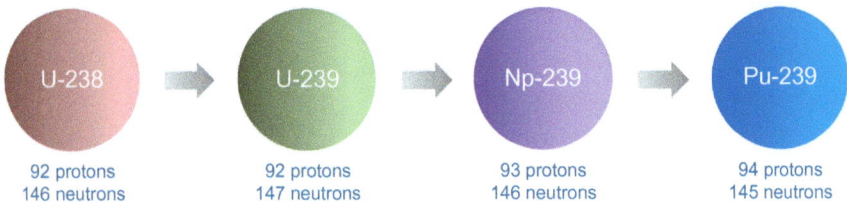

Fig. 1.27 Formation of plutonium in a reactor.
© Bill Collum.

Fig. 1.27 shows the sequence, with the first step being the neutron capture process that we have just discussed, starting here with uranium-238. Once it becomes uranium-239 the efforts to stabilize itself results in one of its neutrons being transformed into a proton. As we know, once the number of protons changes so does the material itself, in this case from uranium-239 to neptunium-239. However, Np-239 has a short half-life of only 2.35 days, so this is little more than a transient stage. As the neptunium quickly decays another of its neutrons transforms into a proton and the material changes again, this time to plutonium-239. Actually it does not end there, as plutonium can continue capturing neutrons right up until it becomes Pu-246, but we shall stick with Pu-239 as it has a comparatively long half-life of 24,100 years.

Isotope	Percentage	Half-life (y)
Pu-239	53	24,100
Pu-240	25	6560
Pu-241	15	14.4
Pu-242	5	375,000
Others	2	Various

Fig. 1.28 Distribution of the approximately 1% plutonium in irradiated fuel.
© Bill Collum.

Creating plutonium within a reactor's uranium fuel is significant (Fig. 1.28) because just as with U-235, its Pu-239 and Pu-241 isotopes are highly fissionable. In other words capable of generating a chain reaction. The big difference though is that they have a higher fissionable content than uranium, so harbor the potential to power reactors much more efficiently.

The problem, though, is how to unlock plutonium that is bound up within a reactor's fuel so as to capitalize on the power it holds. But then only around 1% of the U-238 in spent irradiated fuel will have transitioned into plutonium isotopes, not all of it fissile, so at first glance it may appear that it is hardly worth the effort involved in retrieving it. However, over the operating life of a reactor that 1% equates to several tonnes of plutonium with a fissile content of around 68%, which compares very favorably with the meager 0.7% of fissionable U-235 that we find in naturally occurring uranium.

If we were to dwell on the subject for a little while, the thought would occur that if it were possible to isolate the highly fissile Pu-239 and Pu-241 from spent uranium fuel and blend it with the refined UO_2 powder we discussed earlier, it would be possible to boost uranium's fissile content without recourse to the centrifuge enrichment process. And having figured it out, we would be very pleased with ourselves to hear this is exactly what happens.

1.12   MOX fuel

The plutonium is separated out and retrieved through a reprocessing system that we shall examine later, and which results in the creation of a powder that contains the Pu-239 and Pu-241 isotopes. This plutonium dioxide (PuO_2) powder is then blended with refined UO_2 in a ratio of around 6% plutonium to 94% uranium, and subsequently formed into MOX fuel pellets. MOX by the way is simply an abbreviation for *mixed oxide*, so is a name that captures the heritage of this particular fuel very well.

MOX fuel assemblies are fabricated to the same geometry as PWR and BWR fuels, so that they can be used in those reactors alongside their standard fuel. Not only that, during the irradiation process fissile Pu-239 and Pu-241 are substantially burned up;

so apart from its ability to power a nuclear reactor, using plutonium in fuel can also be a very effective way of decreasing stockpiles and therefore, if deemed appropriate, contribute to a national reduction strategy.

So far it appears that using plutonium to power nuclear reactors is all pros and no cons; there are, however, two notable drawbacks. As we have seen the manufacturing process for AGR, PWR and BWR fuels is free of any significant radiological hazard, to the extent that no harm will come to us even from handing newly fabricated fuel rods. However, in this regard MOX fuel is not quite so obliging. The main culprit is its Pu-241 content, which decays to become americium-241, and as it subsequently decays Am-241 emits both beta and gamma radiation. We shall return to this theme in later chapters, so I will just say here that once americium forms in plutonium shielding will be needed to protect those nearby.

If it were possible to guarantee MOX fuel would only ever be fabricated from fresh plutonium, then americium would have little chance to form and penetrating radiation would not be a major issue. Unfortunately though such a proposition is impractical, so MOX fuel must always be fabricated with at least some degree of remote intervention, which of course adds several layers of complexity to the more hands-on processes employed in the manufacture of uranium-based fuels.

Fig. 1.29 Gloveboxes.
Science Photo Library.

The other major downside to handling plutonium, no matter what its age or isotope, is the harmful effects that ingesting even minute particles can have on our health. Again, this is a topic we shall discuss in the next chapter. For the moment the thing to bear in mind, is that in order to ensure a workforce cannot possibly breathe in its invisible dust particles, MOX fuel fabrication must be conducted within high integrity containment. For example, where processes can be performed manually, then they are conducted within the kind of sealed glovebox arrangement that is shown in Fig. 1.29.

You will recall that when we looked at the operation of gas centrifuges earlier, we found that DU is created as a by-product of the enrichment process. It does have some limited applications, such as the shielding I mentioned earlier, but now we know how plutonium forms within irradiated fuel rods it opens up the possibility of elevating the use of DU to a much higher calling.

1.13 Fast breeder fuel

This is where a FBR enters the scene (Fig. 1.30). They differ from the more familiar reactors in two fundamental ways. Firstly, their fuel is highly enriched with PuO_2 powder, typically in the order of 25–30% rather than the 5–6% for thermal reactors. For the rationale behind such high enrichment, we need to think back for a moment to our earlier discussion on neutron capture and the role of a moderator in slowing neutrons down. Armed with that knowledge, we can see that since FBRs do not have a moderator they need a higher fissile content to achieve criticality.

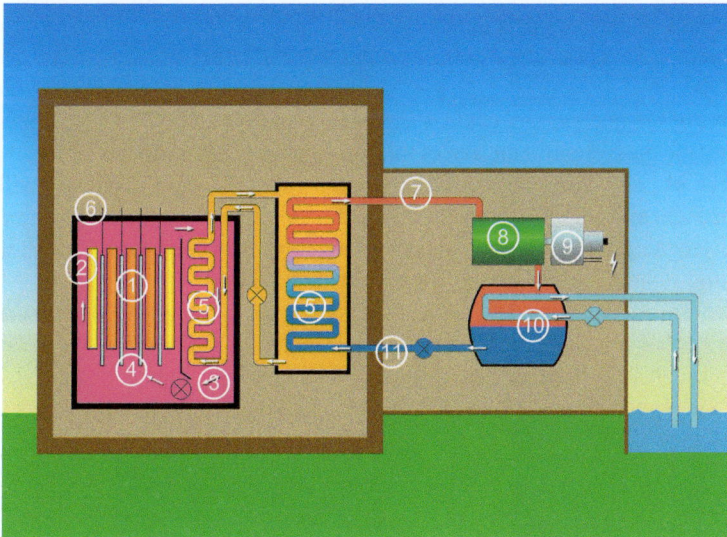

1 Fuel	2 Breeder blanket	3 Liquid metal
4 Control rods	5 Heat exchangers	6 Charge floor
7 Steam	8 Turbine	9 Electricity generator
10 Condenser	11 Water	

Fig. 1.30 Fast breeder reactor (FBR).
Science Photo Library.

Secondly, rods of DU (U-238) are placed around the inside of a reactor, in an arrangement known as a *blanket*. At first it sounds like a strange thing to do, because as we know U-238 is not fissile so will not contribute in any way to generating a chain reaction. However, the U-238 is there to soak up neutrons emanating from the reactor's fissile fuel and in so doing create plutonium, primarily Pu-239, hence their name.

When you put the whole thing together, it really is such an elegant concept. The depleted U-238, or tails, that might otherwise be discarded is put to meaningful use. And while plutonium is being burned up within a reactor's core, more is being created around it.

1.14 Spent fuel removal

If we go back now to the irradiated fuel in a reactor—and this applies to all fuel types—there comes a time when its fissile isotopes are burned up to the extent that their ability to generate heat begins to wane. Initially partially spent fuel is relocated, moving closer to the core of a reactor, while fresh fuel takes its place on the periphery. This rotation, which occurs at around one and a half year intervals, helps to balance the chain reaction which is always hottest at a reactor's core, while at the same time maximizing the efficient use of fuel as its power deceases during its operating, or *burn-up* cycle. Finally though the day dawns when fuel is so exhausted that it must leave a reactor for good; durations vary, but typically fuel remains in a reactor for around 5 years.

In terms of the nuclear fuel cycle, the point at which irradiated fuel leaves a reactor is arguably the most significant step of the entire journey. After all, the early stages are largely free of penetrating radiation and when it first occurs, it is initiated within the confines of a heavily shielded reactor. From this point on, though, things enter a very different league of complexity, because from now on we shall be handling or transporting radioactive materials, some of which are very active indeed.

From here on then, we leave reactors behind and instead need an assortment of facilities that are capable of processing, packaging and storing the radioactive materials which arise from numerous operations that follow. In later chapters, we shall examine the type of equipment found in such facilities, but first we need to get fuel out of the reactors.

1.14.1 Charge floors

As we have seen, fuel loading for AGR and Magnox reactors is conducted via a *charge face*, typically in the form of a *charge floor*, located above their core. The floor is peppered with an array of steel plugs which, as you would expect, are positioned directly above fuel channels in the core down below.

It can be an odd feeling the first time you step onto a charge floor, and ponder the fact that intense radiation and fierce heat is practically within touching distance from the soles of your shoes. Of course the whole floor matrix is heavily shielded so you can wander around quite safely. Nevertheless it is a pretty unique situation so is bound to get your attention until you become accustomed to it.

Fig. 1.31 Reactor refueling machine.
© EDF Energy.

Irradiated fuel is withdrawn from these reactors by use of a heavily shielded refueling machine. Their design can vary quite a bit from one reactor to another, but a not untypical example is shown in Fig. 1.31. They are equipped to handle both shield plugs and fuel assemblies, so can conduct the refueling exercise without allowing radiation to escape into the surrounding area. In fact you could even stand nearby while fuel transfers are taking place.

1.14.2   Cooling ponds

Fig. 1.32 Fuel cooling pond.
© Nuclear Decommissioning Authority.

Irradiated fuel leaving a reactor is still exceptionally hot, in the order of several hundred degrees centigrade and continues to self-generate heat for years to come, so the first destination for all fuel is a cooling pond (Fig. 1.32). AGR and Magnox reactors utilize their refueling machine to transfer spent fuel to a pond nearby, whereas the submerged PWR and BWR are configured to transfer fuel directly to an adjoining cooling pond.

Fig. 1.33 Initial fuel cooling periods.
© Bill Collum.

Reactor type	Duration in cooling pond
PWR and BWR	3 years
AGR	1 year
Magnox	3 months

These deep ponds serve the dual purpose of providing shielding and dissipating heat emanating from irradiated fuel. Recognizing a possibility that the integrity of fuel cladding may be compromised by corrosion, pond water is purged by circulating it through an effluent treatment facility which removes any contaminants before discharging it back to the environment. The cleanup processes, which we shall examine in Chapter 6, utilize ion exchange for soluble substances and other processes such as ultrafiltration to remove particulate matter. After a suitable period has elapsed (Fig. 1.33) the spent fuel, although still hot, will nevertheless have cooled sufficiently to be removed from its temporary pond and transported to a longer term home, one which is often located at a different nuclear site.

1.15   Spent fuel routing

Fig. 1.34 Fuel transport flasks.
© Nuclear Decommissioning Authority.

Before being sent on their way fuel assemblies are loaded, several at a time, into shielded flasks such as those shown in Fig. 1.34. Due to their length PWR and BWR assemblies are transported in long cylindrical flasks, while shorter AGR and Magnox assemblies travel in flasks with more of a cuboidal profile. We shall discuss flasks in Chapter 7, so I will just say here that the ones you may have seen traveling by road, rail, or occasionally by sea, are exceptionally robust and subjected to a series of tests before being licensed to carry such a hazardous payload.

Once fuel is loaded up and ready to travel there are broadly speaking two routes open to it, disposal or reprocessing. However, it does not necessarily need to head for its final destination straight away. In fact this is quite unlikely, so fuel will probably be routed first to an interim home. The main thing to be aware of here is that fuel can take either of the interim routes without foreclosing options for its

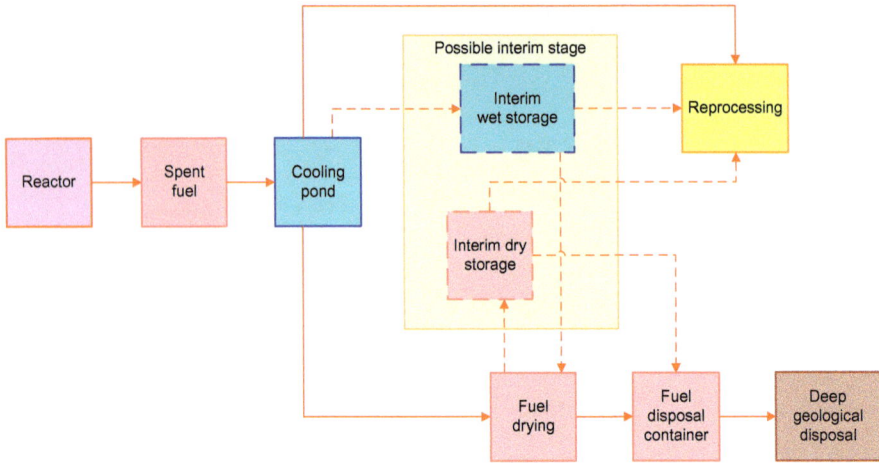

Fig. 1.35 Spent fuel routing options.
© Bill Collum.

ultimate destination. As we can see in Fig. 1.35, fuel is either moved to a pond at another location, or dried and stored. It may seem like an odd concept to take fuel from one pond only to move it to another one, but reactor site ponds normally have enough capacity to hold fuel from up to 10 years of operations. So once it has cooled sufficiently fuel must be moved so as to free up some space.

If, rather than dispatch spent fuel to another pond it is going to be dried out and stored, then the actual drying process is aided considerably by irradiated fuel's inherent heat. The crucial factor here is that fuel must be absolutely bone-dry before being stored, otherwise the combination of heat and moisture in a confined space would create a humid atmosphere that is just perfect for accelerating the corrosion of fuel cladding. This is clearly something to be avoided.

Several fuel assemblies are loaded together into a stainless steel container and, after a lid has been sealed in place, thoroughly dried to ensure any remaining traces of moisture have been driven off. Drying is typically carried out either within a vacuum or by a forced gas technique, probably using helium. To further inhibit any possibility of corrosion, the container is deprived of oxygen by backfilling it with an inert gas such as argon, after which it is ready to be stored.

Dry fuel that is destined for interim storage may be housed in a purpose built vault. But the most common practice is to place the fuel-filled containers, described earlier, into heavy concrete casks, such as that shown in Fig. 1.36. Whether it be in a vault or storage casks, dry fuel may remain in its interim home for up to 100 years before moving on to be reprocessed or disposed of.

Where spent fuel is destined for a permanent disposal repository, maybe a mile underground, then it is loaded into copper containers (Fig. 1.37). With a wall thickness of around 50 mm, they are both robust and geologically stable so are ideally suited to their long term role. Actually gold and platinum are equally stable, but are not the kind of materials you want to be purchasing by the tonne. The main constraint here is that

Fig. 1.36 Dry fuel storage cask.
© EDF Energy.

Fig. 1.37 Copper disposal canister.
© Posiva Oy.

fuel needs to cool until its temperature falls below a set threshold, still measured in several 100 °C, before it can safely be dispatched underground. Otherwise, heat arising from an accumulation of fuel could compromise the integrity of its cladding, or result in thermally induced spalling of a repository's chambers.

Drying fuel and disposing of it does have the attraction of being relatively straight-forward, but on the downside spent fuel still harbors tremendous energy that could be unlocked and used to manufacture new fuel. So maybe discarding it is not such a good idea after all? This is where spent fuel's reprocessing route comes in.

1.16   Reprocessing

You will recall that fresh thermal reactor fuel typically contains around 5–6% of uranium-235 and is enriched to that level by either a centrifuge process or, as in the case of MOX fuel, by blending plutonium dioxide with refined uranium dioxide. With such a small percentage, it would seem reasonable to assume that spent fuel leaving a reactor will have had all of its U-235 burned up, to the extent that disposal is the pretty much the only route it can take. However, it turns out that around 1% of fuel's U-235 still remains, along with another 1% of the newly created Pu-239 that we discussed earlier. In other words approximately 2% of spent fuel is still fissionable, and over the operating life of a single reactor this adds up to an awful lot of potential energy, so we need to think carefully about what happens next.

Of course, gaining access to that energy isn't exactly achieved in a click of the fingers. It takes quite an investment to build a facility that can do it, and generates a greater volume of radioactive waste than simply drying fuel and dispatching down the disposal route. Nevertheless, recycling in this way eases the burden on the world's uranium resources and makes plutonium available for more electricity production. All in all then, it is an option well worth considering.

We shall discuss the various types of nuclear waste in Chapter 13, so I will just mention here that it divides broadly into three categories. *Low level waste* (LLW) which presents minimal radiological hazard and seldom needs shielding, so can therefore be handled manually. *Intermediate level waste* (ILW) which can require 500 mm or more of concrete shielding around it, and *high level waste* (HLW) which typically needs to be contained by a meter or more of concrete. And in addition to its intense radiation, HLW also emits heat as a result of the decay processes we discussed earlier.

In nuclear circles the single word *reprocessing* (Fig. 1.38) is taken as a short form for the entire process of converting spent fuel back into products that can be used to manufacture new reactor fuel, so that one word will suffice. The first step in the process is to liberate fuel pellets by separating them from their outer cladding.

In the case of Magnox fuel, the rods are de-clad by feeding them through a powerful machine which strips away their cladding (Fig. 1.39) and turns it into a swarf, whereas oxide fuels rods such as AGR, PWR, and BWR are simply chopped into small sections by a shearing machine. Each segment, or *hull* as they are known, is around 35 mm long and when first sheared will still have fuel pellets within them.

The reason fuel rods are broken up in different ways, is that the next step in the reprocessing cycle involves dissolving fuel pellets, of any type, in nitric acid. This is fine for the stainless steel rods used to fabricate AGR fuel and the zirconium alloy which encases PWR and BWR fuels, as both of these materials are immune to attack from nitric acid so remain intact when submerged in it. On the other hand, the magnesium alloy used to fabricate Magnox fuel rods would react with nitric acid, so

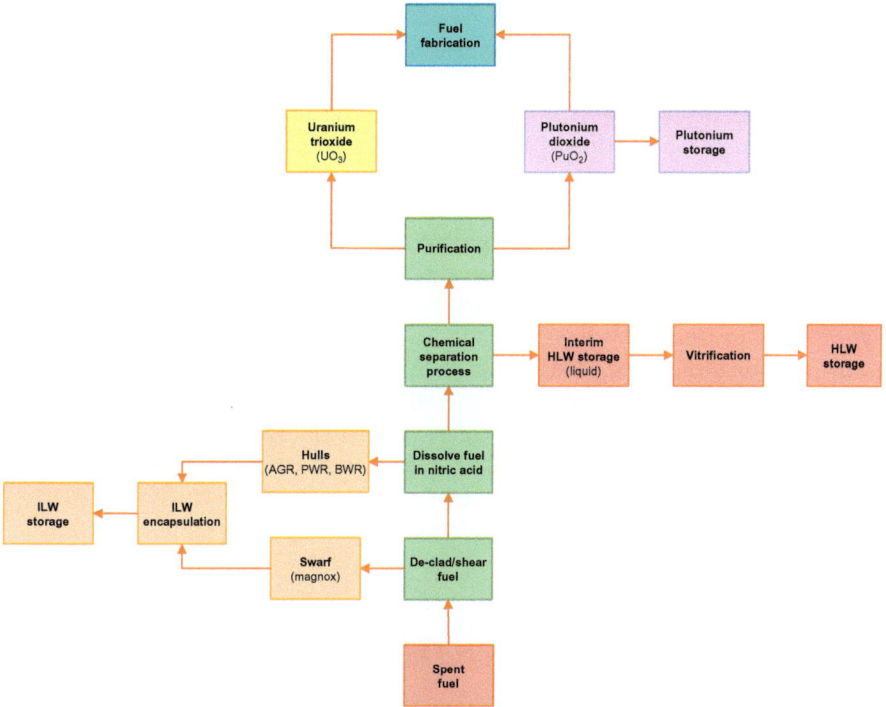

Fig. 1.38 Reprocessing.
© Bill Collum.

Fig. 1.39 De-cladding Magnox fuel rod.
© Nuclear Decommissioning Authority.

the swarf created when they are de-clad must be dispatched via a different route to the fuel. For the moment though we shall stick with the fuel and come back to swarf and hulls later.

Once pellets are dissolved, the fuel bearing nitric acid is routed through a chemical separation process which removes fission products such as caesium-137, strontium-90 and so on, which were created by a reactor's chain reaction. This leaves the uranium and plutonium which are separated into their individual streams, then further purified to remove any remaining traces of fission products. With that done, the two liquid streams are converted into a powdered form: namely uranium trioxide (UO_3) and PuO_2.

At this stage, the powders are ready to begin the fuel cycle again and be fabricated into new reactor fuel: either uranium-based (PWR, BWR, AGR), MOX, or fast breeder fuel. It is worth mentioning, though, that while they are waiting for that to happen both products must be carefully stored. UO_3 will still contain some slight impurities, which although fairly benign, are sufficiently radioactive to rule out handling it quite so freely as the enriched uranium powder originally created via a centrifuge process. In the case of PuO_2 there is the in-growth of americium-241 which we touched on earlier, so some shielding will be required. And because of its potential weapons applications plutonium must also be stored in an exceptionally secure environment.

Fig. 1.40 Constituents of spent fuel.
© Bill Collum.

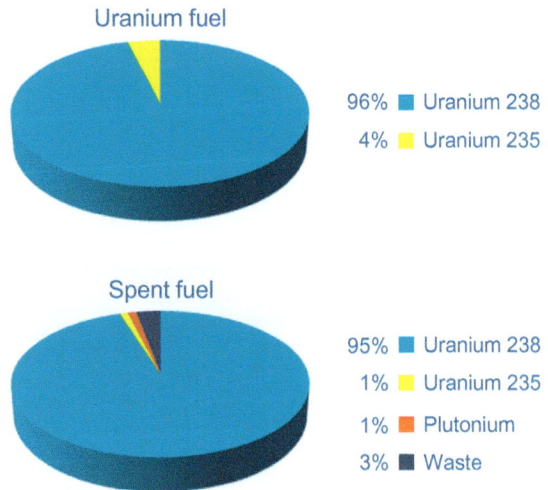

Uranium fuel

96% ■ Uranium 238
4% ■ Uranium 235

Spent fuel

95% ■ Uranium 238
1% ■ Uranium 235
1% ■ Plutonium
3% ■ Waste

As we can see in Fig. 1.40, reprocessing recovers around 97% of spent fuel for reuse, but this does leave around 3% of waste fission products which need to be processed and disposed of. This liquid waste is highly radioactive, particularly if concentrated by an evaporation process, and therefore generates significant heat. So to keep it well below its boiling point and stop it crystallizing the liquid is typically routed, initially at least, to double-skinned stainless steel holding tanks, equipped with cooling coils; the double-skin is a safety measure by way of providing additional containment. This is fine as an interim measure, even for a few decades, but is not sustainable for the thousands of years it takes for the hazards of this particular liquid to subside.

1.17   High level waste

1.17.1   HLW conditioning and packaging

Fig. 1.41 Vitrified radioactive waste.
© Nuclear Decommissioning Authority.

To make it more easily manageable the liquid is turned into a solid form via a process known as *vitrification* in the UK or *glassification* in the United States. Behind heavy concrete shielding it is first dried, or *calcined* to be precise and blended with frit. Following this stage, it is either melted and poured into stainless steel containers, such as that shown in Fig. 1.41, or melted within the container itself. Of course, it is still highly radioactive and will continue to self-generate heat in a similar way to irradiated fuel for several hundred years; importantly, though, the danger of leaking has been dealt with and it is in a form that can be stored indefinitely. If rather than concentrating the liquid it remains in its original state, then the option exists to immobilize it within concrete rather than glass.

1.17.2   HLW interim storage

As an interim measure, which could be 100 years or so, vitrified waste will probably be stored above ground in a shielded vault such as that shown in Fig. 1.42. In this particular arrangement a steel flask is used to stack containers in long tubes which are located below a heavily shielded floor. In time the containers are retrieved and loaded into copper canisters, similar to those used for the disposal of irradiated fuel, and ultimately dispatched to a deep geological disposal site where they remain indefinitely.

That covers the route taken by pellets of spent fuel, so we can go back now to follow the fuel cladding.

Fig. 1.42 Vitrified waste store.
© Nuclear Decommissioning Authority.

1.18   Intermediate level waste

1.18.1   ILW conditioning and packaging

I must admit when you look through a shielding window at Magnox swarf or oxide fuel hulls, it is hard to imagine that such benign looking pieces of metal could possibly do you any harm, other than maybe a cut from careless handling. The reality, though, is that innocent as they appear these metal fragments have been activated while in a reactor, and also harbor trace elements of the fuel they once encased.

Although these fragments do not reach the intense levels of radiation inherent in irradiated fuel, they are nevertheless extremely radioactive and categorized as ILW. In fact it takes around a meter of concrete shielding to contain their radiation and protect those in the vicinity. In addition, the decay of their isotopes results in self-generation of heat that will continue for many decades to come.

All in all then, dealing with fuel cladding presents quite a significant challenge. As with liquid waste, it is important to get the swarf and hulls into a form that is as easy as possible to handle and that will ultimately be suitable for deep geological disposal. Although there are examples of vitrifying ILW the preference in the UK is to encapsulate it, normally in concrete. The containers used to store encapsulated ILW are fabricated from stainless steel and are typically, although not exclusively, either 500 L drums (Fig. 1.43) or boxes with a capacity of 3 m^3 (Fig. 1.44).

To simplify handling and enhance their ability to remain standing in an earthquake, 500 L drums are loaded into four-drum stillages. And to streamline operations even further, the handling features and dimensions of a stillage are identical to those on a

Fig. 1.43 Five hundred liter drum.
© Nuclear Decommissioning Authority.

Fig. 1.44 Three cubic meter box.
© Nuclear Decommissioning Authority.

3 m^3 box; so both can be picked up, moved, and stored together using the same handling apparatus. We shall examine how such packages are handled in Chapter 9.

Depending on the product being dealt with, encapsulation within concrete divides into two different processes. Swarf, hulls, and larger objects such as those that arise from decommissioning are placed in a disposal container, say a 500 L drum (Fig. 1.45). This is then filled with a free flowing grout, and vibrated to distribute the waste more evenly and dislodge any pockets of air. Slurries and sludges are poured into the same type of container but with a paddle mounted inside (Fig. 1.46). Then, depending on their water content, either grout powders or liquid grout is introduced.

Fig. 1.45 Magnox swarf encapsulated in a 500 L drum.
© Nuclear Decommissioning Authority.

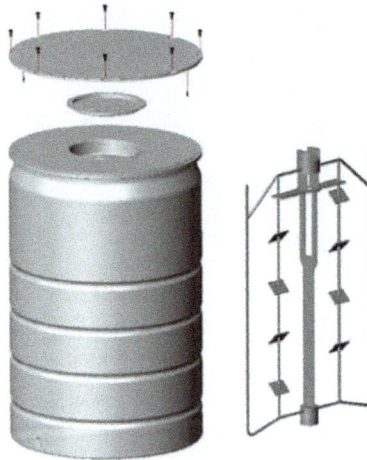

Fig. 1.46 In-drum mixing.
© Nuclear Decommissioning Authority.

To create a homogenous mix, the paddle is slowly rotated, but with its job done is far too messy, not to mention radioactive, to remove. So instead the drive mechanism is disengaged, leaving the paddle to be solidified inside the drum. Unsurprisingly, this particular technique is known as *in-drum mixing*, although it is often referred to as either the *lost paddle*, or *sacrificial paddle* process.

1.18.2 ILW interim storage

As with vitrified containers of HLW, packages of ILW are first routed to an above ground interim storage facility. Where large numbers of packages are involved, they will no doubt find themselves in a building known as an *engineered* store. Clearly all buildings need to be "engineered," so getting the word into their title gives a hint that these deceptively simple looking structures need a lot more of it than might be obvious from their appearance. And on a more practical note, in regular conversation it does help to differentiate them from the conventional stores on nuclear sites, which are no different to those found at any industrial complex.

These stores, although heavily shielded by maybe a meter of concrete, are a somewhat simpler affair than those used to house the hotter and more radioactive vitrified waste containers. Typically boxes and stillages are simply stacked neatly in vast open vaults which can hold several thousand packages each. As I say, they do not look too complicated, in fact not much more than a glorified shoebox. However, once you overlay the constraints of riding out severe earthquakes, not dropping packages, the possibility (depending on exact design) of a sophisticated ventilation system and so on, they do attract a surprising level of complexity.

Fig. 1.47 Shielded overpacks.
© Nuclear Decommissioning Authority.

Where relatively small numbers of ILW containers are involved, say a few hundred or less, it would not be cost effective to house them in such a splendid store, so instead there are broadly speaking two alternatives. Firstly, containers can be placed in hefty, stackable overpacks, normally concrete, which provide the necessary shielding, Then, depending on prevailing details, they could be stored in a shielded enclosure such as that shown in Fig. 1.47. Alternatively, overpacks may be placed in an inexpensive industrial building, with a simple construction not too dissimilar to a large hardware store. In fact, in the right climate they can even be stored outside on a concrete pad.

Fig. 1.48 Modular ILW storage vault.
© Nuclear Decommissioning Authority.

The second option sits half way between a fully engineered sore at one extreme and overpacks at the other. It comes in several guises but all have an ability to store ILW containers within a shielded matrix, one which is often modular (Fig. 1.48) so that it can be expanded as the need arises. The common attribute of these systems is that shielding plugs (Fig. 1.49) are used to cover tubes, or individual compartments, which can hold several stacked containers.

Fig. 1.49 ILW drum being lowered into storage tube.
© Nuclear Decommissioning Authority.

In selecting an interim ILW storage strategy, the challenge is to weigh up all of the competing alternatives, including the feasibility of converting existing buildings into stores, along with environmental impact, whole lifecycle costs and so on. Then with that done, we can make our selection and most importantly be able to justify it.

Ultimately then, all high and ILW finds itself processed, or *conditioned* as it is known, packaged into a form that is relatively easy to handle, is transportable in the public domain and satisfies constraints imposed for its indefinite geological disposal.

1.19 Permanent disposal

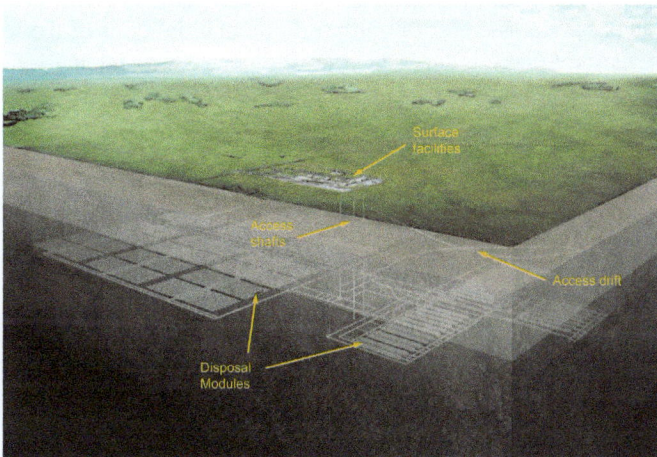

Fig. 1.50 Deep geological disposal.
© Nuclear Decommissioning Authority.

1.19.1 HLW and ILW

Once waste packages have been placed deep underground (Fig. 1.50) the vaults which house them are backfilled with concrete, forming enormous monoliths that entomb the waste for eons to come. There is no doubt, though, that given sufficient time even the most robust stainless steel or copper container will corrode and eventually allow their contents to come into contact with the concrete that encases them, which itself may be also be eroding in some way.

The rationale goes that by the time radioactive materials do come into contact with their surroundings, their activity will have subsided significantly and also that any potential to cause harm will be smothered by a mile or so of rock above the disposal vaults. Clearly though the right geological conditions are all-important, with factors such as susceptibility to earthquakes and minimal presence of groundwater coming to the fore.

Reflecting on deep geological disposal, it might seem that storing waste indefinitely above ground could have more going for it, but then tens of thousands of years is a long time to be faced with managing this type of waste and keeping it safe. The structure of engineered stores might last for say 100 years or so, but their external wall cladding and roofing would only be good for around 50 years and mechanical and electrical equipment for maybe 25 years. Even waste containers would be susceptible to degradation, raising the prospect of ongoing retrieval and repackaging of their contents.

For above ground storage then, there would be an almost perpetual cycle of repackaging waste and refurbishing or replacing storage buildings themselves, not to mention the ongoing security provision that always accompanies facilities housing radioactive materials.

Clearly with such a complex subject and an almost infinite timescale to consider there is bound to be an ongoing technical, financial and political debate on how best to dispose of radioactive waste. For our purposes, though, in terms of the nuclear fuel cycle, all we can note here is that stated government policy for the UK and come to that most other nuclear nations, is that ultimately all high and ILW will be disposed of in vaults deep underground.

1.19.2   LLW

Fig. 1.51 LLW disposal drum.
© Nuclear Decommissioning Authority.

Apart from the HLW and ILW produced from some nuclear processes, practically every nuclear facility generates LLW. Although shielding is rarely required, it nevertheless presents a radiological hazard so must be carefully managed and disposed of. Ordinarily, LLW items are packed into bags or drums (Fig. 1.51) which are perfectly safe to handle, so, with some mechanical assistance, they can be manually loaded into transport skips or ISO containers.

Fig. 1.52 LLW storage vault.
© Nuclear Decommissioning Authority.

In the UK LLW is stored in large concrete vaults (Fig. 1.52) where loaded ISO containers are filled with a cement grout before being placed in their final location. Once the vaults reach their full capacity they are mounded over with several meters of earth, gravel and so on then placed under a regime of ongoing monitoring and site management.

It is quite a journey, but with admirable symmetry a process that was set in motion by mining materials from the ground, ends decades later with all of their transformed descendants being placed back there again. It turns out then, that the term *cycle* wraps up the whole endeavor very neatly indeed.

Radiation

2.1 Radiation and contamination

Before we delve into this subject, we need to familiarize ourselves with the difference between radiation and contamination. Understandably these companions are often thought of as being pretty much the same thing, but actually they each have their own distinct characteristics so must be managed in very different ways. Let's start with radiation.

Fig. 2.1 Confining light rays. Shutterstock.com.

Take the example of placing a light bulb inside a box (Fig. 2.1). With its lid open, light would spread from the box and illuminate the surrounding area. If however, we secured the lid, and it was a good fit, the light would disappear. It would still be on, beaming away, but its light rays would be contained and unable to escape. Radiation is exactly the same, it is just that its rays can have a penetrating power that light does not possess. In addition, as with light, radiation travels in straight lines and although there may be a little *scatter* where it encounters obstacles, it certainly cannot snake around corners, a trait that will work to our advantage when we look later at ways to guard against it.

So if you enclose a solid chunk of radioactive material within sufficient shielding, such as the transport flask shown in Fig. 2.2, or behind thick concrete, its rays cannot get to you and you are perfectly safe. The thing to keep in mind is that of itself radiation does not leave any kind of residue or aftermath hanging in the air, so that once it is contained the surrounding atmosphere reverts back to normal immediately.

Nuclear Facilities. http://dx.doi.org/10.1016/B978-0-08-101938-2.00002-7

Fig. 2.2 Transporting radioactive material.
© Nuclear Decommissioning Authority.

The one exception to the rule is *neutron activation*, which we discussed in the previous chapter, where we saw it is a phenomenon that occurs in the uncommon confines of a nuclear reactor. Ordinarily then, once a radioactive source is adequately contained the environment it occupied moments before is perfectly safe to enter. Contamination on the other hand is not quite so accommodating.

Airborne radioactive contamination results from various processes within nuclear facilities. We shall see shortly that its ingestion poses a serious health hazard: then in the next chapter look at the subject in more detail, particularly how nuclear facilities are configured to corral contamination and guard against its uncontrolled migration around a building. For the time being it is sufficient to understand the crucial difference between radiation, which can be contained, and contamination which unconstrained can waft around, leaving a whole host unwelcome hazards in its wake.

2.2 Electromagnetic spectrum

Radiation is like sound in that it occurs at varying levels of intensity, with its *volume* indicator, so to speak, known as the electromagnetic spectrum. It is made up of three elements: frequency, wavelength, and energy, which in turn are split into two main types, nonionizing radiation and ionizing radiation. The one we are primarily interested here is the ionizing variety.

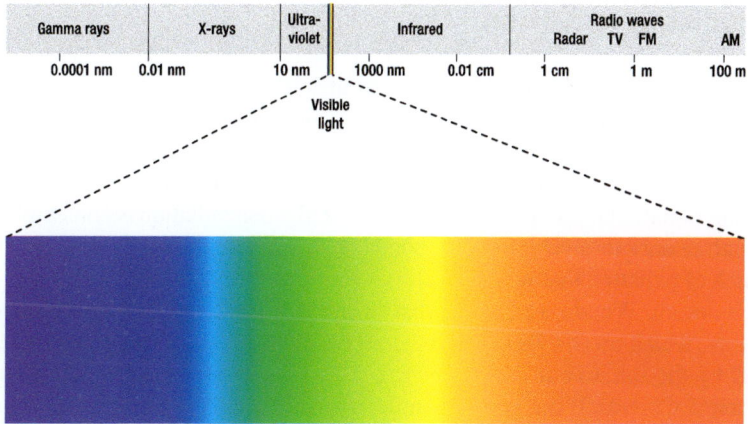

Fig. 2.3 Electromagnetic spectrum. Shutterstock.com.

Looking along the electromagnetic spectrum (Fig. 2.3), we discover how utterly diverse radiation is. At one extreme there is much about it that we rush to embrace, while at the other we are equally determined to keep our distance.

Everything around us reflects visible radiation, which we can see and differentiate as various colors. Having said that, the visible light we perceive sits between infrared and ultraviolet, which are just outside our range of vision. Apparently dogs, if they are to be believed, can hear higher pitched sounds than we do. Nevertheless, even they cannot view the full range of colors that are out there, and neither can we. So radiation contributes to how we see the world around us, but this is just one facet of something with properties so strange it is barely believable.

It is quite remarkable that a single entity, *tuned* in different ways, can carry a radio program or TV picture, cook our dinner, enable us to differentiate colors and use mobile phones. It can give us a suntan or X-ray, be harnessed to make weapons or provide electricity, and either cause cancer or help to cure it. Offhand I cannot think of anything else that is quite so diverse, particularly when we consider how utterly intangible it is.

2.3 Nonionizing radiation

Nonionizing radiation does not have the same energy levels as ionizing radiation, so it does not have the power to do physical damage. It surrounds us all the time and includes infrared and ultraviolet radiation, visible light, microwaves, and radio waves.

2.4   Ionizing radiation

Ionizing radiation is so called because it has sufficient energy to cause an electron to leave the orbit of an atom. As an electron hurtles away at the speed of light, two things happen very quickly indeed: it instantly changes its name to an *ion*, and electrical balance within the atom left behind shifts to become positively charged, or *ionized*.

Ionizing radiation is the one normally visualized when radiation is contemplated. It has enough energy to break chemical bonds or to strip electrons from atoms. It is in this branch of radiation that we find alpha and beta particles, gamma rays, and X-rays, but that's not all. We are well aware that over exposure to direct sunlight can cause skin cancer, which means we can add the higher frequencies of ultraviolet radiation, given off by the sun, to our list of ionizing radiations. It too has sufficient energy to break chemical bonds. It is a good example, if we needed one, that visible light really is part of the same family as other types of radiation.

In the nuclear industry we are primarily concerned with three different types of ionizing radiation, alpha, beta, and gamma, and how to shield against them (Fig. 2.4). The materials responsible for generating these types of radiation and ways in which they are handled are quite different; nevertheless, in operational terms we can group the last two together. In practice then, nuclear facilities tend to be designed and operated as one of two generic types, either *alpha plants* or *beta-gamma plants*. There are some exceptions but for the most part the two do not really mix.

Fig. 2.4 Shielding against ionizing radiation.
© Bill Collum.

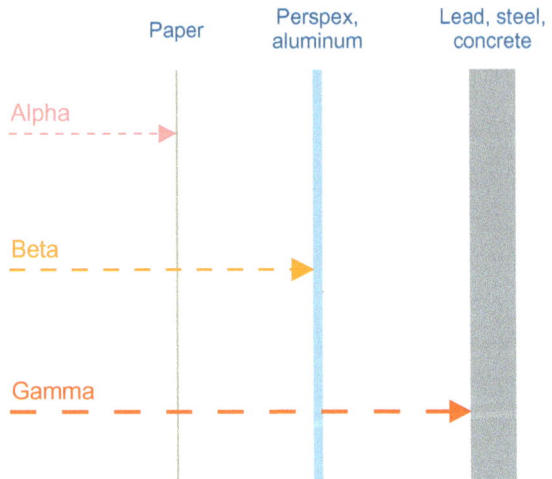

2.4.1   Alpha

Alpha radiation originates from what are classed as relatively heavy particles, more specifically helium atoms without electrons, which can only travel a short distance of a few centimeters through the air. Typical alpha emitters include radium-226, radon-222, uranium-238, plutonium-236, thorium-232, and polonium-210. Generally radiation tends to be perceived as something with considerable penetrating power and an ability to pass unhindered through pretty much anything. However, this particular radiation can easily be stopped in its tracks by a sheet of paper and certainly does not have the energy to pass through our clothes or our skin.

2.4.2 Beta

Beta radiation originates from what are classed as relatively light particles, actually electrons or positrons, which are smaller and more penetrating than alpha particles and can travel a few meters through the air. Typical beta emitters include strontium 90, technetium 99, caesium-137, carbon-14, sulfur-35, and tritium. Beta has more penetrating power than alpha radiation but would still struggle to get through our clothes. It does though have sufficient energy to enter our skin but not enough to pass through it. So unlike alpha, beta particles can penetrate a sheet of paper, but can easily be stopped by a thin sheet of either Perspex or aluminum.

Crucially, though, in situations where beta radiation is not accompanied by gamma radiation, materials such as steel and lead are not suitable as shielding. Beta particles striking such unyielding materials are stopped so suddenly that their energy is converted into gamma radiation which can pass straight through. So although it may seem counterintuitive, when it comes to halting beta radiation a thin sheet of Perspex is more appropriate than several centimeters of steel.

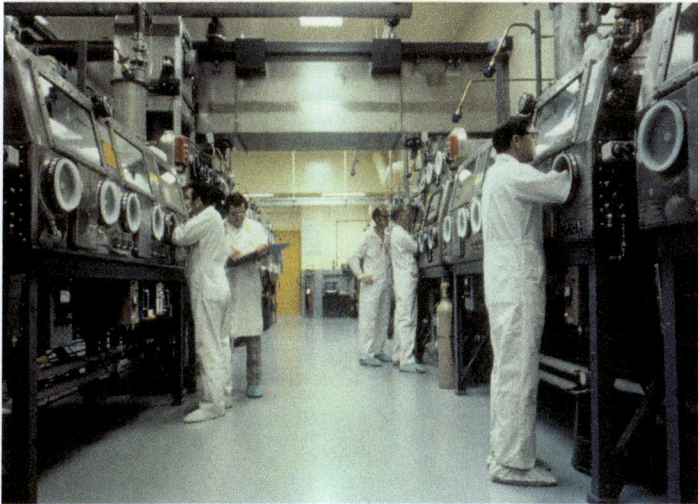

Fig. 2.5 Gloveboxes.
Science Photo Library.

The behavior of alpha and beta radiation explains something that puzzled me as a young boy. I would see people on TV or in the movies, standing at what looked like a glass box with ports cut in it. They would be handling nuclear material of some sort inside the box, but wearing only rubber gloves for protection. I now know those enclosures are called gloveboxes (Fig. 2.5) and that materials inside them are from the alpha and beta families, so do not need the heavy shielding which I assumed was necessary for all things nuclear. I also discovered that the reason for enclosing such materials within a box is that their particles of contamination, particularly alpha, can be exceptionally harmful if we inhale them, as once inside our body can cause biological damage which increases the risk of cancer.

Since it does not travel too far, one of the ways in which we can protect ourselves from beta radiation is to simply keep our distance. This works very well if the beta emitter is a solid object but if it is in the form of dust, which is so often the case, it can easily become airborne and pose a similar inhalation risk to that of alpha contamination. However, beta is not quite as harmful as alpha, due to the way it loses energy while colliding with cells in our body, so is less likely to cause permanent damage. Nevertheless, unprotected contact with both types needs to be strenuously avoided. That being said, many of us will have been in close proximity to small amounts of tritium, a beta emitter, as it is a constituent of the paints used to illuminate some wristwatches.

2.4.3   Gamma

Gamma radiation needs materials such as lead, steel, or concrete to halt its travel, so is definitely the one with characteristics most commonly associated with radiation. It differs fundamentally from both alpha and beta, in that gamma radiation does not come in the form of particles but is instead pure energy. So it is from the same family as radio waves, visible light, ultraviolet light, and so on that we see on the electromagnetic spectrum. Typical gamma emitters include iodine-131, barium-137m, cobalt-60, and radium-226. It has tremendous penetrating power, travels at the speed of light, and in some cases can cover hundreds or even thousands of meters through the air.

2.5   Exposure to radiation

Ordinarily we can depend on our senses to warn us of potential dangers, but in the case of radiation they let us down. We cannot see it, hear it, or smell it, so it is not surprising that radiation is generally eyed with some suspicion. There is, though, more to it than assuming if radiation *zaps* you that the consequences are invariably fatal; in fact, we are all bathed in radiation every minute of every day, and it has always been so. Our task here is to understand the subject, in that way we can separate the benign from the fearful. More importantly, we shall learn how to corral that which might harm us and treat it with the respect it deserves.

2.5.1   Background radiation

At some time or other you must have been in a building that was palpably silent, maybe in an office block, a factory, or a warehouse, late at night and on your own. You are going about your business, when all of a sudden a noise you were unaware of stops. For a moment, it is just a little eerie. It may be that a distant refrigerator or vending machine has switched off, or the hum from a light fitting ceased as it was automatically extinguished. Whatever it was, you were not conscious of the sound until it stopped and the silence deepened. Radiation is a bit like sound in that it ranges from the almost silent (a butterfly flapping its wings) to the incredibly loud (a jet engine at full power). The thing about radiation, though, is that its imperceptible hum is ever-present; unlike those barely perceptible background sounds it can never be switched off.

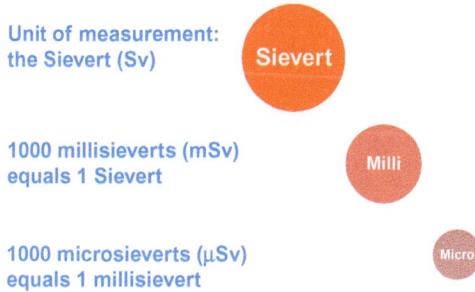

Fig. 2.6 Quantifying exposure to radiation.
© Bill Collum.

The unit of measurement for radiation (Fig. 2.6) is the Sievert (Sv), which is equal to 1000 millisieverts (mSv), or 1,000,000 microsieverts (μSv). Across the United Kingdom, the average annual radiation dose for residents is 2.7 mSv (Fig. 2.7) with 84% resulting from naturally occurring sources and the balance coming from manmade activities. The bulk of this, 15%, arises mostly from medical procedures, namely X-rays, CT scans, and radiotherapy, with the remaining <1% coming from other sources. This includes discharges from nuclear sites which account for a small portion of that 1%.

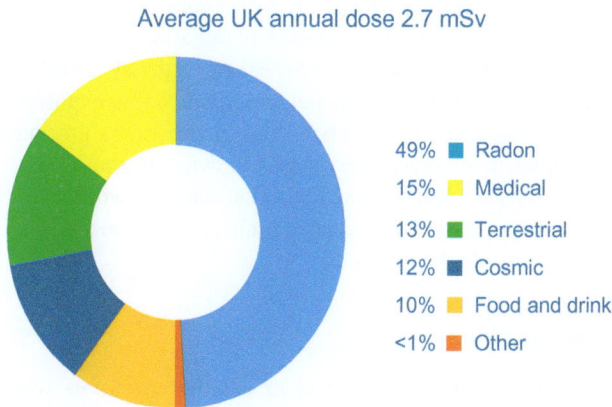

Fig. 2.7 Radiation exposure in the United Kingdom.
© Bill Collum. Data from Public Health England.

As we can see, the major contributors to naturally occurring radiation are quite diverse, ranging from cosmic radiation bombarding us from outer space, to ingestion of traces in our food, to radon gas, which forms in rock as a result of radioactive decay of uranium and seeps from the ground around us. I was in my teens when I first learned that radiation occurs naturally in our foods and must say I was a bit perplexed at the thought. I noticed that bananas were regularly mentioned as containing more than average levels of radiation, but having looked into it discovered they are just one example from a very long list. In fact, just about all foods are to some extent naturally radioactive and it has been that way since time began. At any rate, I am a big fan of the humble banana and seldom a day goes by that I don't have one. It is all about perspective isn't it.

With an average annual exposure of around 2.7 mSv, you might reasonably expect that we would all receive a dose within a few percentage points of that, but actually there are wide variations. Our actual exposure to natural radiation depends on where we live, with radon gas, or radon-222 to be precise, being the main contributor to regional variations. In the United Kingdom for example, people living in Cornwall can receive around three times the national average. This is mainly due to radon seeping from granite rock which contains trace amounts of uranium and predominates in that area of the country.

Of course, having recognized the issue, we take steps to stop radon accumulating in our homes and places of work. If you happen to live in Cornwall, you will be pleased to hear that such pockets of background radiation have always been with us, so our bodies have evolved to cope with the experience. In fact several parts of the world experience naturally occurring radiation at levels in excess of 50 mSv per annum, some even exceeding 100 mSv.

Having established that our bodies have an inherent ability to tolerate some exposure to radiation, the onus falls on the nuclear industry to guarantee that its workers, and members of the public beyond its sites, are not subjected to levels that could do them harm. To ensure consistency in this crucial area, the International Commission on Radiological Protection (ICRP) established what are commonly referred to as "safe working limits" for exposure to radiation. Their recommendations have been adopted by the International Atomic Energy Agency (IAEA) and therefore form the basis for worldwide regulations governing allowable exposure.

With the IAEA's more than 160 member states incorporating this guidance within their national legislation, pretty much everyone is following the same protocol. Actually, the practice on many UK nuclear sites and others around the world is to limit worker exposure to half, or less, of the prescribed legal limits, so they are taking a conservative approach by being at least doubly cautious.

2.5.2   Dose equivalent

I should mention here that whereas most of us use the word *dose* when discussing radiation exposure, strictly speaking the correct term is *dose equivalent*. Furthermore, a person's annual dose equivalent is actually made up of two elements: *external exposure*, which is unambiguous enough, and *committed dose*, which applies to that which is received by inhalation, ingestion, and skin absorption: in other words radionuclides entering the body. Ordinarily then, when we discuss radiation exposure we are speaking of *whole body dose*, which is a combination of external exposure and committed dose. We shall return to this theme in the next chapter when we examine the Sievert versus the Becquerel.

2.5.3 Nuclear workers

Those who work on nuclear sites tend to be known collectively as nuclear workers, but the reality is that many such workers are employed in roles which do not entail dealing with radioactive materials or even entering a nuclear facility; they may work in finance, human resources, legal, catering, or any of the myriad of roles necessary to support day-to-day running of any large industrial complex. That said, many organizations sensibly minimize the number of workers based within the footprint of their licensed site boundary, simply because it keeps down the volume of traffic which must pass through their security gates and releases sometimes limited space for nuclear facilities that could not be built elsewhere.

Employees who do operate in the nuclear sphere are covered by a regime that keeps a check on their exposure to radiation. Those whose occupation entails minimal exposure are categorized as *monitored workers*; this includes people like myself who are not based on a nuclear site but visit nuclear facilities on a regular basis. On the other hand, personnel whose role carries potential for higher radiological exposure are defined as *classified radiation workers* and subject to more comprehensive health surveillance.

For workers who are permanently based on the same site, it is a relatively straight-forward matter to keep track of their exposure through the site's in-house dosimetry systems. Those of us who visit multiple sites are issued with a passbook in which a radiological specialist logs our exposure at the end of each site visit, a bit like the bank books which prevailed before plastic cards became so popular. I have used my passbook many times around the United Kingdom and overseas, and must say it is the kind of simple yet effective system that I like very much.

Trainees age 16–17	6 mSv per year
Men from age 18	20 mSv per year
Women from age 18	20 mSv per year Maximum 13 mSv to the abdomen in any consecutive 3 months

Fig. 2.8 Radiation exposure limits.
© Bill Collum.

To conform to IAEA guidance, the maximum whole body exposure to radiation, or dose, that a radiation worker may receive divides into three categories (Fig. 2.8). In rare cases it is permissible for an adult male to receive a dose of 50 mSv in a single year, but when this does happen they must still stay within the maximum of 100 mSv that a classified radiation worker may receive over a 5-year period. Crucially, though, it is not permissible to design a new nuclear facility on the basis that workers going about their normal duties may occasionally receive a dose of 50 mSv in a year.

The principle is that day-to-day exposure must be as low as reasonably practicable, or ALARP, a concept we shall examine in Chapter 14 when we discuss safety. Ideally then exposure should be zero and certainly not up at a multiple millisievert level over a short period. The thought of exposing a worker to 50 mSv in a year would only be countenanced when all other alternatives had been examined and shown to be implausible. Typically this might be for a one-off operation which is necessary to recover from mal-operations or an accident. Outside of this it would be difficult, if not impossible to demonstrate that such exposure was justified.

Recognizing that it is far from ideal to expose any one individual to more radiation than is absolutely necessary; then where potential minimization solutions have been exhausted, the practice may be employed whereby several workers perform different elements of a task which is deemed to be radiologically hazardous. It is known as staff rotation or dose sharing, and is an effective way of spreading exposure and keeping those involved within the ceiling of 20 mSv per year. As I say, many sites around the world self-impose a limit which is well below the permitted maximum, to the extent that the vast majority of UK classified radiation workers receive less than a quarter of the annual legal limit.

2.6   Protection from radiation

2.6.1   Shielding

In nuclear facilities that handle gamma emitting materials, it can take quite a lot of shielding to protect workers from the effects of its overwhelming energy. The usual material of choice is concrete, as apart from providing shielding it also has excellent structural properties. Typically walls are around a meter thick, but they can get up to 2 m or even more, depending on the inventory being shielded.

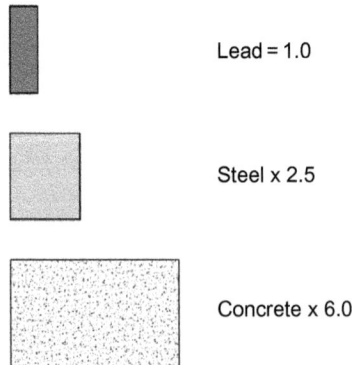

Lead = 1.0

Steel x 2.5

Concrete x 6.0

Fig. 2.9 Indicative shielding equivalents.
© Bill Collum.

Occasionally steel plate or lead is used, as being denser materials they do not need as much thickness. As a rule-of-thumb, the ratio of thickness to provide the same level of shielding is as shown in Fig. 2.9. However, I should stress that comparative performance of shielding materials can vary considerably, depending on exact characteristics of the shielding material and to some extent materials being shielded, so the ratios shown should be used only as a useful approximation.

Lead comes into its own in situations such as decommissioning tasks, where temporary shielding may be required. As we know though it is toxic, so the proposed use of lead must be weighed against potential health and environmental consequences. Apart from its minimal thickness, lead can be formed into conveniently sized bricks which are easy to assemble fairly quickly, a factor that can be all-important when radiation is present. In addition to a range of solid shielding materials we can also use water, such as the fuel storage ponds discussed in the previous chapter, or alternatively keep our distance.

2.6.2 Distance

When we consider the notion of using distance to protect us from harmful effects of radiation, particularly gamma radiation, it may seem at first to be a rather vague concept, and certainly feels like considerable distances would be needed before we could begin to feel safe. After all it would seem reasonable to expect that if you double the distance between yourself and a radioactive source, the best you could hope for is that the radiation's energy would have diminished by half of the power it would have if you were stood next to it.

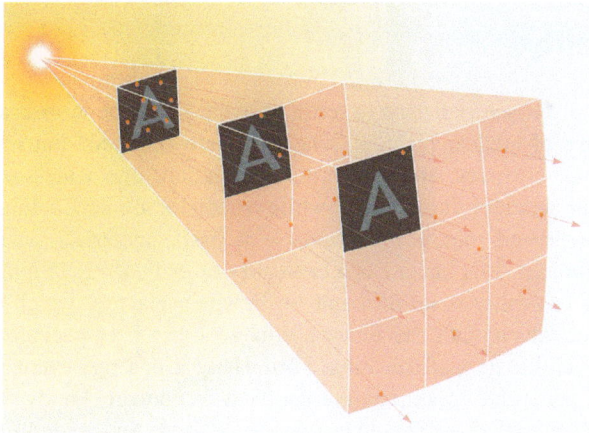

Fig. 2.10 Spherical dispersion.
Science Photo Library.

The way it works, though (Fig. 2.10), is that as you move further and further from a source of radiation its energy decreases due to *spherical dispersion*, or to be more precise, in accordance with the *inverse square law*. Simply put, each time you double the distance, or radius, from an object, intensity reduces fourfold. In fact the same rule applies to sound and light, so we witness this law in action every time someone calls out to us, or we switch on a flashlight.

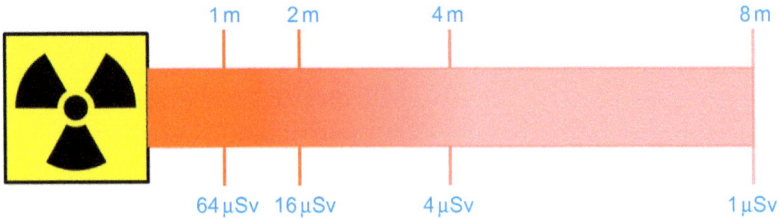

Fig. 2.11 Reduction in radiation energy over distance.
© Bill Collum.

As an example (Fig. 2.11), we can see that exposure of 64 µSv at a distance of 1 m from a source of radiation, will reduce to 16 µSv by doubling the distance to 2 m, and so on, continually reducing by the inverse square. That being said, when dealing with large radioactive items, effects in their *immediate* vicinity may not quite conform to the rules. Essentially, spherical dispersion kicks in at a distance of around two and a half times an object's longest dimension. However it is a point of detail, one for the specialists. The crucial point, one that has a bearing on the design and operation of nuclear facilities, is that although gamma radiation may be able to travel quite a long way its power reduces rapidly during the journey.

2.7 Criticality

So far we have discussed radiation in terms of what might be termed a steady state or, as in the case of nuclear reactors, when they are "excited" but in a controlled manner. There is, however, another form of radiation, one which results from a particular type of accident. It can occur while personnel are conducting what would normally be considered routine operations with fissile materials such as enriched uranium or plutonium. It is known as a criticality or *critical excursion*.

In essence, a criticality is no different to the chain reaction which occurs inside a nuclear reactor, albeit with the major distinction that it is uncontrolled and not contained by several meters of shielding. Fortunately it is a rare event which is only possible if two particular conditions are satisfied in concert. Firstly there must be sufficient fissile material, or a *critical mass* in one location. And secondly the materials, whether they are solid, powdered, or liquid, must be held in a geometrical configuration, such as a sphere, which enables neutrons to interact in ways that can lead to a chain reaction.

If that exact coalition is realized, then it can lead to a penetrating burst of both neutron and gamma radiation which can be lethal to those in the immediate vicinity. Beyond that local area, spherical dispersion dissipates energy to the extent that those a few tens of meters from the scene are unaffected. However, if a criticality does occur there is a possibility that its initial burst may be followed by subsequent pulses of radiation. It is therefore imperative that if a criticality alarm sounds, everyone within earshot urgently follows the nearest evacuation route until they reach a designated place of safety.

But, you may ask, how can we be sure criticality detection systems are operating? If they malfunctioned or were accidentally switched off, how would we know? If you are new to the nuclear world, the first thing you notice on entering many facilities is a sonar type sound, a watery ping that echoes every few seconds. Normally we get accustomed to the noise and after a while hardly notice its presence. However, there are systems that continually make slight adjustments to its modulation, which although quite imperceptible are enough to keep its pings at the forefront of our consciousness. Whatever the sound pattern, it is linked to criticality detection systems.

So there you have the reason for that annoying noise: it confirms criticality monitoring systems are operating and checking the environment on our behalf. If its ping stops or the system goes into alarm mode, we know straight away that a malfunction has occurred, a practice drill is underway, or there is a genuine emergency.

It is sometimes assumed that a criticality, being a chain reaction, could result in an explosion of the type produced by nuclear weapons. However, such explosions require exacting conditions, including specific characteristics of the fissile material itself and extremely precise engineering features. As these conditions do not occur in civil nuclear facilities, including reactors, we can be confident a criticality will not inadvertently initiate a nuclear explosion.

2.8   Personal dose measurement

I have mentioned radiation exposure and personal dose quite a bit in this section, so the question that follows is: how it is measured? How can a classified radiation worker, or, just as importantly, visitors to a facility, be sure their personal dose is correctly recorded?

There are many kinds of measuring devices, or *dosimeters* available, but they all divide broadly into two generic types. The *passive* variety, which record a person's exposure over multiple site visits, with accumulated dose being evaluated after the event so to speak. And direct readout dosimeters, which display exposure in real time and are reset to zero after each use.

2.8.1   Thermoluminescent dosimeter

A thermoluminescent dosimeter (TLD) is one of the most popular passive personal dosimeters and is generally issued to users for a period of 1–3 months. Straightaway the notion of long-term use is bound to raise questions of safety and particularly risks of being "over exposed" without even realizing it.

The thing to keep in mind here is that the environment in and around nuclear plants is continually being monitored by static devices and not just for the criticalities

I mentioned a moment ago, but for all radiological conditions. TLDs can therefore be relied on in stable environments that are known to have very low levels of radiation, the kind of place that is safe to occupy for hundreds, or even thousands of hours.

Fig. 2.12 Thermoluminescent dosimeter (TLD).
Science Photo Library.

A TLD (Fig. 2.12) contains crystals of thermoluminescent (TL) material such as calcium fluoride, lithium fluoride, calcium borate, or one of several others. When exposed to ionizing radiation, electrons within TL crystals behave in a way that creates "holes" in the material, or to be more precise *electron–hole pairs*, a phenomenon we shall discuss in Chapter 11. Once a TLD has been worn for its specified period, it is returned to a site's dosimetry department and placed in a unit, or *reader* which heats the TL crystals. This heating, or *annealing* process, causes electrons to revert to their original condition and in so doing release energy in the form of light. And crucially the intensity of light is directly proportional to the degree of radiation exposure received by the TLD's wearer.

To obtain maximum data from a TLD they normally contain several crystals, each shielded, or *filtered* by various types of metal foil. In this way some crystals can be designated to measuring the shallow beta dose, in other words, to a person's skin, and others to the deep gamma dose received by internal organs.

In addition to what we might call the standard TLD there is also a much smaller version that can be worn on the fingers (Fig. 2.13). They are used to measure the extremity dose received by those using gloveboxes, an item of equipment we shall discuss in Chapter 7. When in use TLDs do not become radioactive in any way, so after appropriate annealing can be reused.

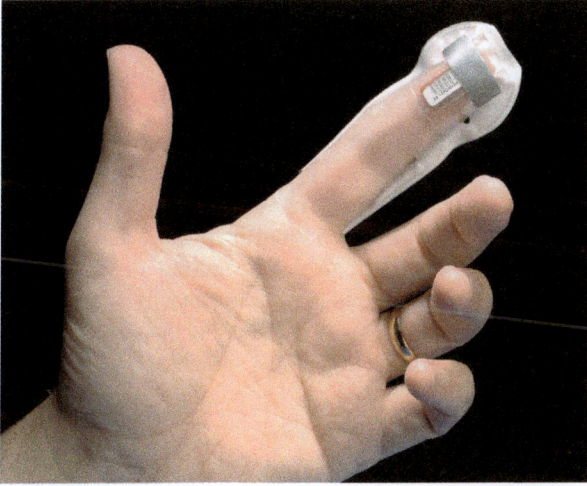

Fig. 2.13 Fingertip TLD.
Science Photo Library.

2.8.2 Electronic personal dosimeter

So a TLD is fine where there is no immediate imperative to quantify radiation exposure, but in high dose, or *high-field* situations, where residence time is necessarily tightly controlled, we need real time information. Enter the electronic personal dosimeter (EPD) (Fig. 2.14). Their operation is based on semiconductor technology, which we shall discuss in Chapter 11, so for the moment we can concentrate on what they do.

Fig. 2.14 Electronic personal
dosimeter (EPD).
Science Photo Library.

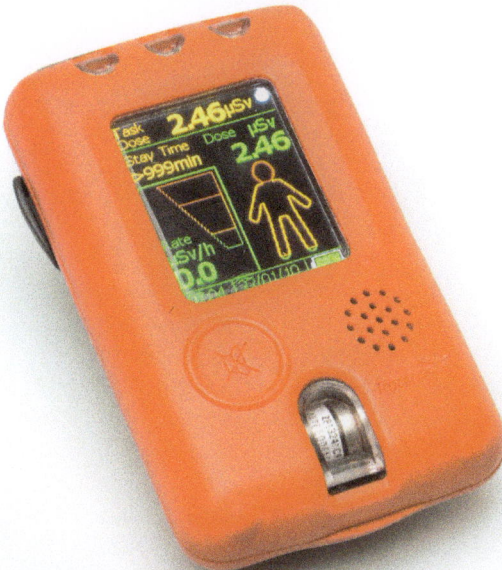

An EPD can be used as a person's only dosimeter, but they are often supplemented by a TLD. In this way a TLD records dose accumulated over several weeks, maybe up to 3 months, while an EPD records the specific dose for each entry into a high-field area. As with TLDs, an EPD can detect and quantify a person's shallow dose and penetrating dose, along with keeping a running total. Crucially, EPDs can be programmed to trigger alarms, both audible and visual, at preset levels and even audibly "count down" to a specific dose measurement.

In common with TLDs, an EPD is ordinarily worn at chest level, say attached to a pocket, or alternatively clipped to a belt around the waistline. However in some situations, say where radiation sources are primarily at floor level or at a high elevation, radiological specialists may advise that dosimeters are worn in a different position. After each use, data from an EPD is uploaded into a person's dosimetry record, and with that done they can be set back to zero and reused multiple times. All in all then, an EPD is quite indispensable.

So we have seen something of the different types of radiation, permissible exposure limits, how we guard against it and how it is measured, but in the context of nuclear facilities radiation can only be fully fathomed when we group it with its partner, contamination. For that we must examine *radiological zoning*, so this is where we shall go next.

Radiological zoning

3

As we have seen, radiation places unique constraints on the design and operation of nuclear facilities, but in many ways the contamination which often accompanies it is even more problematic. I suppose the basic objectives of constraining this duo are fairly obvious; make sure radioactive materials have plenty of shielding around them, and do not let particles of contamination drift uncontrolled around a building. Sounds fairly straightforward, but the trouble is we need to move active materials around and in certain situations operators need to handle them. Doing these things, whilst at the same time keeping people safe and not harming the environment is quite a challenge, in fact is sits right at the heart of what it takes to design and operate a nuclear facility. Having examined radiation, let's look now at how we deal with its almost ubiquitous accomplice: contamination.

3.1 Zoning rationale

If you were to walk into a nuclear facility, off the street so to speak, you would have left an area of regular fresh air, the environment where we all live every day. In the case of a beta gamma plant, we know that somewhere at its heart there will be highly radioactive materials behind heavy concrete shielding. Similarly, for an alpha plant there are gloveboxes where materials such as plutonium are handled. And no matter what the building there will be support activities, such as maintenance or decontamination, which involve handling radioactive substances and dealing with the contamination that often hovers about them. These will be on a scale from high radiation operations which must be performed remotely, to being less and less active, until we reach a level where classified radiation workers can manually handle the materials.

Even if we put aside security considerations, it is unthinkable that we could simply walk into such a facility, wander around and then leave without any kind of safeguards to ensure we were not placing ourselves in danger of exposure to radiation, or unwittingly being enveloped in contamination. In such a situation we could find ourselves inhaling contamination, carrying it about our person from one area of a building to another and even taking it home at the end of the day. Clearly we need a system.

The starting point for restraining this twosome is to recognize that operations with different levels of radiation and contamination need different barriers and operating procedures to keep them safe. And what this tells us is that nuclear facilities need to be divided into areas, from the normal environment of an entrance lobby, right up to the most hostile of radioactive environment's where it would be unsafe to tread even for a moment. The system used to segment nuclear facilities is called *radiological zoning*, commonly referred to as rad zoning, and it has an overwhelming influence on how these facilities are designed and operated.

Nuclear Facilities. http://dx.doi.org/10.1016/B978-0-08-101938-2.00003-9

Crucially, radiological zoning is not just about physical barriers. It is equally reliant on a facility's ventilation system which operates in ways quite unlike anything we find in office blocks, shops, factories and the likes, and plays a crucial role in keeping contamination in its rightful place. In fact, ventilation is so specialized a subject that it needs a chapter of its own. For the moment, it is worth keeping in mind that radiological zoning is just one facet of a facility's containment philosophy, and can only function when seamlessly integrated with its ventilation system. Furthermore, in the nuclear world this subject is so fundamental that it must be addressed from day one of a project's life and expounded in a *radiological zoning philosophy* which is one of a project's most important documents. So how does it work?

3.2 Naming conventions

All areas within nuclear facilities are designated two radiological zoning references: one signifies the highest level of radiation that could possibly occur as a result of its day-to-day operations, and the other a maximum level of contamination. The contamination factor can sometimes be divided into airborne and surface, but for the most part we are combating floating particles so for the moment will concentrate on those.

By the way, when we talk about "zones" in this context, it normally refers to an area which is enclosed by walls and therefore segregated from other spaces around it: for example, a room, or corridor, or the shielded caves and cells where highly radioactive operations take place. In other words, it is not normally the case that we assign different radiological zones within the same airspace. It does happen, but only in the particular "dual classification" circumstances that we shall be discussing later.

Naming conventions for radiological zones vary a little around the world and even within the same country. The United Kingdom is typical in this regard, with different nuclear sites and organizations adopting their own particular preferences. Fortunately, though, no matter what their name-tags they all boil down to the same thing, with four radiation categories and a matching number for contamination.

In the United Kingdom some nuclear sites name their radiological zones white, green, amber, and red, with red signifying the highest level. Others tend to favor an alphanumeric system where contamination zoning is categorized in four levels from C1 to C4, and radiation in the same way from R1 to R4. In both cases 1 denotes the lowest classification and 4 the highest. Other sites opt for designations such as clean, low, medium, and high, or similarly themed variations.

I should add that in the United Kingdom, a number of organizations subdivide their radiological zones, particularly contamination categories, but it is a matter of fine detail and strictly speaking outside standard practice. One custom for example, is to employ five zones but when this happens the two highest levels, in both categories, are managed in much the same way. Similarly, several countries adopt conventions with 10 or more radiological zones, but once again they are essentially subdivisions of the standard nomenclature. For our purposes then, we shall stick with four zones for both radiation and contamination, and adopt the alphanumeric system. Along the way I will also make regular reference to zone colors which are an equally popular alternative. As I say, there are quite a few naming conventions but their labels are easily enough transposed.

3.3   Radiation zones

The limits on radiation and contamination which apply to individual radiological zones can vary a little from one nuclear site to another, with variations resulting mainly from local custom and practice, particularly where zones have been subdivided. However, figures are always broadly the same and of course legal limits, discussed in the previous chapter, are applied uniformly to all nuclear installations.

Fig. 3.1 gives an overview of the four standard radiation zones, along with examples of exposure limits which typically apply to each of them. In the previous chapter we saw how background radiation, along with exposure limits for classified radiation workers, is generally described in millisieverts (mSv) per annum. However, for

Zone	Example areas	Max levels Microsieverts (μSv) h
R1	Non-active areas with radiation broadly comparable to normal background conditions eg, roadways, administration buildings, and restaurants	2.5
R2	Routinely accessible eg, Corridors, crane halls, and electrical equipment rooms	7.5
R3	Radiation present to levels that require special access arrangements eg, Workshops for maintenance of lightly contaminated items and rooms housing primary ventilation HEPA filters	100
R4	High to extremely high levels of radiation. Areas locked off, access infrequent and only with special controls. eg, Ranging from hands-on maintenance or dismantling of heavily contaminated items, to areas such as shielded caves housing highly radioactive materials	Unrestricted

Fig. 3.1 Radiation zones.
© Bill Collum.

radiological zoning this is far too crude a measure, so instead we adopt its junior sibling, the microsievert (μSv), which is one thousandth the scale. And rather than annual exposure we adopt an hourly measure, so microsieverts per hour (μSv/h).

3.3.1   R1 (white)

The lowest classification for radiation (white) is broadly equivalent to normal background conditions, so is assigned to areas which are entirely free of a radioactive inventory, or indeed any other form of radiological hazard. In practice then, it applies to offices, restaurants, stores, workshops and a long list of other non-nuclear buildings. In addition it encompasses anywhere within the confines of a site's perimeter fence, including the open air. It might seem a bit odd to allocate a radiological zone to the regular environment, but it is needed so that we can identify and manage interfaces between the inside of nuclear facilities and the world around them.

Incidentally, the R1 designation could legitimately be applied to many areas within nuclear plants. However, the normal practice is not to employ it in radiologically controlled environments, electing instead to adopt R2 as the lowest on-plant level.

3.3.2   R2 (green)

R2 levels of radiation are broadly equivalent to air travel and apply to most of a nuclear plant's routinely accessible areas, such as corridors, the majority of plant rooms, non-active workshops and so on.

Even though the 7.5 μSv/h ceiling for this category is relatively low, it is still incumbent on designers and operators to ensure any on-plant exposure is maintained at its lowest possible level. We shall see more of this when we discuss ALARP principles in Chapter 14. The acronym stands for *as low as reasonably practicable*, which rather gives the game away in terms of what it is all about.

3.3.3   R3 (amber)

R3 covers a considerable spectrum from its entry level at a smidgen over 7.5 μSv/h, all the way to 100 μSv/h. It applies to active workshops, decontamination areas and the likes.

You will recall from our discussion in the previous chapter that the maximum exposure for a classified radiation worker is limited to 20 mSv (20,000 μSv) per annum which would equate to working in an area at the upper R3 level of 100 μSv for 200 hours. However, with many UK sites operating to half legal limits, or less, this reduces to a period of no more than 100 hours a year. Then again if an R3 zone was subject to say 40 μSv/h, an operator could occupy the area for 500 hours in a single year, reducing to a maximum of 250 hours on most sites. They are just a couple of examples, but give an indication of how the system works.

In addition to site operators self-imposing lower limits on annual exposure, many also elect to reduce the standard maximum of 100 μSv/h which applies to R3 areas.

The reduction is typically around 50% but varies from site to site. When applied, it results in areas above the voluntary level being categorized as R4 and therefore attracting the stricter controls which accompany that classification.

3.3.4   R4 (red)

At its lowest levels of somewhat above 100 µSv/h (unless a local reduction is in place), it is possible for classified radiation workers to enter R4 areas. However, there is a constant proviso that operator exposure must be kept to the lowest possible levels, and always with an eye to staying within maximum exposure limits, whether it be 20,000 µSv/annum or more likely a site's self-imposed lower limit. For the most part, though, radiation levels within R4 areas are so high that access is out of the question, even for a moment.

This is the category most often associated with radiation and ranges from areas housing heavily contaminated items, to zones where levels are so intense that all operations must be conducted remotely. Even areas at the "accessible" end of the spectrum will be surrounded by at least some form of shielding. And of course as we move further and further up the R4 scale, to areas where radiation is measured in thousands of microsieverts per hour, shielding keeps pace until it reaches one, two, sometimes more than 3 m of concrete. Apart from its containment function, the equally important role of shielding is to ensure exposure levels on its "cold" side are reduced, ideally to R2 and certainly no higher than the lower levels of R3.

3.4   Contamination zones

Normally we only become aware of dust particles when sunlight, maybe streaming through a window, catches them in a particular way, so at home and practically everywhere else these invisible companions only get our attention when they accumulate and we need to get a duster out. In nuclear facilities many processes involve handling solid materials in a way that causes minute particles of radioactive dust, or *contamination*, to be released into its surrounding environment, maybe from cutting or shaping, or all manner of operations. Similarly, radioactive liquids can evaporate and dry out, with particles being gradually released into the air.

Once airborne, particles of contamination take on a mind of their own and unconstrained would go pretty much where they please. Almost as troublesome, they settle on exposed surfaces where accumulations can result in a significant build-up of radiation. In addition, as we have seen in the previous chapter, ingestion of such particles poses a very real risk to our health. All in all then, contamination presents a significant challenge, one that design teams and plant operators need to diligently guard against.

Imagine though trying to stop dust particles in your kitchen wafting through to the lounge, or drifting outside through an open door; it would be impossible. Clearly nuclear facilities need some kind of compartmentation system, so this is where the second "C" or contamination element of radiological zoning comes in. However, before we get into the zones themselves we must take a slight detour.

3.4.1 Sievert versus Becquerel

In the previous chapter, we saw how radiation is quantified in terms of the Sievert, which can be used to describe a person's external exposure. However, when it comes to contamination the hazard is more to do with risks of inhalation, so we need a different approach. The unit of measurement in this context is the Becquerel; unlike the Sievert, though, it does not lend itself quite so easily to assigning relatively straightforward limits within particular radiological zones.

The Becquerel is a measure of radioactive decay, expressed as disintegrations per second. In other words it quantifies, isotope by isotope, how radioactive particular particles are and how long-lived their radioactivity is going to be. Armed with this information it is possible to deduce how damaging individual isotopes, such as strontium-90, plutonium-239 and others could be to our internal organs. So we can think of the Sievert in terms of measuring *dose* and the Becquerel as measuring *activity*. So far, so good.

Complexity kicks in when we consider that before assigning a radiological contamination zone to any given area, there must be a thorough assessment of all isotopes that could possibly be present, what their particle size might be, how densely they may be congregated, consequences of combining dissimilar isotopes and so forth. It is a desperately complicated exercise and certainly one for the specialist radiological protection advisors (RPAs) to wrestle with. The good news for the rest of us, is that having done their analysis, RPAs can advise, not only on contamination zoning, but exactly what personal protection must be worn within individual zones. So for example, if a respirator is needed they will spell out the specifics of its filters, a component that is influenced by potential isotopes, their particle size, whether gas may be present and so on.

As I say, unlike the Sievert we cannot simply assign a Becquerel limit to C2, C3, and C4 areas; unfortunately there are just too many variables. Normally then, in everyday parlance radiological contamination zones are discussed in terms of the protection they demand, so this is what we shall do. Fig. 3.2 gives an overview of the four standard contamination zones, along with an indication of the protective clothing and equipment generally worn within them.

3.4.2 C1 (white)

As with radiation the lowest level, C1, is broadly equivalent to normal background conditions, or simply put, is reserved for zones which are free of contamination. And again, although it could legitimately be applied to many areas within nuclear plants, the normal practice is not to employ it in radiologically controlled environments, electing instead to adopt C2 as the lowest on-plant level.

Zone	Example areas	Protective clothing and equipment
C1	Non-active areas with no potential for contamination eg, roadways, administration buildings, and restaurants	None required
C2	Routinely accessible Airborne contamination is not generated in these areas, so minimal potential for its presence eg, corridors, crane halls, and electrical equipment rooms	Change of clothes and footwear
C3	Contamination present to levels that require special access arrangements eg, workshops for maintenance of lightly contaminated items and rooms housing primary ventilation HEPA filters	Respirator Plus additional protective clothing
C4	High to extremely high levels of contamination eg, ranging from hands-on maintenance or dismantling of heavily contaminated items, to locked off areas such as shielded caves housing highly radioactive materials	Air-fed PVC suit

Fig. 3.2 Contamination zones.
© Bill Collum.

3.4.3   C2 (green)

When you step from the outside world into a nuclear facility, this is the zone you will enter. In fact for visitors, and indeed some plant-based personnel, it would be unusual to venture beyond its confines. Having said that, it is ordinarily by far the largest radiological zone in a facility, often accounting for maybe half or three quarters of a building's total volume. It comprises corridors, circulation areas, electrical equipment rooms, primary services distribution routes, non-active workshops and so on.

Significantly, and unlike the two higher zones, these areas are designed and operated to be entirely free of airborne contamination, so the inhalation risk is very low indeed. Nevertheless, air in C2 zones is constantly being sampled and monitored, with alarms poised and ready to instigate safety procedures if undesirable particles invade its airspace.

It is worth bearing in mind that even though C2 (green) zones are radiologically speaking rather clean, potentially squeaky clean in this regard, all activities within them are conducted on an assumption that contamination is present. To make the point they are often referred to as "suspect active," so no eating, chewing gum or needlessly meddling with anything that does not concern us. It is a prudent and mandatory approach; after all, with radioactive materials in the vicinity, *caution* is ever the watchword.

3.4.4   C3 (amber)

If we think forward for a moment to the next and highest category, C4 (red), we shall already have surmised that it must be capable of constraining the most severe levels of contamination. Furthermore, we could also figure that maintainable items, withdrawn from such an environment, will need to be decontaminated before anyone can handle them. Similarly, many items leaving a C4 zone to be stored or disposed of elsewhere may also need to be decontaminated; otherwise we risk needlessly spreading contamination to other areas or even, after packaging and transport, to distant receiving facilities. What all this tells us, is that we need a contamination zone which stands between the almost pristine conditions of C2 (green), and often grossly contaminated environment of C4 (red). This is where C3 (amber) comes in.

A C3 zoning classification is assigned, for example, to rooms where decontamination processes take place, and to *amber* workshops where lightly contaminated items can be handled by maintenance personnel. In addition, any other area that serves as a transition between C2 and C4, or for storage of items at C3 levels, is assigned this classification.

Environmental conditions in C3 areas can vary considerably from dry to wet, and from very little contamination, to potentially quite high levels. As a result, personnel entering these areas always need to change their attire, more specifically, donning some form of protection such as an additional layer of clothing, changing their footwear and maybe wearing gloves and a respirator.

3.4.5   C4 (red)

The term most often employed when discussing zones with this classification is "gross contamination," which pretty much sums it up. It can apply to heavily contaminated items of equipment, but ordinarily when we come across this zone it is in the context of a shielded cave where contamination levels are exceptionally severe. If their radioactive inventory is removed, then it is possible to enter C4 areas, albeit with the protection of an air-fed PVC suit. It is, however, a rare occurrence and as we might expect the whole process is very tightly controlled. So for example, lone entry is not permitted, operations are meticulously planned so as to minimize durations and a support team is always in attendance.

Particulars of C4 (red) areas can vary considerably from one to another, but to give a feel for what is involved we can take a "typical" situation where two operators need to gain access. Straight away we can assume their support team will comprise at least six, but more likely eight personnel. If we add time needed to dress and prepare prior to entry, followed later by decontamination and disrobing procedures; then during an 8 hour period we could expect the two operators to spend somewhere between 1 and

2 hours in the C4 (red) environment. So apart from radiological issues it is quite a labor-intensive and expensive exercise.

3.5 Guiding principles

To recap, so far we have built up a picture of why nuclear facilities need to be divided into various radiological zones, their naming conventions and something of the criteria that apply to individual radiation and contamination classifications. Before we can move on to examine how zones are designated in real buildings, we need to visit the principles which, to a large extent, dictate how they are assigned.

3.5.1 Designations need not match

As we have seen, all areas within nuclear facilities are assigned both a radiation and a contamination classification. Crucially, though, they are evaluated independently so there is no expectation that "C" and "R" numbers, or colors for that matter, should match. In an extreme, although not entirely uncommon example, it is possible to categorize an area as C1/R4, where there is no airborne contamination but radiation levels are very high. As long as there is appropriate shielding in place then such a combination is perfectly acceptable.

3.5.2 Apply to normal operations

When it comes to establishing a facility's radiological zones, classifications are ordinarily assigned against the highest "potential" level of contamination or radiation that may occur during "normal" operations, and not against every conceivable fault or accident condition. Having said that, we do not ignore such possibilities, quite the opposite, but adopting them as the norm would add unnecessary complications to a facility's operation. So we need to get the balance right.

In practice, a thorough hazard analysis is performed which postulates every conceivable fault or accident, including loss of radiological containment, along with what the consequences would be. And with that done, measures are implemented to ensure, should the worst occur, nuclear workers, the public and environment are protected from harm. We shall discuss this crucial process in considerable detail when we get to Chapter 14. In the meantime we just need to be aware that, unless overridden by higher safety considerations, rad zoning is always harmonized with normal operating conditions.

3.5.3 One step at a time

This particular principle is all to do with minimizing the risk of spreading contamination, so applies irrespective of radiation levels. When moving between contamination zones, things should be arranged so as to move, in either direction, from any given zone to its immediate neighbor. So between C1/C2, C2/C3, and C3/C4 is fine, but not for example C1/C3. A study of Fig. 3.3 makes it clear why this objective is so important.

Fig. 3.3 Containment zone interfaces.
© Bill Collum.

If we concentrate initially on the standard contamination zones, specifically what they invoke in terms of potential airborne levels, then first up we have C1 (white) which is akin to a regular non-nuclear environment, followed by C2 (green) where there is no airborne contamination, then C3 (amber) where we can expect to encounter at least some, and finally C4 (red) where airborne contamination levels can be very severe indeed.

Clearly we cannot simply step, or move equipment, from one zone to another without some form of safeguarding procedure, otherwise we risk spreading contamination across zones and undermining their segregation role. Logically then, we need some way of safely transitioning from one zone to the next, so sure enough this is what we find.

In the case of personnel, transitions are effected via radiological changerooms which we shall discuss in the next chapter. For items of equipment, along with a wide variety of containers which hold radioactive materials we need something more vigorous. Here we find decontamination processes, linked to monitoring systems which ensure cleansed items satisfy criteria of their destination zone before being allowed into it. More on this shortly.

So one way or another, whether it be people or kit, there are trans-zone protocols and systems which enable us to effect transfers between zones without trailing contamination around a building. However, moving between contamination zones is best achieved in steady increments, particularly when moving from a high zone to a lower one, otherwise we run the risk of allowing contamination to jump between areas. Monitoring systems would detect such errant particles, so we would know there was a problem and be able to do something about it. Nevertheless, we could face the prospect of a potentially unending cycle of halting operations, in order to clean up contamination that has managed to invade another zone's airspace; it is far better to observe the guiding principle and stick to moving from each zone to its immediate neighbor.

Now, having said all that you will not be surprised to hear there can be the odd exception to the rule, but not for personnel, so we can discount that. It comes about because occasionally we assign a C4 (red) classification to zones which are destined to experience mostly C3 levels, albeit with infrequent forays into the lower echelons of C4. In these situations we *may* elect to transition items between C4 (red) and C2 (green) but only if detailed analysis confirms, case by case, that such an approach is viable. Otherwise, step by step.

3.6 Dual classifications

Having established the principle of transitioning between zones, it becomes clear that transit areas must be assigned dual classifications and not just for contamination. That said, we need to keep in mind that containment, in terms of shielding and ventilation, is always arranged to satisfy the most onerous of potential "operating" conditions. Let's look a little closer, once again excluding, for the time being, ways in which personnel negotiate comparable transitions.

Fig. 3.4 depicts the classic example of adjoining decontamination and monitoring areas, which stand guard between C4 (red) and C3 (amber) environments. You may also come across arrangements where both activities are conducted in the same space, but when this happens the dual zoning we are about to examine follows similar principles. In this particular case, decontamination is needed so that operators can conduct hands-on maintenance of items which normally reside in a heavily contaminated cave. For this discussion, we shall assume decontamination and monitoring are both performed by remotely operated equipment, as is usually the case.

Prior to the sequence beginning, the decontamination and monitoring areas will be in their dormant condition, which corresponds to completion of the previous cycle; namely both will be C3/R3.

3.6.1 Start sequence

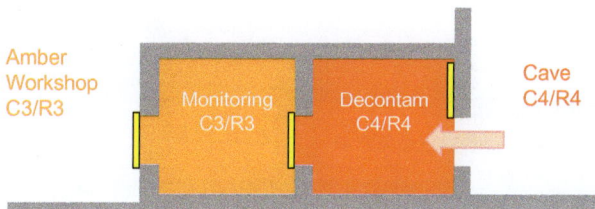

Fig. 3.4 Start sequence.
© Bill Collum.

The steel shield door between the decontamination bay and adjacent cave opens and an item enters (Fig. 3.4) bringing along its aura of contamination. The door closes. With a contaminated item aboard we must elevate the decontamination bay to C4; and since contamination levels are potentially very severe, we must also designate the area R4. Decontamination is carried out, say by multiple jets of high pressure water; then once the cleansed item has dried to a predetermined level, we are ready for the next stage. But before that, we must look back for a moment.

Several years before any of this happens, the facility's design team, designated plant operators and specialist equipment manufacturers, will have put considerable effort into developing decontamination techniques which are best suited to this particular situation, including full-scale trials with contamination simulants. Having gone to so much trouble, we can reasonably expect the chosen decontamination process to have a high success rate. So we shall assume this is the case and come back later to what happens if decontamination fails to achieve specified cleanliness levels.

3.6.2   Monitoring—pass

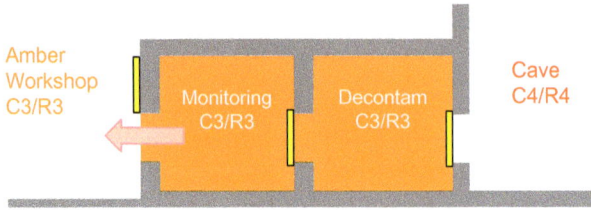

Fig. 3.5 Monitoring—pass.
© Bill Collum.

With decontamination complete, the cleansed item and decontamination bay itself should revert to C3/R3 levels; however, at this pre-monitoring stage we cannot be certain so must assume both are still C4/R4. The shield door between the two bays opens, the item moves to its monitoring position and the door closes behind it. Monitoring confirms that decontamination has been successful, to the extent that the item and both bays are within C3/R3 parameters. With that the final shield door opens (Fig. 3.5) and our now-cleansed item is transferred to the workshop where suitably attired operators can safely handle it.

3.6.3   Monitoring—fail

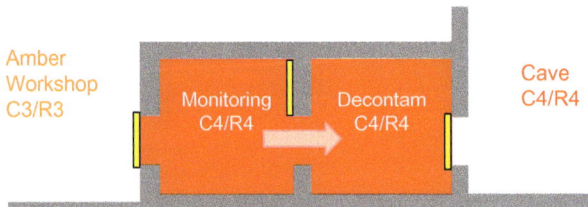

Fig. 3.6 Monitoring—fail.
© Bill Collum.

As above, decontamination is conducted and the item moves to its monitoring position; in this case, though, checks reveal it is still at C4/R4 levels, so we must assume both bays are similarly contaminated (Fig. 3.6). The item retraces its steps and is decontaminated again. In the meantime, a specialist cleanup team enters the decontamination bay and removes any traces that may have been deposited whilst it was occupied by an overly contaminated item. With that done, the recleaned item is monitored again, if necessary repeating the cycle several times until it passes and is transferred to the workshop.

Clearly the prospect of failing monitoring is not something to be taken lightly, so we can see why there is so much emphasis on conducting timely trials and selecting the right decontamination technology. To further reduce risks of contaminating a monitoring bay, some design teams elect to install monitoring sensors in the decon-tamination area. This *background* monitoring is not as sophisticated as the thorough processes which take place in a purpose-made monitoring bay. However, where levels are sufficiently elevated, it can be helpful in signaling further decontamination is required before opening the door to an adjoining monitoring bay.

3.7   Surface contamination

Up to this point, we have assumed contamination is airborne, but as I mentioned earlier it can just as easily settle in one location without getting into the air, and when this happens it needs to be managed in different ways. The first thing to mention in the context we are considering here, where surface contamination can have a bearing on how radiological zones are determined, is that we are only talking of very small amounts. In situations where surface contamination may accumulate to any degree, it will be deemed sufficiently hazardous to be considered airborne and therefore fall within the regular "C" classifications we have discussed so far.

Having established that contamination is not always airborne, the next logical step is to conclude that areas which are subject exclusively to surface contamination need not necessarily be indoors. Granted the notion of assenting to "outdoor" contamination may seem counterintuitive, but as we shall see, it can be safely accomplished. Let's take indoors first.

3.7.1   Indoors

We may come across this type of area where the occasional drop of a radioactive substance falls onto a surface, such as when a filling head is decoupled from a drum. Alternatively it can happen when a lid is tightened onto a container of some type, say a shielded transport flask, and a little moisture seeps out as the two surfaces are squeezed together. Importantly we are not talking here of gross contamination, so in this context are restricted to that which complies with C3 conditions. And as I say, just small amounts, not airborne and not something that can be expected to happen routinely.

Being nether wholly C2 (green) or wholly C3 (amber), these areas do not conform to standard zoning criteria so are not covered by any of the regular naming conventions; in fact, they do not have a formal designation. In essence, though, they are largely analogous to C2 zones, albeit with the additional characteristic of harboring traces of C3 surface contamination. For this reason my personal preference is to name them C3 (S), or amber (S).

We know that ordinarily when zone C3 (amber) is assigned to an area, it is so we can put measures in place to stop airborne particles escaping. The thing about surface contamination, though, is that it is not going anywhere so does not pose the same kind of hazard and therefore does not warrant being zoned in the same way.

The upshot is that in situations where operations may occasionally generate a little C3 surface contamination, it is permissible to conduct them in a C2 (green) zone. Albeit with a proviso that the area is cordoned off, say with handrailing. In addition, entry must be tightly controlled, otherwise there is a risk that surface contamination could inadvertently be carried into adjacent C2 areas. In practice, if there is a possibility that surface contamination may be encountered, then access can only be gained via special rooms where particular procedures must be followed. We shall look at how these rooms function when we discuss radiological changerooms in the next chapter.

3.7.2 Outdoors

At first it can feel a little strange to think you can be out in the open, and yet at the same time governed by radiological controls. However, once you consider the logic of how this type of environment functions, particularly not being susceptible to airborne contamination, you quickly become accustomed to the idea. Ordinarily these zones are found in older areas of nuclear sites, but not exclusively as from time to time they can also be linked to modern nuclear facilities, say a flask marshalling area.

If you were to have a good look around an outdoor area that has surface contamination zones within it, you will find it is a securely fenced enclave with no direct access to the nuclear site that surrounds it; for this reason, you sometimes hear them referred to as an *island site*. Furthermore, there may be not just one, but several nuclear facilities dotted around these enclaves.

Radiologically speaking, these enclaves are classed as radiation controlled areas, which acknowledges their potential hazards and imposes appropriate controls. So as with indoor C3 (S) areas, personnel can only enter via a radiological changeroom, although not of the same type. Access for vehicles is restricted to a needs-must basis, and those that do enter must pass through a checkpoint where they are monitored for contamination before being allowed to leave again. In addition, if there are areas within an enclave where surface contamination or radiation levels may present a particular hazard, they will be cordoned off from general access.

The difference between indoor and outdoor "surface contamination" areas is that air within indoor versions is contained, so can be monitored before being released back to the environment, whereas air associated with outdoor enclaves is unconstrained, so is free to waft around and ignore fencing. In this context, therefore, the watchword is "surface"; if contamination does not satisfy that clear definition, then any thoughts of creating this type of enclave are out of the question.

As with their indoor counterparts, the enclaves we are considering here do not conform to standard radiological zoning criteria so do not figure in regular naming conventions. My own preference is to name them C2 (S). It reinforces the message that contamination cannot be airborne, so by implication affirms that air can come and go without risk to those on a nuclear site or way beyond its boundaries. In terms of rad zoning, the story does not end there.

When applying radiological zones to facilities within C2 (S) enclaves, we must be mindful that although C2 (green) figures in their nametag, it is not what we might call a true C2 of the type we find indoors. For zoning purposes, "air" within these enclaves is by definition C1 (white) as it is free to go where it pleases. It is an important distinction and particularly relevant when we consider the principle of moving one step at a time between contamination zones. So whereas transitioning between C2 and C3 is perfectly acceptable, plans to transition between C2 (S) and C3 are not. Effectively it would constitute a C1/C3 interface, which as we know is outside the rules.

3.8 Depicting zones

When it comes to rad zoning, the normal practice is to lead with an area's contamination zone. So as we might expect, when floor plans are prepared which enable us to view all zones at a glance, each area will be labeled C2/R3, C3/R3, C3/R4 or whatever the combination may be. When these plans are overlaid with helpful color coding, then where "C" and "R" levels do not match, the usual practice defaults to depicting contamination colors. It is not a rule of any sort so there is the odd exception, such as when radiation is the sole or primary topic of discussion. Apart from that, colors signify contamination with radiation classifications picked up from the text. It is a convention which highlights the fact that, in rad zoning circles, contamination is most definitely the category which demands most of our attention.

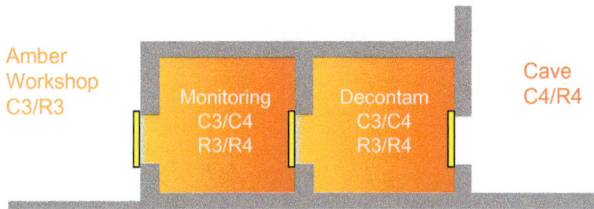

Fig. 3.7 Dual radiological classification.
© Bill Collum.

In the dual zoning example we discussed above, both areas would have a classification of C3/C4 and R3/R4. When depicted on color-coded drawings, they would appear as in Fig. 3.7, or alternatively be hatched with an amber/red pattern. Either way the significance of their dual classification will be highlighted.

Now that we have seen how nuclear facilities are divided into radiological zones and how equipment, radioactive materials and so on move between these zones, we are ready to examine how people do the same thing.

Radiological changerooms

Most factories and industrial premises have changing facilities of some sort, where the workforce can swap their regular clothes for something much less fashionable but more appropriate for the environment in which they will spend the day. In nuclear plants these areas are configured very differently, as apart from the usual clothes changing functions they also have a role in stopping the spread of radioactive contamination. For this reason they are known as radiological changerooms.

Fig. 4.1 Nuclear facility with administration building attached.
© Nuclear Decommissioning Authority.

Large nuclear facilities often have an administration building attached (Fig. 4.1) where, following entry processes, personnel can gain access from the street outside. Here you find the usual offices, conference rooms, dining facilities, and so on. For many workers this is as far as they will go, since their duties do not entail entering the adjoining nuclear plant. Changerooms may be located in a separate building but are quite often found within these administration blocks, maybe on their own floor, so there is direct access for those who divide their time between office duties and a role within the adjoining nuclear facility.

Nuclear Facilities. http://dx.doi.org/10.1016/B978-0-08-101938-2.00004-0

4.1   Generic types

The first thing to note about radiological changerooms is that they fall into two distinct categories. A *changeroom* segregates the regular external environment from the inside of a nuclear facility, so where we cross between C1 (white) and C2 (green), while *sub changerooms* are the ones we find inside facilities, and are used by personnel to move back and forth between higher contamination zones. In fact sub changerooms are further divided into those used to move between C2 (green) and C3 (amber) areas, and a slightly different type between C3 (amber) and C4 (red). We shall begin with changerooms and then take a look at how the usually smaller sub changerooms differ.

4.2   Changerooms

The steps followed on passing through a changeroom are pretty much the same whether you are a visitor, or a worker based permanently in the facility it serves. For our discussion, we shall assume a changeroom is of the fully comprehensive variety, those which serve large nuclear facilities, and for completeness assume a dedicated change area is needed for visitors, something which is not always the case. We shall follow the permanent worker route and along the way take in any slight variations for visitors.

4.2.1   Locker rooms and basics

The first area you encounter in any radiological changeroom is similar to changing facilities in any other large industrial complex, with lockers, showers, and so on (Fig. 4.2). Those based in the building are assigned their own locker, where they can leave their clothes and personal belongings.

Clothes worn within the facility are commonly referred to as basics (Fig. 4.3) but can also go by names such as blues, browns, greens, etc. depending on a site operator's color of choice. Whatever its name, this fresh attire is collected from a room adjacent to the locker area.

Selecting your clothes does not compare to the experience of shopping on the high street but all of the essentials will be there, neatly laid out, in size order, on stout wooden shelves: everything from underwear to trousers, shirts, and socks. Shoes come later, so at this stage you move around in your stocking feet.

Once through this regular changing area you will be heading towards the nuclear plant itself, where all areas are assigned a contamination classification of at least C2 (green). Clearly in this kind of environment it would be quite reckless to handle and eat food, or even to chew gum, as it would needlessly incur a risk of ingesting contamination. For this reason there is always a strict policy of not allowing food products of any type to pass beyond the locker room. In fact it is so serious a matter that, for our own wellbeing, contravention of this particular rule is normally considered a disciplinary matter.

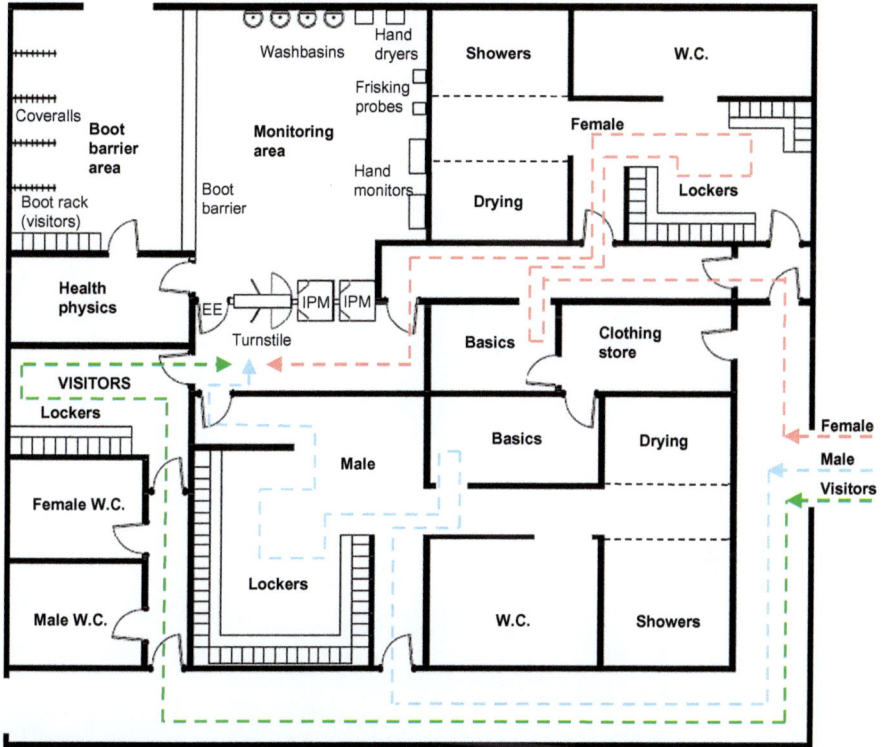

Fig. 4.2 Nonradiological access routes.
© Bill Collum.

4.2.2 Access to C2 areas

On leaving the locker room you will be confronted by a robust security turnstile (Fig. 4.4) which is a sure sign you are about to enter a nuclear environment. Once through a turnstile the nuclear world beckons, so access arrangements at this point are necessarily very tightly controlled.

For visitors to get as far as the turnstile it is not always necessary to change into basics, as actual attire depends on specifics of a visit and the facility itself. The minimum requirement will be to leave jackets, socks and shoes in a locker and don a pair of long socks identical to those worn by permanent workers. Sometimes, depending on local practice or conditions, you may be asked to tuck trouser bottoms into your socks.

4.2.2.1 Monitoring and coverall areas

On passing through a turnstile, the first thing that strikes you is that the room you are standing in is divided into two sections by what looks like a long smooth box or a bench of some sort, about half a meter high. It is known as a boot barrier (Fig. 4.5) and innocuous as it looks plays a crucial role in the successful operation of a changeroom.

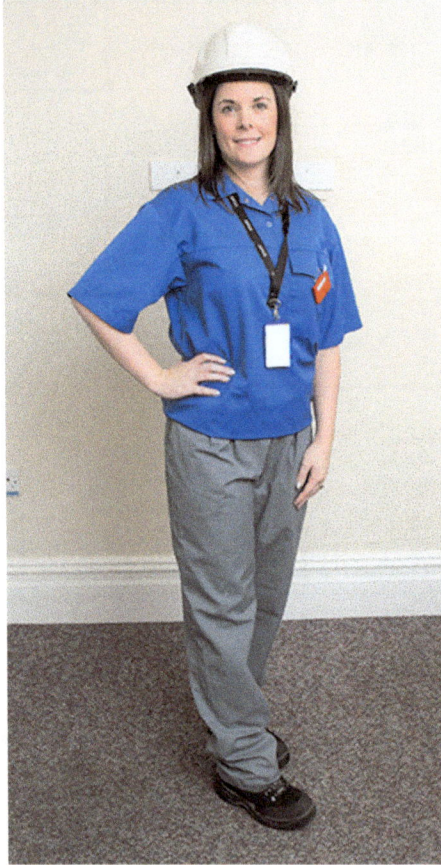

Fig. 4.3 Nuclear worker "basics."
© Nuclear Decommissioning Authority.

The area between a turnstile and boot barrier houses an array of monitoring equipment, along with several washbasins; however, these are not used on the inward journey, so we shall save discussion on those for later. Across the boot barrier there will be racks stacked with visitor work boots in an assortment of sizes.

I must admit, the first time you pass through a changeroom it does feel a little disorientating, particularly on the return journey where there are more procedures to follow. Fortunately, as a newcomer or visitor you will be accompanied by an experienced guide who has undergone the necessary training, and after a couple of visits you start to get the hang of it. Actually when you get to the stage of understanding what is going on and the context of why it is happening, you do feel yourself marveling at how very simple and yet effective the whole process is. But as I say, this is definitely not the way it feels on your first visit.

Fig. 4.4 Security turnstile.
© Nuclear Decommissioning Authority.

Fig. 4.5 Boot barrier.
© Nuclear Decommissioning Authority.

As I mentioned earlier, arrangements in changerooms do vary from one to the other but that said they all follow a similar pattern. One notable exception is the location of coverall racks, which in some facilities are across the boot barrier but in others, as Fig. 4.5, before it. This apparent anomaly can stem from the practice of some facilities to issue clean coveralls for each plant entry, in which case racks tend to be situated prior to the boot barrier. On the other hand, it might simply be a matter of space constraints or local preference which, bearing in mind there is no right or wrong arrangement, does permit some flexibility in where they are positioned.

The area of a changeroom between its turnstile and boot barrier is normally referred to as a *monitoring area*, while the zone across its barrier is known rather unimaginatively as the *boot barrier area*, or if racks are present then it may be referred to as a *coverall area*. Those are the names we tend to see on drawings or referenced in documents. However, during informal discussions a monitoring area is often referred to as the *clean side* and boot barrier area as the *plant side*, which is a good tag in that it gives the clue that once across a boot barrier we can access the facility beyond.

Interestingly, with a monitoring area being the buffer between a facility and outside world, it is neither C1 (white) nor C2 (green). Strictly speaking it is categorized as C2, but recognizing its role as a transition area, I often refer to it as C one and a half. Of course, this is not the kind of term that can be referenced officially, but if used informally, it is a handy tag that highlights its radiological role.

4.2.2.2   Boot barrier—inward

A boot barrier is so called because its plant-side it is divided into open-fronted compartments where work boots are stored, with each compartment individually numbered and assigned to a member of the facility's permanent workforce.

Fig. 4.6 Boot barrier procedure—step one.
© Nuclear Decommissioning Authority.

To cross the barrier a worker reaches over it, retrieves their boots and places them on the plant-side floor (Fig. 4.6).

Fig. 4.7 Boot barrier procedure—step two.
© Nuclear Decommissioning Authority.

With that done they sit on the barrier and swing around (Fig. 4.7) while ensuring their stockinged feet do not touch the plant-side floor.

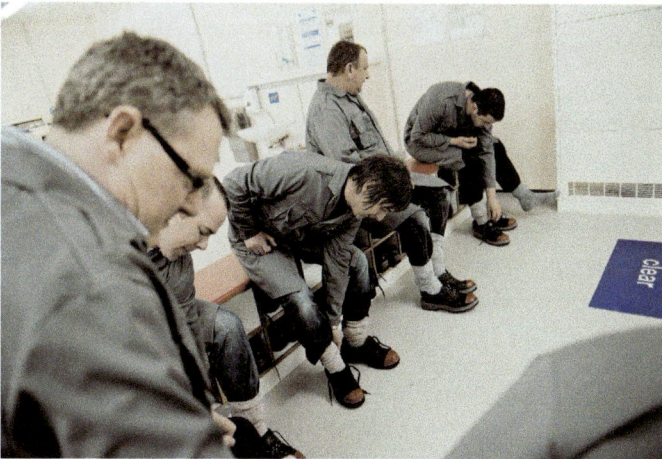

Fig. 4.8 Boot barrier procedure—step three.
© Nuclear Decommissioning Authority.

Finally they slide their feet into the waiting boots (Fig. 4.8) and tie the laces. This procedure ensures that footwear worn within a nuclear facility always remains on the plant-side of a boot barrier, and therefore mitigates any risk of trailing contamination onto the floor of a monitoring area, or beyond.

Visitors crossing a boot barrier wait for their guide to cross first and retrieve a pair of right-sized boots for them from racks on the other side. Boots are placed on the plant-side floor beside the barrier; then visitors sit down, swing around and don their boots in exactly the same way as resident workers.

Local practice will dictate whether those wearing basics also don coveralls before entering a plant. However, visitors attired solely in their own clothes and newly acquired work socks, will definitely be asked to don some form of coverall. If hard hats are needed, then they too are picked up in this area. Once dressed everyone must clip on dosimeter, along with any other monitoring devices which may be required, and with that we are on our way.

4.2.3   Egress from C2 areas

Once on-plant, most workers spend either all of their time or the vast majority of it in a facility's C2 (green) areas, with just a small number of specialists entering zones categorized at C3 or above. For visitors, it would be quite unusual to venture beyond a C2 environment. Those who do will almost certainly have had some additional training, including *sub changeroom* procedures, which are used when entering areas categorized as C3 (amber) and C4 (red). We shall look first at the return journey through a changeroom and come back later to how we navigate sub changerooms.

You will recall from the previous chapter that C2 areas are designed and operated to be free of airborne contamination. However, under certain conditions traces of surface contamination are permissible, with a proviso they are limited to C3 levels and confined to areas which are identified and segregated. All of this means that when we arrive back at the plant-side of a boot barrier, there is a low probability that we shall be carrying any particles of contamination with us. Of course we must be absolutely certain, so this is where a changeroom really comes into its own.

4.2.3.1   Airflow

The first line of defense is to take no account whatsoever of an assumption that C2 areas should be free of airborne contamination. On the contrary, the ventilation system is configured in such a way that air is continually drawn across a changeroom from its monitoring area towards the active, or C2 side (Fig. 4.9). We shall examine this in more detail in Chapter 8 when we discuss nuclear ventilation. For the moment it is sufficient to note that air in changerooms, and sub changerooms for that matter, always flows towards the side which is potentially more contaminated.

It is unlikely you will notice these airflows but they are ever-present and ensure that, if airborne particles were in the vicinity, they would be unable to drift through a turnstile and invade the building beyond. This leaves potential surface contamination, which is dealt with through procedures that must be closely observed by everyone passing through a changeroom.

Fig. 4.9 Airflow direction.
© Nuclear Decommissioning Authority.

4.2.3.2 Coveralls

Workers wearing either overalls or a lab coat hang them back on their personal numbered peg on the coverall racks. The same number is assigned to their boots compartment in the boot barrier and also to their personal locker. Likewise, visitors return coveralls to a nominated group of pegs. And regardless of which side of a boot barrier the coveralls were originally collected from, once used they are left on its plant side. This is a sensible precaution which guards against unwittingly carrying contamination across the barrier.

When approaching the coverall racks there is a very simple action to be carried out, one which is disproportionately embarrassing if you should happen to forget. It is simply to unclip your dosimeter and be sure to keep it about your person when leaving the area. It might not sound like much, but if someone has to re-enter a changeroom to retrieve it on your behalf the story is almost certain to be retold, so it's best to stay alert.

4.2.3.3 Boot barrier—return

The procedure for re-crossing a boot barrier is essentially the reverse of that followed on an inward journey; again, care must be taken to ensure stockinged feet do not touch the floor on a barrier's plant side. In this instance visitors cross before their guide, who places borrowed boots back on a rack which is usually out of reach from the barrier, whereas permanent workers deposit boots back in their own numbered compartment.

Knowing that every permanent worker has a personal slot within a boot barrier, explains why some barriers appear much longer than they need to be. It is simply a direct correlation to the number of boots that must be stored, a factor compounded in facilities which operate a two- or even three-shift working pattern. As a consequence, years before a changeroom is built, designers and operators must work together to determine how many workers will need to avail of it. It is a task which sounds straightforward enough but in reality is quite a major undertaking, particularly when you bear in mind how important it is to get the numbers right.

4.2.3.4 Monitoring

With barrier procedures complete we have left the main plant behind, so there is just the monitoring area to traverse. Fig. 4.10 gives an overview of the scene, with an inward security turnstile in the background, hand monitoring and washing to the right and whole-body monitoring, prior to exit, on the left. We shall take each of the monitoring procedures in turn.

Fig. 4.10 Monitoring area.
© Nuclear Decommissioning Authority.

Before we move on, I should mention that monitoring procedures can vary a little from one site to another and even between facilities on the same site. It all depends on local practice and exact conditions within the facility in question, so I have gone for what we might call a standard sequence (Fig. 4.11) certainly one which is very commonplace. Normally then, the first task is to wash our hands. Recognizing there is potential for contamination, it would not be sensible to handle taps and risk passing it on to the next person, so water flows are automated by use of an infrared sensor or similar device.

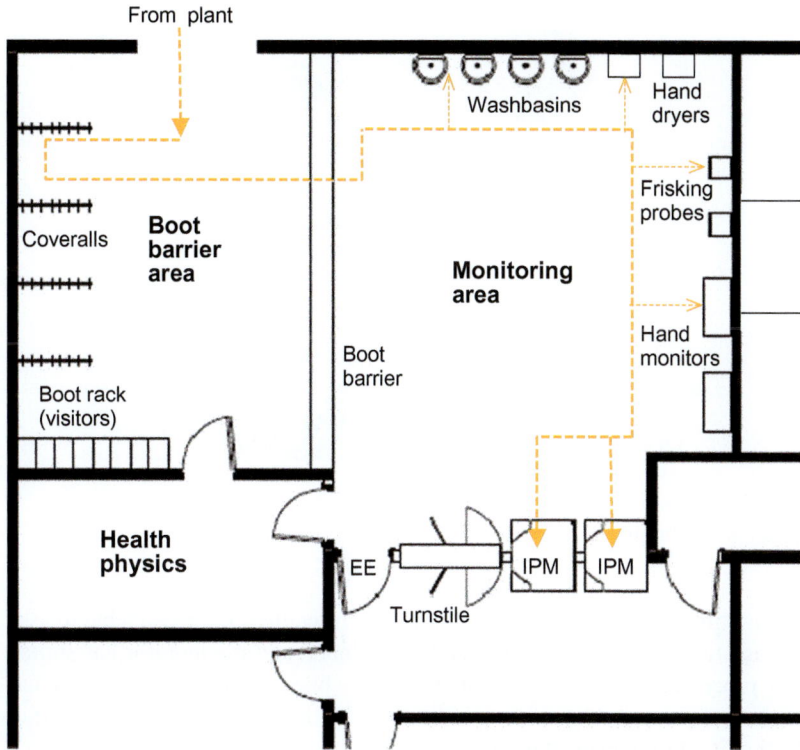

From plant

Washbasins

Hand dryers

Coveralls

Boot barrier area

Frisking probes

Monitoring area

Boot barrier

Hand monitors

Boot rack (visitors)

Health physics

EE

Turnstile

IPM

IPM

Fig. 4.11 Monitoring sequence.
© Bill Collum.

Incidentally, it must be assumed that any water discharged from a changeroom could be carrying particles of contamination within it; the term used is *trace active*. It is therefore channeled through monitoring and, if necessary, cleanup processes, before being declared ready for return to the environment.

We know from discussions in Chapter 2 that it only takes a sheet of paper to halt alpha radiation; indeed, alpha particles can even be masked by a thin film of water, making it impossible for monitoring equipment to detect them. After washing, it is therefore essential that hands are dried thoroughly before moving onto the monitoring stage.

Hand monitoring

First up is a hand monitor (Fig. 4.12) where our hands are inserted into open slots on the unit's front face. The inside of these slots tapers inwards, in such a way that we are forced us to keep our fingers nice and straight. Once your hands are inserted as far as the wrists, you will feel your fingertips touching a solid surface that closes the far end of each opening.

Fig. 4.12 Hand monitoring.
© Nuclear Decommissioning Authority.

The first time you use one of these monitors, it would seem reasonable to assume at this stage that your hands are perfectly positioned and the monitor has started its scanning process. However, if you were to wait a while and then withdraw your hands, the monitor would sound an alarm to indicate its work was not yet done, or to be more precise that it had not even started.

It turns out the plate at the far end of each slot moves forward if you push it hard enough with your fingertips. The technique is to place your hands deep inside the slots and then push firmly. Personally I like to lean in a little to make it easier. As you do this your fingertips push the plates forward and immediately a chime signals the start of a monitoring sequence, which takes about 10 seconds. Assuming there is no contamination present, the sequence ends with a very satisfying ping which confirms you have passed the first test.

Frisking station

Next up is a scintillation detector (Fig. 4.13), which in this particular guise is known as a frisking station or frisking probe. Unlike hand monitors there is no start-up technique, as they begin scanning the moment their flat pad is lifted from its holding bracket. The procedure here is to slowly run its detachable pad over your arms, hands, face, hair, clothes, and stockinged feet.

Using a frisking pad reminds us that it is inadvisable to carry anything into a nuclear facility unless we really need it, as there is always a possibility it may become contaminated. At best this could delay your departure, but may conceivably result in the loss of a treasured possession. However, if objects such as a pen or notebook have been taken across a boot barrier, then they must be carefully scanned with the pad. Notebooks by the way must be scanned a page at a time, which is another good reason

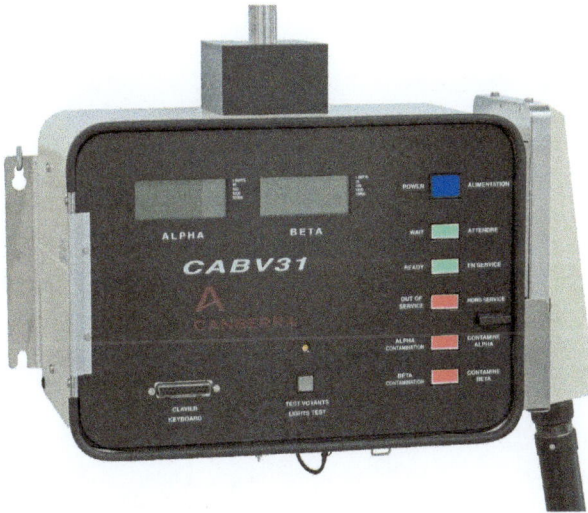

Fig. 4.13 Frisking station.
© CANBERRA.

for trying to manage without. Personally I leave my watch behind and either carry nothing at all, or at most a pencil and single sheet of paper.

While it is scanning, a frisking station crackles now and again like the Geiger counters you might have come across or seen on TV; however, this is part of their normal operation and nothing to be anxious about. I do wish someone had told me that before I used one for the first time. I was quite startled when it started crackling and found myself envisioning a night in quarantine being tended by a team of specialists, but then I looked around and noticed others using the same crackling equipment and not showing the slightest concern. Reassured, I regained my composure and feigned all was going exactly as I had anticipated. Assuming no alarm sounds, the second test has been passed.

For facilities harboring particularly low levels of contamination, this is all the monitoring that is required, possibly with an additional foot monitor; for most, though, there is an additional stage, one that must be followed very precisely. Furthermore, if it does not confirm a person is entirely free of contamination, they will literally be unable to leave the monitoring area.

Installed personnel monitor

This apparatus goes by the name of an *installed personnel monitor* widely referred to as an IPM, or in plain English it is a whole-body monitor. They operate by combining scintillation detectors, as used in hand monitors, and gas flow detectors. Alternatively they may use scintillation detectors on their own. We shall examine how this type of equipment works when we discuss radiometric instruments in Chapter 11, so for the time being can concentrate on how to use it.

Fig. 4.14 Installed personnel monitor (IPM).
© CANBERRA.

IPMs come in several configurations, each of which is used slightly differently. All have full-height detector panels. And while some have hand monitoring slots, at elbow height, on each side of the panels, others (Fig. 4.14) have an additional detector to one side. The other main variation is that while some are entirely open, as shown here on the left, others have a panel attached, so forming a cubicle. And very often cubicle versions have a personnel access door fitted to each side. As I say, some slight variations.

The turnstile through which we enter a changeroom is usually alongside an IPM. However, it does not have a pass reader on the monitoring side so cannot be used to exit the area. The only exception would be in an emergency, when a turnstile will automatically reset to freewheel towards the locker rooms. In addition there is always an emergency exit door nearby, which if used illegitimately will trigger an alarm. So in normal conditions the only way out of a changeroom is through an IPM.

For our purposes here we can follow the procedure adopted when using a fully enclosed IPM cubicle, one with built-in hand monitoring slots, but keep in mind that there are some slight differences when using other configurations. There are just two steps.

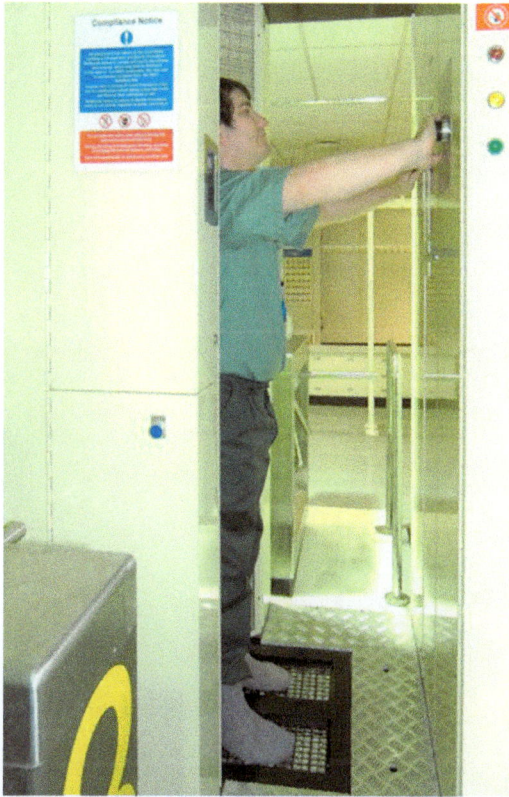

Fig. 4.15 IPM in use—step one.
© Nuclear Decommissioning Authority.

First, stand with your back to the concave area so that its monitoring panels wrap around you slightly (Fig. 4.15) and with both feet placed on the floor panels. Facing you at about shoulder height there will be two large buttons with hands outlined around them, showing how to position your palms over the buttons. Once in place, push your hands firmly on the buttons, which forces the back of your body against panels behind you. A chime will sound as the first scanning stage begins, followed about 5 seconds later by a ping, or prerecorded voice, indicating it is time to turn round and face the other way.

For the second stage (Fig. 4.16), place your hands into slots in the same way as a hand monitor and lean your body into the monitoring panels, as if someone was trying to squeeze past behind you. Again it takes around 5 seconds and, assuming you are free of contamination, concludes with another ping. With that, an adjacent door or exit turnstile will sound a metallic click, confirming you are cleared to leave the changeroom. You step through the exit, retrace your steps to a locker room and if you wish take a shower before retrieving your own clothes.

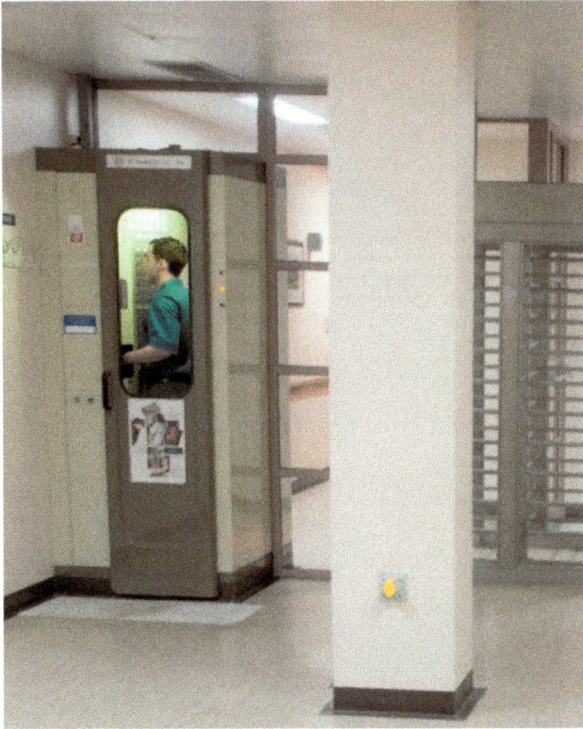

Fig. 4.16 IPM in use—step two.
© Nuclear Decommissioning Authority.

4.2.4 Health physics

If any monitoring equipment sounds an alarm while it is being used, there are straight-forward procedures which must be followed. In the case of a hand monitor or IPM, you simply wash your hands and scan yourself a second time. If the system alarms again then it is time to summon assistance. If a frisking station signals it has detected contamination about your person, further washing would serve no purpose so it is also time to send for help.

Every changeroom, and sub changeroom for that matter, has a telephone with a dedicated line to the facility's health physics office. It is manned by a team of specialists, headed by a radiological protection advisor (RPA), who know all there is to know about radiation and contamination. In fact the role of an RPA so important, it is a legal requirement that they advise project teams during the design process, and once a facility begins operating they must always be present or on call.

If we do need to call on their assistance, health physicists are trained in techniques employed to locate particles of contamination and gingerly remove them. Old hands will cheerfully inform newcomers that the process involves a wire brush and one of

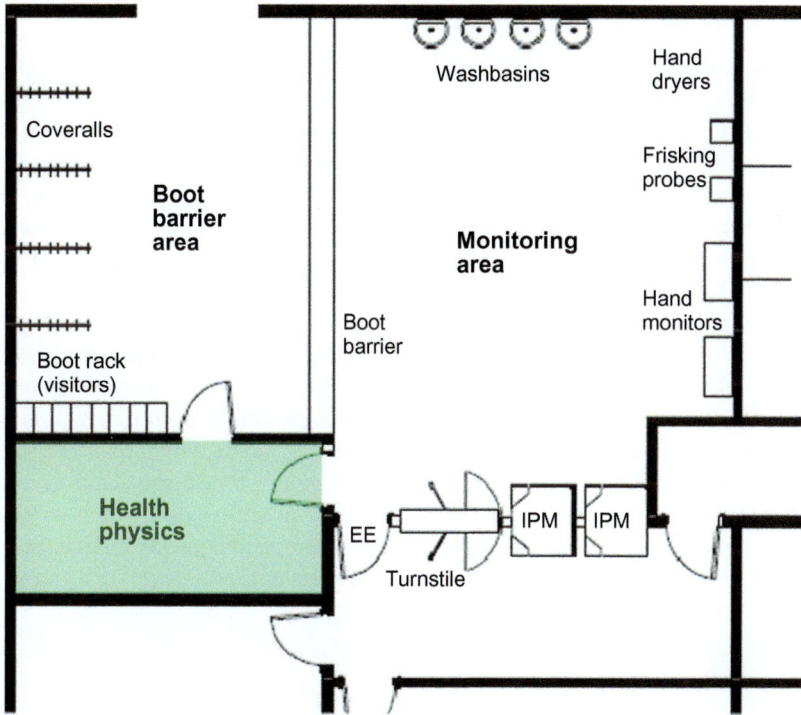

Fig. 4.17 Health physics office.
© Bill Collum.

those powerful jet wash machines that are so effective at cleaning driveways, but happily it has more to do with adhesives, special chemicals and knowing how to use a swab.

The main base for a health physics team is normally located deep within a nuclear facility, close to where most of their monitoring and surveillance duties are performed. For changerooms that cater for high numbers of personnel, it is also a good idea to locate a satellite health physics office adjacent to the boot barrier. In fact the ideal position is as shown in Fig. 4.17, where the office has direct access to both sides of the barrier and enables specialists to work most effectively.

Incidentally, with all this talk of personal monitoring and the prospect of a health physics team swinging into action, it may appear that the probability of becoming contaminated is almost routine. The thing to keep in mind though is that nuclear plants, particularly C2 (green) areas, are designed and operated with an eye to radiologically cleanliness, so changeroom procedures are precautionary. That being said, possibilities are real enough, so I have no wish to play down the situation. What I can say, in the interests of balance, is that while finding oneself contaminated is not inconceivable, it would be unusual.

4.2.5   Duplicating equipment

When contemplating changeroom planning, it might be assumed that separate monitoring areas and boot barriers are required for male and female personnel, and also for visitors if they have a dedicated changing area. Clearly each must have their own locker room, showers and so on, but when you think about it everyone is fully clothed around a monitoring area so there is no need for segregation. Yes, such replication is an option but should not be seen as a foregone conclusion, for two reasons. First is the straightforward use of space which is much more efficient when these areas are combined, but the second is not quite so obvious.

From time-to-time monitoring equipment is going to be out of service, due to either routine maintenance or breakdown. In busy changerooms queues would quickly build up, with little hope of an instant remedy. The solution is to duplicate each item of equipment, so there is a high probability that at least one of each pair will always be working. However, if we opt for dedicated male and female monitoring areas, we must purchase four IPMs, hand monitors, frisking probes, and hand dryers; if catering for segregated visitors, then six of each item are needed. When we factor in how expensive some of this equipment can be, segregation quickly loses its appeal; it is far better to purchase pairs of the necessary kit and route everyone through a common monitoring area.

4.2.6   Restrooms

You may already have guessed that with a possibility of encountering contamination within nuclear plants, it would be inadvisable to locate toilets or restrooms within them. Apart from anything else monitoring procedures would necessarily become quite intrusive, so a combination of safety considerations and common sense dictate it is a nonstarter.

Of course there are always toilets adjacent to locker rooms but, if partway through a shift, it can be quite time consuming to pass back and forward through a changeroom in order to use them. The only slight improvement that can be made to the situation is to provide restrooms which are directly accessible from a monitoring area. Personnel must still follow boot barrier procedures, wash their hands and use a hand monitor and frisking probe, but it is not necessary to pass through an IPM. It is a fairly slight difference and one normally only encountered in large changerooms, but is helpful in speeding things along and freeing up access to IPMs during busy periods.

4.2.7   Access to outdoor radiation controlled areas

Occasionally you may come across a changeroom in a location which at first seems to contravene the normal rules of radiological zoning and containment. It happens where we need access to a radiation controlled area, or outdoor radiological zone, of the type discussed in the previous chapter. In one variation the changeroom itself is no different to any other, with the usual lockers, security turnstile, boot barrier, monitoring equipment, and so on. The difference only becomes apparent when you cross to the plant side of a boot barrier and expect to make your way into a nuclear facility. You walk a short distance, pass through a few doors and then step onto the street outside. With other variations there is no boot barrier, so you simply pass through a monitoring area and then step outside.

The first time it happens it can be a bit of a puzzle, as you find yourself wondering where the nuclear plant went. And even more bewildering, what was the point of passing through a changeroom only to step outside and look up at the sky again? Well, the answer hinges on the fact that changerooms are used to access not just regular C2 radiological zones, but also outdoor radiation controlled areas, described in the previous chapter as C2 (S). And as we know these areas are not prone to airborne contamination: surface contamination maybe, but nothing floating in the air. When you put the whole thing together, it becomes clear that changerooms do not necessarily need to serve areas enclosed by four walls and a roof; of course, the vast majority do, but not all.

If you pass through a fully equipped changeroom, with a boot barrier, to get into a radiation controlled enclave, C2 (S), then you may not encounter any other boot barriers as you enter facilities located within it. On the other hand, if you gain access via a changeroom without a boot barrier, then you will find them located at the entry point to individual facilities. In addition, each building may have local security measures that restrict access to authorized personnel only. However, this relates to protection requirements so is a separate matter to the radiological controls we are considering here.

On entering an actual facility within a radiation controlled enclave, its environment will be C2 (green) with which we are familiar. If access is required to C3 (amber) or C4 (red) areas within the building, then personnel follow procedures identical to those which apply in any other nuclear plant, namely via the sub changerooms that we are about to discuss.

4.3 Sub changerooms

4.3.1 Generic types

As with changerooms, there is a fair degree of variation in the way sub changerooms are configured, but once again they all adhere to the same basic principles. The majority are classed as C2/C3 (green/amber) and enable personnel to cross between those two radiological zones. A smaller number are categorized as C3/C4 (amber/red) and facilitate entry into the more highly contaminated areas of a nuclear plant.

In addition to their distinct radiological classifications, sub changerooms also divide into structures which are either temporary or permanent. Permanent versions, which are by far the most commonplace, are housed in recognizable rooms, whereas their temporary counterparts are accommodated within either tented enclosures or demountable partitioning. As a user the most notable difference is that permanent sub changerooms, while not palatial, will certainly be adequately sized, whereas the ad hoc nature of temporary rooms dictates that they are often located in some handy space, so can be a bit tight on elbow room. That being said, they work just fine.

As a rule, sub changerooms which are needed during a facility's operational phase are permanently built-in, whereas temporary enclosures tend to be a feature of decommissioning, so do not appear until maybe decades after a facility was first built. There can be exceptions, such as where a particular sub changeroom is needed so infrequently during normal operations that expenditure on a permanent fixture would be unwarranted. In these cases it makes more sense to allocate appropriate floor space and simply erect a temporary enclosure on the rare occasion that one is needed.

The way in which both permanent and temporary sub changerooms are equipped and operated is broadly similar, with main differences arising from whether they provide access to C3 (amber) or C4 (red) areas. The one notable difference for temporary enclosures is that they are accompanied by a mobile filtration unit, which is needed to maintain containment and keep air moving in the right direction, a subject we shall examine thoroughly in Chapter 8. For our discussion here, we shall assume sub changerooms are permanent structures and take them in radiological order, starting with those designated C2/C3.

4.3.2   Sub changeroom—C2/C3

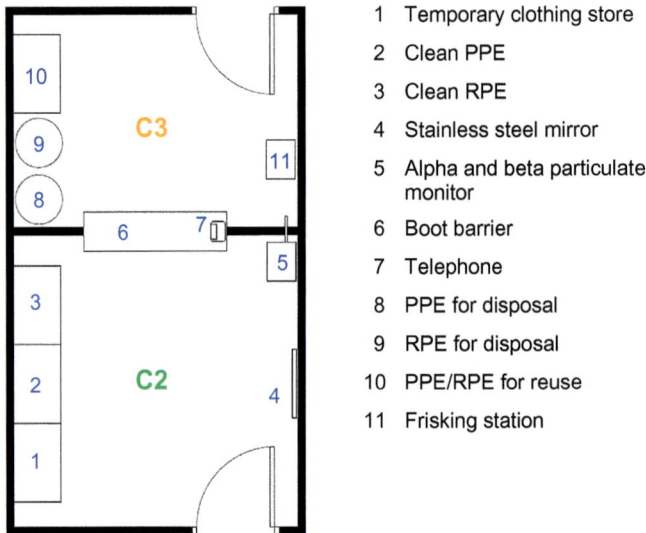

1	Temporary clothing store
2	Clean PPE
3	Clean RPE
4	Stainless steel mirror
5	Alpha and beta particulate monitor
6	Boot barrier
7	Telephone
8	PPE for disposal
9	RPE for disposal
10	PPE/RPE for reuse
11	Frisking station

Fig. 4.18 C2/C3 sub changeroom.
© Bill Collum.

As we can see in Fig. 4.18, a sub changeroom has the familiar boot barrier located approximately across its center. The procedure for inward crossing is similar to that followed in changerooms, but this time involving different footwear.

4.3.2.1   Clothing and protective equipment

Often it is just a matter of selecting a pair of disposable covers with an elasticated opening (Fig. 4.19) and, while sitting on a boot barrier, pulling them on over the boots you are already wearing.

Fig. 4.19 Disposable boot covers.
© Nuclear Decommissioning Authority.

In the United States, they are known as *booties*, a name I like very much, simply because it is so unexpectedly genteel for the rather austere industrial surroundings in which they are worn. For those wearing a lab coat the normal practice is to leave it on a boot barrier's C2 side and exchange it for one on the C3 side. If you are not already wearing a lab coat when you enter a sub changeroom, you may need to don one over the basics or overalls you have been wearing up to this point. To complete the minimum C3 attire, you normally pull on a pair of latex gloves.

If you are about to enter a C3 (amber) area that is potentially wet or even a little damp, then you will exchange work boots for a pair of short rubber wellingtons, again leaving your boots on the C2 (green) side. And in line with the usual procedure, wellingtons are slipped on while sitting on a boot barrier and facing towards the room's C3 side. In addition to revised footwear, a lab coat, and gloves, if it is deemed that there are particular radiological hazards due to airborne contamination in a C3 area, then you will don additional protection such as the respirator shown in Fig. 4.20.

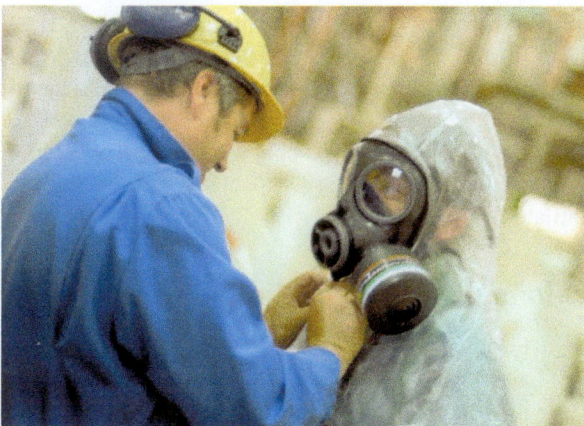

Fig. 4.20 Full face respirator.
© Nuclear Decommissioning Authority.

It is not possible to give definitive descriptions on exactly what one might be expected to wear in a C3 area, as specifics are dependent on characteristics of the particular environment being entered. We can be sure, however, that some form of personal protective equipment (PPE) or respiratory protective equipment (RPE), or both, will be necessary.

4.3.2.2   Personal monitoring devices

Whatever you need to wear, the one thing to remember is to unclip your dosimeter, discussed in Chapter 2, and transfer it to new clothing. If you are entering an area classified as C3 or above, then you may well be wearing both a thermoluminescent dosimeter (TLD) and an electronic personal dosimeter (EPD).

Recognizing there is a fair bit to think about when getting dressed and ensuring the necessary paraphernalia is in its proper place, many sub changerooms have a full length mirror on their C2 side, where you can admire yourself and double-check that everything is in order before moving on. Actually the mirrors are polished stainless steel, which has the dual benefits of making them more robust than a regular mirror and easier to dispose of when the time comes.

4.3.2.3   Disposable items

Day-to-day disposable items, such as booties and gloves, are normally picked up while on a sub changeroom's C2 side; then on return they are deposited into designated bins on the C3 side. These bins are located close to a boot barrier, so that we can remove used items and drop them into their appropriate bin while sat on the barrier, or standing on its C2 side. Once again, the emphasis is on ensuring C3 contamination does not get an opportunity to migrate from the confines of its designated area.

4.3.2.4   Containment

It is worth mentioning here that the containment principles applied to changerooms, which involves using airflow to stop contamination absconding, applies equally to sub changerooms. However, in this context contamination is a real rather than theoretical possibility, so air moves across sub changerooms with greater speed. At this point it is just a matter of being aware that airflows play a role in the operation of all types of changeroom. We can save the details for Chapter 8, where we shall see how nuclear facilities are ventilated.

4.3.2.5   Monitoring procedure

If you think through the logic of how a C1/C2 changeroom operates, much of it hinges on the fact that C2 (green) areas are designed and operated to be free of airborne contamination. In addition, if surface contamination is present in a C2 area, it must be limited to small amounts and restricted to locations where it can be confined. With this in mind, on returning to the plant-side of a changeroom boot barrier we can confidently cross with little expectation of trailing contamination onto its clean side. Of course we thoroughly monitor ourselves before heading back to a locker room and the outside world, but in the normal run of things we can consider this to be a confirmatory process. For sub changerooms, though, this is not the case.

On returning to the C3 (amber) side of a boot barrier, there is a real possibility we may be carrying contamination about our person; so, unlike a changeroom, the practice here is to monitor ourselves *before* crossing the barrier. That said, personal monitoring within sub changerooms is indistinguishable from that adopted in changerooms, so is normally conducted with a frisking probe, occasionally supplemented by a hand monitor.

If monitors do detect contamination, we contact the health physics team. For this reason all sub changerooms have a telephone, which is normally attached to the wall and directly above one end of its boot barrier. This is the ideal spot as it affords direct access to the phone from both C2 (green) and C3 (amber) areas.

As a further reminder of potential contamination, air on the C3 side of a boot barrier is continually sampled by an *alpha and beta particulate monitor*. We shall discuss this type of equipment in Chapter 11, so I will just say here that in order to simplify maintenance the unit itself is best located on the C2 side of a sub changeroom, but then fitted with an extension to its *sniffer probe* so as to enable it to sample from the C3 side.

4.3.3   Sub changeroom—C3/C4

Entering a highly contaminated C4 (red) area is not something to be undertaken lightly, so there is a regulatory responsibility on those designing nuclear facilities to ensure it is very much a last resort. For example, if a task within a C4 area is deemed to be in any way routine, say replacing the seal on an item of equipment, it is simply unthinkable to proceed on a basis that it will be performed by workers entering the area on a regular basis. A solution must therefore be developed whereby such tasks are performed by remote means, or alternatively make provision to retrieve items and maintain them in a C3 environment. In practice then, entry is only permissible for ad hoc or unforeseen circumstances.

For new facilities a design team's task is not without its challenges, but at least the starting point is a blank sheet so there are opportunities to develop proposals which comply with this particular constraint. When it comes to decommissioning aging nuclear plants it gets much more complicated, but the onus is still on designers to, as far as possible, keep the workforce out of C4 (red) areas.

Of course by its nature decommissioning entails a much higher degree of *manual entry,* as it is termed, into contaminated areas, so this is where we tend to see the majority of C3/C4 sub changerooms (Fig. 4.21). Having said that they are still pretty thin on the ground, and those that we do come across are housed almost exclusively in temporary structures.

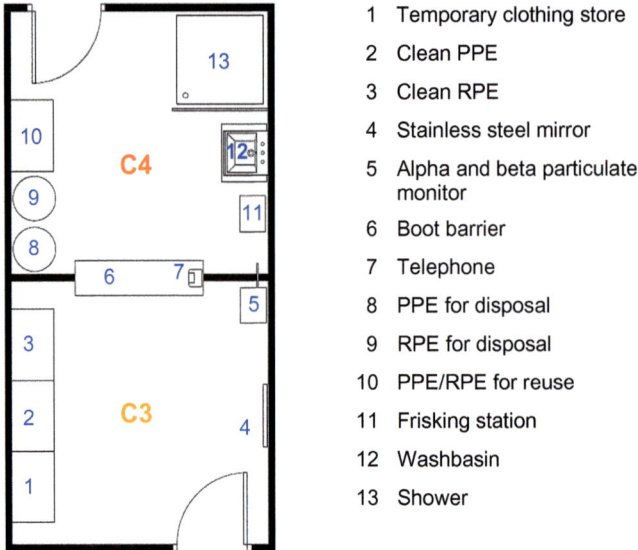

1 Temporary clothing store

2 Clean PPE

3 Clean RPE

4 Stainless steel mirror

5 Alpha and beta particulate monitor

6 Boot barrier

7 Telephone

8 PPE for disposal

9 RPE for disposal

10 PPE/RPE for reuse

11 Frisking station

12 Washbasin

13 Shower

Fig. 4.21 C3/C4 sub changeroom.
© Bill Collum.

4.3.3.1 Trans-zone protocol

You will recall that in the previous chapter we examined transferring items between contamination zones and found that the normal practice is to move from each zone to its direct neighbor. Well exactly the same constraint applies to sub changerooms. In practice then, it is not normally permissible to use a single room to facilitate access between C2 (green) and C4 (red) zones; instead we must pass first through a C2/C3 sub changeroom and then cross a second boot barrier to get from C3 to C4.

4.3.3.2 Clothing and protective equipment

C3/C4 sub changerooms are quite similar to their C2/C3 counterparts, but with the additional features needed to accommodate workers changing in and out of more extensive PPE and RPE, including the kind of air-fed PVC suit we see being fitted in Fig. 4.22.

So a washbasin or even a shower may be required to aid decontamination before a worker removes their suit. In addition, space is also needed for the coiled breathing line which carries air to a suit and can be tens of meters long. Actually an air tank can be carried on the back, much as firemen do, but they are pretty heavy so breathing lines tend to prevail. Where operators are performing cutting operations, there is a risk that hot apparatus or sparks may puncture their PVC suit, so they protect themselves by wearing a long leather apron over the top. Furthermore, if it is deemed that protection is needed from radiation the apron will be lead-lined.

Fig. 4.22 Air-fed PVC suit being fitted.
© Nuclear Decommissioning Authority.

As you can imagine, there is quite a lot of preparation needed before workers can enter a C4 area. And when they emerge again there are procedures which must be followed in removing their PPE and RPE. Quite sensibly then there is always a support team, including a health physicist, on hand to give assistance. Actually, this even occurs in many C2/C3 sub changerooms where there is less attire to contend with. One of the consequences of entry and exit procedures is that actual working time in C4 (red) areas is quite limited, added to which the cumbersome nature of protection being worn dictates it is inherently slow-paced.

4.3.3.3 Monitoring procedure

Personnel returning to the C4 (red) side of a boot barrier will undoubtedly be carrying contamination about their person so, as with a C2/C3 sub changeroom, must be monitored *before* crossing the boot barrier (Fig. 4.23). And again, air on the C4 side of a boot barrier is continually sampled by an alpha and beta particulate monitor, which is on the lookout for tell-tale signs of airborne contamination.

No doubt there will always be a need for personnel to enter C4 areas. However, when we consider what it entails, it does focus the mind on doing all we can to minimize its occurrence.

Fig. 4.23 Health physicist
conducting monitoring.
© Nuclear Decommissioning
Authority.

4.3.4   Open sub changeroom—C2/C3 (S)

As is so often the case, there is an aspect of this particular discussion which does
not quite conform to the normal pattern. There are rooms which appear identical,
and perform a similar function to sub changerooms, but on closer inspection are slightly
different and could not accurately be described as conforming to rules which govern
their design. We could say they are part of the same family, but from a different branch.

It all comes about because of the C3 (S) areas mentioned in our discussion on radiological zoning, which are located in C2 zones yet accommodate items that may harbor
traces of C3 contamination on their surface. We know such contamination will stay
put so we can see the logic of simply cordoning these areas off, say with handrailing,
but this still leaves the problem of how we gain access without unwittingly carrying
C3 contamination into the C2 area around them. This is where a quasi or *open* sub
changeroom comes in.

In appearance, they are practically indistinguishable from a standard C2/C3 sub
changeroom, to the extent that those passing through must follow exactly the same
procedures. The major difference being that both sides of its boot barrier are in the
same C2 (green) zone, the same airspace. True sub changerooms always have air
flowing towards the side which is potentially more contaminated, which means that

for their ventilation system to function they must be fully enclosed. The room we are considering here is a single airspace, so in radiological terms forcing its air to move in a particular direction would serve no useful purpose. The tell-tale sign is that they are often open-topped, unless located in a space that happens to provide a ceiling by default, and there is no need for a door at either entrance.

These rooms go by various informal names, so to differentiate their purpose I normally refer to them as *open* sub changerooms. Whatever we call them it is important that they are distinguished in some way, as it highlights subtle but important differences in their design and operation. Most notably, unlike sub changerooms, airflows need not be constrained and radiologically speaking we are not dealing with a C3 (amber) environment, with all that entails in terms of airborne contamination.

4.3.5 Equipment transfer lobby

By their nature, sub changerooms often provide access to maintenance workshops or areas where decommissioning tasks are being performed. Invariably this brings with it a requirement to transfer not just people but also equipment between radiological zones. Small items can be carried across boot barriers but larger, heavier items need to be transported on trolleys which would be unable to negotiate a boot barrier.

If the need for such transfers is fairly infrequent, it is possible to design boot barriers to swing out of the way and allow a trolley to pass. However, where heavy items will regularly cross between zones, it is prudent to locate an equipment transfer lobby adjacent to a sub changeroom. As we know from our rad zoning discussions, such lobbies attract a dual classification of either C2/C3 or C3/C4.

An RPA or health physics officer must be on hand to oversee transfers. In particular they monitor equipment exiting through a lobby, so as to ensure contamination is not inadvertently carried from its designated area. To help in this endeavor, the usual practice is to bag contaminated items before they begin their journey.

In common parlance, most of us would refer to this type of lobby as an *airlock*. However, strictly speaking airlocks are airtight chambers which separate very different environments, extreme examples being seen on craft out in space or beneath the ocean. In truth there is some differential air pressure between radiological zones, but not of a magnitude that would demand an airlock. It is good to know this because use of the wrong term will incur either puzzlement or smart remarks from specialists, such as the ventilation folks, who know exactly what an airlock is for. So *lobby* it is.

Up to this point, we have explored the nuclear fuel cycle and examined the essential constituents of all nuclear facilities, namely radiation, contamination, and radiological zoning, and also had first indications of the important role played by ventilation systems (more on this later). In addition we have seen the constraints these factors impose on how personnel and equipment can move around. With this done, we are now ready to look at the engineering which, based on these fundamentals, turns nuclear facilities into a reality.

Structural

5

With so many nuclear projects being one of a kind, there can tend to be an emphasis on the innovative processes that will ultimately deliver a bespoke nuclear solution. We may find an understandable focus on what might be termed the high profile aspects of a facility, such as remote mechanical handling or some groundbreaking chemical processes. It may appear to some that a project's entire challenge is tied up in developing new robotic techniques or processing exotic substances; believing that once those *challenging* areas are defined, all the structural engineer needs to do is wrap some concrete and steel around the nuclear heart and everything will be fine. Not so.

We shall see in this chapter that it is those selfsame nuclear materials and processes which place so many demands on the structural engineer. These are the people who must guarantee that, come what may, a facility's radioactive containment will not be breached and equally important, work with other disciplines to ensure equipment which is crucial to safety will continue to perform after, say an earthquake or a severe storm. Unsurprisingly, much of a facility's all-important safety case and ultimately its license to operate is in their hands.

Fig. 5.1 Nuclear facilities.
© Nuclear Decommissioning Authority.

There is no such thing as a standard nuclear facility; even reactors can differ from one to another, with not all being housed beneath the seemingly ubiquitous dome-shaped roof. Just as with schools, hospitals, and international airports, the external appearance of nuclear plants can vary considerably. There is, however, a fair element of commonality. Just as we would have little difficulty recognizing the examples I have just mentioned, we would also find it easy enough to identify major nuclear facilities

Nuclear Facilities. http://dx.doi.org/10.1016/B978-0-08-101938-2.00005-2

(Fig. 5.1). Typically, if I can use that word for a moment, they occupy a large footprint, sometimes very large indeed; they can be just a story or two high, but very often at least part of a facility will be 20 m or more in height. It is also quite unusual to come across a nuclear facility that is a straightforward shoebox shape; it does happen, but ordinarily they have stepped roofs and smaller building blocks appended to their sides. And finally, there is often a fair sized ventilation stack attached.

I mention all this because as we walk through the following sections, it will be useful to have an image of a building that occupies a pretty big footprint, in some areas is at least 20 m high and is not an entirely regular shape. For those nuclear facilities that are much smaller, and there are plenty of examples around, the issues I am about to describe still apply, it is just that they are scaled down somewhat. We shall start with construction materials and aspects of their design, then later on examine how robust nuclear facilities need to be.

5.1   Steelwork structures

5.1.1   Structural grid

If you are not too familiar with the concept of a structural grid, just think of any sizeable building you have ever seen being constructed. It does not matter if it was a tower block or a new supermarket. When all the steelwork is erected you can visualize that no matter which side of a building you view, the vertical columns always line up one behind the other (Fig. 5.2). The underlying principle of arranging structural steelwork is based on adopting a regular pattern, hence the term grid.

Fig. 5.2 Steelwork grid.
© Nuclear Decommissioning Authority.

Ordinarily a structural grid indicates locations of a building's main steelwork columns and interlinking beams. However, the same principles can also be applied to large facilities constructed entirely from concrete, or indeed a mixture of the two materials. Where a facility is constructed exclusively from concrete, a grid is still needed to organize locations of its main walls and columns. For our purposes, we can assume a facility is constructed from rolled sections such as the stanchion shown in Fig. 5.3, and just keep in mind that concrete could be covered by a similar discussion.

Fig. 5.3 Steel stanchion.
© Bill Collum.

The imaginary lines connecting steel columns together are shown on virtually every drawing produced by a design team, as they are the datum from which just about everything in a building is dimensioned: walls, floor penetrations, equipment, and so on. Grid lines are the perfect anchor from which to establish positions of everything else, for two reasons. They cover a whole building in neat rectangular segments and crucially, once frozen provide an unchanging matrix which is easy to identify, both on drawings and within a real building. It is also much easier than, for example, giving measurements to the nearest outside wall which could be a considerable distance away. In fact when preparing drawings for very large buildings, it is often the case that the nearest outside wall is literally "off the page" in relation to an object being dimensioned.

5.1.1.1 Grid arrangements

Each grid line (Fig. 5.4) is numbered consecutively in one direction, 1, 2, 3, etc. and alphabetically in the other, A, B, C, etc. So each grid intersection can be referred to alphanumerically, for example, C-3. The space between rows of stanchions is known a structural bay but unlike graph paper it is not absolutely necessary that all bay spacings

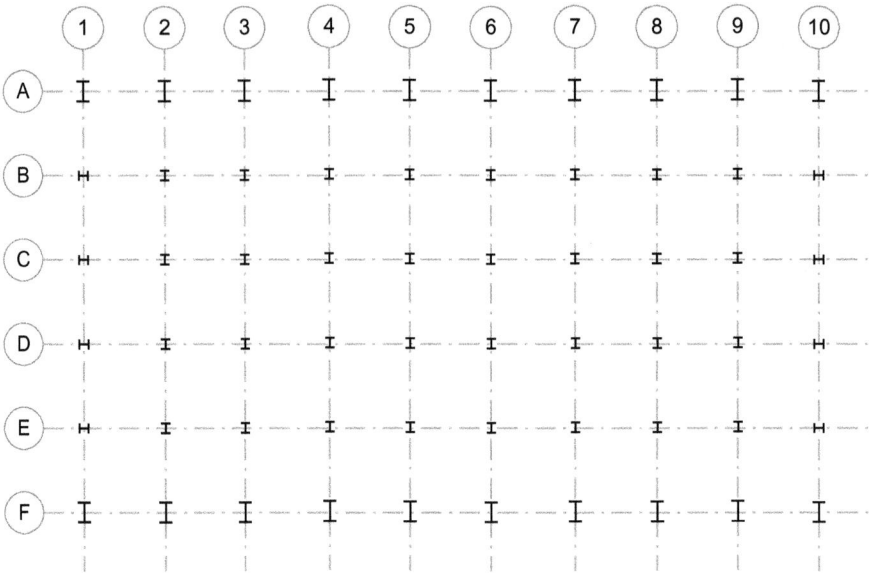

Fig. 5.4 Structural grid.
© Bill Collum.

are identical. If they are it can make structural calculations a little easier and the repeatability of a uniform grid can simplify construction a touch, but it does not have to be.

This brings me to a grid's dimensions. Structural engineers prefer grid spacings of between 5 and 6 m. It is not a hard and fast rule, but when faced with a blank sheet of paper it is not a bad place to start. Spans of this sort can comfortably be achieved, so there is no structural imperative to strive for, say, 2 or 3 m. A building's configuration may ultimately demand all sorts of grid dimensions, but that is a different matter and can be accommodated in the event that a design solution demands it.

5.1.1.2 Load paths

No matter what the building, there is always one abiding problem for the structural engineer: how to support the weight of everything within a building, including the building fabric, and how to safely carry those loads down to the ground. I suppose we could say that is two problems, but they are so closely related as to present a combined difficulty. The vehicle for addressing this particular challenge is a structural grid.

Fig. 5.5 illustrates how very effective this type of matrix is at directing loads into the grid's beams and stanchions, and from there down into the foundations. It is also plain to see that if an element of a grid was removed, then loads would need to take a more convoluted route, which would increase forces in adjacent steel members. Even worse, loads may be unable to find a path to the ground, which could result in steelwork becoming overloaded and ultimately failing.

Fig. 5.5 Steelwork structure. Shutterstock.com.

5.1.1.3   Grid adjustments

Even though it is such an important element of the design development process, I do feel that a grid's ranking in project priorities can sometimes be underrated. I know this because at various stages in the life of every project, design team members often ask for a grid line moved, or, to be more precise, a piece of steelwork on it. It will turn out that the ideal location for kit they are designing just happens to be spoken for by a hefty chunk of steel. So how do such requests come about?

If you are involved in the design of say ventilation plant rooms, items of mechanical equipment, and so on, you may find that as a design matures more space is required than previously envisaged. We know this is not entirely unexpected, after all early schemes will have been based on provisional information which may itself have changed.

The problem is that adjacent building areas will already be spoken for and almost certainly fully utilized; but you need space, so where are you going to get it? If steelwork happens to be passing through an area which is troubling you, surely it can be moved. After all, there are truckloads of steel all over the building so why not nudge some bothersome chunk out of the way? Well, there are repercussions.

If a single column, say on the ground floor of a building, is moved off its grid (Fig. 5.6) it would no longer be part of the network delivering forces directly into the ground. Let's look at one of the consequences. At least one of the horizontal beams that previously connected into the top of the column must now be increased in length, most likely spanning two bays. The strength and therefore dimensions of that beam must now be increased, because the column above is now delivering loads to a point where the removed steel column had previously supported it. As it is subjected to loads the extended beam will want to deflect like a tightrope beneath a circus performer. In addition, the enlarged beam may be supporting an adjacent floor or connected to

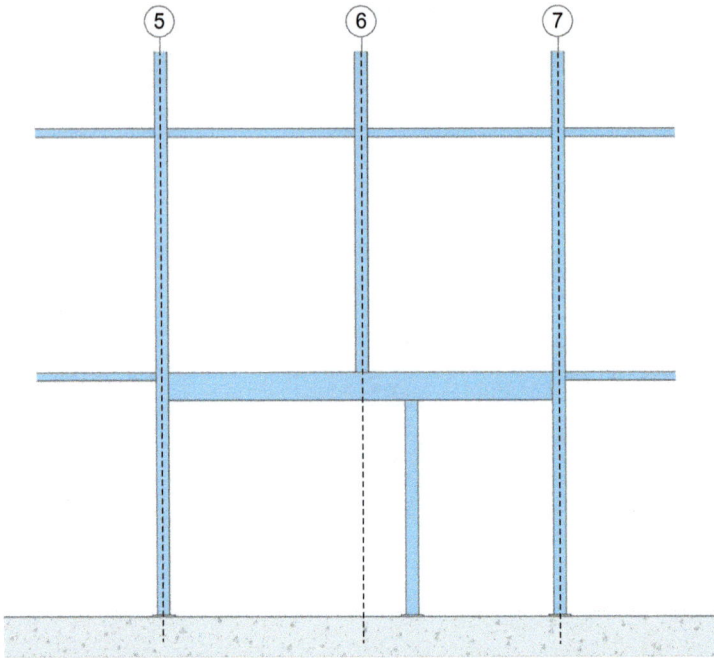

Fig. 5.6 Steel column off grid.
© Bill Collum.

bracing which, as we shall see shortly, plays an important role in fortifying a steel framework. So a bit of tinkering can have widespread implications.

The structural team will do their best to accommodate change requests, but very often they must disappoint. Unfortunately the reality is that if they accepted every request they receive for a change to the grid, the whole building would literally fall apart.

5.1.2   Floor construction

Horizontal beams must be positioned a little below the final floor level, as the actual floor will be constructed directly on top of them. The floor deck itself may be formed from precast concrete panels (discussed later) or steel plate, but for nuclear facilities is often constructed from in situ concrete. Where this is the case, concrete is poured onto metal decking which is fixed on top of the horizontal beams.

To support the concrete slab, which is typically around 150–200 mm thick, along with equipment which will be placed on it later, smaller steelwork beams are inserted between the main beams which follow the grid lines (Fig. 5.7). If it were not for these intermediate beams then a floor slab would need to span the full width of each bay, which could be 6 m or even more, so we need to support the load with additional steelwork. That's it, apart from one vital element: bracing. It is bracing that stops a steelwork structure swaying and falling over.

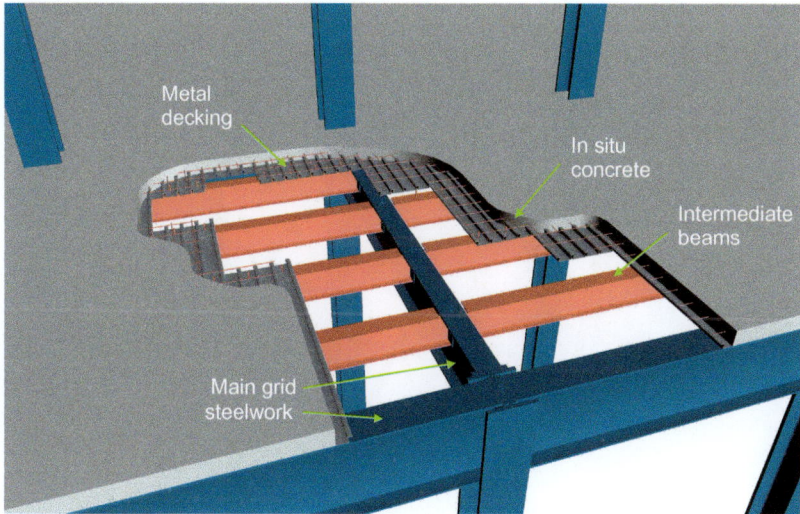

Fig. 5.7 Steel beams supporting floor construction.
© Nuclear Decommissioning Authority.

5.1.3 Bracing

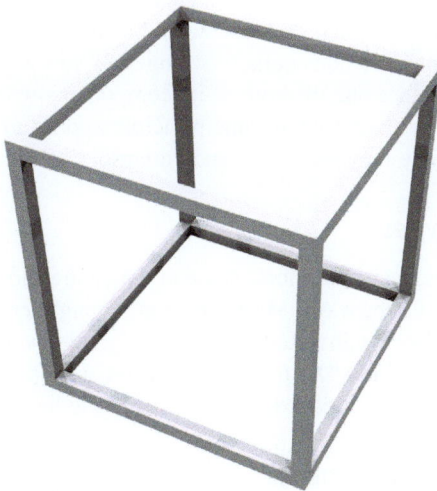

Fig. 5.8 Unbraced cube.
© Bill Collum.

Imagine you were to build a small plastic framework, say it is in the shape of a cube (Fig. 5.8) with each piece around 5 mm square and 200 mm long. If you were to glue the whole thing together and fix it to a base it might look quite firm. However, if you gently prodded the assembly and pushed any of the top *beams* sideways, the structure would easily lean and collapse. A frame of this type is unable to withstand sideways

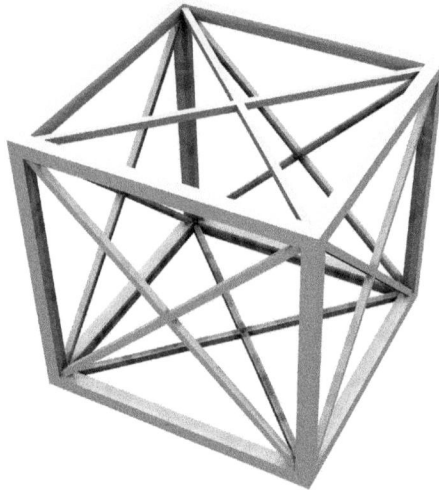

Fig. 5.9 Braced cube.
© Bill Collum.

forces, so if constructed in steelwork would be incapable of riding out an earthquake or braving the fierce winds that nuclear facilities are often designed to withstand.

If you made the same frame but this time attached more members, corner to corner and on all six sides (Fig. 5.9), then each side would be described as *cross braced* and the whole assembly would be considerably stronger. If you were to prod it again it would not lean over, in fact one or more of its components would break before the structure would give way. Exactly the same principle applies to real buildings: they must be braced, otherwise they will sway and collapse.

5.1.3.1 Bracing types

There are various bracing configurations, some of which are shown in Fig. 5.10, each distributing loads in different ways. Most common among them is the "X" or *cross* bracing, but there is also the inverted "V" the "K" and several others. However, we shall stick with the common "X" for our discussion, as it is more suited to dealing with the level of seismic performance we are considering here.

5.1.3.2 Bracing arrangements

If it is to perform its function efficiently, bracing in real buildings must be positioned in a particular way, although it is not necessary to brace every structural bay. If you studied the steelwork for a large industrial building before its cladding was installed (Fig. 5.11) you would see how it works. Arrangements can vary quite a bit, but bracing is often grouped into pairs of adjoining bays, leaving clear bays in between. Each "X" spans floor to floor and is stacked one on top of the other, from the ground to the roof. The shorter gable walls will also be braced, again not necessarily in every bay.

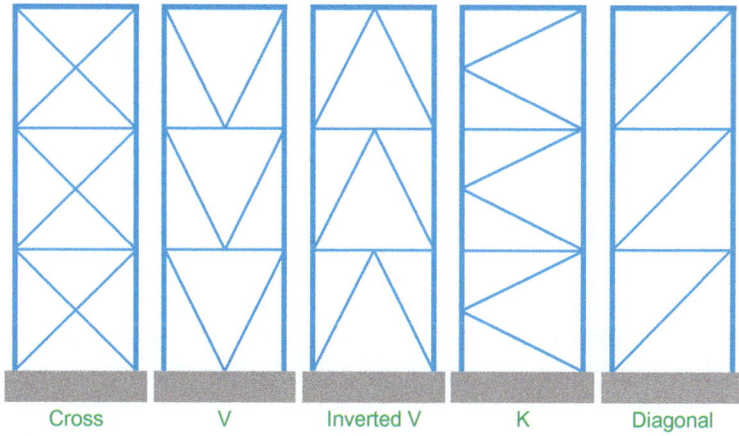

Fig. 5.10 Bracing types.
© Bill Collum.

The good news, from a design perspective, is that there is some flexibility in selecting which bays are braced. However, once locations are agreed then bracing must be distributed over the full height and entire width of a whole structural bay. Well, I say *must*. If it becomes absolutely necessary, our structural friends may be able to work a bit of magic and tinker with the odd bracing location. Really, though, their options are tightly constrained, particularly when grappling with extreme seismic criteria. So it is best to assume bracing, shown here in red, should follow the same pattern across an entire structural bay.

Fig. 5.11 Typical bracing arrangement.
© Nuclear Decommissioning Authority.

If internal bracing is overlooked during a project's early design evolution it inevitably leads to clashes later on, for when bracing is belatedly overlaid on a mature scheme it is bound to cut through machinery, door openings, ductwork, and so on. Similarly, a room that looks to be comfortably sized for its plant and equipment may not be quite so spacious when it is divided here and there by cross bracing. It is always prudent, therefore, to identify potential bracing locations as early as possible in the design process.

The same sort of thinking applies to recognizing that a network of steel beams is needed to support concrete floors, otherwise it is tempting to view their open expanse as an apparently blank canvas, one to be divided up and perforated pretty much as we wish. Nuclear facilities in particular always demand lots of floor penetrations to accommodate their various processes and service distribution networks, so it helps to remember that just below their floors there is an interlinked matrix of steelwork holding them up. Once again there are some limited opportunities to reposition beams, but wholesale modifications would undermine a floor's structural integrity.

5.1.4 Portal frame

So far, we have concentrated on structures where steel stanchions are located at every grid intersection across a whole facility. The arrangement is sometimes referred to as a *stick build,* although strictly speaking this term applies to the way in which a facility is constructed, namely in situ and piece by piece, rather than modular. That being said, the context within which the term is used usually makes its meaning clear enough.

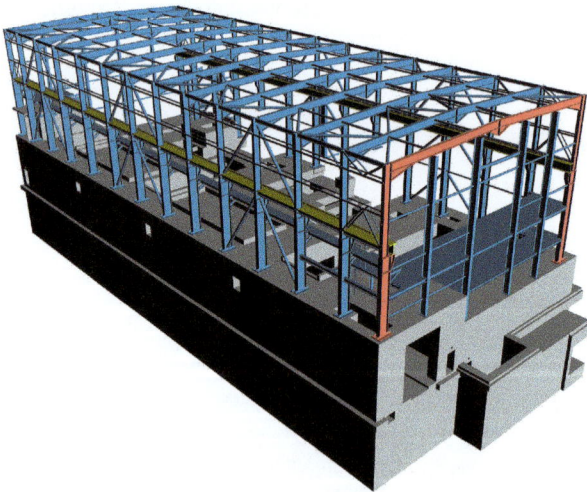

Fig. 5.12 Portal frame.
© Nuclear Decommissioning Authority.

There is, however, an alternative structural arrangement, one which is often utilized for nuclear facilities: the portal frame (Fig. 5.12). Steel columns are again erected along the longest external sides of a rectangular building, with steel beams spanning across the top and joining them together. This time, though, internal steelwork, or concrete for that matter, can be structurally independent, which literally opens up more opportunities.

Fig. 5.13 Overhead traveling crane.
© Nuclear Decommissioning Authority.

For designers and operators of nuclear facilities, one of the main benefits of a portal frame structure is its ability to accommodate an internal crane across the full span of a building (Fig. 5.13), which of course is impossible where a network of steel stanchions is holding up the roof.

5.2   Concrete structures

In many ways, reinforced concrete is a wonderful material for large heavy-duty buildings. It is very robust, terrific for carrying hefty loads, relatively inexpensive and in the case of nuclear facilities provides excellent shielding from radiation. Of course it cannot match steel for ductility, but all in all concrete is a material you want to have around.

5.2.1   Precast concrete

Much of the following discussion is based on using in situ concrete, which is the type mostly favored by the nuclear industry. So before we go on I need to set the discussion in context with a few words on the difference between precast units and in situ concrete.

Fig. 5.14 Precast concrete wall panel.
Istock.

Precast concrete arrives at a construction site not wet or as a dry mixture of cement and aggregate, but already formed or *cast* in the desired shape (Fig. 5.14) hence the name. You may have noticed large precast blocks being used on roadways to temporarily segregate traffic, or panels with a decorative finish adorning the external walls of a building, so we know how it works.

Precast units come in all sorts of shapes and sizes, but when used for construction they are normally in the form of planks (Fig. 5.15) which are commonly used for roofing and flooring. The attraction of precast concrete is that its installation is almost instant. For example, planks being used for roofing will arrive on the back of a truck, be hoisted into position by a mobile crane, butted against each other and fixed in place. Hey presto, a roof. All that is needed to complete the job is to apply a weatherproof finish.

Fig. 5.15 Precast concrete planks.
Shutterstock.com.

Attractive as it may be in other settings, the use of precast concrete within nuclear facilities is fairly limited. Standard precast units arrive on site with their steel reinforcing bars embedded to a predetermined pattern, but in buildings such as these, that need lots of floor and wall penetrations, this imposes some severe limitations. Just drilling a small hole, say 100 mm in diameter, is almost bound to cut through one or more reinforcing bars and weaken a unit's strength. Making several holes, even small ones, in a single plank can completely undermine its structural integrity, while large penetrations for services or ventilation ductwork are really out of the question.

Having said that, for nuclear sites modularization of a building's structure is always an attractive proposition; this is mainly because security considerations, and in some cases radiological working practices, can add time to construction activities and increase costs. Occasionally then precast concrete is used, say for a flat roof, or in the form of bespoke units with their particular penetrations cast in place and appropriate reinforcement bars embedded around them.

So to some extent it is possible to deal with penetrations, but the thing that really rules out widespread use of precast concrete within nuclear facilities is coping with the demands of extreme seismic events. Structural engineers need concrete walls to direct loads downwards and concrete floors to act as a continuous diaphragm. That is to say, floors need to transfer lateral loads to concrete shear walls or steel columns and from there down into the foundations.

Individual concrete units, butt jointed together, are incapable of acting in this way. In a severe earthquake you can just imagine a floor of precast planks rattling away, doing nothing to hold a building together and certainly not helping to direct forces where they need to go. If planks are used, then in order to get some kind of diaphragm action it becomes necessary to pour an in situ concrete floor screed on top of them. This works fine, but unless it is needed for shielding purposes rather defeats the object of using precast units in the first place.

5.2.2 In situ concrete

Unlike precast units, in situ concrete is poured in its permanent location (Fig. 5.16) and offers a much more robust solution. If constructing a wall, for example, its reinforcing bars are assembled first, then shuttering or *formwork* erected around them and finally concrete poured into whatever space remains. Shutters are positioned to line up with a wall's face, so that once concrete has cured they can be removed to reveal a wall in exactly the right location. If you look at a concrete wall when it is finished, you may imagine it to be a mass of concrete embedded with a little steel mesh or a few reinforcing rods, or *rebar* as it is commonly known; but for structures designed to withstand particularly ferocious earthquakes, the size and density of rebar is quite remarkable (Fig. 5.17).

I still remember very clearly the first time I stood at the base of a wall that, when its concrete was poured, would be 1.5 m wide and 25 m high. I had seen drawings and photographs of such walls and hadn't dwelt on them overmuch, but with my nose pressed up against the steel matrix I was struck by a realization that erecting rebar

Fig. 5.16 In situ concrete pouring.
© Nuclear Decommissioning Authority.

Fig. 5.17 Shear wall reinforcing bars.
© Nuclear Decommissioning Authority.

on this scale is a major construction project in its own right, one with its own stability and support issues. Yes, it was quite a sight.

When concrete for walls such as these is finally poured, there may be just 50 mm between the face of a wall and the first row of rebar buried within it, or 50 mm of *cover* as structural engineers like to say. Once I realized what was going on behind their surfaces, the thought crossed my mind that simply calling them *concrete* walls tells only half the story, and rather belies the monumental effort it takes to design and construct them. Whatever they are called they certainly look up to the challenge of riding out a severe earthquake, and I can vouch they remain absolutely immovable if you attempt to prize the bars apart, no matter how hard you try.

5.2.3 Wall penetrations

When structural engineers set about designing a facility they obviously need to establish exactly where all of its walls and floors will be located, what width they will be and so on. No surprise there. Following that first exercise things get more complicated, as they also need to know exactly where all the penetrations are going to be. Clearly rebar must be routed around holes, which is quite a complex design and fabrication exercise, so it is essential this information is to hand in good time. If openings change size, are moved, or new ones introduced, it is time consuming to revisit both a design and calculations that have already been performed.

The challenge therefore, for the rest of the design community, is to have penetration information fixed and frozen when their structural colleagues need it, and it is quite a challenge. This is one of those seemingly routine matters which actually demands an enormous amount of cooperation and coordination across all the engineering disciplines.

I should mention here that when discussing penetrations in concrete you will normally hear design teams refer to *holes in walls* information, when really they mean holes in walls and floors. It is not that they are ignoring penetrations in floors; it is just that everyone uses *holes in walls* as a short form for all penetrations. Normally we could expect a clever acronym to cover all penetrations no matter what their geographical location, but someone must have decided, quite rightly, that we already have enough of them. So in this particular case we get by with a partial phrase: "holes in walls…"

You may be wondering why we should make any fuss at all about a few holes. The thing is we are talking here of at least hundreds, more likely thousands of holes scattered around just about any major nuclear facility you care to scrutinize. Granted a fair proportion will be through building fabric such as partition walls which are not associated with structural loadings, but within any large nuclear facility there are bound to be a myriad of holes passing through its concrete walls and floors. The issue here, is that concrete is there to perform a structural or shielding function, so we need to be careful in deciding where so many holes are going to go.

Where do all these holes come from? I suppose a hole is a negative so they don't actually come from anywhere, but you know what I mean. The list is long: door openings are an obvious one, but the majority of penetrations are required to accommodate services such as ventilation ductwork, banks of electrical cables, pipework, and so on, all running around a building and looking for the most direct route to their destination.

For the type of facility we are considering here, each of these services will be measured not by the meter but by the mile, or kilometer. In fact, for many buildings are measured mile upon mile, all making their way from point A to point B, and heading for thousands of holes scattered around a building. Against this, structural engineers will be valiantly guarding against a plethora of uncontrolled penetrations, which could easily destroy the integrity of concrete that may be binding a building together and keeping it standing.

5.2.3.1 Concrete banding

Hefty concrete walls, and floors for that matter, look so robust it appears they could shrug off a considerable number of penetrations before their strength would be seriously compromised. However, not only is it inadvisable to pepper concrete with holes, there are also constraints on where they can be located.

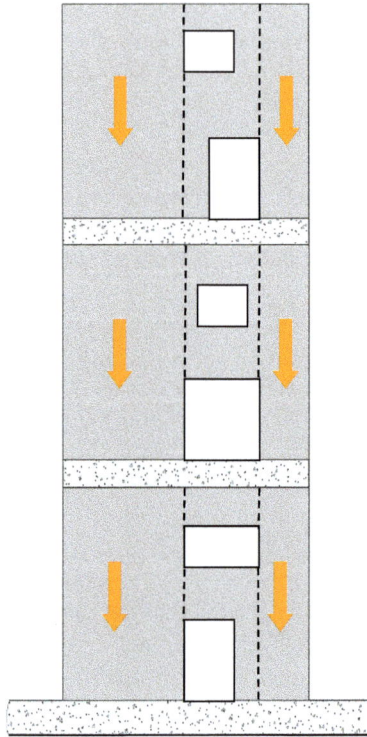

Fig. 5.18 Concrete banding.
© Bill Collum.

Take the case of a concrete shear wall (Fig. 5.18) that rises through several floors of a building and plays a crucial role in directing horizontal forces down into the ground. Individual design teams are understandably only concerned with sections of wall that inhabit their area, probably on just one floor. Structural engineers, however, must view the wall as a whole, from top to bottom. If we ignore for a moment floor slabs and all of the paraphernalia attached to a wall, then the guiding principle is that if you were able to stand back and view an entire wall, you must see bands of concrete running from top to bottom with no penetrations whatsoever. The minimum width of these bands varies from case to case, but to fulfill their structural role must be measured in meters or significant parts of a meter and certainly not in centimeters.

As we have seen, walls like these must direct forces down into the ground, but where a load path encounters a penetration then its path is interrupted, which effectively nullifies the entire band of concrete above and below a hole. In terms of structural performance the strip of concrete running vertically through a hole almost ceases to exist. Actually it is quite analogous to the consequences of moving a steel column off grid, which we discussed earlier.

If penetrations are arranged without due regard to the principle of concrete banding, then forces generated within a building will have no satisfactory route into the ground and an apparently substantial structure will be unable to fulfill its purpose. For design teams engrossed in the challenges of a particular zone, the thought of not being able to make holes where they please may feel like a grievous interference. Structural engineers, however, are mandated to take their view from a more inclusive perspective, so now and again must get tough in enforcing the rules.

5.2.4 Phased release of structural information

It takes a great deal of time—years maybe—to get from pouring a foundation slab to constructing the walls and floors situated high in a building. With this in mind, structural engineers are kind enough not to demand every last detail from the rest of us until it becomes absolutely essential to their work. What they are interested in, initially in broad terms, is the maximum weight which will be carried by a building's upper floors. With that proviso they can accept *frozen* design information incrementally, starting with the lowest levels.

This allows them to finalize their detail drawings in sequence, beginning with the foundation slab, then moving onwards and upwards through a building's various levels. These drawings can then be released in the same sequence to construction teams who are fixing rebar, pouring concrete, and erecting steelwork out on a site.

When it comes to really huge buildings, each floor level is divided into zones with their construction tackled an area at a time. These buildings are simply too big to contemplate erecting the whole thing en masse. So there is further flexibility to issue information not just a floor at a time, but also in packages for the zones that subdivide them. However, we cannot ignore penetrations high in a building until just before construction teams get to them, as there are other imperatives that drive us to scrutinize even the highest elevations long before that. We shall come back to this theme shortly, when we discuss seismic analysis and look in more detail at how structural information, particularly in relation to penetrations through concrete, is developed and released.

5.3 Combined concrete and steel structures

Through recent pages we have looked separately at steelwork and then at concrete structures. However, in the nuclear industry it would be quite unusual to come across a building that relies exclusively on one or other of those materials. Most often we find both nestling side by side under the same roof (Fig. 5.19), so need to consider how they react when found together.

Fig. 5.19 Concrete structure within
a steel framed building.
© Nuclear Decommissioning
Authority.

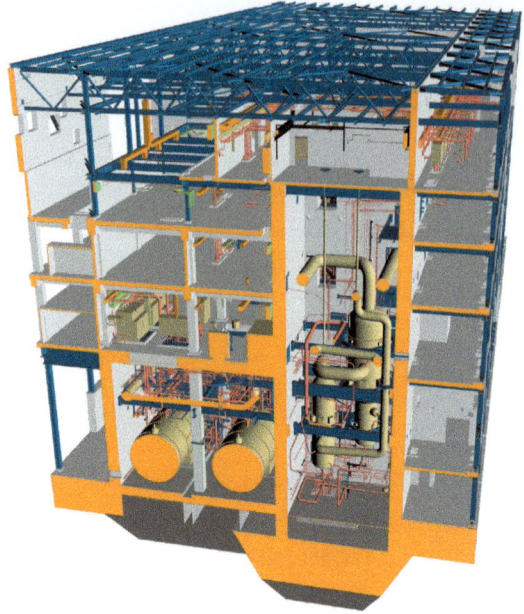

If you think about the way a concrete structure will behave in an earthquake you can visualize it being quite stiff, whereas steelwork by comparison will be much more flexible. If a building is constructed entirely from steel, with its beams spanning between vertical columns, then the whole assembly is steel-to-steel and moves in harmony if an earthquake occurs. In the type of construction we are considering here those same beams may run from a steel column at one end to a concrete wall at the other, so we need some way of allowing concrete and steelwork structures to shake to their own rhythms without damaging each other or falling apart.

5.3.1 Differential movement

One of the classic solutions is to stop floor beams short of the concrete walls they adjoin, thus providing a movement gap between the two types of structure. Of course the beams cannot simply be left hanging in mid-air, so various ways have been devised to support their unconnected ends. Essentially they all entail fixing some kind of a hefty steel bracket to concrete walls; the crucial detail is that beams are supported vertically on the brackets but not bolted or welded to them, so that horizontal movement can still be accommodated.

A neat example comes in the form of a sliding bearing (Fig. 5.20) which allows a beam to move during an earthquake and absorb differential movement without causing any structural damage.

Fig. 5.20 Sliding bearing.
© Bill Collum.

5.4 Seismic analysis

When it comes to conducting a seismic analysis for nuclear facilities, there are always several factors pulling in opposing directions. The analysts want frozen, unchanging information; they also want it as soon as possible and in a logical sequence, which all sounds reasonable enough. Meanwhile, however, other members of the engineering community will be feverishly developing designs and coordinating interfaces, so understandably want to stretch out the time they have to finalize exact positions of walls, floors, and penetrations, not to mention the exact size and location of all structural steelwork. They really do not want to nail down such information until they are good and ready, so to keep everyone happy there are rules of engagement.

5.4.1 Phased release of wall penetration information

Drawings and supporting design information are issued to seismic analysts in three main phases. They are referred to by all sorts of nametags, none of which stand for anything in particular, so we shall call them S1, S2, and S3 (Fig. 5.21).

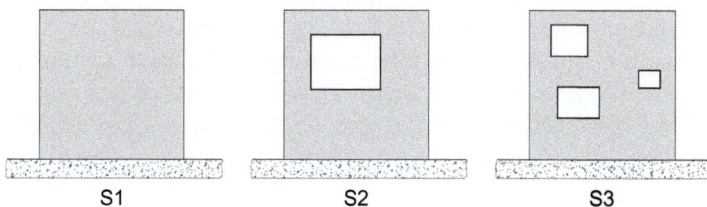

Fig. 5.21 Phased concrete penetration information.
© Bill Collum.

S1 drawings must show a frozen structural grid along with all major walls and floors in their final locations, but at this stage it is not necessary to include any information on penetrations. But remember analysts do need information on the mass within a building, particularly on its upper floors. S2 drawings contain more detail but need not show the exact dimensions and location of all penetrations; instead, it is permissible to show a single hole representing the approximate cumulative size of several holes that may be needed in a particular wall or floor area. S3 is the final issue and must spell out every nuance with millimeter-perfect precision.

The beauty of phasing the release of information in this way is that it maximizes an engineering community's window of design time, yet still allows analysts to make progress by giving them sufficient information to advance their work incrementally and in a systematic manner.

If a project is on a particularly tight timetable, as most of them seem to be, it can be a good idea to schedule a preliminary release of S1 information, enough to get the ball rolling for our seismic folks. No doubt they will need to do a little reworking when the formal S1 issue arrives later, but at least they can steal a few valuable weeks to get ahead of the game. Of course it costs money to have them revisiting their analysis, so that expenditure needs to be weighed against the schedule advantages it delivers. I would not advocate contemplating an advance S1 issue as a regular routine, but it can be a useful maneuver when time is especially precious.

5.4.2   Seismic evaluation

You may be wondering what happens if a seismic analysis demonstrates that a proposed building configuration is incapable of withstanding its particular design basis earthquake (DBE) (discussed shortly), maybe because a structure is too top-heavy, for example. Apart from the obvious impact a redesign would have on a project's planned schedule, it would also be quite embarrassing to say the least. Such repercussions focus the mind, so that in practice controls are in place to ensure engineering teams do not simply develop proposals then hand them to specialists in the hope that everything will be okay.

As with so much of the nuclear facility design process, the key here is iteration, in that the design progresses in incremental steps. In this particular case, structural engineers continually evaluate evolving proposals in order to ensure they embody appropriate seismic design principles, such as striving for a symmetrical configuration and distributing loads as low down and as evenly as possible. Perfection is seldom realized, but under the structural team's watchful eye, other designers can be confident that their solution will be seismically robust, and ultimately pass the test of a detailed analysis. Results may reveal the need for a few tweaks but early cooperation, and those incremental steps, will deliver an engineering solution which is inherently sound.

5.4.3   Response spectra

One other factor to be aware of in this context is that a seismic analysis is not conducted simply to prove a building can withstand the stress and strain of its DBE. It is an important consideration but by no means the only reason for so much arithmetic. The analysis also generates response spectra for local areas throughout a whole facility. In other words it shows how individual areas of a building will move when subjected to various seismic conditions. It is important, because accelerations surging through a building during an earthquake can vary considerably between its lowest and highest levels, and even within different areas on the same floor.

If you are tasked with designing vessels, pipework, mechanical equipment, and so on, which will be dotted around a building, you are not terribly interested in seismic accelerations down in its foundations or up on the roof. You want to know exactly how walls and floors will behave at your particular point of interest. After all, crucial plant and equipment (P&E) must be capable of shutting a facility down after a severe earthquake. And in addition, other equipment which is deemed important to safety may need to remain fully operational. Equipment designers therefore need details of response spectra in neighborhoods that interest them—what we might term their *local earthquake*—so that they can set about satisfying the demands of its signature at that particular point in a building.

5.5 Extreme environmental events

Nuclear facilities are subject to much tougher design standards than commercial or other industrial developments; most of us would assume just that. After all, it is reasonable to expect that buildings housing nuclear materials will be built to more stringent regulations than your local supermarket. In the United Kingdom, for example, most industrial buildings and down town office blocks are built to withstand the worst environmental events that might conceivably occur on average once every 50 or maybe every 100 years; this is known as the 1 in 50 or 1 in 100 year return period. This covers the worst of wind, rain, snow, extremes of temperature and seismic events, or earthquakes.

We know from our own experience that some years will, for instance, bring much more severe storms than others, so can foresee that over a 50-year period there is almost bound to be some particularly ferocious weather, or even an earthquake which is beyond the norm—not that any earthquake could be considered routine, but you get my meaning. For nuclear facilities, the return period can be as much as 10,000 years, so you can just imagine how appalling things may get at some stage during such a prolonged spell.

As you will no doubt have surmised, the characteristics of extreme environmental events are not plucked from thin air. There is indeed quite a lot of science involved in coming up with the numbers which draw largely on historical data, along with some carefully considered extrapolations of how severe things might become. It is important to have these predetermined figures on which to base our designs, as it ensures consistency in envisaging extreme scenarios and gives engineering teams a common datum on which to base their calculations. I shall take us through each of the environmental events in turn and explain what Mother Nature might conceivably get up to over such a long period.

The specific extreme environmental criteria which an engineering team must satisfy are determined by a facility's safety case, so vary from one project to another. Precise figures depend on the function of a building, its radioactive inventory and geographical location. Having said that, across the United Kingdom, for example, geographical conditions are broadly similar, so it all comes down to specifics of individual facilities What we shall see is on those rare days that Mother Nature chooses to remind us of her awesome power, it is not good to be out and about.

Before moving on, it is worth mentioning that although our discussion here focuses on specific *design basis* events, there are additional factors that must be taken into consideration and incorporated into a nuclear facility's design and operational planning. They come under the separate heading of *resilience* although in truth the way they are dealt with can have some considerable overlap with ways in which we guard against design basis events.

Broadly speaking, the difference between the two is that design basis events relate to predefined criteria, normally environmental, whereas resilience is more of a catch-all for "what if?"—Essentially having an eye on potential man-made hazards, or even accidents. In practice, such events are identified during the development of a facility's safety case, which we shall discuss in Chapter 14. The reason I mention it here is that very often it is the structural community who are called on to provide the primary resilience, or *defense in depth* demanded of a safety case, and in some situations this may be over and above the measures we are considering here.

5.5.1   Seismic

If we take the United Kingdom as an example, it is in the fortunate position, literally, of being some distance from tectonic plate boundaries where most of the world's severe earthquakes tend to occur. However, it is not immune to the possibility of experiencing significant tremors, particularly over a 10,000-year period; consequently seismic events figure high on the list of priorities that nuclear facilities must be designed to combat.

The stresses and strains a building will experience during an earthquake, the same earthquake, can vary significantly, with factors such as local ground conditions, a building' shape and height and exact configuration of its concrete and steelwork coming into play. That said, I can outline some rule-of-thumb type numbers to give a feel for how it works, but we do need to keep in mind that every case is unique so will face its own bespoke seismic experience.

5.5.1.1   Design basis earthquake

For most of the United Kingdom, the specified or DBE which an engineering team must address typically assumes a peak ground acceleration of $0.25g$ which can equate to, say, $1.0g$ at the top of a building, or an average of $0.625g$ across the building as a whole. So what do those numbers mean and why are accelerations at the top of a building different to those at the bottom?

The "g" stands for gravity, so at ground level accelerations experienced by our example building are equal to one quarter that of gravity, while the top of the building is subjected to accelerations which are four times greater, or equal to gravity.

In theory then, with accelerations being equal to gravity, we may hear these earthquakes described as equivalent to tipping a building on its side, and indeed it does count as a partial explanation of what is going on. In truth, though, the image is a bit extreme. For one thing the situation at ground level is much less onerous than that at the roof; in addition, gravity continues to push a building downwards, plus accelerations in any given direction only last about one tenth of a second, and so on.

Typically, then, and remember these are broad numbers, during a DBE we can expect a building to move 80 mm in any direction at ground level, and 300 mm up at the roof, again in any direction. Actually the ground might be moving in one direction while the top of a building is moving the opposite way, all with some haste, so clearly these buildings must be designed to withstand quite a pounding.

5.5.1.2   Operating basis earthquake

In addition to a DBE there can also be an *operating basis* earthquake (OBE) of a much lower magnitude, say, $0.05g$. This is the seismic threshold above which a plant's operations may be temporarily suspended until it has been inspected and declared safe to resume its operations. In other words, although in seismic terms an OBE is quite minor, we may need to check that no harm has been done before continuing operations. That being said, there is considerable variation in the response required to an OBE, so particulars are tailored to the specifics of individual facilities.

5.5.1.3   Seismic categories

When we consider an earthquake, particularly a DBE, we may picture a large building being shaken around like a rag doll and just managing to remain standing. Everyone around breathes a sigh of relief and life goes on; but for a nuclear facility, just managing to hang in there until the ground stops shaking is not good enough. Not only must its radiological contents not be allowed to escape into the surrounding environment, in many cases it must also continue to be operable, or at least parts of it must. So it is not just the fabric of a building that needs to withstand a severe tremor, but also much of the equipment within it. Just how well a building and its equipment must perform after an earthquake, will depend on which seismic category it is designed to, a factor which is dictated by its safety case.

The number of seismic categories and their exact performance requirements can vary from one type of facility to another, and also depend on whether the category pertains to the building itself or equipment within it. It really all comes down to safety. So what needs to be protected and contained and what equipment needs to be operable in order to maintain a facility in a safe state, or to shut in down safely. Broadly speaking, though, if we include nonnuclear standards, structures must conform to one of four seismic categories, with generic examples shown in Fig. 5.22. Incidentally, site owners may opt for a higher category as a means of investment protection, a factor

Category	Performance
1	Structures must maintain their integrity and continue to provide radiological containment. Equipment must remain operable
2	Structures must maintain their integrity and continue to provide radiological containment Equipment must remain operable, at least sufficient to perform recovery and shutdown procedures
3	Structures and equipment must maintain their integrity sufficient to not compromise (by collapse etc.) structures covered by categories 1 and 2
4	Structures and equipment must be designed, as a minimum, in accordance with national nonnuclear standards

Fig. 5.22 Seismic categories.
© Bill Collum.

which links back to the resilience I mentioned earlier, but in terms of safety, one of these categories will apply.

5.5.1.4 Displacement

Let's go back to the seismic accelerations on a building. Remember, in the United Kingdom it is typically $0.25g$ at ground level (or at *grade* in the United States) and $1.0g$ at the top of a building. Why the difference? You could even imagine it being the other way around. On the face of it, it might seem reasonable to assume that the further you are from the ground, the less movement there will be in floors and walls around you. But no, it doesn't work like that.

Let's say we had a pile of sticks all 10 cm long and representing steel beams. Assume the sticks are made from rubber, the kind used to make pencil erasers, so not too soft and not too hard. Each rubber stick has a little joint at each end so they can be clipped together. Armed with the sticks we could build a meter-high tower, glue its base to a dinner plate and place the whole assembly on a table. With everything complete, it we were to prod the top of the tower with a finger it would sway a little. This is what real buildings do.

If we were to take hold of the dinner plate and gently slide it back and forward, just a centimeter or so in each direction, we would notice the top of the tower would not move in quite the same way as the base. Movements at the top would be amplified, with a result that the top would sway a little further than slight moves being made at the base. You can just picture it happening.

If we were to speed up movement of the plate and introduce a few sideways changes of direction, we would be simulating an earthquake and the top of the tower would move around quite energetically. It is not difficult to imagine that if the plate suddenly became motionless, the top of the tower would continue to sway even though the ground, so to speak, had stopped moving. Seismic specialists refer to these exaggerated movements as displacements, which often have enhanced accelerations associated with them. And it is these enhanced accelerations which can amplify $0.25g$ at ground level into $1.0g$ at the top of a building.

5.5.1.5 Subterranean construction

Processing constraints or the straightforward functionality of a facility may dictate that a substantial part of it needs to be located below ground. On the face of it, enveloping part of a building within ground that delivers an earthquake may appear to bring more challenges for seismic specialists; indeed sometimes it does, but not always. A lot depends on the depth and extent of the basement area relative to footprint of the building above it. Along with that, other factors such as the proximity of other buildings, the type of soil, and whether excavations have disturbed it will come into play. Clearly there are a lot of variables to consider, but the thing to bear in mind is that seismically speaking it is not always bad news to locate part of a large facility below ground. In fact in some situations, softer soil surrounding a basement area can dampen the effects of an earthquake and help to minimize accelerations experienced by a facility.

5.5.1.6 Steelwork

The next time you visit your local supermarket or hardware superstore take a look at the steel columns, or *stanchions*, around the perimeter of the building and, assuming it is a *stick build*, those dotted around inside. All will be arranged in the grid pattern we discussed earlier. Typically the steelwork used to fabricate these columns, their webs and flanges, is around 12 mm thick and the columns themselves say 400 mm × 150 mm. As you stroll around your local supermarket think about this. In some nuclear facilities those same dimensions could be 2.5 m × 1.5 m and the steel used to fabricate them 60 mm thick.

The first time you see structures of this magnitude I guarantee you will stand and stare, I know I did. You find yourself adjusting your perspective and wondering if maybe you have stepped into a scene from *Gulliver's Travels*, but this is the reality of what it can sometimes take to ensure facilities housing nuclear materials will safely withstand a 1 in 10.000 year earthquake.

5.5.2 Wind

5.5.2.1 Building movement

I once visited a friend who was working close to the top of a very tall office block in Paris. You didn't need to know it was in Paris, but it sounds so good I just had to mention it. Their office was more than 50 floors up, so high it often looked out on clouds down below. As if that wasn't surreal enough the Eiffel tower could be seen in the distance, poking through the clouds. So all in all it was quite an enviable location. Someone on the same floor had taped a long sheet of paper to a wall; it had a line drawn vertically down the center and a pendulum hanging over the line. It was just a simple cord with a weight at the bottom, but it worked just fine.

Tall buildings like these sway in the wind, but unless it is a particularly stormy day you may not feel a thing. So the pendulum was there to illustrate the building's sway for the amusement of those around it. There was, however, an unusual twist with this particular arrangement; here the pendulum stayed perpendicular, more or less, and the building did the swaying. If I were working there, I do not believe I would feel the need for such a strong visual image of how my surroundings were moving back and forward through the sky. Happily on the day I visited it was absolutely calm, even at 50-odd floors up. In the more extreme cases it may be a little unsettling, but steelwork and even concrete will flex a little when the occasion demands, so there really is no problem with this kind of movement. Nuclear facilities may not have that kind of height to contend with, but they do need to be capable of toughing out a fierce wind.

5.5.2.2   Design basis—1 in 10,000-year wind

The 1 in 10,000-year wind which many of the United Kingdom's nuclear facilities must withstand, equates to a gust zipping by at 216 km/h (134 mph), equivalent to a Frisbee passing you and covering 60 m in just 1 second. Ideally I would like to tell you how that equates to a hurricane and other similarly named fierce winds. The thing is, though, there are several conventions used to describe them, each relating mainly to their geographical location and adopting different names such as tropical storm, cyclone, typhoon, and so on. Added to that the number of hurricane categories can vary from one convention to another, and just to keep us on our toes each has its own range of wind speeds to denote how appalling the situation is.

Having said all that, in broad terms the weakest hurricanes form when sustained wind speeds reach around 100 km/h (62 mph) and the most catastrophic kick in when sustained speeds hit 250 km/h (155 mph) or more. Furthermore, I can tell you that in the right conditions, or maybe that should be wrong conditions, it is possible for winds to exceed 320 km/h (200 mph) which really does not bear thinking about.

You will notice that at 216 km/h, the United Kingdom's 1 in 10,000-year wind is below the 250 km/h entry level for the most severe of hurricanes. However, the thing to keep in mind here is that the United Kingdom circumstance applies to *gusts* of wind rather than the very different challenge of *sustained* wind speeds which characterize hurricanes. So although a 1 in 10,000-year wind is severe to say the least, it does not compare to the relentless pounding of *sustained* winds that power through the world's most merciless hurricanes. The United Kingdom may not have a too delightful a climate, but on the up-side it can reasonably be assumed that the turmoil of 216 km/h (134 mph) *gusts* of wind will be unleashed on it very rarely indeed, in fact on average just once every ten thousand years.

5.5.2.3   Protection from airborne debris

For the designer, there are a couple of problems inherent in winds of the hurricane variety, the first being airborne debris. No nuclear facility stands in isolation, as there is always a need for support facilities nearby. The list is endless but could include office accommodation, maintenance workshops, stores, restaurants, and so on, none of which will be designed to withstand the ravages of extreme environmental events. The problem is that if cladding or roofing were to be torn from one of these ancillary buildings, it would effectively become a missile that could penetrate the external skin of an adjacent nuclear plant and potentially cause some serious damage inside. There is plenty of evidence (Fig. 5.23) to demonstrate the possibilities are very real, so this is something we need to guard against.

Fig. 5.23 Aftermath of a hurricane.
Shutterstock.com.

As a consequence, the outer walls of some nuclear facilities may be called upon to withstand impacts from airborne debris, and not just any debris but a predetermined *design base* projectile such as a scaffolding plank. On the same theme, I recall working on the design of facility located on a nuclear site in the southern United States, which as we know is an area of the globe that endures much fiercer winds than the United Kingdom. In this particular situation, the design basis called for external walls to withstand a tornado initiated impact from a large car tumbling into them, so quite a challenging circumstance.

Building cladding, or siding, may not appear to be very robust but it can make a contribution to slowing airborne debris down. Of course on its own a single sheet of metal cladding is pretty flimsy, but the panels specified for most industrial buildings, including nuclear facilities, are normally assembled in layers (Fig. 5.24). Panel faces are formed from aluminum or steel with a filling or backing made up of insulation, typically between 50 and 100 mm thick. So although panels do have some inherent strength, it is not enough to contend with the challenges we are considering here.

It is unlikely that the entire outer skin of a nuclear facility would need enhanced protection from flying debris, so there is a prescribed methodology for identifying areas that do indeed warrant special treatment. The safety case for a facility identifies any equipment which is classed as *important to safety*, that is to say, those items which

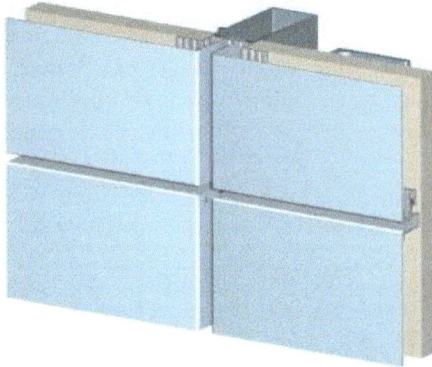

Fig. 5.24 Cladding panel.
© Nuclear Decommissioning Authority.

must be protected in order to maintain the facility in a safe state, or shut it down safely. If such equipment is close to an external wall which is deemed susceptible to puncture from flying debris, then depending on performance requirements, a blockwork or even concrete wall will be constructed inside the outer skin, one robust enough to protect safety-critical equipment.

Of course there is another way to mitigate the possibility of airborne debris perforating the cladding of a facility: simply build the entire outer shell in concrete. Even architects and structural engineers argue amongst themselves over which is the preferred approach, with cost, life expectancy, appearance and ongoing maintenance all entering the debate. Some favor unadorned concrete, whilst others expound the value of cladding attached to a steel framework. Clearly in those parts of the world that are regularly subjected to fierce winds there will be minimal debate, and the answer is almost bound to involve a fair amount of concrete. However, in the United Kingdom and many other parts of the world the answer is not quite so straightforward. In these situations then, design teams must evaluate a building' elevational treatment on a case by case basis.

5.5.2.4 Suction forces

We tend to think of wind causing destruction because it pushes so hard against things. In the case of trees, caravans and other unsupported or lightweight objects, it is absolutely true and they stand very little chance of surviving the onslaught of a particularly determined wind. For buildings, however, particularly those that are clad in some way, it is a different story. When you think about it, there is a substantial structure behind a building' facade which will stop it being pushed around, something which is especially true of nuclear facilities, where seismic demands could well dictate that a structure is far more robust than it would need to be if only countering the effects of strong winds.

It would be quite unusual, therefore, for parts of a building shell to be pushed inwards, even by the most violent of storms, so instead it is suction that creates the main challenge. In large part it results from the Bernoulli principle (discussed in

Chapter 10) which gives aircraft their lift, although its consequences for buildings are not quite so appealing.

If wind is blowing directly onto one face of a rectangular-shaped building, coming in at a right angle, its other three faces and the roof would experience suction forces. What happens is that wind hitting the leading face is slowed down considerably and pushed out towards the side walls and up over the roof, so that when it rejoins passing wind which has not been slowed down it must accelerate violently to catch up. And it is this turbulent interaction that can create powerful eddies on the lee side of a building.

The phenomenon is always at its most extreme on corners of buildings and along their parapets, where negative pressure created by these swirling winds tries to pull or suck the outer skin away and send it skywards. Where wall cladding and roof decking is fighting against this type of suction force, the system of bolts and brackets securing it must be designed to both hold it in place and transfer wind forces into the main structural elements. The problem is so acute that architects and structural engineers often need to work closely with manufactures, to develop bespoke systems which are up to the job of confronting a 1 in 10,000-year wind.

5.5.2.5 Wind simulation modeling

The presence of other buildings nearby can be a mixed blessing, with a lot depending on wind direction and local topography. They may provide shelter, but then again if wind is forced into a narrow gap between two high sided buildings it will speed up and swirl around, creating pressures which can amplify suction forces considerably. Clearly it is important to understand how the most ferocious winds will behave so that engineering teams can come up with solutions to confound them. Predicting wind behavior is not quite as tricky as accurately forecasting next month's weather, but is nevertheless a complex and time consuming business.

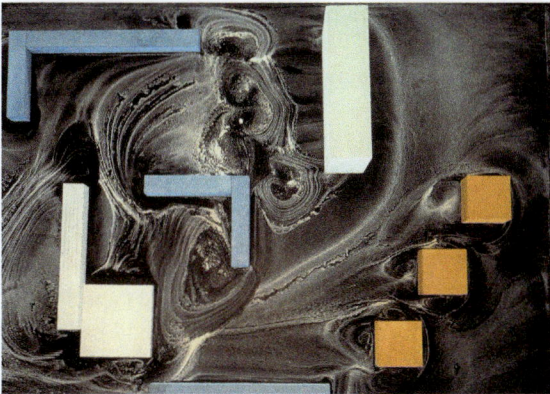

Fig. 5.25 Wind tunnel test. Science Photo Library.

When new facilities are planned, the effect their presence is destined to have on wind behavior is simulated by powerful computers, wind tunnel tests (Fig. 5.25) or even both. If results demonstrate that a planned profile will be detrimental to existing neighboring facilities, or indeed the facility itself, then modifications can be made to a design before it enters the real world. It just goes to show that countering the ferocity of a 1 in 10,000-year wind is not an exercise that could be described as being a breeze.

5.5.3 Snow

Snow might look very pretty on the side of a mountain or on a Christmas card, but if you get too much of it on a roof it can cause considerable damage. Most large industrial buildings have a flat roof. There will be a shallow slope to help rain runoff, but they are still flat enough to accumulate the maximum amount of falling snow. Many roof decking materials are in themselves quite flimsy. Profiled aluminum for instance might only be in the region of 1 mm thick, after all there is very little activity on a roof so a covering which is lightweight and durable makes perfect sense. However, it does need to be very well supported if it is to withstand the loadings of a 1 in 10,000 year snowfall.

5.5.3.1 Design basis

We know from our own experience that snow drifts in the wind and piles up against vertical faces, maybe against the side of a house or a fence running through a field. So we can be sure that the greatest loadings will come from drifting snow, particularly where there is a change in roof level or behind high parapets. The United Kingdom's 1 in 10,000 year design basis for drifting snow, typically assumes it could be up to 5 m deep and maintain that depth over say two bays of a building's roof.

As we have seen, the distance between structural bays varies from one building to another and may even differ within a single building; however, for our purposes we can assume two bays equates to a width or around 12 m. So although a roof may be completely covered in snow, we can visualize it being most concentrated to one side and in a strip 12 m wide and 5 m deep.

The weight of snow depends on its density, with factors such as how fresh it is and whether it has been compacted coming into play. Once again the characteristics of what we might call our *design basis snow* have been predicted, enabling engineers to come up with a loading. If we were to take a square meter of this particular snow and it was the specified 5 m deep, then it would impose a loading of over 7 kN/m^2. This is quite a burden, especially when you consider that if environmental factors were not an issue, then the most significant load a roof would experience would be that of personnel carrying out maintenance duties, equivalent to around 0.8 kN/m^2.

5.5.3.2 Shifting snow

Snow, even the 1 in 10,000 year variety, is at its least destructive when sitting quietly on a roof looking pretty. If our industrious team of structural engineers have done their jobs correctly, and we can assume they have, then a roof's deck along with its supporting steelwork will absorb the heavy load and direct it down into the ground where it can do no harm. The real drama starts when a thaw sets in.

Fig. 5.26 Shifting snow.
Istock.

The volume of drifting snow we are considering here is equivalent to parking family cars tightly together on a roof and then placing at least one more layer of cars on top of those. When that kind of weight starts to shift (Fig. 5.26) it has the potential to rip parapet flashings and cladding panels off a building, not to mention the considerable damage it can do by simply falling onto lower parts of a facility. In this arena architects and structural engineers must work in concert to thwart Mother Nature' sometime frosty behavior, together ensuring mountains of snow can come and go without leaving their mark on a building, or harming anyone in the vicinity.

5.5.4 Temperature

When it was in service Concorde illustrated the effects of thermal expansion very well indeed. At supersonic speeds its increase in temperature, due to air friction, caused the airframe to grow in length by over 100 mm, shrinking back down again as the aircraft slowed. It may not be quite so pronounced in buildings, but it happens all the same. At the other extreme, I have visited places that are so cold the tires on parked vehicles develop flat spots which only fade after they have been driven for a while.

5.5.4.1   Design basis

For the United Kingdom, 1 in 10,000 year extreme temperatures typically range from a sizzling 35°C to a glacial minus 20°C both of which are pretty harsh to say the least. And of course other parts of the world can be subjected to even tougher conditions. A facility's ventilation system has a role to play in dealing with such temperatures and maintaining an agreeable internal environment. Although it has to be said, it may not be designed to maintain perfect conditions in circumstances which are predicted to be exceptionally rare.

5.5.4.2   Embrittlement

From the structural perspective, the greatest challenge arises from embrittlement. Standard carbon steel begins to lose some of its toughness as temperatures dip below zero, becoming brittle and therefore susceptible to cracking at around minus 10°C. To counter this problem, structural engineers specify higher grade steels which are capable of retaining their strength under such extreme conditions.

5.5.5   Rain

Strictly speaking, combating the effects of extreme rainfall is not really a structural issue, but is a responsibility jointly discharged by architectural detailing and civil infrastructure; however, for completeness I shall include it here.

5.5.5.1   Flood planning

You will have noticed that if 20 or 30 mm of rain is due to fall in the same location within a 24 hour period, there is a fair chance it will warrant a mention in the weather forecasts, and when 50 mm falls it invariably causes sufficient disruption to make the news bulletins. Flooding in any environment is dangerous whilst it is happening and in populated areas leaves a particularly distressing trail in its wake.

On nuclear sites flooding brings an additional complication, as uncontrolled contact between nuclear materials and water is never a good idea. As a consequence, some considerable energy is expended on designing systems to capture rainfall and direct it away from a site. Even with all this effort, it is still recognized that during truly horrendous conditions, some sites will experience a certain amount of flooding. The challenge in these situations is to accurately model potential flood routes across a site, calculate the maximum depth of water, including any surge, and ensure the safety of nuclear facilities and their radioactive inventory will not be compromised by the worst that could possibly occur.

Incidentally, although I mention flooding here in the context of extreme rainfall (Fig. 5.27) exactly the same analysis should go into determining and mitigating the consequences of a tsunami. Sadly, we know from historical events how catastrophic the situation can become if a tsunami' reality is either underestimated, or adequate defenses are not put in place to guard against its onslaught.

Fig. 5.27 Extreme rainfall.
Shutterstock.com.

5.5.5.2 Design basis

For the United Kingdom, the 1 in 10,000 year rainfall is defined as 100 mm in a single hour or a remarkable 200 mm during a 24 hour period. Furthermore, such rain-bearing clouds cannot be expected to suddenly appear from a clear blue sky, so we can assume rainfall either side of these periods, particularly a single hour, could be almost as dire.

5.5.5.3 Rainfall management

To harness such torrents, architects calculate the volume of water that would surge across flat roofs during such abnormal conditions and ensure gutters and rainwater pipes can cope with the deluge. On large nuclear facilities, with their vast expanse of near flat roofing, it is not at all unusual to come across gutters that are a half a meter wide and 300 mm deep (Fig. 5.28) with rainwater pipes maybe 200 mm in diameter.

Fig. 5.28 Box gutter.
© Nuclear Decommissioning Authority.

Down at ground level civil engineers must design a network of drainage pipework, tunnels and open trenches, capable of carrying the cascading water and delivering it beyond a site' perimeter. All in all it is quite a considerable challenge, so I must say I do feel sorry for architects and civil engineers when I reflect on how little visibility there is of their handiwork. Still, they do have the satisfaction of knowing that should a biblical flood happen by, then the site will not be found wanting.

We began this chapter with an observation that the role of a structural engineer, even in facilities that must withstand particularly severe environmental conditions, can be perceived as almost routine, maybe an add-on to the main event. However, once we peel back a few layers and take a look inside their world, we find they play a pivotal role in the whole enterprise. Not only do they support everyone else, in more ways than are immediately obvious, but in following their lead we can rest easy when Mother Nature is having a bad day.

Process engineering

6

If you are ever looking for process engineers in a large office block filled with every assortment of engineer, you could for your own amusement decide not to ask for directions. They will not be difficult to find; process or chemical engineers tend to occupy desks loaded with the biggest reference books in the building, always to hand and ready to consult. Happily for the rest of us, we do not need to know the minutia of whatever it is that occupies their minds. What we are interested in is knowing enough of what they do to contribute to our understanding of nuclear facility design and to appreciate how their activities interact with those of the wider engineering community. So this is where we shall concentrate our efforts. What we shall discover, is that once a nuclear project is launched it is process engineers who set its direction, to the extent that the rest of us must await results of their early deliberations before we can even set about making our own contribution.

6.1 Closed cells

The first impression you normally get on entering a nuclear processing facility, particularly those based on the closed cell concept (Fig. 6.1) so popular in the United Kingdom is that there is very little going on.

They can be almost eerily quiet and in some areas give little indication of any interaction with operations personnel, so not exactly a hive of activity? Well, no.

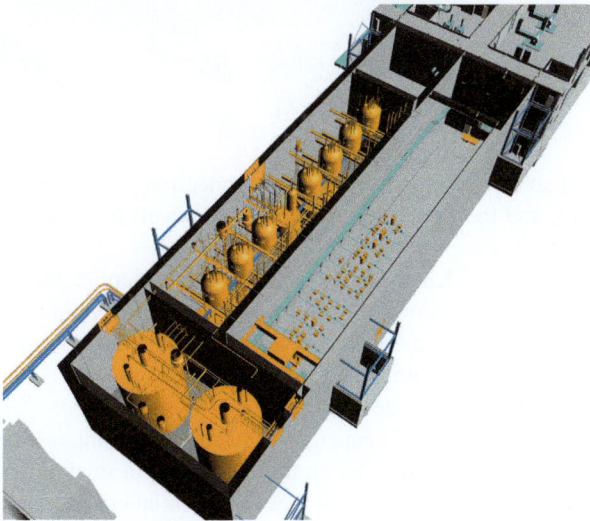

Fig. 6.1 Closed cell processing facility.
© Nuclear Decommissioning Authority.

Nuclear Facilities. http://dx.doi.org/10.1016/B978-0-08-101938-2.00006-4

You see apart from maintenance activities, just about everything in these plants is either wholly or partially automated. In addition, almost all of the liquors being processed are radioactive and therefore surrounded by hefty shielding, probably concrete, which means there may not be that much to see. The elusive operators will be there alright, overseeing the goings on, but they may not get out too much and certainly will not need the level of involvement required in facilities more biased towards mechanical handling.

This illusion of inactivity belies the fact that behind those hefty walls, liquors, sludges, and slurries, often measured in tens of thousands of liters, are being processed every day and dispatched to their various destinations. So if you do get the opportunity to look around this type of facility, do not be taken in by superficial appearances. A little investigation will show they have much to reveal. Just one example will illustrate the point.

We tend to think of vessels as hollow shells with maybe a couple of pipes connected, so quite a simple affair. However, vessels in the type of facility we are

Fig. 6.2 Internals of a maintenance-free processing vessel. © Nuclear Decommissioning Authority.

considering here can have so many pipework connections that it is difficult to find space for them all, and their concealed internals may well look like the example in Fig. 6.2. Clearly there is a lot going on.

As noted, in the United Kingdom the usual, although not exclusive practice, is to adopt the *closed cell*, sometimes referred to as *dark cell* or *passive cell* approach. It is an appropriate name because almost all of their equipment, which must be maintenance free, is housed within an environment that is sealed up, normally in total darkness and untouched for decades.

The relatively small number of items that do need to be maintained are positioned in a way that allows them to be withdrawn through concrete shielding and dealt with outside the cell. The notion of effectively locking essential equipment away like this may sound like a strange thing to do, but it is nevertheless a concept which works well, and the way it operates will become clear as we make our way through the coming pages.

The main alternative to a closed cell is the *canyon* concept, which is more popular in the United States. The way they operate is quite different and more akin to the mechanical handling caves (discussed in the next chapter) that we see in the United Kingdom. Essentially the main difference is that closed cells are divided into concrete compartments where their equipment is effectively locked away for its entire operational life. Whereas canyons are a single space and have the wherewithal to retrieve plant and equipment in order to carry out maintenance, repair, and even a full exchange. However, even though their operation and maintenance regimens are quite different, much of the processing equipment used in cells and canyons is based on similar principles.

If you do get the opportunity to step inside a closed cell before it is sealed up (Fig. 6.3) I would recommend the experience very highly indeed. The first thing that

Fig. 6.3 Closed cell during commissioning.
© Nuclear Decommissioning Authority.

strikes you is how little space there is to maneuver, closely followed by wonderment at how such a tangle of unyielding stainless steel pipework, not to mention vessels and other equipment could possibly be squeezed so tightly into the area. Knowing it is made possible by sophisticated computer modeling which, apart from generating pretty pictures, also feeds millimeter-perfect information into automated pipe cutting and bending machines, does not diminish the occasion.

By way of illustration, Fig. 6.4 shows a snapshot from the computer model which resulted in the real world image seen in Fig. 6.3. It certainly justifies its name tag of *intelligent modeling*.

By the way, you will notice I have been using the terms *cell* and *cave*, which are often thought to mean pretty much the same thing and therefore be interchangeable.

Fig. 6.4 Closed cell computer model.
© Nuclear Decommissioning Authority.

However, they have quite different meanings. The term *cell*, as we have just seen, refers to a sealed up concrete structure that houses stationary equipment which, apart from some components on its periphery, cannot be retrieved. Whereas *cave* is used to describe a shielded concrete structure that houses operating mechanical equipment. So unlike cells, caves are well illuminated, have the wherewithal to retrieve their equipment, normally include a crane operating along their entire length and probably have viewing windows embedded within their perimeter walls. Quite often the terms are transposed and it does no harm, but strictly speaking their meanings are as I have described.

Liquor processing equipment may contribute to a nuclear facility's primary purpose, such as supporting the operation of a reactor. Frequently, though, it will instead provide a crucial role in dealing with by-products, such as radioactive liquors and sludges, which may be present when decommissioning an aging facility. Wherever they originate, having performed their various tasks, the resulting solutions must be dealt with in some way.

Liquors not treated during a facility's operations phase are invariably transferred to stainless steel holding vessels to await processing at a later date. This is satisfactory as an interim measure, but long-term storage of active liquors is unacceptable as there is always a possibility that it may seep from aging vessels, contaminating the local area and even the environment beyond. So whether it happens straight away or after a period of interim storage, active liquors will need some form of additional processing.

As we saw in Chapter 1, to stabilize liquors, sludges, and so on and make them safe for long-term storage, they really need to be converted into a solid form such as glass or concrete. However, before we can even start to think about solidifying a solution it needs to be prepared by routing it through a sophisticated processing plant. As with so much in the nuclear industry, similar equipment and processes can be found in other spheres such as the petrochemical and pharmaceutical industries. Of course they are dealing with entirely different products, so shielding, for example, is not an issue. Nevertheless, many of the fundamental principles are the same. So if you are familiar with similar non-nuclear processes you will recognize some of the concepts, but then you will also discover that the radiological dimension complicates matters by several orders of magnitude.

Nuclear engineers are often asked an entirely legitimate question, "Why don't you just build a process plant identical to the one you built the last time?" Sounds reasonable, until we consider that each plant must be tailored to suit the constituents, or *radionuclide fingerprint* of the inventory—liquors, sludges, and so on—that it is destined to process. The chances of finding two radioactive solutions formed from entirely different processes and yet with the same fingerprint are as close to zero as makes no difference. You will get an idea of just how bespoke a processing plant can be when we discuss the range of equipment which is available to treat active liquors.

Now that we have set the scene, we can examine how process engineers go about their task; then with that done, we shall take a look at the array of equipment that is available to them. So where do they start?

6.2   Mass balance

Mass balance, or mass and energy balance as it is more correctly known, is where it all begins. No matter what the facility something will come in at one end, be processed and go out at the other. It is absolutely crucial to determine exactly, and I mean exactly, what will be involved at each step along the way. If you are embarking on a decommissioning project, then the *something* that came in probably did so several decades earlier. The principle is still the same: understand what is destined to be handled and then draw up detailed plans for how it will be accomplished.

With any sort of processing facility, the first questions to be addressed on the very first day are, "Exactly what are we dealing with?" and "What form should it be in when it leaves our facility?" That gives rise to a third question, "What must we do to answer the first two?" This is where mass and energy balance comes in. Mass refers to the weight and constituents of the original feed liquors, along with new additives blended along the way, whilst energy refers to any heating, cooling, and propulsion that may be required on the journey through a facility. Balance is the key word here and believe it or not is analogous to baking a rather splendid fruit cake.

Fig. 6.5 shows how it works, with main process steps running left to right across the middle, materials and energy coming in from the top and waste products along the bottom. Process engineers will no doubt be alarmed by the diagram's simplicity, but for the rest of us it captures the principles of mass balance handily enough. You will probably have guessed that apart from radiation, contamination, sophisticated ventilation systems, and so on, there is a fundamental difference between baking a cake and handling substances in a nuclear facility. Whereas a cook can throw

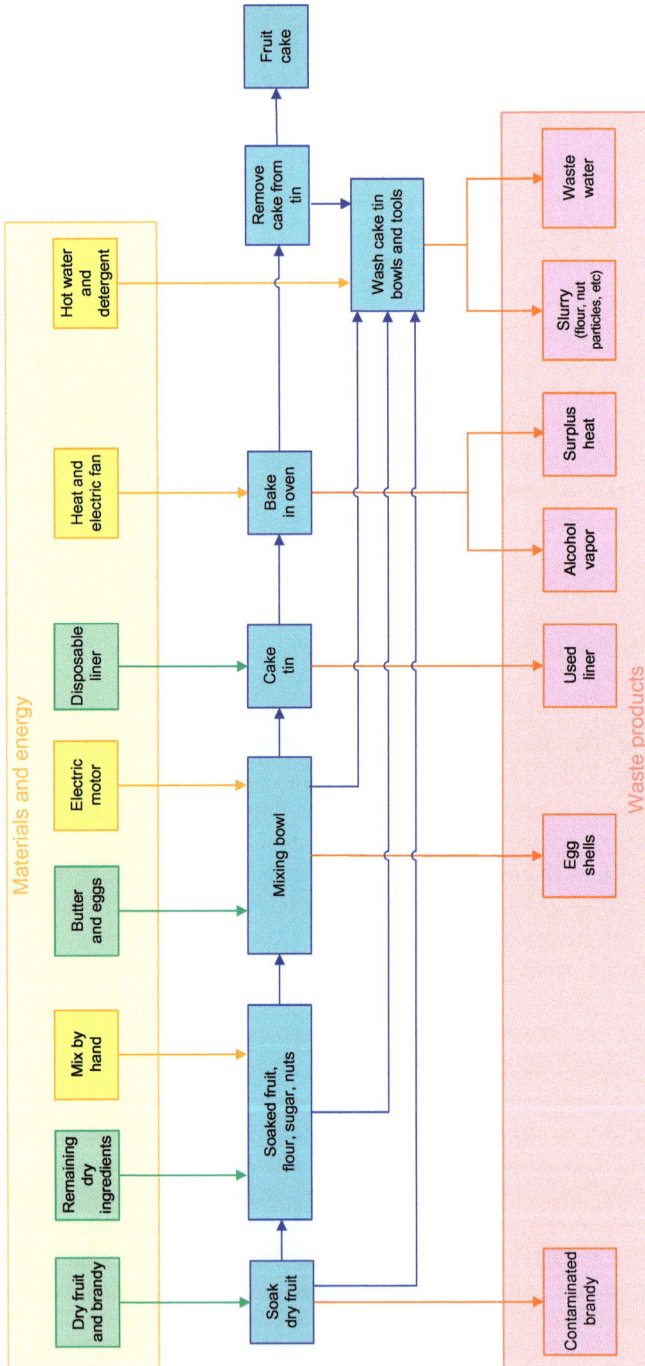

Fig. 6.5 Mass balance of a fruit cake.
© Bill Collum.

eggshells away, flush waste water down the sink and perhaps drink the "contaminated" brandy; their radioactive equivalent must undergo additional processing, and maybe packaging, before it can safely be disposed of or stored. So with the kind of mass balance we are considering here, every thread must be followed until it reaches a safe and regulatory compliant endpoint.

In the nuclear world then, mass and energy balance looks at the precise makeup of feeds entering a facility, how they are heated up here, cooled down there, additives blended in and substances drawn off and sent on their way. Ultimately it demonstrates that, in terms of constituents and energy, what goes into a building at one end is exactly equal to what comes out at the other, albeit that (as with our cake) material character-istics may change considerably along the way. Nothing is unaccounted for; if it were, then the required processing energy would be unavailable and even more problematic overlooked substances would accumulate in the system with nowhere to go.

Ideally, when compiling a mass balance sequence, the first steps should be focused on removal of the most radioactive elements, because if they can be segregated early, this will minimize the volume of highly active waste generated by a facility. Cesium, for example, is not something you would want to route through umpteen processes before finally getting around to removing it. It is highly radioactive and needs an awful lot of shielding, so is best isolated before it contaminates more equipment than is necessary.

There is, however, an exception to the philosophy of removing highly active constituents first, one which can be brought on by the presence of particles, or *solids* as they are known in a waste stream. We shall discuss solids removal later, so I will just say here that they would clog up the processing equipment used to remove cesium or other species from a liquor stream, which pretty much dictates that when solids are around it is usually prudent to remove them as soon as possible.

The format of a mass and energy balance, as with so much of what is produced in this arena, is known as a *process flow diagram* (PFD). And as with other PFDs they are always supported by reams of calculations, explanatory text and all manner of design justification. Once completed, a mass balance tends to look like an intricate car wiring diagram, because it no doubt tells a very complicated story. When well prepared, they will, for example, highlight that a liquor is being heated up by one process, only to be cooled soon afterwards by another. Whilst the balance is still on paper, there would be opportunities to amend the process, if indeed it is feasible, in order to avoid such potential wastes of energy.

A mass balance also documents the fate of something which is not quite so obvious: gas. It is impossible to operate processes where liquors are simmering and fermenting, without vapors and gases, or *off gas* as it is known, being generated. And these gases need somewhere to go, otherwise they could build up and cause a rupture in the system, or even explode. The thing is there will be particles bound up within these vapors and gases, radioactive particles at that, so they must be scrubbed, filtered, or both, and finally monitored before being directed to a ventilation stack and released into the environment. It just shows how thorough the exercise of creating a mass and energy balance must be, when even gases, along with microscopic particles borne within them, must be identified in advance and their ultimate destinations determined.

A mass and energy balance is the cornerstone on which a project's entire design evolution is built, so much so that very little meaningful design can commence without at least a first draft; beyond that, a project team is guided by further more detailed iterations. In fact, as the process folks are only too keen to remind us, it is they who get the show on the road and keep it on track thereafter. Obviously they overplay their hand now and again, but there is some truth in what they say.

Clearly producing a mass and energy balance demands a mighty effort. However, once it takes shape a project team holds a wonderful route map, one showing exactly what must happen within their facility if it is to deliver on its objectives. This is not a bad place to start.

6.3   Feedstock analysis

Imagine this situation. You are starting out on a new project, one which is destined to treat liquors that have been stored for many years, and asking yourself the fundamental questions: "What are we dealing with?" and "What must we produce?" If we expect crystal clear answers to both of those questions on the very first day, the chances are we are in for a big disappointment.

6.3.1   Conditions for acceptance (CFA)

Getting an answer to the second question is the more straightforward of the two. It may be that the task is to convert a highly active liquid or sludge into a solid form, to eliminate the danger of it leaking, or giving off gases that could cause a danger if they were allowed to accumulate. We should also be able to readily determine what form that solid product should be in—maybe to lock up the liquor in a concrete matrix or within glass blocks, so as to prepare it for safe storage.

In addition, on that second question of, "What must we produce?" there should also be pretty good guidance on the specification or *CFA* as it is known, for the final packaged product. So, for example, the type of container, no doubt stainless steel, into which the product must be poured, how much voidage space is permissible and how well the external surface of containers must be decontaminated. All of these things should be spelt out clearly enough, after all the containers will ultimately be destined for a long-term storage facility which will have set out its CFA years in advance.

Again on that second question, it may be a little more difficult but still possible to establish the CFA for containerized concrete or glass at a reasonably early stage. So we can determine exactly how many parts per million will be permissible for species such as cesium, americium, technetium, strontium and many others, until finally the full spectrum of acceptance criteria for a particular project are known.

All of this relates to that "what do we produce" question, which although it is a difficult one to crack, is in the grand scheme of things relatively straightforward. If key players from all the interested parties, which may include regulators, get their heads together, they can sort it out and give project's a sound basis on which to proceed.

Facilities that elect to include liquor processing within their primary operations phase, will be in the happy position of having a precise answer to the issue of what they are dealing with, the mass balance exercise sees to that. However, for facilities processing substances that have been stored for many years, the question of "What are we dealing with?" can be in a different league of complexity altogether. It cannot be solved by technical discussions and coming up with assumptions that everyone feels are reasonable. Ideally we want rock solid information on the full range of constituents in an aged feedstock, and there can be scores of them, all locked up in a stew which is often highly radioactive and therefore very difficult to get at.

In fact sampling and fully understanding such liquors could even be a major project in its own right, conceivably with a level of complexity and costs not too dissimilar to extracting and processing them. To make meaningful progress, what we need is sufficient knowledge of a feed inventory to mitigate unwelcome surprises and allow us to move on. This is an extremely complex area, one where many a project has faced an immense challenge, even before they get properly started. Let me give you an example.

6.3.2 Stratification

Let's say we are tasked with processing the aging contents of a large vessel holding thousands of liters of active liquor. Unless it has been regularly agitated, which is not always the case, it will have stratified and within those strata there will be distinct layers (Fig. 6.6). If the liquor is a weak slurry, then sitting on top there may be a spongy crust, at the bottom, heavier particles will have settled into a sludge, and in the middle, will be the main body of the liquor.

Fig. 6.6 Stratification.
© Bill Collum.

The easiest area to sample, which could be difficult enough, is obviously the top crust. Armed with a sample or two from that area, along with historical records, skilled analysts could extrapolate the available data and deduce what the other constituents of a vessel may be. The problem is that science in this area is extremely complicated, particularly when so little real data is available, so it really is a gamble on how close the analysts' predictions will turn out to be. Of course it is not a gamble in the sense of win or lose; it is all about confidence in designing a facility that can cope with the full spectrum of potential constituents that may come its way.

It gets even more complicated when we consider that the ideal processing route for different strata within a vessel will almost certainly be quite different, a situation further compounded by the variable characteristics at transition zones between one strata and another. So, for example, we could be happily drawing off a free flowing liquid from the main body of a vessel when it starts to change characteristics, carrying more and more suspended particles until it thickens into a sludge at the bottom. In some cases, the sludge may have hardened into a claylike consistency, which presents even more of a challenge as it will have to be fluidized before it can be liberated. The variation in a liquor's viscosity is bad enough, but potentially more challenging are changes in its radionuclide composition. This occurs because different species, cobalt, americium, and so on will also have stratified throughout the vessel.

6.3.3 Organics

As if all of that were not enough to contend with, there may also be some doubt as to the presence and quantity of organic materials in the feedstock, such as oil, lubricants, or solvents. It is a crucial factor, as organics generate hydrogen which can become trapped in a sludge and, if uncontrolled, risk initiating a detonation of some type: anything from a small "pop" to something much more serious. This is one of the reasons it might not be such a good idea to stir up a mixture before processing it. A design team can introduce measures to disperse hydrogen or inert a vessel where it may be present, say with argon, nitrogen or helium. It certainly mitigates the risk of a detonation but can cost a great deal of money. So ideally we must be sure a risk is real and not imagined before drawing up plans to expunge it. And so it goes.

Even from that condensed summary, you get a feel for how terribly difficult it can be to pin down the exact characteristics of an aging feedstock. No doubt carefully targeted sampling campaigns will be instigated to gather as much real data as possible, but when you get right down to it there comes a point where a design team must proceed with the information available. One way or another, calculated assumptions must be made about the physical characteristics and spectrum of species within a process feed and how they might fluctuate across its various strata.

6.3.4 Orphan streams

The big problem here is that any species falling outside assumptions of what we might call an "artificial" process envelope may potentially be untreatable. These *orphan* streams as they are known, would then need to be stored or even left where they are to await treatment at a later date. But then you can imagine the expense of building or modifying a plant, just to treat a few constituents that were not anticipated during the original assessments.

You may think, "Why not just design a facility to cope with the worst-case feed envelope that could possibly come its way? This would eliminate any uncertainties and the plant could take anything thrown at it." Even if the worst-case could be identified, a project would risk spending a fortune, a considerable fortune at that, on plant and equipment that turned out to be unnecessary, which would be embarrassing to say the least. So there is considerable pressure on process engineers to get their analysis right, create an unassailable mass and energy balance and justify their recommendations. Maybe this is why they like to keep such hefty books to hand.

6.4 End product

Before we move on to look at the various types of processing equipment, it is worth taking a moment to contrast the challenge faced by nuclear process engineers with that of their colleagues in other industries. It is quite revealing, because when you look closely you find that processing demands in the nuclear industry are quite unlike those encountered elsewhere. Ordinarily the objective is to create something, say a pharmaceutical product or a bathroom cleaner, which means that in the normal run of things the task facing engineers and chemists is to come up with a recipe that will deliver a new concoction of some sort. Their role is to figure out the ingredients along with how to blend and cook them, so to speak. The whole impetus therefore, is to start with a blank sheet and conjure up a product that people will be happy to pay good money for.

In the nuclear industry, process engineers do not always set out to make a product; sometimes they do, maybe nuclear fuel for a reactor, but very often they are presented with redundant radioactive equipment or a not very well defined brew of some sort and asked to sort it out, to make it safe. So the job of a nuclear process engineer tends to be diametrically opposed to that of their colleagues in other industries, in that they must unravel an existing product rather than create a new one. It turns out that the end product for a nuclear engineer is very often nothing more than a means to an end, and certainly not something that would be eagerly snapped up on the high street.

6.5 Transfer devices

6.5.1 Mechanical pumping

Strictly speaking we would expect to see mechanical pumps discussed in Chapter 7 (Mechanical). However, their deployment is inextricably linked to operation of the processing equipment we are examining here. In addition, and as we shall see shortly, mechanical pumps are not the only way to move liquors around. For these reasons it makes sense to discuss all forms of pumping here, that way we can get the whole picture in one place.

As we know from Chapter 2, the intensity of radiation can vary considerably, from background levels that surround us every day, to the kind of penetrating energy that needs a meter or two of concrete to contain it. The characteristics of this sliding scale dictate that where mechanical pumps are employed, the nuclear industry utilizes two distinct types. The first are akin to those found in any other sphere, but modified somewhat so as to facilitate their use in a nuclear environment. They are widely used to pump low level liquors, plus many from the lower end of the intermediate level waste (ILW) category. Liquors at the top end of the ILW spectrum, along with those categorized as highly active, are moved along by an altogether different kind of device that we shall get to shortly. Let's start with a look at standard mechanical pumps, then examine how they are deployed in a nuclear environment.

On the face of it, the job of a mechanical pump, transferring liquors or gases from one location to another, is straightforward enough, so we might reasonably expect to find maybe a handful of variations in their basic design. It turns out, though, that there are a myriad of generic types, each with even more sub-types, all satisfying a particular niche. The nuclear industry employs quite an assortment of them, but two in particular are more popular than most.

6.5.1.1 Progressive cavity pump

The progressive cavity, or mono pump, as it is more commonly known is simple, robust and particularly adept at pumping slurries and sludges, a trait which makes them particularly well suited to oil and gas exploration, sewage treatment, and so on. They are really a modern version of the Archimedes screw that has been around for centuries, where liquors are moved along by the revolving motion of a helical thread. For the nuclear industry, where maintenance needs to be as slick as possible, such simplicity is very attractive indeed.

6.5.1.2 Centrifugal pump

The centrifugal pump (Fig. 6.7) is easily the most popular of all pumps and widely used in the water, chemical, and oil and gas industries, along with food and textile production and many more.

They operate by drawing liquid into the eye of a spinning impeller which accelerates incoming fluid and hurls it towards its discharge route. This rapid rotation creates

Fig. 6.7 Centrifugal pump.
Shutterstock.com.

a vacuum at an impeller's eye which has the effect of sucking more liquid into it, so once up and running the flow through these pumps is wonderfully balanced.

As we know mechanical pumps can fail, so where they are being used the first item on the agenda is a decision on whether or not a standby pump is required. Having a standby pump does not mean keeping a shiny new one in a box ready to install, it means having one permanently plumbed in and ready to go at a moment's notice. It may sound extravagant but it can easily take several hours, maybe even a few days to change a pump, so there are production losses to think about and in some cases damage to associated processing equipment. Either one of those, especially the last one, could potentially add up to more than most of us earn in a lifetime, so an extra pump, although expensive, might turn out to be a shrewd investment.

Suction head

Before a pump can be fired up, most must have a good head of pressure, or a positive *suction head* behind them. In other words it is not a good idea to start, say a centrifugal pump, whilst it is dry and wait for liquor to start making its way through. In fact, if this type of pump is activated with insufficient suction head, liquor can bubble up and vaporize, which can cause considerable damage.

This is where the self-priming pump comes in (Fig. 6.8). All it takes is a water supply vessel above the pump to feed it with a guaranteed head of liquid. Unfortunately, the additional water dilutes process liquors so this option may not always be viewed too favorably. Ideally then, it is preferable to design a system so as to use the actual liquors being processed to provide the suction head.

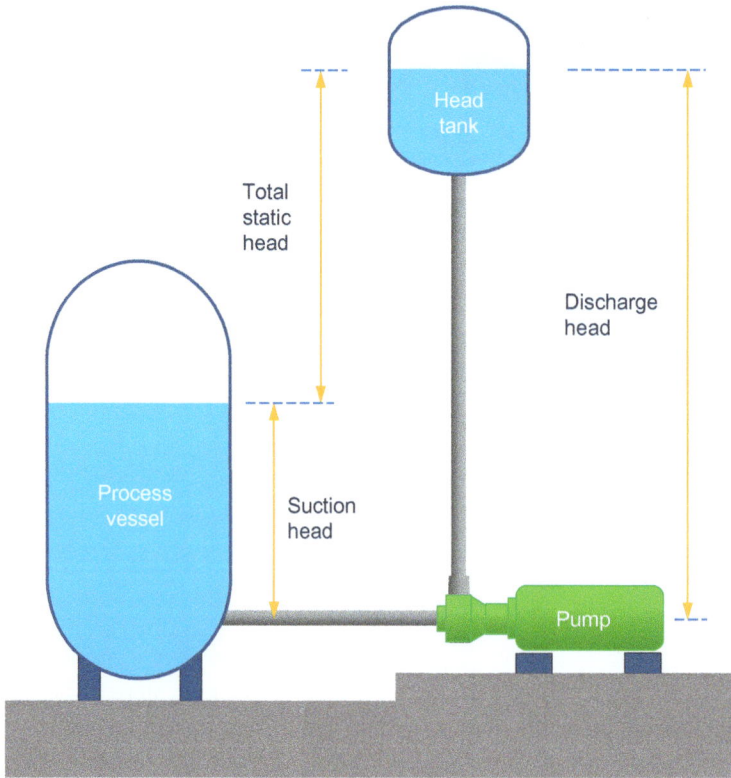

Fig. 6.8 Self-priming pump.
© Bill Collum.

When you think about it, it becomes obvious that the easiest way to guarantee a positive suction head is to simply place pumps below the vessels they serve; that way they are always primed. However, the downside with this arrangement is that before a pump can be removed for maintenance, or exchanged for a new one, it must be isolated from the rest of the system, otherwise active liquors will flow out the moment a failed pump is removed.

Double block and bleed

Where a standard mechanical pump is located below its feed vessel, it is accompanied by valves which can be shut off whilst the pump is removed from service. In most industries we would expect to see two valves, one on the suction line entering a pump and one on its discharge line. However, with a significant head of pressure behind the entry point it would be too risky to depend on a single valve to hold back radioactive liquids. The arrangement used to defeat what would otherwise be a weakness in the system, goes by the rather unglamorous title of *double block and bleed* (Fig. 6.9). All it amounts to is two valves on the pipeline which delivers liquid to the item of equipment being isolated, say a pump, with a bleed leg between the two.

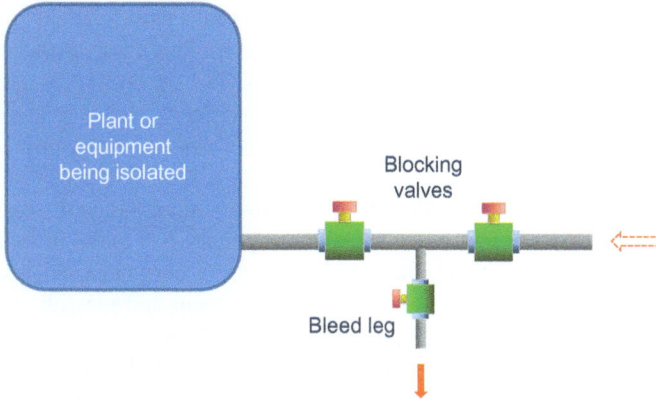

Fig. 6.9 Double block and bleed.
© Bill Collum.

In the case of a pump, before being removed from service it is flushed through with water, or maybe nitric acid or a solvent, and then all three valves closed off. With that done liquor trapped within the pump and blocking system is drained away via the bleed leg, probably to a local sump and from there sent for onward processing. With draining complete it is safe to work on the pump or remove it for maintenance or replacement. It is a good example of simple yet reliable engineering that does a good job without any fanfare.

Shielding bulge

Where standard type pumps are serving processing equipment which is located in an inaccessible C4 cell, they will be located just outside the cell and surrounded by local steel shielding. This type of shielding is known as a *bulge* and it can be removed to permit access. When used in this setting, the most notable difference in *nuclearized* versions of standard pumps is that their drive shafts are extended. This enables pumps, or valves for that matter, to be housed within the shielded bulge, whilst drive motors are positioned outside and therefore more readily accessible. Where pumps are enclosed in a bulge there are two options available regarding how they are positioned and maintained.

With option one, which tends to be more prevalent in the United States than the United Kingdom, pumps, along with their local shielding are located in C2 (green) areas, such as corridors, and normally adjacent to the more hazardous C4 (red) areas occupied by vessels they serve. In this situation it is not permissible to go to work on a pump right there in a corridor, as contamination may be released and spread uncontrolled into the surrounding area. Instead temporary containment is erected around the bulge, in effect creating a new room which has its own mobile ventilation and air filtration system.

Typically these rooms are constructed from demountable partitioning, or some form of lightweight sheeting secured to a frame. Once the room is in place it can be temporarily classified to the higher C3 (amber) radiological level and work on a pump can begin. With maintenance complete, the room is monitored and if necessary decontaminated to ensure it complies with C2 (green) conditions, then the structure removed and the area returned to its regular duties. The beauty of this option is that we can locate pumps pretty much where we please; all that is needed is space to erect a temporary enclosure and a suitable clear route for trolleys to haul equipment to and from the scene.

Option two, which is more widely favored in the United Kingdom, takes significantly more trouble to configure but is easier to operate. The main difference is that rooms housing pumps and their shielding bulges are a permanent feature. So their ventilation is integrated with the building's main system and the sub changeroom, needed to move in and out of a C3 pump room, is built-in.

Here's the challenge, though: there is no way you would want to build lots of permanent rooms like these, around pumps dotted all over a building. The cost, including space needed to accommodate them, would be far too prohibitive. To make this option viable, pumps really need to be congregated into a small number of rooms, ideally just one. Realizing that goal will take some effort because each pump may well serve several vessels, and there are constraints on how the relationship between them is configured.

To achieve this option, the arrangement of vessels within shielded cells and positioning of pumps that serve them must be developed in parallel, each influencing the other until ideally all pumps are gathered together in a common location. If designers persevere and manage to deliver an arrangement with just one or two permanent pump rooms, then it opens up the possibility of fitting them out in a way which is not really feasible for their more transient counterparts.

The typical arrangement for a permanent C3 (amber) pump room allows personnel to gain access via a fully equipped sub changeroom, whilst equipment is transferred to and from the room via a separate C2/C3 lobby. In addition, lifting beams can be permanently installed above pumps to assist in dismantling their local shielding bulge. In some cases the whole floor is lined with stainless steel as it is very robust and, should the need arise, easier to decontaminate than alternative finishes. Of course this option does not have the locational flexibility of temporary enclosures and it costs more to lay on the additional fittings and improved access. However, it does have the advantage of speeding maintenance and repair tasks by turning them into activities that are more routine in nature. As always, the choices on offer are much easier to identify than an *obvious* solution.

The mechanical pumps we have considered so far are of a standard variety, but quite often nuclearized in a way that enables them to operate behind local shielding. In addition there is a breed of pump that operates on similar principles, but is specifically designed for nuclear applications and is particularly useful in dealing with more active ILW liquors.

6.5.1.3 Remotely maintainable pump

These pumps are so named because their internal parts are specifically designed with remote maintenance in mind, a task that is more problematic with standard pumps, even when they are nuclearized. Having said that, many of them operate in radiological conditions that are compatible with a regimen of hands-on maintenance. So their name does not quite tell the full story.

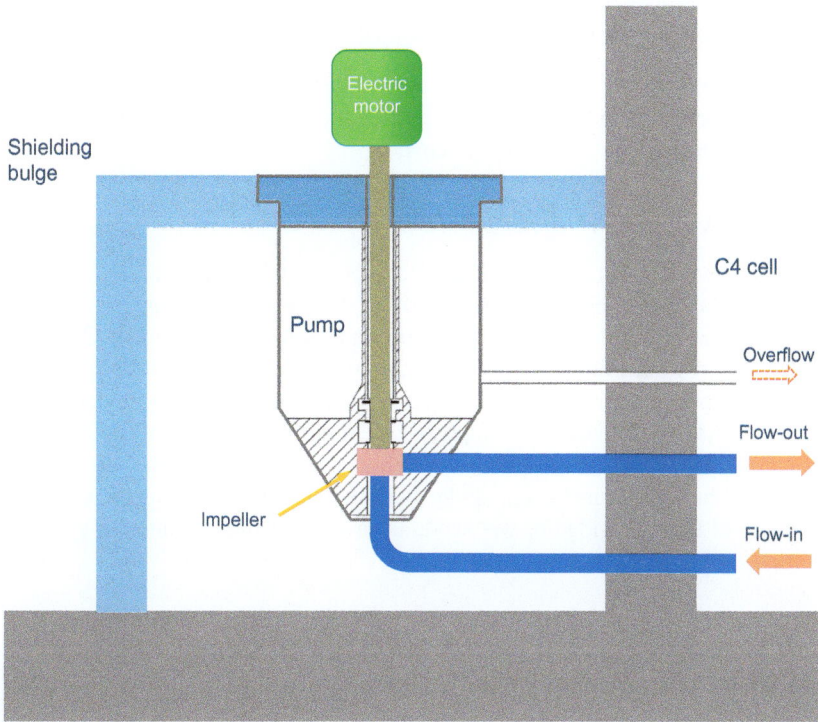

Fig. 6.10 Remotely maintainable pump.
© Bill Collum.

These splendid pumps are normally deployed in a bulge (Fig. 6.10) with their drive shaft and body extended, but can equally well be embedded within the concrete roof of a C4 (red) cell. In common with nuclearized versions of standard mechanical pumps, their drive motor is mounted outside the bulge, or cell, where it is more easily accessible.

Prior to maintenance, they are flushed out in the same way as other mechanical pumps and their electric motor removed; after this the shaft, along with its impeller and various seal components, is withdrawn. If, after flushing, a pump's radioactivity levels are still high, then the mechanism is hoisted into a shielded flask (discussed in the next chapter) and taken to an area equipped for remote maintenance. Where activity levels are sufficiently low the mechanism can be loaded into a stout transport bag and taken to a C3 (amber) workshop for hands-on maintenance.

So mechanical pumps, in their various guises, are entirely appropriate where one way or another it is possible to retrieve them for maintenance, repair, or even disposal. However, where highly active liquors are being processed we need an altogether different approach. This is where fluidic pumps come in. In fact if it were not for the fluidic pump and its derivatives, closed cells would not be viable.

6.5.2   Fluidic pumping

Process engineering is strewn with examples of elegant solutions to seemingly intractable conundrums, the epitome of which must be the reverse flow diverter (RFD) that sits at the heart of fluidic pumping systems.

6.5.2.1   Reverse flow diverter (RFD)

Let's say we were given the task of developing a maintenance free liquor transfer device, one that had to be exceptionally robust and reliable enough for decades of trouble free use in a hostile and inaccessible radioactive environment. Given such a challenge, we would do very well to come up with anything half as good as the RFD. Simply put they have no moving parts, which for a pumping device is pretty remarkable. As if that were not impressive enough an RFD can cope with just about any "liquor" you care to pump through it, not just those at the watery end of the spectrum but also slurries and sludges. They can even pump substances with a consistency of thick porridge or wet sand; all in all, it is quite a piece of apparatus.

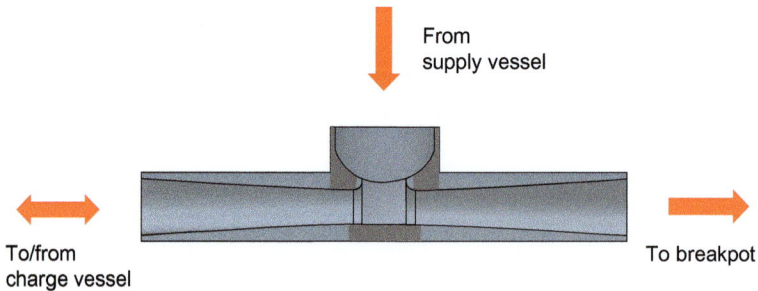

Fig. 6.11 Reverse flow diverter.
© Elsevier.

The RFD itself (Fig. 6.11) is really nothing more than an innocuous looking piece of "T" shaped pipework with regular openings at each end. The clever bit lies in how its internal diameters are tapered and sculpted, and how simple compressed air movements can be used to turn it into a very effective pumping device. Crucially, although RFDs are invariably deployed in a C4 (red) environment, their control systems are located in a C3 (amber) area, so are quite readily accessible for maintenance and repair. This configuration enables fluidic pumping systems to meet that all-important objective of having no moving parts whatsoever within a highly radioactive environment.

6.5.2.2 Fluidic pumping system

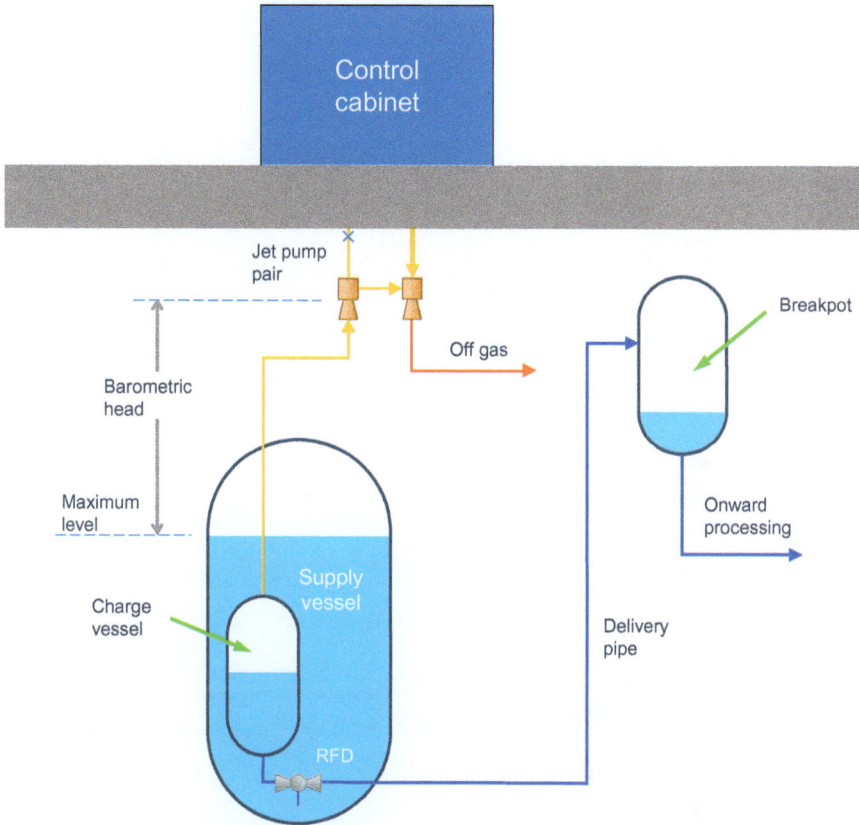

Fig. 6.12 Fluidic pumping system—RFD internal.
© Bill Collum.

Fig. 6.12 shows the main components of a typical fluidic pumping system. As air is drawn upwards through the charge vessel it pulls a depression on the RFD, which in turn draws liquid through its open ended pipe and into the vessel. Straightaway there is an obvious concern that radioactive liquor could be pulled too far up the pipework above a charge vessel, escape from the closed cell and spill into an area occupied by the RFD control cabinet. To overcome this possibility, the system's suction is calibrated in such a way that it can only pull liquor so far up the charge vessel. Added to that the jet pump pair, which is responsible for directing airflows, is always located inside a cell and at least a barometric head above the supply vessel, so the system is "safe by design" and it is impossible to draw radioactive liquor into the area outside a shielded cell.

Barometric head

Fig. 6.13 Pipetting.
© Bill Collum.

Before we move on, I need to explain how a barometric head fits into the equation. If you put a drinking straw into a glass of water, until it touches the bottom, place a fingertip over the open end and then lift it out of the glass (Fig. 6.13) water will remain in the straw until you raise your finger to release it. We can visualize, though, that there is a limit beyond which a straw would be too long to replicate the trick, simply because the weight of water within it would be too great to be held back.

The same principle applies even when liquid is being sucked up into a tube by some force. There comes a point when no amount of suction can raise liquid any higher than a given point; it happens at a height of around 10.2 m and is known as a *barometric head*. Strictly speaking this height applies a true barometric head which pertains to water. More importantly there is also a working, or practical barometric head which varies depending on the density of liquid being handled. Heavier liquids, say a slurry, will not rise as far, whilst lighter ones, say a solvent, will travel further. There are ways of circumventing dependence on a barometric head, such as ensuring a suction device is simply not powerful enough to draw liquors beyond a certain point, but the most straightforward way of ensuring active liquors are not inadvertently sucked from a cell is as I have described.

Pumping sequence

Getting back to the pumping sequence, a valve in the control cabinet that was open for this first suction operation now closes, and another valve opens. Air is then forced down a pipe via the jet pump pair, which is effectively a static switching mechanism, and into the charge vessel, driving its stock of liquor across the RFD and up into a breakpot.

You would expect liquor to be driven back out of the RFD's open-ended pipe, where it was drawn in—particularly as the alternative route, the one it actually takes, is upwards. But an RFD is founded on the principle that liquid, and air for that matter, will always take the path of least resistance, so shaping within an RFD ensures the path to a breakpot is the least demanding. In truth some liquor will flow out of an RFD's open end and back into the supply vessel, but the bulk of it will go where it is needed.

As the transfer sequence nears completion, air pressure within the charge vessel will begin to drop off, until it reaches the stage where it is no longer able to push liquor through the RFD. When this happens, liquor in the delivery pipe falls back across the RFD and into the charge vessel, until liquor levels in the delivery pipe and charge vessel are both the same. Air which is pushed out of the charge vessel by this backflow, travels back across the jet pump pair, from where it is routed to the off gas treatment system. The sequence settles down and the entire cycle can begin again.

RFD arrangements

As an alternative to the arrangement shown in Fig. 6.12, where an RFD is mounted within a supply vessel, the RFD, charge vessel and associated pipework can all be mounted externally (Fig. 6.14). Apart from being able to empty a supply vessel completely, which is not possible when an RFD is installed inside it, some designers prefer this approach because it allows access to the whole fluidic pumping system. Others favor the enclosed arrangement because it minimizes any possibility of leaks within the surrounding cell. Both are equally effective so selection comes down to the specifics of a given situation.

There is a particular side effect inherent in the use of RFDs which designers need to consider. The system is cyclic in that it delivers liquid one batch at a time, each followed by an interlude, then another batch, and so on. Consequently, if batches are dispatched directly to their final receipt vessel they would arrive in spurts, which does not suit some processes. If a steady supply is required then the flow can be smoothed by the introduction of a breakpot, which we shall discuss shortly.

The main downside to using RFDs, and maybe the only significant one, is the quantity of air consumed by their operations. Each cycle dispenses a fair sized volume which must be cleansed and monitored before it can be discharged back to the external environment. In practice then the system is ventilated via a dedicated route, the first stage of which is an *off gas* treatment process which we shall discuss later. Once it has been cleansed by that process, air is routed through a ventilation system which we shall examine in Chapter 8.

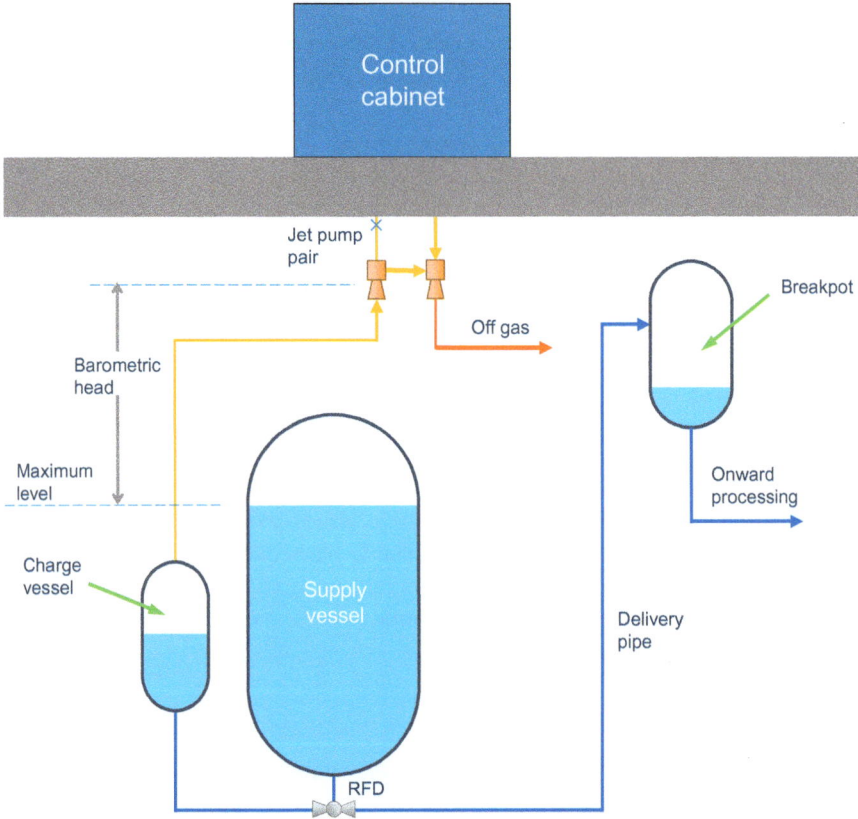

Fig. 6.14 Fluidic pumping system—RFD external.
© Bill Collum.

Propellant

Throughout this section I have referred to air as the driving force for RFD operations; however, I should mention that it is sometimes necessary to employ a different propellant. If volatile liquors are being transferred it may be prudent to use an inert gas to mitigate the risk of explosion. So, for example, solvents, which give off potentially explosive vapors, may be driven through RFDs by nitrogen. Of course it costs more than air, but is a small price to pay for a system that is inherently safe by design. Incidentally, dispersing the cleansed nitrogen via a facility's ventilation stack is not a problem, since it already makes up almost 80% of the Earth's atmosphere.

In addition to RFDs, other air-activated equipment is also employed to move highly active liquors around; among them is the vacuum operated slug lift which is often used in sampling systems. Its operation is based on using a vacuum to suck liquid into a collection pot, from where it falls to its next destination by force of gravity alone. For closed cells, RFD based systems are easily the most common.

6.5.3   Steam ejectors

Apart from air, nitrogen, and so on, there is another force that can be used to transfer liquids: steam. If you are familiar with open fires, maybe at home, you will have noticed that the harder the wind blows outside the more suction there is at an opening to a fireplace. And on a particularly stormy day wind roaring by a chimney stack will suck air out of a flue with some vengeance. Steam ejectors work on the same principle.

Steam being driven across an ejector creates a suction action known as *striking*, which is easily capable of drawing liquor up through a pipe submerged in a supply vessel. The force of steam behind a liquor then drives it along to its next destination. Although steam ejectors are not as versatile as fluidic pumping systems, they are capable of delivering much more force. However, they do have a couple of side effects that need to be kept in mind. For a start, heat within steam raises the temperature of any liquor being transferred. It is not going to boil or come close to it, but there will be a slight rise in temperature which may not suit some processes. Additionally, as steam condenses it dilutes liquors with which it has been in contact, increasing their volume by anything from 5% to maybe 10%, so designers need to consider potential process consequences due to liquor heating and dilution.

6.5.4   Gravity flow systems

So we can use mechanical pumps, fluidic systems and steam ejectors to move liquors along, but there is another method that has been around for very much longer: gravity. Process engineers are always on a quest for an easy life, not for themselves of course but for plant operators who will ultimately inherit their handiwork, so wherever possible they employ Mother Nature to take the strain. Gravity may appear so obvious a means of moving liquors that it barely warrants a mention, but it is worth a few words for two reasons. Firstly, it is a remarkably simple "system," if it can be called that, so if it can be employed why use something more complicated? And its other great benefit, in certain configurations, is guaranteed delivery.

A good example is the introduction of non-active chemical solutions to a processing system. If they were conveyed by a mechanical pump on a "just in time" basis, then if that pump failed the process would need to be halted until repairs were performed and the system was up and running again. However, if the same chemicals were pumped to a head tank capable of holding several days, or even several weeks supply, and it was located above its receipt vessel, then it is simply a matter of opening a valve to guarantee a flow to where it is needed. Failure of the pumping system which delivers chemicals to a head tank would have no impact whatsoever on operations, unless the outage lasted for quite some time. Installing a head tank, along with the space needed to accommodate it, will come at a price but it is an excellent way of laying on a very dependable supply.

6.5.5   Breakpot

There is an essential element of any closed cell processing system, which I mentioned earlier, and is inextricably linked to the use of liquor transfer devices, particularly RFDs and steam ejectors: the breakpot.

We once had a top loading machine at home which after 5 or 6 years of valiant service broke down. The water pump gave up just as it was due to empty the drum, which of course was full to the brim. I don't suppose they ever break down when the drum is almost empty. I removed the wet clothes, and studied the deep water the way you do, as though it might have gone away, but no it was still there. So I set to work. I cut off 2 m of garden hose and placed two empty buckets beside the machine, then secured one end of the hose inside the drum with its open end touching the bottom.

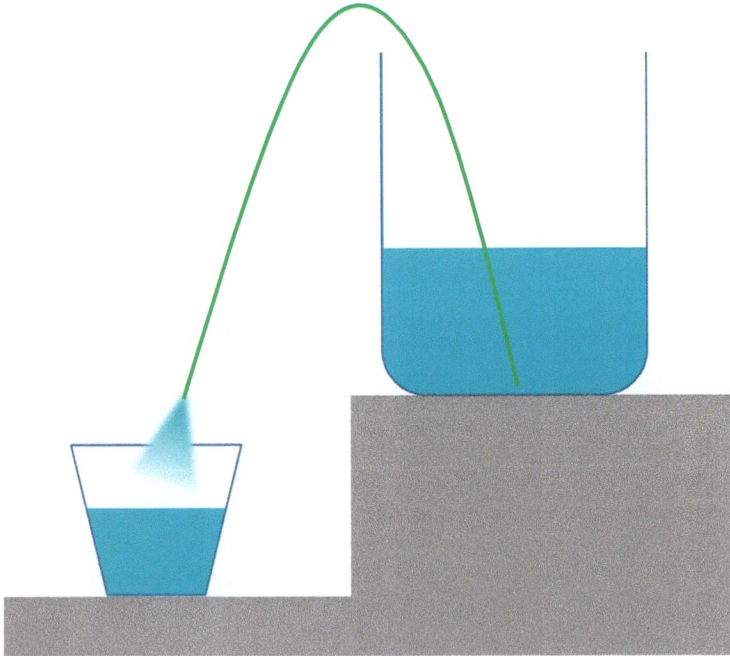

Fig. 6.15 Siphoning.
© Bill Collum.

With preparations complete I knelt down beside the buckets and sucked hard on the end of the hose that was still free. Sure enough water came gushing out and kept on pouring. As long as the loose end of hose was kept below the water line within the drum, water would continue to flow. I emptied several buckets one after the other, pouring each away as the next was filling. As if that wasn't enough, I managed to cap this marvelous achievement by installing a new water pump. Of course I shared my triumph with friends, but only got lots of smart remarks about how my engineering know-how had *finally* been put to some useful purpose.

The trick with the hose is called siphoning (Fig. 6.15) and it can be a bit of a problem for process engineers. Imagine you want to transfer a 500 L batch of liquor from a vessel holding 5000 L, but siphoning kicks in and sucks out the vessel's entire contents—definitely the kind of thing that can spoil your day. In a processing facility liquors are continually being transferred from one vessel to another, often in batches, with the size of batch no doubt important to chemical processes taking place. However, with some transfer systems, particularly fluidic pumps and steam ejectors, there is a possibility that siphoning may occur and thwart what should be a steady batch process. This is where a breakpot comes in. Although I am using the term *breakpot*, they are also well known, particularly in the United States as *disentrainment vessels*.

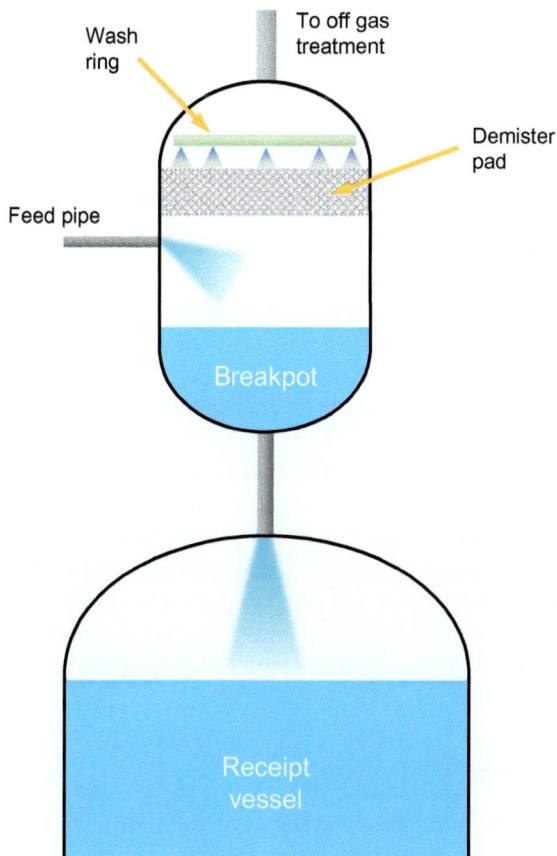

Fig. 6.16 Breakpot. © Bill Collum.

Instead of fluidic pumps and steam ejectors transferring liquid directly from one vessel to another, it passes through a breakpot (Fig. 6.16) above the intended receipt vessel. As liquid enters the pot it cascades downwards, breaking away from its feed line and therefore terminating any possibility of a siphoning action. Then once within a breakpot liquid can continue on its journey by force of gravity alone, so delivery could not be simpler. By the way, it is inadvisable to skimp on diameter of the pipe leaving a breakpot's base. If liquid is cascading into the pot at a fair rate, a narrow pipe could inhibit its exit so much that it may back up beyond its point of entry and could even end up back where it started.

As an added bonus, breakpots can also solve a problem which occurs fairly commonly on the top of processing vessels, where there can often be so many pipes vying for space that it becomes extremely difficult to fit them all in. When this happens the sidewalls of a breakpot can provide additional surface area for making pipework connections and getting feeds into the vessel below.

Although breakpots are primarily designed to cope with the siphoning that can occur when moving liquids with steam or compressed air, the nature of such transfer systems gives rise to another problem. The force behind these delivery systems inevitability leads to liquid being sprayed into a breakpot, particularly as it splutters at the beginning and end of a transfer cycle. The problem is there can be a lot of contaminated particles held within that spray, which if not moderated could overburden a ventilation system. Breakpots must therefore be designed to filter out the bulk of contaminated particles within a liquor spray, leaving the facility's off gas treatment systems to deal with any remaining contamination further downstream.

Close to the top of a breakpot we find a demister pad inserted like a plug across the pot's full diameter. As vapor rises up through the pad, which is essentially a metal filter, it coalesces, forming beads of moisture which have active particles trapped within them. These droplets grow in size until eventually their own weight causes them to fall from the pad, carrying their contamination back to the liquid below. The wider a breakpot's diameter, the better the disengagement of particles will be.

At the very top of a breakpot a wash ring is fitted. It is used to spray the demister pad from time to time and dislodge any lingering contamination. If a filter were not rinsed in this way it would gradually fur up and lose its efficiency; in fact it could even become completely blocked.

If a malfunction occurred whilst a steam ejector was driving liquor into a breakpot, the transfer line would run dry and steam alone would be blown into the pot. Safety systems recognize the drama by sensing the rise in temperature caused by a rush of steam, and can then shut the system down.

6.5.6   Transfer device selection

When I first started to become familiar with various pumping devices employed by the nuclear industry, I assumed each would be appropriate for a particular application, so it was just a matter of selecting the right device for each task and then getting on with the details. In my innocence, I figured that if only I knew as much as my process engineer colleagues, then decisions on equipment selection would become obvious to me. However, it turns out that the more you know, the more difficult it gets. I will take you through a few issues to give a flavor of how the selection debate goes.

Let's say the task is to feed highly radioactive liquors into equipment that needs its flow to be under some pressure in order to operate efficiently. If the option was to go for a gravity supply then it would probably be necessary to position a feed vessel some distance above the equipment so as to generate a significant head of pressure. On the other hand, a mechanical pump could push liquor through equipment at pretty much any pressure you like, which on the face of it makes it look like a clear winner. But no, each liquor delivery system has its inherent pros and cons.

A gravity feed system is definitely the low-cost option; it is incredibly reliable and virtually maintenance free. However, on the downside, getting the necessary pressure from a gravity system, if indeed it is feasible, requires the height I just mentioned. On the other hand, remotely operable mechanical pumps can easily deliver the all-important pressure so crucial to many processes but as we have seen they cannot be guaranteed to last forever. So a debate must ensue on what to do with pumps that have failed.

In this example, we are assuming the components of our remotely operable pump are so radioactive they cannot be handled manually, even after decontamination, so its maintenance must be done remotely. But then providing a remote maintenance facility is not something to be undertaken lightly. It can turn out to be full blown mechanical handling cave, which costs a great deal. Not to mention the problem of getting pumps to and from such a cave, either along a shielded route, or more likely by moving them around within shielded flasks. Then again, just as the specter of a remote maintenance facility begins eroding the appeal of mechanical pumps, it transpires that maybe we shall not need to maintain them after all.

Considering the environment they operate in, remotely operable pumps can be very reliable indeed, so it is not inconceivable that a splendid remote maintenance facility could be built only to find itself with very little to do. The thought occurs that pumps should last for years, for some facilities maybe their entire operational life, so maybe we could skip the maintenance provision altogether and spend the money more wisely?

The strategy associated with that sort of thinking can be dressed up in all sorts of ways, but what it comes down to is a gamble. We can gamble that pumps will be so reliable they hardly ever fail, and in the unlikely event that they do we can just dispose of them. Of course we cannot exactly throw them away; they must be processed and disposed of as radioactive waste. Nevertheless, it is an option that can appear quite attractive, particularly for facilities that will only operate for a few years. In other cases, where pumps will be subjected to a tough operating regimen that lasts for many years, the potential cost of replacing pumps becomes a major consideration. Remotely operable pumps are expensive, very expensive indeed; think of a small city center apartment, so a nuclear facility that operates for several decades could easily get through enough pumps to fund a remote maintenance cave several times over. So now maybe disposing of failed pumps is not such a good idea after all. It's not easy.

We have covered just a little of the discussion that may occur when considering gravity flow versus remotely operable pumps. In reality, of course, these debates will center on the specifics of each case, and may also consider other nuclearized mechanical pumps along with air- and steam-driven devices, plus the consequences that various options would have on a facility's off gas and ventilation systems.

As if that were not enough to contend with, the debate must be repeated for each liquor transfer situation, and indeed each agitation process. Ultimately, a plan must be formulated that joins the kaleidoscope of often conflicting requirements into a coordinated liquor transfer strategy for a whole facility.

It is not possible to say here that this or that option tends to be preferable for particular sets of circumstances, as it really does come down to the specifics of each situation. The crucial factor here is that the debate must be rigorous and certainly not based on familiarity or a historical preference. And it must happen early, after all philosophies adopted here will set the scene for so much of what follows, so it is a serious business. Factors such as the planned operational life of a facility, radiological characteristics of liquors in the system, space required for a maintenance cave, balancing the initial capital cost of providing maintenance facilities against the cost of replacing an unknown number of pumps during plant operations, their disposal costs and on and on. The only thing we can be certain of is a lively and exceptionally complicated debate.

6.6   Services distribution

To round off this section on transfer devices and all our talk of services such as water, steam, and air, we shall turn our attention to how these, and other services, make their way around a building before finally arriving at the point of need. It is a topic with fingers in so many pies that it could legitimately be discussed within several other chapters. However, since the process folks have a direct interest in wet services and can legitimately claim a close affiliation with all others, it is appropriate that we cover the subject here.

Service	Derivatives
Ventilation ducts	C2 supply C3 and C4 extract
Electricity	400 V 220 V 110 V
Instrument cables	Control systems Safety systems Alarms Measurement
Gas	Argon Nitrogen Helium P-10
Compressed air	Breathing Instruments Tools
Water	Drinking Fire fighting Demineralized
Steam	High pressure Low pressure
Reagents	Nitric acid Sodium hydroxide

Fig. 6.17 Services.
© Bill Collum.

Up to this point I have mentioned a few of the usual services, but then this being *nuclear* the full list is a good deal longer. Fig. 6.17 provides a more comprehensive tally, but even here we could add further reagents, gases, water types, and so on. There is, however, enough to be going on with, and more than enough to illustrate the breadth of this particular challenge. I would not want to give the impression that all nuclear facilities demand the full gamut of potential services, but even those with moderately complex operations can draw on close to the full list.

Fig. 6.18 Services distribution.
Shutterstock.com.

Coordinating the distribution of services around a major nuclear plant (Fig. 6.18) is so complex a task that a team of space management specialists will be dedicated exclusively to overseeing it. In the early stages of a design's evolution, whilst things are still at the concept stage, they establish the primary arteries which will be common to all distribution networks. Essentially these services highways tend to follow a facility's main operations corridors and take the most advantageous route across its open spaces which, particularly if there are electric overhead travelling (EOT) cranes in the vicinity, are principally close to the walls.

In terms of horizontal distribution, we find similar routes duplicated on each floor. And whilst some vertical distribution occurs locally within the body of a building, primary routes are often via services risers, some the size of a stairwell, dotted around a building's perimeter. As a minimum we could expect to see a handful of these risers, but for some plants their numbers could easily extend into the high-teens.

Apart from the inherent difficulties of routing literally miles of piping, cabling and the likes around a large facility, not least of which is avoiding clashes between them, the space management team will be exercised by a number of more specific constraints.

6.6.1 Service risers

First up are service riser locations. Knowing that a facility may need several of them around its perimeter, it can be tempting to select a long featureless wall and position maybe four or five risers alongside each other. The difficulty here is that multiple services, water, steam, reagents, ductwork, and so on, emerging from conjoined risers, will all be heading for the same distribution artery as they snake towards their individual destinations. Inevitably this leads to serious congestion, or *bunching* as it is known, and makes it virtually impossible to develop efficient routing proposals. With this in mind some careful planning must go into selecting locations for a facility's primary service risers, and we can assume it would be inadvisable to place more than two in close proximity to each other.

6.6.2   Access

It would be very satisfying if we could route miles of services around a nuclear plant and then forget about them until it came time for decommissioning. The trouble is, there will be all manner of associated valves, sensors, dampers, and other paraphernalia that need to be adjusted from time to time, and in some cases regular inspections are required to ensure all is well. As a result there must be sufficient space, adequate lighting and in many situations, permanently installed walkways, which afford operators access to anything they need to view, maintain or variously tinker with. It may not sound like too much of an issue, but in reality it is one that can devour an awful lot of space.

6.6.3   Services hierarchy

A further constraint relates to a predetermined hierarchy by which different service types are layered within a shared distribution zone (Fig. 6.19). It is the kind of logical

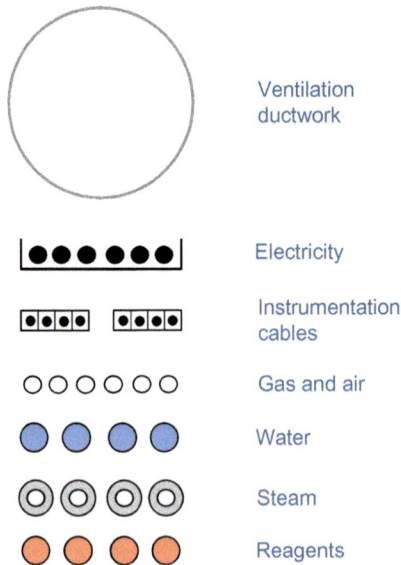

Fig. 6.19 Services hierarchy.
© Bill Collum.

protocol that, given time to think about it we could have a crack at ourselves, but then the specialists have managed to get there before us.

Ventilation ductwork takes the top position, simply because it has a large cross sectional area, sometimes very large indeed, so its main runs are best kept out of the way. Below that we find electricity and instrumentation cable trays, for two reasons. Cables are practically maintenance free, so being at height does not present a significant access problem. And, most important we need to avoid any possibility of cables getting wet, which is why the various types of pipework are always below them. There is a little flexibility on positioning gas services, but a desire to simplify access to valves normally sees them taking the slot above wet services. And finally, the specter of acid leaks dictates that, among wet services, reagents should always occupy the lowest level.

Of course this protocol is not totally prescriptive, but where services run one above the other we can expect to see an arrangement along these lines. One area that demands particular attention to detail, is the relationship between electrical and instrumentation cables. Essentially if they are too close together it can result in instrument signals being disrupted by electromagnetic interference.

6.6.4   Joggle boxes

The final point on services distribution concerns how pipework and cables are routed through hefty concrete walls. As we know radiation travels in straight lines. There may be a little scatter as it strikes objects in its path, but essentially radiation's power is linear. The issue here is that if services entering a highly radioactive area did so via a straight tube-like penetration, then radiation would be able pass through the hole, including its services, and invade the adjoining area.

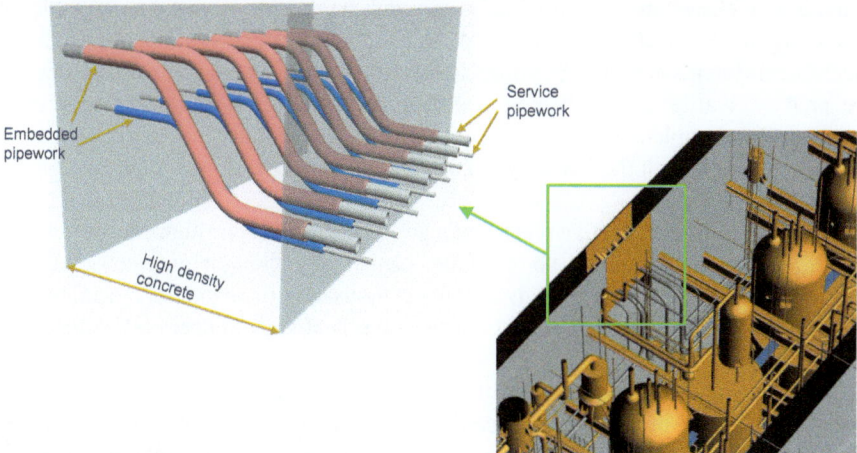

Fig. 6.20 Joggle box.
© Nuclear Decommissioning Authority.

Fortunately all that is needed then to thwart radiation's escape is to route pipework and cables through the jovially named *joggle box* (Fig. 6.20). They sit within wall boxes which are cast into the hefty concrete that bounds high radiation areas, with each box capable of accommodating several penetrations. By bending the embedded pipes and surrounding them with high density concrete, radiation is unable to find a route through, so is confined to its rightful place within an R4 (red) environment.

6.7   Agitation systems

Liquors handled in nuclear facilities exhibit a wide range of characteristics. Some need no mixing at all, some a gentle stir now and again, whilst others demand fairly continuous and intense agitation to keep them from stratifying or practically solidifying. Some are eager to mingle, whilst others need much more coaxing. And whilst some are fiercely abrasive, others can slide along pipework for years without leaving a mark in their wake. Added to these variations we find a wide spectrum of liquor types, from very light and viscous solvents to sludges with a consistency of thick porridge. All in all then, there are variations to suit every conceivable processing challenge, and unsurprisingly the technologies used to keep these fluids stirred share many similarities with those used to pump them around.

6.7.1   Mechanical agitation

As with pumping, the guiding principle when agitating liquors is that there should be no requirement for maintenance within a closed cell and ideally it should be kept to a minimum elsewhere. Unfortunately, air and steam driven systems capable of delivering those objectives cannot satisfy the full spectrum of agitation demands, so mechanical systems must also be utilized. Having said that, air and steam driven devices are pretty versatile, so mechanical agitation in closed cells is fairly uncommon. However, for completeness we shall cover it here. Incidentally mechanical agitation does fit very well with the US canyon concept, where retrieval of equipment is more easily accomplished.

From the safety perspective it is never a good idea to combine high energy systems with radioactive substances, so we must banish any thought of electric motors and mixers whirring active liquids around at thousands of revolutions a minute. If mechanical systems are used then the imperative is always to operate them at slow speeds.

The closed cell concept dictates that when electric motors are driving agitation systems they must be located on a cell's roof, in areas where operations personnel can gain direct access to them. It is possible to arrange agitation motors at ninety degrees to their drive shaft, but ideally a motor should be centered above the vessel it serves, with its drive shaft passing through a cell roof and into the vessel below. To minimize drive shaft lengths it is important that vessels are located as close as possible to the underside of a cell's roof, as the longer a shaft is the more torque, or twisting motion it will experience as it turns. The effects of torque can be overcome by increasing the diameter of a shaft, but there are limits on how far it is sensible to push this. A total of 150 mm might be a good place to stop and consider other options.

It sounds so simple to say that vessels should be located *just* under the roof of a cell, but in practice designers will have to work hard to achieve this objective. For a start, there will almost certainly be a multitude of pipes crowded around a vessel's top, all demanding space. Then there are the constraints of interfaces to other vessels and out-cell equipment, hydraulic issues, barometric head, and so on, no doubt all conspiring to position vessels in very unhelpful locations. So it can take quite some time to coax a system's various components into a compliant location.

If mechanical agitation can be achieved, even the most optimistic engineer would have to concede the possibility that over time drive shafts and agitators may fail, particularly if they are dealing with corrosive or abrasive substances. Precautions can be taken, such as specifying heavier gauges of stainless steel, coating agitators with hardwearing materials like titanium, or manufacturing components and even whole vessels from exotic metals such as zirconium alloy. But even with all these defenses in place, it is still impossible to guarantee that components will not fail. Therefore, provision must be made to remove and replace failed drive shafts and mixing paddles, essentially by following a routine similar to that employed when changing a remotely maintainable pump.

6.7.1.1   Pump-round

If mechanical pumps are being used for liquor transfers then this system comes more or less for free, which is not something you hear every day. All that's needed is to add an extra length of pipe to the pumping network, one which terminates open ended at the bottom of the vessel being stirred. As we can see in Fig. 6.21 most of the pump-round pipework can be shared with other operations, in this case a liquor sampling system, hence the low cost. In my example I have shown valves to the sampling connection closed off, so to keep liquor agitated we just continue running the pump on tick over.

As usual there is a downside to think about, namely that agitation must stop for a while when a pump is being used for another purpose, in this case, for sampling operations. However, this is only a problem for those liquors that need continual stirring, either to mix dissimilar constituents or to keep particles in suspension. The main difficulty here is that some liquors need to be agitated whilst sampling is taking place, otherwise an incorrectly blended product will be dispatched. And unfortunately, for those liquors, the very time you need to maximize agitation is the time that agitation is not possible.

The effectiveness of this system sits about halfway between that of air driven systems and full blown mechanical agitation, so for many liquors it can work very well indeed. You have to admit that when it comes to process engineering and the mechanical systems that support it, there are always plenty of options.

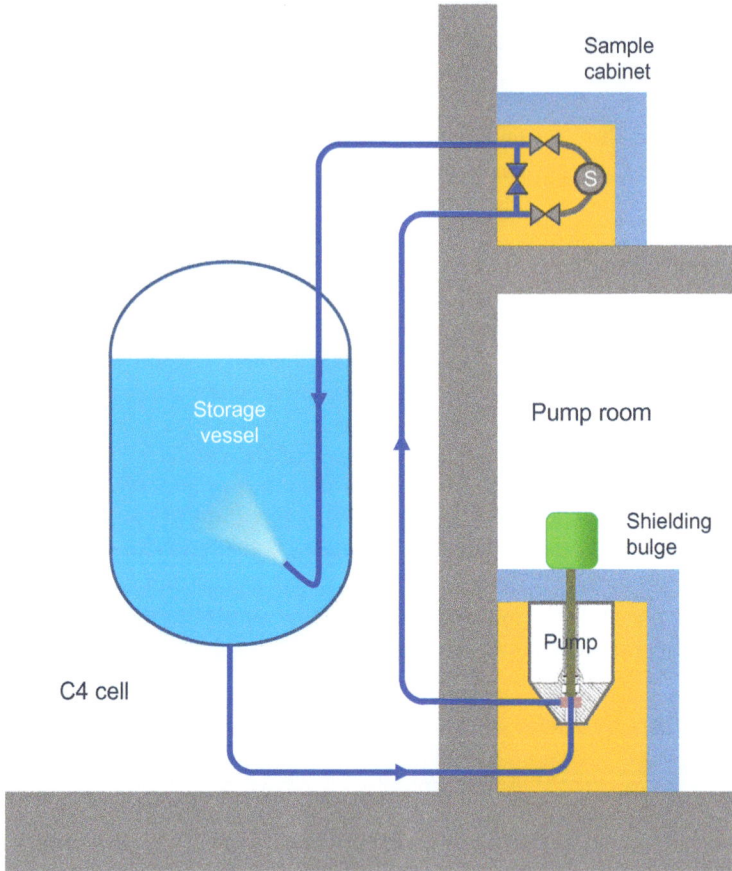

Fig. 6.21 Pump-round arrangement.
© Bill Collum.

6.7.2 Fluidic agitation

6.7.2.1 Pulse jet mixer

Where a vessel is holding liquor that just needs a gentle stir now and again to keep it happy, then it opens up the opportunity of using less energetic non-mechanical agitation systems. A pulse jet mixer, sometimes referred to as *pulsed jet* (Fig. 6.22) is simply half of the equipment used in a fluidic pumping system. It utilizes the same type of compressed air control cabinet, a jet pump pair, charge vessel and a length of pipe. In this case, there is no RFD or breakpot; instead the pipe from a charge vessel is open ended and terminates in a sweep at the bottom of the vessel whose contents are being stirred. Knowing how a fluidic pumping system works, you will no doubt have figured out how this gentle agitation system operates.

Air is drawn up through the charge vessel pulling liquid behind it; then when liquid within the vessel reaches a predetermined level, air suction stops and one of two things

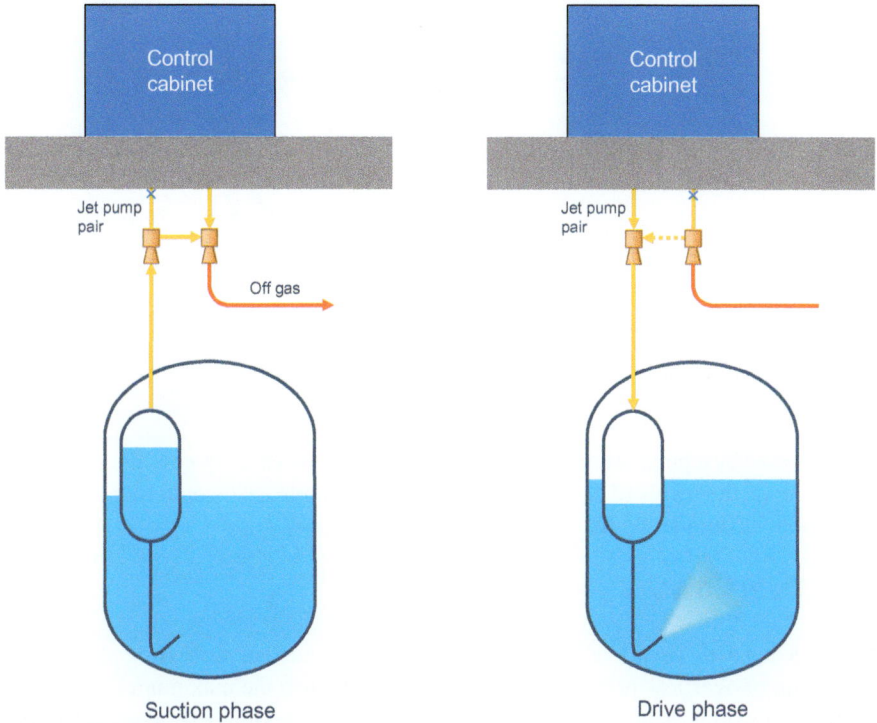

Fig. 6.22 Pulse jet mixer.
© Bill Collum.

happen. Liquid flows back out of the open ended pipe by force of gravity alone, or air is forced into the charge vessel to drive its contents out at a faster rate. Either way, liquid exiting the pipe stirs up surrounding liquor as it re-enters the vessel. To maximize agitation, there can be several pulse jet mixers arrayed around the inside of large diameter vessels. A pulse jet's action may not be very dramatic but it is simple, very effective and best of all maintenance-free.

6.7.3   Air sparge

For this system, all we need is a pipework ring like a doughnut, perforated with small holes (Fig. 6.23) which is placed across the bottom of a vessel and has a length of pipe connecting it to a compressed air cabinet on top of its cell. To stir a vessel's contents, air is driven down the pipe, then as it exits through holes in the sparge ring, it bubbles up and causes agitation. It is so simple that it is almost too good to be true, in fact for some applications it is just that.

The downside to air sparging is that bubbles can carry hydrogen or active particles up through a liquor and into the air space, or *ullage*, at the top of a vessel. From there the ventilation system must be designed to cope with both the volume of air generated by sparging and the newly liberated gas or contamination, which by the way would be

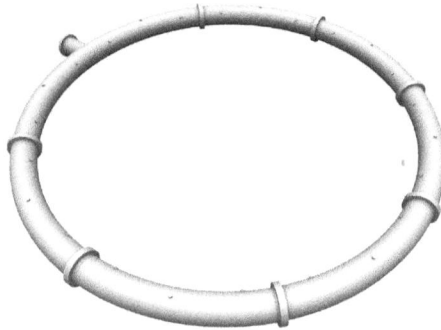

Fig. 6.23 Air sparge ring.
© Bill Collum.

less disturbed by a pulse jet mixer. This is another of those case-by-case situations; for some liquors, the levels of hydrogen or contamination released would rule out use of this simple technique, whilst for others it is perfect.

6.8   Overflows

As with a domestic situation, many tanks and vessels in nuclear processing plants need some form of overflow protection, so we need to establish the maximum permissible liquor level within them and install a drainage line at that point. However, unlike the domestic setting, we cannot allow liquors to flow out at some convenient spot and make a mess until the situation is rectified. Instead, overflows may cascade through other vessels, or even across a cell's floor, on their way to a *drains tank*, which as you would expect is located at the lowest point in the system. Once captured, liquors in the tank can be pumped back into the main system for onward processing.

It is worth mentioning here that drainage lines, and not just those to a drains tank, must be laid to a particular fall, or incline. The preferred minimum fall for lines carrying process liquors is normally taken to be around 1 in 20, although it depends what is flowing through them, so in practice there can be a fair degree of variation. The reason for an incline is that drainage relies on gravity for its flow and if a fall is too shallow, particles or *solids* within process streams may settle out and get left behind. Even though lines may be flushed from time to time the deposition of solids could eventually result in a blockage, so it is an issue worth keeping an eye on.

Although drains tanks are normally quite a small affair, say a cubic meter or two in volume, they can sometimes be quite a substantial vessel in their own right. It depends on the capacity of other vessels in a system and how much day-to-day duty the tank is expected to perform. Locating a drains tank is one of those seemingly inconsequential matters that in reality invokes quite a significant challenge, one that stems from what would happen if a major vessel above the tank were to rupture.

The safety case for a nuclear processing plant will dictate an assumption that any vessel could fail completely and spill its entire contents into the enclosing shielded cell; understandably, designers need to show what they are going to do about it. There is no point arguing that a concrete cell structure is so robust that it will contain the spilled liquid, maybe tens of thousands of liters, until it can be pumped out, even if the floor construction is 2 or 3 m thick, which is not that uncommon. No amount of reinforcing bar can guarantee that concrete will not experience hairline cracks, especially following an earthquake. So cells are normally lined with stainless steel sheeting which covers their entire floor area and extends partway up the walls.

There is a simple rule of thumb for how far up cell walls the liner must go. Simply take the largest vessel in the system, calculate how deep liquor would be if its contents plummeted into the cell and then add 10%. The additional percentage is to accommodate a surge that would result from the initial deluge. Having said that, it would seem reasonable to question why a cell's liner need only accommodate the contents of a single vessel rather than the entire system. Where it is important to maintain their integrity, these vessels will be seismically designed, maybe to withstand the forces of a 1 in 10,000 year earthquake. So it is considered quite improbable that one would fail and even less credible that several would fail together. Even the ultraconservative nuclear industry concedes that such an occurrence is implausible, so that the largest vessel will therefore satisfy the worst (or in this case, deepest) case.

What does all this have to do with a drains tank? Picture the scene: if a large vessel were to rupture and dump its entire contents into a cell, the newly released liquor would immediately cascade around a drains tank which will be sitting at the cell's lowest point. This sounds fine until you think about trying to hold a ball under water in a swimming pool. Most of us have tried it and found that once you let the ball go it shoots out of the water with some urgency. A drains tank suddenly submerged in liquid will want to do exactly the same thing, especially if it is empty when a flood drowns it. It might be fun in a swimming pool, but not here. Designers must mitigate the danger by either building a robust structure to hold the tank down, or ensuring spilled liquors are routed directly into a drains tank to quickly neutralize its buoyancy.

6.9 Volume reduction

6.9.1 Evaporation

If you were to leave a shallow dish of muddied water outside on a sunny day, much of it would steadily evaporate leaving a scum behind that, if left long enough would dry out completely. If we were able to capture the resulting vapor and cool it down, it would turn back into clean water. Process engineers use pretty much the same technique to draw cleaned-up liquid from contaminated solutions. It has the dual benefits of producing a stream which is more readily disposable and reducing volume of the originating liquor, which in turn eases the burden on processing equipment further downstream.

There are quite a few variations on equipment used and how it is configured, but Fig. 6.24 captures basics of the most common themes. With this configuration liquor is heated by passing high pressure steam through a coil submerged within it, although routing fluids through a heat exchanger is equally popular.

The evaporation process is not quite as simple as Mother Nature's use of the Sun, but is nevertheless straightforward enough. As vapor rises up from its heated parent liquor it passes through a demister pad, which performs a role similar to those found in

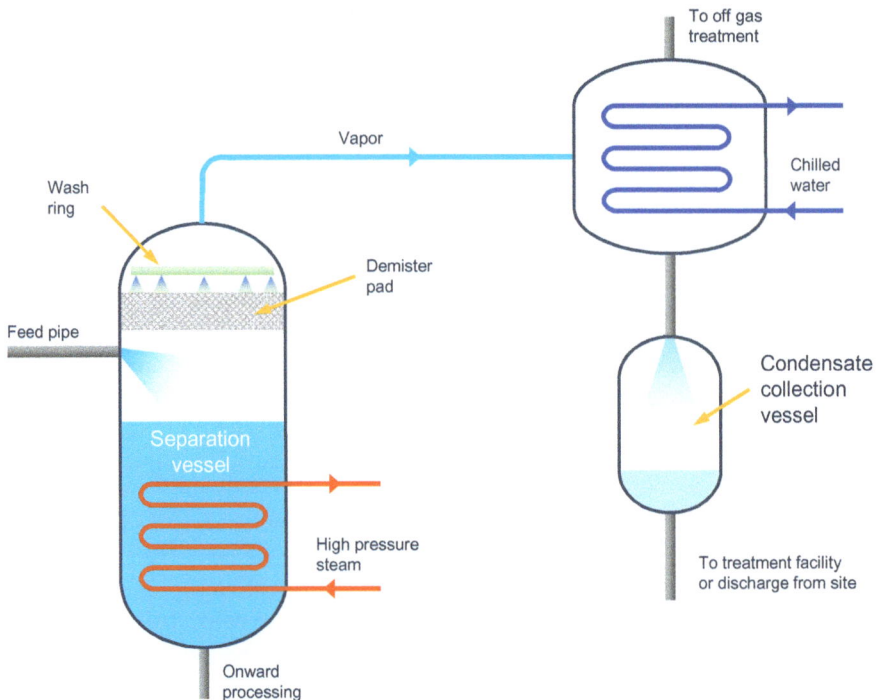

Fig. 6.24 Evaporation process.
© Bill Collum.

the breakpots we discussed earlier. Any contaminants entrained within the mist will gather in water droplets which coalesce in the pad and fall back into the liquid below. The filtered vapor flows into a condenser, which squeezes its cleansed moisture back out and routes it, by gravity, to a condensate collection vessel. Meanwhile the gas, or *off gas* to be precise, is directed to a separate clean-up process which we shall discuss later.

Having been disengaged from its parent liquor the condensate may be routed to a facility which treats low active effluents. However, in many cases it can achieve a level of cleanliness that permits discharge from a site without need of further processing. Unsurprisingly, with an ability to clean up liquors so effectively evaporation technology has many other applications, most notably in the desalination plants that transform saline rich seawater into drinking water.

If you had plenty of thinking time, you would, after a few cups of coffee, conclude there are two problems inherent in any evaporation process. Heating sizable quantities of liquid is going to take a lot of expensive energy, and some chemical solutions will not take too well to the prospect of a significant rise in their temperature. These nuisances can both be assuaged by holding the ullage space within a separation vessel under a vacuum. With the chamber's pressure reduced its contents will boil at maybe half the temperature required under normal atmospheric conditions, so it keeps heat sensitive chemicals happy and reduces heating bills, both of which are always welcome.

6.10   Solids removal

Given the task of processing a murky liquor, say a sludge or a slurry, it makes sense to remove any solid particles first, otherwise they will only clog up equipment which is used later in the cleanup cycle.

Broadly speaking there are three ways to isolate suspended particles, or entrained solids as they are known in the United States: gravity, spinning, or passing them through a filter. The choice of method depends on the size of particles being removed. The first stage, the one used to remove a liquor's heaviest particles, is so simple that it barely qualifies as a "process."

6.10.1   Settling

The settling process provides a simple method of dividing an existing feed into two separate streams, one thickened into a sludge and the other much more free flowing than the original liquor. Ideally it would be helpful here if I could provide precise definitions on what constitutes a slurry and a sludge. However, there are so many variables in the properties of particles suspended within them—size, weight, buoyancy, and so on, not to mention characteristics of parent liquors themselves—that such clear definitions are impossible. For our purposes, we can make the not unreasonable assumption that a weak slurry will have around 5% of solids suspended within it, and a sludge 20% or more.

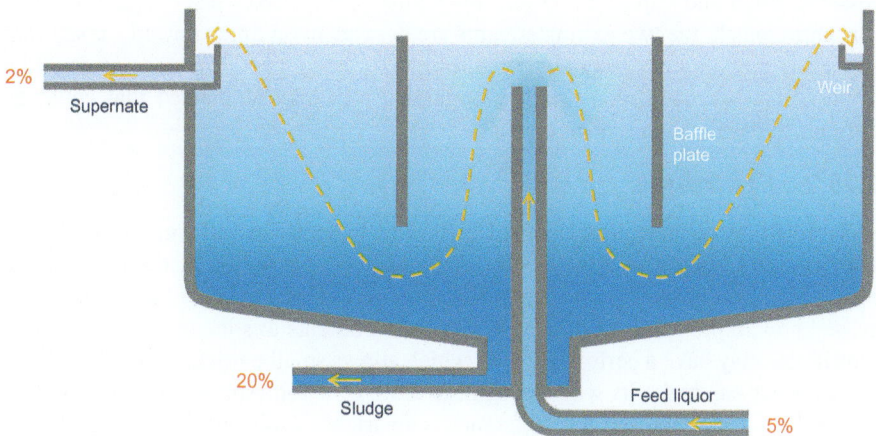

Fig. 6.25 Settling tank—continuous flow.
© Bill Collum.

Fig. 6.25 shows how two streams are created, in this case following a continuous flow process. The feed liquor, with its particle concentration of 5%, is fed slowly into the settling tank and cascades downwards inside a cylindrical baffle plate. Entering in this way minimizes any disturbance to settling that has already taken place. The primary constraint on particle concentrations within both exiting streams is the length of time they are given to settle, something which is easily regulated by adjusting flow rates on the inlet pipe. Typically, though, and bearing in mind the variations

I mentioned a moment ago, we can expect durations to be such that sludge, or *concentrate*, will contain at least 20% solids, and the more free flowing liquor, or *supernate* in the order of 2%.

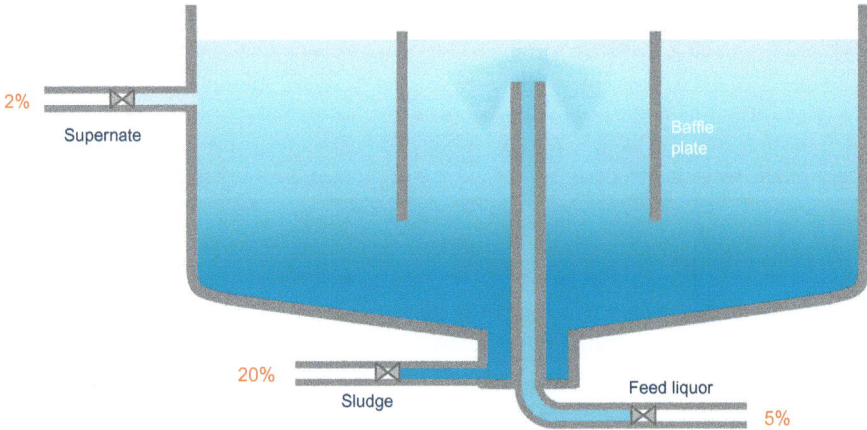

Fig. 6.26 Settling tank—batch process.
© Bill Collum.

A similar result can be achieved by adopting a batch process (Fig. 6.26) where a vessel is filled and valves are closed off whilst settling takes place. Again particle concentrations in the two exiting streams are dependent on time allowed for settling. However, as with any process where a solution is being concentrated, care must be taken not to overdo it, otherwise a coagulated mix may be created that could be difficult to mobilize and move on.

6.10.2 Centrifuge

So settling may be about as simple as it gets when it comes to removing solids from a solution, but close on its heels comes the centrifuge. Operation of the type we are considering here is quite akin to the spin drying function of a washing machine, where, as we know, rapid spinning of a bowl forces its contents to stick to the sides. Some centrifuges may have a perforated bowl which allows small particles to push through, just as water can make its way through the drum of a domestic washing machine.

For the nuclear industry, the preference is invariably a solid bowl which can be put to use in 2 ways. With both systems the bowl comes to a stop on completion of its spin cycle and any remaining liquid, or *centrate* is drained away. With that done, the *cake* which has accumulated on a drum can be jetted off, collected and disposed of. Assuming a cake's radioactivity is categorized as ILW, which is often the case, it can be made ready for long-term storage, for example, by encapsulation in concrete.

Where a centrifuge is dealing with a relatively modest throughput, then it makes more sense to use small bowls, say 150 mm in diameter, and simplify the process by disposing of the bowls themselves. With this approach several bowls, with their centrifugal cake aboard, can be loaded into a stainless steel drum and encapsulated together in a concrete mix.

6.10.3   Cross-flow filtration

Cross-flow filtration is the generic name given to systems that pass pressurized fluids alongside a filter, and where the pressure is so great that it forces fluid to taker a sideways detour through the filter. In the nuclear industry the cross-flow technology of choice tends to be ultrafiltration; it comes in various guises and in other spheres has diverse uses that range from fruit juice production to blood purification.

6.10.3.1   Ultrafiltration

The job of an ultra-filter is to take fine particles out of a liquor, send the cleaned up liquid in one direction, and concentrate the remainder, which has not been filtered, into a slurry. It can even thicken a light slurry by squeezing some clear liquid out of it. Once a *dewatered* slurry reaches its desired consistency, it can be drawn off and dispatched to the next process.

There are just three main components to an ultrafiltration system, a supply vessel, a pump and the ultra-filter itself, which is really the business end of the whole process. Stripped down to its basic form an ultra-filter has just two elements, an outer cylindrical casing and a hollow metal mesh filter which runs from top to bottom along its axis; so if you were to cut across an ultra-filter it would look like a doughnut, hollow in the middle.

The process is initiated by pumping slurry into the space between an ultra-filter and the casing which surrounds it. The key to wringing clear liquid from a circulating slurry lies in the fact that it is pumped around the system at high pressure, and that crucially there is a difference in pressure across the filter. This variation is typically around 6 bar, or 6 times atmospheric pressure, so you can imagine what happens within the chamber. Faced with so much pressure slurry will look for any way out, so instead of just passing alongside the filter it pushes through it. Particles within the liquor get trapped in the ultra-filter's fine mesh, leaving clear liquid to complete the journey to the filter's core: the middle of the doughnut. Once liquid makes its way to this inner space it can rise unimpeded and be drawn off at the top; sometimes the liquid is sent onto another ultra-filter to cleanse it even more.

You may hear process engineers using terms like *permeate* and *filtrate* which sound mildly exotic, but are actually a reference to the kind of liquids we are discussing here. Permeate gets its name because it has "permeated" through a filter, whereas liquor that has passed alongside a filter and therefore still "retains" solids is known as *retentate*.

6.11   Ion exchange (IX)

If we were to analyze a few liters of highly radioactive liquid for a while, we would invariably find that the bulk of it could be radiologically very clean indeed. The problem is that highly radioactive elements within the liquid, which only account for a small percentage of its total volume, elevate the entire batch to the high active category. As we would expect, industry regulators and straightforward environmental considerations always dictate a need to minimize high level waste (HLW) volumes. Added to which, treating and storing it is much more expensive than dealing with less active wastes.

If only those troublesome elements could be segregated from their parent liquor, it would open up an opportunity to treat the small amount of highly active waste via one route and the remainder via intermediate level waste (ILW) and even low level waste (LLW) routes. This is where IX comes in; it cleans up liquors by removing chemical elements such as cesium, technetium, cobalt, strontium, and americium. It is so effective that the remaining liquid often complies with site discharge authorizations so, after analysis, can be released to local waterways or the sea.

IX media, which capture the various elements, divide broadly into two different types: organic material such as polymer-based beads, and inorganic material such as clinoptilolite, usually referred to as clino, which is a naturally occurring mineral. We shall assume that our typical column is filled with beads, since they tend to be the medium of choice, and for our purposes here will also assume the column is removing cesium.

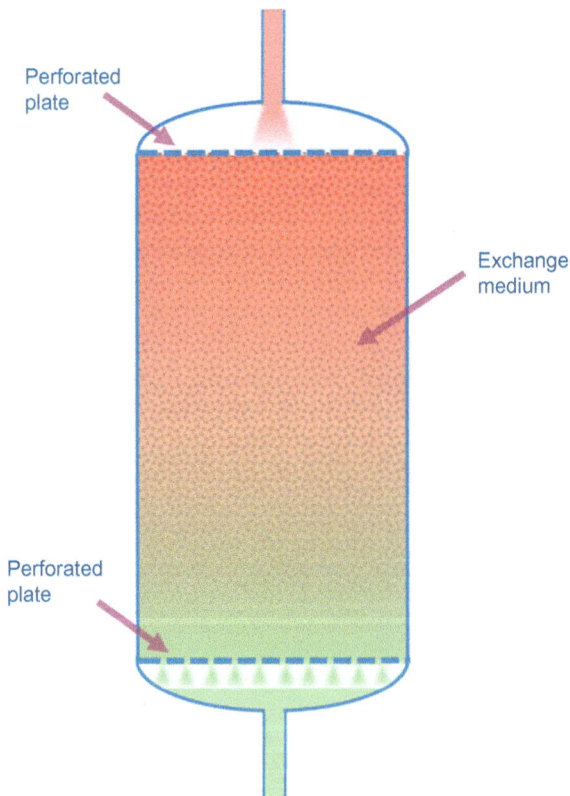

Fig. 6.27 Ion exchange column.
© Bill Collum.

The configuration within IX columns can vary a little, but their basic design sees perforated plates fixed to the top and bottom, with an exchange medium packed in between (Fig. 6.27). Active liquor enters the top of a column, passes through its first plate, carries on through the beads, and leaves via another perforated plate at the bottom. The exact composition of a medium depends on the element being removed, and there are even media capable of removing more than one element at a time; ordinarily, though, there is a different column for the separation of each element.

The easiest way to visualize what goes on inside an IX column is to imagine the beads being covered with tiny indentations. As liquor passes through a column the target element, in our case cesium, becomes trapped in the indentations, so that by the time it exits again at the bottom there are only minute traces remaining within it. The process continues with liquid moving from one IX column to another, each dealing with a different element, until finally all of those being targeted have been removed. With its highly radioactive elements isolated, the remaining liquor is directed to storage tanks and, maybe after further processing, monitored and released to local waterways or the sea.

To get the greatest level of efficiency from a medium, it is essential to maximize surface area where the actual IX process takes place. So the beads used are very small, say 1–3 mm in diameter. If larger beads were used it would increase voidage between them and reduce the efficiency of a column. For example, if you visualize a column filled with beads the size of tennis balls, it is easy to picture lots of unproductive air space and a substantial reduction in the total surface area of the beads which are doing all the work, so the smaller the beads the better.

We return to the IX column, now laden with cesium. Polymer beads are only good for so many campaigns before they become completely saturated and unable to absorb any more of their target element. To give a rough idea of timescales, fresh beads may last anything from a few hours to a week but certainly not for months or years. Unlike clino, though, beads can be rejuvenated. Firstly they are flushed with water, in a process known as *elution*. This does not dislodge the captured cesium but rinses out any remaining active liquor, which can then be routed to the next IX column. Having been rinsed, the cleansed beads are flushed with nitric acid. It displaces the cesium ion and turns the acid into cesium nitrate, which is then drawn off and treated as a high active waste stream. With that done the beads are ready to be used again.

Over time however, the beads lose their rejuvenation abilities so must be exchanged for a fresh batch. This is achieved by either replacing an entire cartridge of beads, or by fluidizing them with water and then pumping the beads away for high active processing. Substituting pristine beads into the newly emptied column enables the process cycle to start all over again. So all in all it is a deceptively simple looking solution to a rather complex chemical engineering challenge.

If we reflect on the IX process for a while and consider the complex chemistry going on inside its columns, it does have the hi-tech feel of something that could only possibly occur in carefully controlled conditions and with the help of a very substantial capital investment. It turns out, though, that similar processes are employed by the water softeners which are built into some of our domestic appliances, it is just that nuclear versions are significantly more elaborate.

6.12   Off gas treatment

As we have seen, radioactive liquors are routed through all manner of equipment, from
fluidic pumps, to breakpots, evaporators, and so on. Along the way many of these
processes generate vapors and gases which must be drawn out of the system and cleaned
up before leaving a building. A facility's ventilation system plays an important role in
filtering out the microscopic particles carried along in these streams but, as with all
gases, those we are considering here would pass through even the most sophisticated
air filter, which means quite a different approach is called for. Considering the job it
does, the process used is remarkably simple; it comes in various guises, with a packed
column (Fig. 6.28) being one of the most popular configurations.

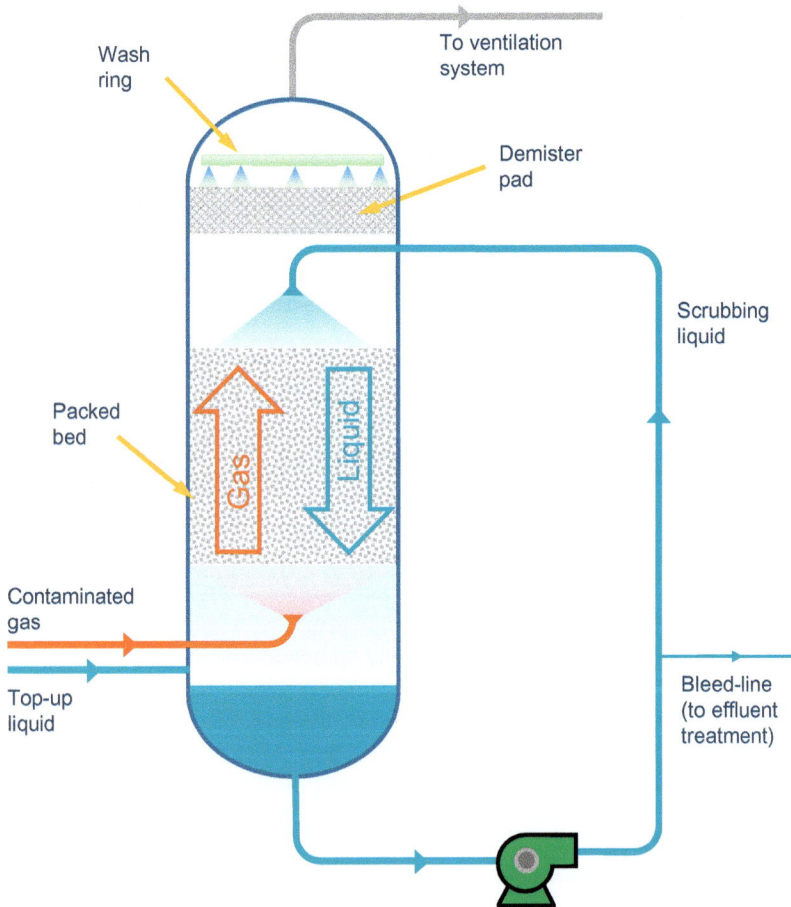

Fig. 6.28 Packed column.
© Bill Collum.

6.12.1   Packed column

Prior to treatment, the chances are that gases from several vessels will be directed to a single pipeline, which is routed to the packed column and enters just above its liquor collection sump. Once inside the gas is drawn upwards, but before it can exit again must run the gauntlet of a packed bed which is the heart of the whole process. The easiest way to visualize a bed is to think of a tightly packed pile of scrap metal, all in small pieces. Packing components can be formed from ceramic materials, but for the nuclear industry they are invariably stainless steel. The way in which a bed is formed can vary quite a bit, but all are porous with some having a regular or *structured* pattern, and others a more *random* assortment of rings and other shapes.

As gas heads upwards through the packing it is met by a scrubbing liquid, such as caustic soda, which is tumbling down in the opposite direction. The job of a packed bed is to make the journey of falling scrubber liquid and rising gas as tortuous as can be, so that the two clash and mingle as much as possible. It is this washing process that scrubs out noxious gases and removes species such as iodine, radon and tritium, dragging them into the sump below and leaving a substantially cleansed gas to arrive triumphant at the top; from here it is directed to a facility's ventilation system, which will deal with any microscopic particles that remain.

At the beginning of this chapter, I suggested that nuclear processing facilities, particularly those based on the closed cell concept, can often appear to be almost dormant. Since then we have looked behind their heavy concrete shielding and discovered the reality is far removed from that initial perception. It turns out that nuclear process engineers are quite adept at hiding their achievements away, forcing the rest of us to search hard if we are to uncover what it is they do all day.

Mechanical engineering

7

When I think of what it is that differentiates mechanical engineers in the nuclear industry from those in other spheres, I am reminded of the words so beloved by presidents, prime ministers and the like, not to mention screenwriters: *"failure is not an option."* But as you might expect, failure of mechanical equipment is indeed an option, or at least a possibility, one that gets an awful lot of attention, particularly in a radioactive environment where you cannot walk up to apparatus and tinker with it. Understandably then, there is considerable pressure to ensure failure does not happen too often. However, it is not uncommon for mechanical and electromechanical equipment to fail—just think of our cars and washing machines. So this particular challenge is very real indeed, especially when we bear in mind that these engineers cannot depend on the reliability gained from multiple prototypes and mass production.

Fig. 7.1 Drum handling grab.
© Nuclear Decommissioning Authority.

In the nuclear industry mechanical engineers spend a fair portion of their time designing equipment such as manipulators, bogies, mechanical grabs (Fig. 7.1) and so on, which we tend to see in facilities that handle solid radioactive materials. Even though this is what we might call *standard* mechanical equipment, it is not unusual to find that an existing design cannot simply be copied from one project to another. Variations in performance requirements can dictate that some redesign, often quite

Nuclear Facilities. http://dx.doi.org/10.1016/B978-0-08-101938-2.00007-6

extensive, is required. However, if something similar has been done before, they may use that as a starting point.

Fig. 7.2 Remotely operated mechanical equipment.
© Nuclear Decommissioning Authority.

When not occupied with this, mechanical engineers are often out there forging a path in developing equipment which is needed to meet some new engineering challenge—maybe designing the type of equipment seen in Fig. 7.2 which must be both operated and maintained remotely. The good news, for our discussion here, is that once we have an understanding of what is involved in designing a range of standard mechanical handling equipment, it gives us a clear insight into what it takes to conjure up an item which is truly one of a kind. So this is where we shall focus our attention. First, however, we need a little context to where much of this kit resides.

7.1 Mechanical handling caves

At the heart of the type of facility we are considering here, there is generally a shielded mechanical handling cave such as that shown in Fig. 7.3. As with process cells, discussed in the previous chapter, the majority of caves have their floors and at least part of their walls lined with stainless steel. Although in the case of caves it is less to do with containment and more about decontamination and an ability to withstand impact or abrasion. In addition, in-cave cranes and just about any item of equipment to be found within a cave will also be manufactured from stainless steel.

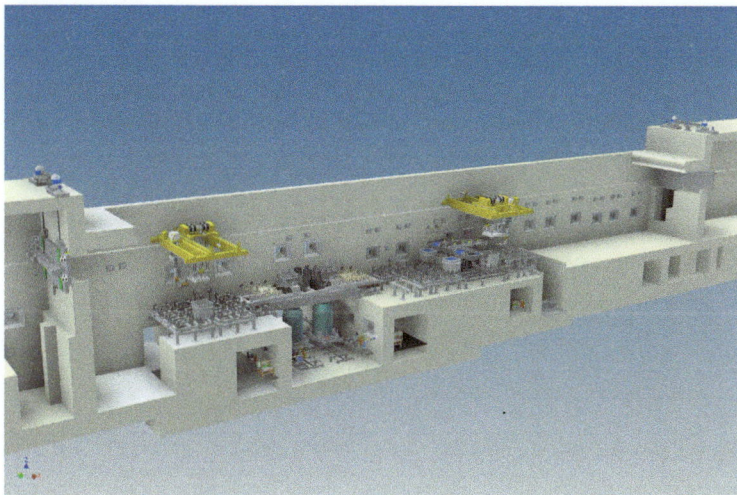

Fig. 7.3 Mechanical handling cave.
© Nuclear Decommissioning Authority.

The hostile radiological environment within operating caves makes it impossible for personnel to enter, so invariably there is an adjoining area where retrieved equipment can be safely handled and maintained. In Chapter 2 we looked at the difference between radiation and contamination so we know that in caves which are free, or virtually free of contamination, the issue of decontaminating their equipment does not arise. In those situations, in-cave equipment, such as cranes, can move directly from a cave to its maintenance bay; once the shield door is closed, maintenance personnel can enter the area. However, this arrangement is not suitable for many caves, since they not only have radiation to contend with but also contamination. In these cases, anything leaving a cave passes through a decontamination bay before moving to an area where it can be maintained.

There are several permutations for ways in which caves and their adjoining decontamination and maintenance bays can be configured, but we shall save that for later. First we need to set the scene by looking at the kind of equipment found in and around most caves. Let's start on the perimeter and then make our way inside.

7.2 Shielding windows

7.2.1 Composition

You will recall from our discussions in Chapter 2 that when it comes to shielding the effectiveness of a material hinges primarily on its density, so one thickness of lead is equal to approximately 2.5 of steel, or 6 of concrete. Exactly the same is true of glass, a material formed mainly from a type of silica sand, making it more akin to concrete than its appearance would suggest.

Fig. 7.4 Shielding window.
© Nuclear Decommissioning Authority.

Shielding windows (Fig. 7.4) are manufactured from borosilicate glass with anything up to 70% of lead oxide blended within it. The lead further increases its density and so improves shielding performance, to the point that some formulations provide even better shielding than concrete. It is good to know these things because if you do find yourself standing at such a window, embedded in a hefty concrete wall, the question of shielding is bound to cross your mind.

7.2.2   Construction

Fig. 7.5 shows a fairly typical shielding window, in this particular case installed within a wall box which is embedded in a concrete wall. Similar arrangements are used for walls constructed from steel and lead, although in those cases windows need to be deeper than the walls in order to achieve an equivalent shielding performance. As we know radiation directs most of its energy in straight lines, so where a window is embedded in concrete its liner is stepped in order to block any direct *shine path* from within a cave. Whereas when windows are installed in steel or lead, the wall profile itself is stepped so a wall box is not required.

On the face of it, we would expect to see a fairly tight fit between the glass and its steel liner, but if we look closely there is a gap, which can be anything up to 25 mm wide, packed with lead wool. It has two primary purposes, one of which is combating straightforward thermal expansion and contraction; more importantly, the gap is there to absorb movement during a seismic event. If the whole assembly was a tight fit, then shielding windows would almost certainly shatter during the buffeting of an earthquake, instantly breaching a cave's containment. So the gap plays a crucial role in cushioning them from their unyielding surroundings.

Earlier versions of shielding windows comprise a thick pane of glass on both sides, with their interspace filled by a liquid such as zinc bromide, or alternatively a silicone or mineral oil. They perform well, but can be prone to leaking and generally need more

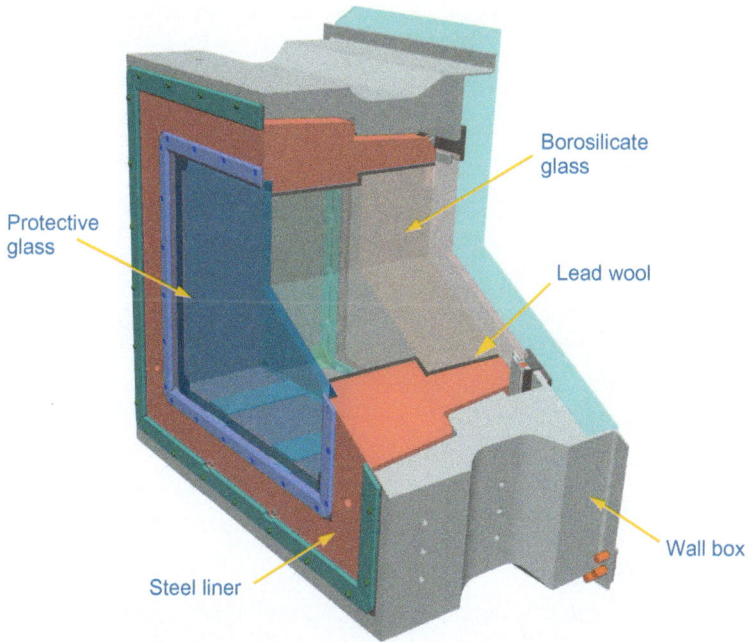

Fig. 7.5 Shielding window—out-cave view.
© Nuclear Decommissioning Authority.

care and maintenance than is ideal. Modern shielding windows are a much more sophisticated affair. They are built up by several panes of thick glass, generally between two and six in number, with the spaces between occupied by air or an inert gas such as nitrogen. These units are very well sealed so ordinarily condensation is not an issue. However, if it does start to form the interspaces can be purged by attaching a gas line to a connector embedded in the window's outer frame.

7.2.3 Refraction

Apart from its shielding properties, the beauty of using several panes is an inherent ability to adjust how much of a cave can be viewed through a window. With one or two panes of thin glass, such as in the windows of our homes, the view we get through them is pretty much the same as it would be if a window was removed. However, just as with water (Fig. 7.6) light passing through thick glass is subject to noticeable refraction, effectively bending the light and therefore changing the scene being viewed. The refractive index of a pane of glass depends on its constituents, so engineers can adjust the path of light through a shielding window by modifying the thickness of its various panes, and to some extent by adjusting their formulation. As you would expect there are computer programs that can perform the necessary calculations instantly, so it is possible to make adjustments to a window's design and simulate an on-screen view of exactly what would be seen in the real world.

Fig. 7.6 Refraction.
Science Photo Library.

7.2.4 Containment

To be doubly sure that they maintain containment at all times, a pane known as *alpha glass* is sealed against a shielding window from the inside of a cave. And where windows may be at risk of impact from materials being handled or from projectile hazards, due to say cutting operations, then a tough polycarbonate screen (Fig. 7.7) or wire mesh is fitted inside a cave to protect the glass.

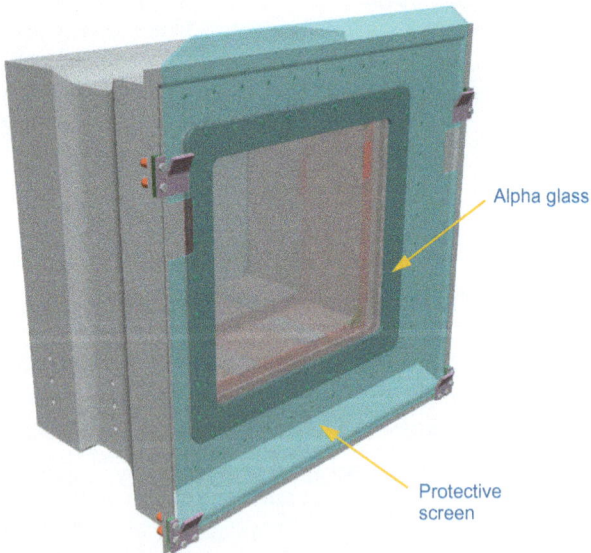

Alpha glass

Protective
screen

Fig. 7.7 Shielding window—in-cave view.
© Nuclear Decommissioning Authority.

However, when plastic is exposed to high radiation it discolors and becomes brittle, actually a speeded-up version of what happens when it is exposed to direct sunlight; so apart from obscuring the view, a plastic's ability to withstand impact is also diminished. For these reasons it is possible to change protective screens with remotely operated equipment.

If in-cave operations are destined to generate significant dust, then it may accumulate on the alpha glass and impair an operator's view. To counter this problem, windows can be fitted with a rinsing system that cascades water down the glass and carries away any offending particles. To minimize liquid effluent arisings, water in such systems is generally recirculated and topped up from time to time.

7.2.5   Tint and degradation

So, mechanical engineers can find ways to protect shielding windows and keep them clean but there is one characteristic of these windows they can do very little about. If you were to look through a regular window at home or in an office block and so on, more or less 100% of the light outside would pass through the glass and give you an exact appreciation of the scene outside. However, with shielding windows chemical constituents of the glass give panes a yellowish tint, to the extent that only around 50% of light within a cave is able to pass through them. In addition optical degradation, which is accelerated by radiation, further erodes visibility by around another 1% every year.

What all this tells us is that for facilities which are destined to operate for several decades, it must be possible to replace a shielding window as it nears the point of becoming too obscured to perform its function. And in any event arrangements must be in place to exchange windows that may be accidently damaged.

7.2.6   Change-out

When you consider that a large family car weighs around 2 tonnes, and compare that to a fully assembled shielding window that can weigh up to five times more, it becomes clear that changing them is a significant task it its own right. In fact there is a lot to do before the actual change-out itself can even begin. Even though windows tend to be manufactured in standard sizes of anything up to 1.6 m wide and 1.5 m deep, their exact configuration and glass formulations can vary so much that each one is effectively a bespoke design for its, particular situation. Unsurprisingly therefore, the process of procurement, manufacture and delivery of a replacement is quite a protracted one, to the extent that a timescale of one year would not be unusual.

By the way, whilst we are on the subject of procurement, it is worth noting that an entry-level shielding window can cost the equivalent of a couple of family saloons, with larger more complex ones coming in at the price of a small city center apartment. That kind of price tag is one of the reasons it is unlikely a replica will be sitting in the stores on the off chance that it might be needed one day. Fortunately failed windows can safely remain in situ until a replacement arrives, so the imperative to change is mainly an operational one.

Before a change-out operation can commence, radioactive materials within a cave are moved to a location where their shine path does not have direct *sight* of the window opening. However if moving materials is not practical, then in-cave equipment can be used to position shielding blocks in front of the window. Contamination, for once, is easier to deal with, as it is halted by the alpha glass which always remains in place. So once the shielding issue has been satisfactorily addressed it is safe to begin the window extraction process.

The outer pane of protective glass is removed, before a trolley is wheeled into place and locked in position by attaching it to anchor points which are already embedded in the cave wall. With that done the trolley's flatbed is raised until it is flush with the bottom of the window box. The window's securing bolts are removed and a withdrawal attachment is bolted to pre-drilled holes in the frame. With preparations complete the trolley's hydraulic ram is activated, slowly withdrawing the window until it is fully extracted and transferred to the trolley.

In situations where windows are located above operator viewing platforms, it is not always feasible to dismantle the platform or remove operational equipment which may be located beneath it. If this is the case, then the platform and surrounding area will have been designed to accommodate a temporary structure, one that provides a similar function to the trolley which is normally used. The main difference with this arrangement is that a crane or lifting beam must be used to lift the extracted window and place it on a waiting motorized trolley.

In both cases the redundant window is moved to an area where it can be packaged and exported from the facility. With a shielding window removed, the priority will be to reverse the procedure as quickly as possible by installing a new window and returning the cave to its original radiological integrity.

7.3 Manipulators

We know from Chapter 2 that some nuclear materials can be physically handled by operators. Such substances tend to be either low active beta-gamma products, or alpha emitting materials which must be contained within a glovebox and only handled through glove ports. So with this type of material some form of what is termed "manual intervention" can be achieved. However, the majority of nuclear materials emit enough radiation to need hefty shielding around them, so any form of direct manual handling is out of the question. This is where manipulators come in; they are an invaluable nuclear industry workhorse, which effectively place the hands of an operator inside a radioactive environment.

They come in two different guises, a common type being the *master-slave manipulator* (MSM), also known as a through-wall manipulator. Although these are largely mechanical devices, MSMs also come in combination electromechanical versions.

Other types are known collectively as *manipulators*, or simply referred to as robot arms, and are driven by either electrical or hydraulic systems. The main difference between the two generic types is that MSMs are effectively a single item of equipment with one end outside a shielded cave and the other operating on the inside. Manipulators, on the other hand, are much more independent from their controls, so need not be anchored to a particular spot and therefore offer more flexibility in where they can be located. Let's look at MSMs first.

7.3.1 Master-slave manipulator

7.3.1.1 Components

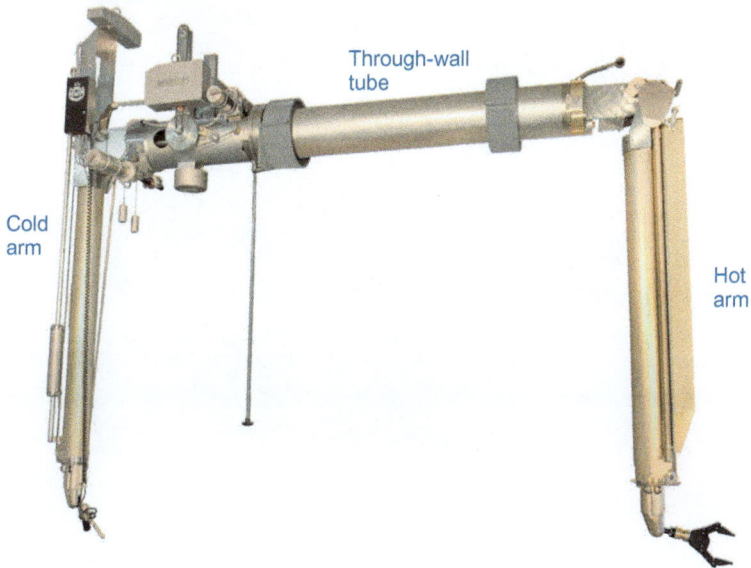

Fig. 7.8 MSM components.
© Wälischmiller.

They divide into three main components (Fig. 7.8): a cold arm or master, which is located outside a cave; a through-wall tube which sits within a shield wall; and the hot arm, or slave, which operates in a radioactive environment.

A system of wires and pulleys, either inside an MSM or mounted externally (Fig. 7.9), transfer movements made by an operator handling a cold arm and replicates them in a hot arm on the other side of a cave wall. It sounds straightforward enough and indeed skilled operators do make it look deceptively easy. Unsurprisingly, though, it does take quite a bit of practice, at a training facility, to achieve the skill which is so evident when we see them being manipulated in operating facilities.

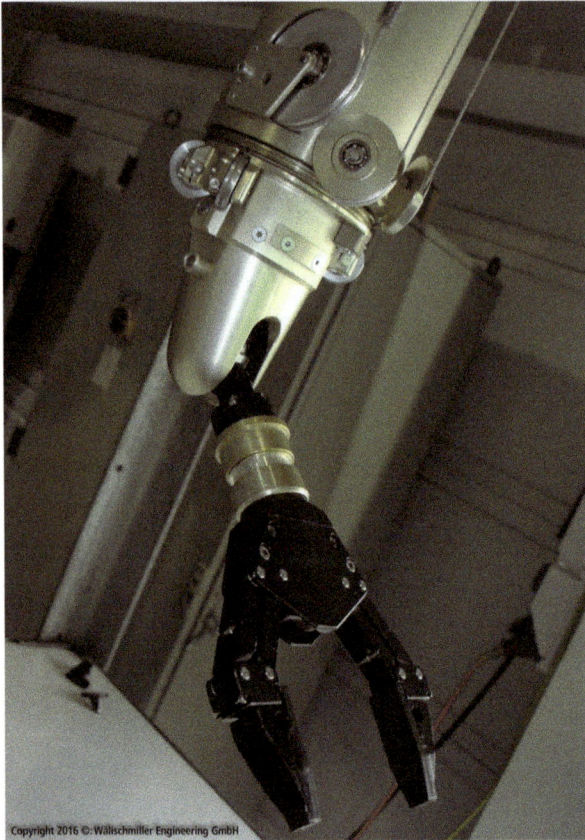

Fig. 7.9 MSM drive wires and pulleys.
© Wälischmiller.

7.3.1.2 Gearing

One of the traits that MSM operators must become accustomed to is the way movements in a cold arm are exaggerated in their hot arm partner. It is brought about by gearing systems in the arms and on the face of it seems like a feature operators could well do without. However, when we look a little closer it is an arrangement that makes perfect sense and makes MSMs much easier to use.

The hot arm of an MSM is typically at least 3 m long so can perform tasks within a wide arc, sweeping both left to right and up and down. If movements in both arms were copied exactly, then operators would need to perform a lot of aerobic style stretching to get a hot arm to do their bidding. In addition, MSMs are invariably located at cave windows so that operators can get a direct view of how arm movements are being replicated. It is the gearing system which makes it possible for operators to remain at the center of a shielding window (Fig. 7.10) and so get the best possible view of the action at all times.

Fig. 7.10 MSM—cold arm controls.
© Nuclear Decommissioning Authority.

There is another benefit of gearing which is quite akin to the mechanisms on most bicycles. Many of us have experienced this first hand and soon discovered that although certain gears move us along more slowly than others, they do have the advantage of propelling our weight up steep hills with relative ease. Just as with a bicycle, the gearing ratios in an MSM are cleverly arranged so as to enable operators to lift 50 kg or more.

7.3.1.3   Human factors

There is an important role here for ergonomics or *human factors* (HF) specialists, those who ensure activities requiring operator interaction are designed with workers' wellbeing in mind. As with so many design activities, the HF team must be consulted early so as to design out issues that could otherwise lead to repetitive strain injury, musculoskeletal problems or operator fatigue. It may seem this is an area where we all know intuitively what constitutes good design, but in truth HF is a highly specialized subject and one most definitely best left to the experts.

7.3.1.4   Reach

MSMs, as seen above the shielding window in Fig. 7.11, can be configured to have a reach of up to 5 m, which gives them a wide swathe within which to perform their duties. In addition their versatility even allows them to reach back to the walls that support them.

Clearly such dexterity comes in very handy indeed, but an ability to swing within what is effectively a large hemisphere could lead to a hot arm clashing with in-cave

Fig. 7.11 MSMs—in-cave view.
© Nuclear Decommissioning Authority.

equipment. To prevent such unwelcome events the movement joints of MSMs are fitted with mechanical stops, which physically limit an arm's reach and confine its operations to within predetermined safe working zones.

7.3.1.5 Change-out

By their nature, MSMs do present a challenge when the time comes to remove them for maintenance or repair, as very often it is not possible to detach a hot arm and remove it from inside a cave. Instead, once a cold arm has been removed, the through-wall tube and hot arm must be withdrawn together through the wall liner that houses them. The cave crane can help out by holding a hot arm in a horizontal position whilst it is pulled through the liner, but it is still quite a delicate operation. Not only that, but with an arm being, say, 3 m long and a tube extending the width of a shield wall, we find that around 5 m of clear space is needed to accommodate the extracted assembly.

As we know hot arms operate in a radioactive environment, so will almost certainly be contaminated and therefore need to be withdrawn into some form of containment. Depending on contamination levels, this will be either a sturdy bag, or a custom made metal container known rather unfortunately as a *coffin*. Once an MSM's components are withdrawn and safely contained, an overhead crane lifts them from a platform that will have been temporarily installed for the purpose. The whole package is then deposited onto a waiting trolley and transferred to a maintenance workshop.

Fig. 7.12 Cave face corridor.
© Nuclear Decommissioning Authority.

Whilst we are on the subject of withdrawing MSMs, it is worth mentioning that cave face corridors invariably have services passing through them. However, because a clear zone is needed for the crane which handles MSMs and maybe shielding windows, these services must be routed along the facing wall (Fig. 7.12). As a result, my personal view is that where a cave is fitted with MSMs then a minimum of 6 m, ideally 7 m, must be left between the cave wall and building structure across the corridor. When early concept designs are being developed such space may seem like an extravagance, but trust me it will be needed to withdraw MSMs, accommodate services and facilitate all manner of routine operations. Not to mention how handy it will be during the decommissioning phase.

7.3.2   Electrical and hydraulic manipulators

The deployment of electrical and hydraulic manipulators share much in common, in that both are operated by joysticks which can be quite remote from the manipulators themselves. And usefully their electrical controls, or hydraulic packs, are located outside a cave where they can be easily accessed by maintenance personnel. These manipulators can be fixed to a cave wall, in similar locations to MSMs, but can just as easily be secured to a cave floor or mounted on rails (Fig. 7.13). They can even be suspended from the crab unit of an in-cave crane.

In situations where they are fixed to a wall, manipulators can be detached with the help of a cave's crane or power manipulator, so do not need to undergo the contortions necessary when extracting MSMs.

Fig. 7.13 Electrical manipulator—rail mounted.
© Wälischmiller.

Fig. 7.14 Electrical manipulator deployed underwater.
© Wälischmiller.

To add to their versatility manipulators can even be designed to operate when submerged in water (Fig. 7.14) which, bearing in mind that this is where spent nuclear fuel spends so much of its time, can come in handy now and again.

Most of the MSMs and manipulators that we see positioned at shielding windows are in pairs. The arrangement provides redundancy if one breaks down, widens the area of coverage on the other side of a window and makes sense because generally two hands are better than one. In fact I once watched an operator use a pair of MSMs to sweep up swarf with a small hand shovel and brush, similar to the kind you can purchase on any high street, and must say it was an impressive feat to say the least.

Unlike MSMs, manipulators can house all of their on-board control paraphernalia such as electric motors, cables and hydraulic pipework, inside the manipulator arms, so apart from looking rather sleek their surfaces are much easier to decontaminate. In addition the improved radiological cleanliness of their concealed parts aids the maintenance process. Manipulators are quite akin to the robots we see working so feverishly in the automotive industry; so where repetitive tasks, such as bolting a lid onto a drum are involved, nuclear versions can be pre-programmed in exactly the same way.

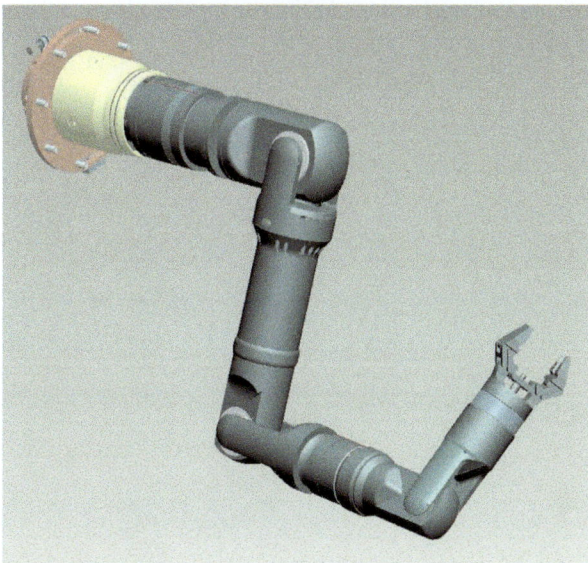

Fig. 7.15 Hydraulic manipulator.
© Nuclear Decommissioning Authority.

In terms of power, manipulators can deliver more than MSMs, with hydraulic variants (Fig. 7.15) being the most powerful of the group. And rather helpfully all three of the generic types, MSMs, electrical and hydraulic, can provide force feedback to operator's at the controls, so there is no danger of powerful jaws crushing a component due to an operator being unable to sense the tightness of their grip.

7.3.3 Power manipulator

7.3.3.1 Deployment

As I mentioned above, manipulators can be attached to a crane's crab unit. The problem is that a cave's crane rails are often located too high for a manipulator to reach down and get involved in the action below. The solution is to attach either a hydraulic or electrically powered manipulator to the end of a telescopic mast which itself is attached to a crane-like crab unit (Fig. 7.16) that way they can easily be deployed to practically any point in space within the confines of a cave. The term used to describe manipulators configured in this way is a *power manipulator*. Strictly speaking the same nametag could be applied to the whole family of manipulators, no matter how they are deployed, but by tradition it is reserved for those that are integrated with a telescopic mast.

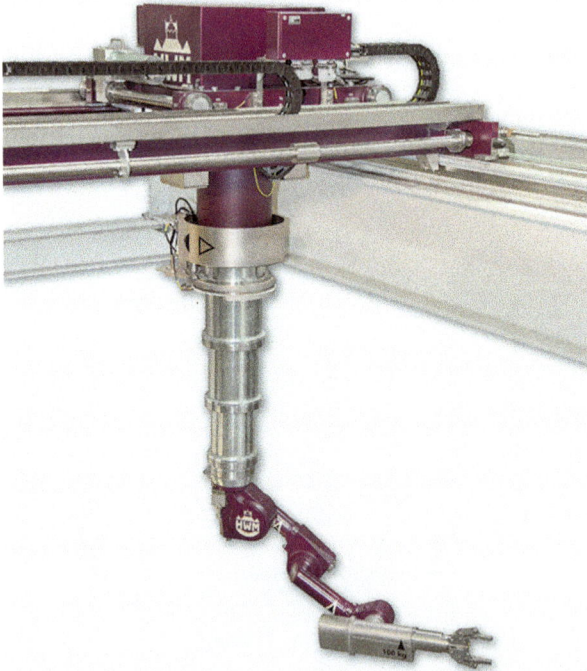

Fig. 7.16 Power manipulator.
© Wälischmiller.

In most respects, the operation of power manipulators is no different to those we have discussed already; there are, however, two important considerations. The first centers on an unavoidable consequence that the longer a telescopic mast becomes, the less stability it has when fully extended. It is all about torsion, so a great deal depends on what duties a manipulator must perform with its mast at full stretch. And just as important, how far the manipulator arms themselves will need to be extended. Unfortunately there are too many variables to make a definitive statement on the maximum length of a telescopic mast, but 5 or 6 m is quite commonplace, and there are plenty of longer examples around. Closely associated with this particular theme, is the conundrum which arises when power manipulators and cranes are vying for space within the same cave. However, this is a separate subject in its own right, one we shall come back to later in this chapter.

7.3.3.2 Decontamination and maintenance

The second issue that pertains to power manipulators, and one that needs to be considered early in the design process, is how a mast will be decontaminated and maintained? A mast's drive apparatus is housed inside its hollow sections so remains radiologically clean or close to it. Its outside surfaces, though, are exposed, so if subject to airborne contamination they will need to be decontaminated before maintenance crews can get to work.

Fig. 7.17 Telescopic mast.
© Wälischmiller.

Ideally there should be provision in a cave's decontamination bay to fully extend a power manipulator's mast, say, into a tube equipped with spray rings or some other decontamination system. However, single sections of a mast can be fully extended in any order (Fig. 7.17), so it is possible to decontaminate them one at a time. Whatever method is used, appropriate equipment and adequate space will be needed in the decontamination bay and to some extent in the area below it.

Decontaminating and maintaining a power manipulator's mast, is a good example of an activity that can appear fairly inconsequential when large design teams are running at full tilt. However, with the passage of time, it does get more and more difficult to accommodate, so is well worth keeping an eye on.

7.3.3.3   End Effectors

Jaws are so commonplace a sight on the MSMs and manipulators operating in nuclear facilities that it may seem they are the only tool which can be deployed. Usefully, though, jaws can be remotely detached from their hot arm and exchanged for quite an array of other tools, or *end effectors* as they are more properly known.

Fig. 7.18 Power manipulator with cutting disc.
© Wälischmiller.

To see these in action you may need to visit a facility which is being decommissioned, where tools can range from cutting discs (Fig. 7.18), to saws, drills, powerful shears and many more. All in all then, whether it be an MSM, a manipulator, or a power manipulator, they play an essential role in the whole lifecycle of nuclear facilities.

7.4 Shield doors

If an engineer told you they specialized in designing shield doors, you might reasonably wonder (to yourself of course) just how they could possibly fill their days. After all shield doors are pretty much large flat lumps of metal that slowly open and close now and again, so how difficult can it be?

It turns out that designing shield doors and turning them into reality involves some mighty challenging engineering. Even deciding on the right door type and drive mechanism is a major exercise in its own right, one that demands close liaison with structural and electrical colleagues, construction specialists, plant operators and many more, so maybe these folks are kept busy after all.

7.4.1 Construction

As their name suggests, shield doors cover accessible openings into areas housing radioactive materials and provide equivalent shielding to the structure, normally concrete, in which they reside. The thinnest doors start at around 25 mm, and from that depth up to, say, 150 mm may be formed from a single steel plate. Beyond that thickness it makes sense to build them up from several sheets bolted together, or construct a steel "box" to the dimensions of a finished door and fill the inside with concrete or lead. By the way, when lead is used it is normally poured in molten form so as to ensure there are no air pockets that would impair its shielding properties.

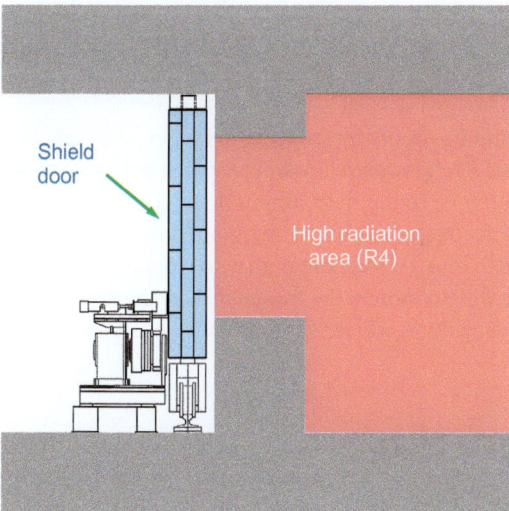

Fig. 7.19 Laminated shield door. © Bill Collum.

Where doors are laminated it is not necessary to use steel plates which are the same dimensions as the finished door; instead each leaf can be formed by assembling several smaller plates in a brickwork type pattern and bolting them to the adjoining lamina. However, the joints do need to be staggered on each successive leaf (Fig. 7.19) otherwise radiation may find a path through them. The beauty of this process is that very hefty doors can be assembled with components that are relatively easy to handle and even better can be built in situ, so neatly sidesteps the challenge of handling the finished article in one piece.

7.4.2 Personnel access doors

The most basic of the shield door types have proportions similar to doors in our homes and are known as personnel access (PA) doors. They are routinely anywhere from 25 to 200 mm thick and are necessarily hung on very substantial hinges.

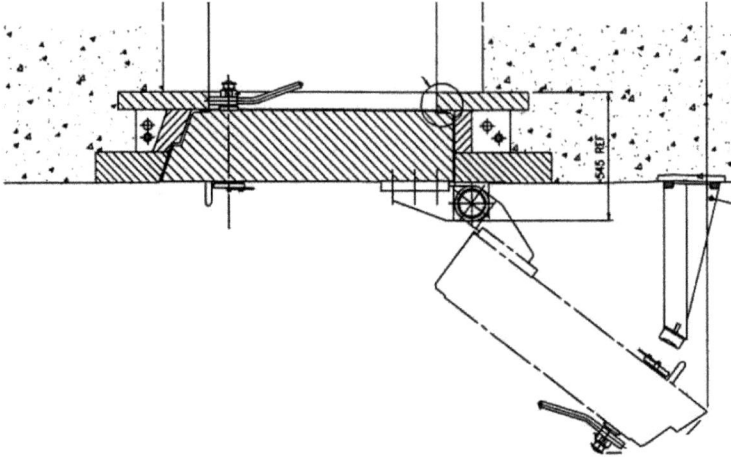

Fig. 7.20 PA door with integral frame.
© Nuclear Decommissioning Authority.

Despite their weight, which for some PA doors may exceed 5 tonnes, they can be opened by hand; alternatively they may be powered. However, the sheer weight of bigger doors does rule out any possibility of fixing their hinges directly to a concrete structure. So instead these doors are accompanied by an integral frame (Fig. 7.20) including hinges, which is embedded into the concrete that surrounds a door's opening. Apart from distributing their weight around the frame, this arrangement also facilitates tight tolerances around a closed door, which is important where radiation is involved.

7.4.3 Pressure relief vent

We shall see in the next chapter that there is always negative air pressure between low contamination zones and higher ones. Furthermore, air travels from C2 (green) areas to C3 (amber) areas at a velocity of 0.5 m per second, and from C3 (amber) to C4 (red) at 1 m per second. All of this is perfect for maintaining containment, but can result in ventilation systems sucking so hard that it becomes impossible to pull a door away from its frame. I have experience this impasse whilst test procedures have been underway, so can vouch that air pressure is so effective a force it can feel as though a door has been welded shut.

The solution is beautifully simple and not unlike the twist caps we see on containers holding flour or talcum powder. PA doors serving openings that are subject to significant differential air pressure have a simple twist vent fitted which can be turned by hand. When opened, air rushes through and evens up the pressure on both sides. In this way a door which moments before was obstinately immovable, will instantly cooperate and swing open with ease. If only everything could be so simple.

7.4.4   Maintenance access

Where their sheer weight, or radiation, dictates that shield doors need to be moved by mechanical means, then they are almost always designed to slide sideways or raise and lower. The robust drive systems employed for such demanding duties are either electrically driven screw jacks or hydraulic rams, both of which will have manual backups that enable operators to maneuver a door if its primary drive should fail.

If we think back to our earlier discussions on radiation and contamination, then the classic situation for a shield door will see its hot side exposed to a R4/C4 (red) area, whilst its cool side faces a zone categorized as R3/C3 (amber). Apart from the obvious objective of ensuring maintenance is only conducted when shield doors are closed, the other goal when configuring them is to locate drive mechanisms on the, radiologically speaking, coolest side of a door. Although it is permissible to conduct maintenance activities in a C3 (amber) area, and it is a regular enough occurrence, engineers always strive to position a door's electrical drive motor or hydraulic pack in an adjacent C2 (green) area.

7.4.5   Vertical shield doors

Large vertical shield doors could twist out of alignment if raised by a single mechanism at their midpoint, so the obvious solution is to provide two drives and position them at both ends of the door. However, if they were to operate independently there is a real possibility that the drives would turn at slightly different speeds, which could misalign a door and jam it in place. Clearly an unhappy situation, so there has to be a better way. A typical solution employs a single electric motor which drives two screw jacks through a common system. As the central gears turn, their movement is transferred into drive rods which revolve in opposite directions and instigate a synchronized up or down movement in the screw jacks.

The configuration of caves housing radioactive materials brings a couple of interesting challenges for shield door designers. The first stems from the way a cave's crane, more specifically its crane rails, must pass through a shield door to get into the adjacent maintenance bay.

One of the ways of achieving this is to temporarily remove a segment of both crane rails so that a shield door effectively slots into them when in its closed position. With this configuration a short section of each rail, directly beneath the shield door, is hinged on a simple pivot. This arrangement enables a *tipping rail* to be pushed into its horizontal position by a crane entering a cave, then raised behind the crane as it passes over. It works just fine but there is a simpler solution, so the tipping rail is not a solution we come across too often. The preferred approach (Fig. 7.21) is to cut notches from the two bottom corners of a shield door so that it nestles around crane rails when in its closed position.

7.4.6 Combination vertical and horizontal shield doors

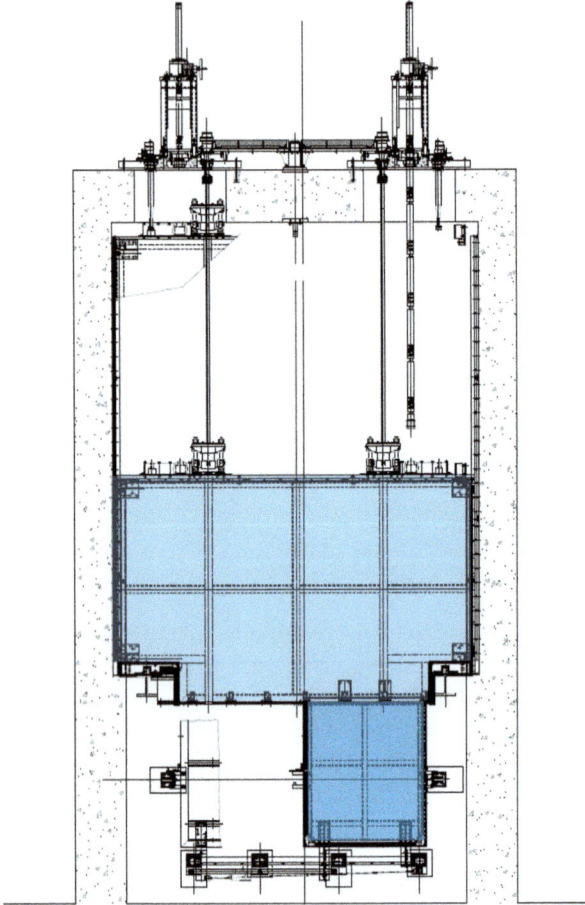

Fig. 7.21 Combination vertical and horizontal shield doors.
© Nuclear Decommissioning Authority.

The second challenge arises from the way a crane's grapple hangs below its bridge. Ordinarily this is of no consequence, but where a crane passes through a shield door, some thought needs to be given to how a grapple will get through the opening. The upshot is that a pair of shield doors is required (Fig. 7.21) one operating vertically, along with another smaller one that can move sideways, allowing the grapple to pass through.

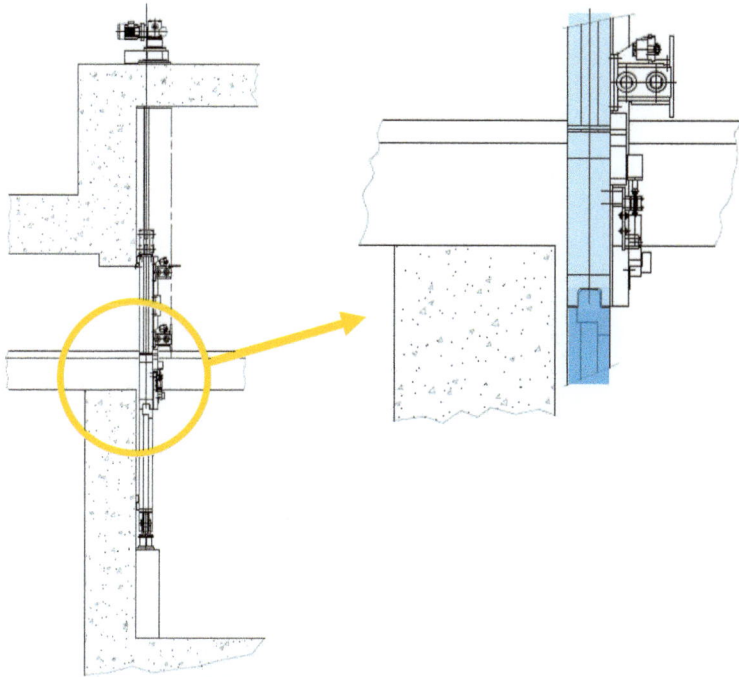

Fig. 7.22 Interface between vertical and horizontal shield doors.
© Nuclear Decommissioning Authority.

It looks straightforward enough until you consider that the horizontally operated door must slide across a clear opening, which leaves it unsupported along its top edge and destined to fall off the first time it is used. The solution is akin to the tongue and groove (T&G) joint that we see on the edge of many floorboards, where an upstand in the bottom shield door is captured by a corresponding channel in the top one (Fig. 7.22). This arrangement literally keeps the horizontally operated door on track, but only works if both doors are opened and closed in the correct sequence. To keep things running smoothly the bottom door must only move when the top door is closed and therefore providing stability through the T&G joint. Happily this can be achieved by incorporating interlocks which will only allow doors to move in accordance with a predetermined safe operating procedure.

7.4.7   Vertical shield door recovery

As always, the problem that looms large with any remote operation in a radioactive environment is how to recover if things do not go to plan. If, for whatever reason, a vertical shield door was to fail whilst closed or partially open, then at maybe 50 tonnes or more it would present a recovery challenge which is substantial in every sense of the word. The solution is necessarily robust, but considering the magnitude of this particular conundrum also manages to be quite an elegant one.

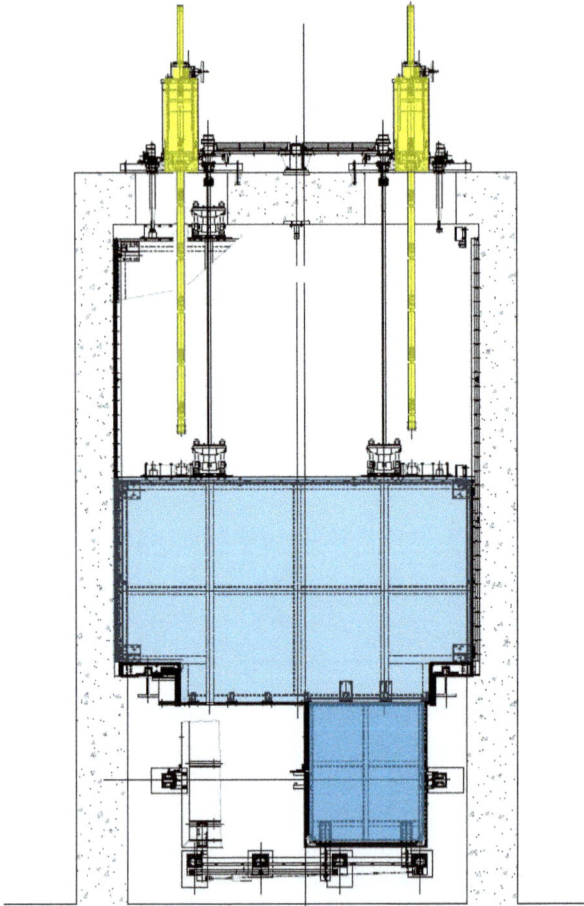

Fig. 7.23 Vertical shield door recovery.
© Nuclear Decommissioning Authority.

Long recovery rods, which are divided into connectable sections, are stored in units that sit alongside the screw jack drives. Should the need arise, the interlocking sections are fed down through penetrations in a cave's roof (Fig. 7.23) until their end effector connects with an attachment on top of the stranded door. With both recovery rods deployed they can be winched up or lowered by hand, taking the door with them as they go. It is a slow process but works well and provides reassurance that recovery can always be instigated.

7.4.8 Pivoting shield doors

Apart from what we might call the traditional vertical shield door, there is also an alternative design which is utilized from time to time. Known as *pivoting*, or *flap* doors, they notch around crane rails in the same way as their vertical counterparts, but rather than sliding up and down swing open, a bit like a regular door but hanging from the ceiling instead of a wall. They are moved by hydraulic rams which are operated from the cave's roof. To be honest it always looks as though no amount of force could possibly swing such hefty doors upwards, but they operate very smoothly indeed which just goes to show how powerful hydraulic systems can be.

7.4.9 Concrete filled shield doors

I mentioned at the beginning of this section that shield doors can be formed by filling a steel "box" with concrete. This technique is not employed nearly as much as the ubiquitous all-steel door, so when we do come across them they can be quite an arresting sight. The largest I have come across had an empty shell weighing over 100 tonnes and when filled with concrete has a gross weight of around three times that. It was impressive, not least because it moved on air skates, which introduced yet another dimension to the challenges faced by its designers. In fact, it neatly illustrates the point we began with: that there is much more to shield doors than their superficial appearance suggests.

7.5 Bogies

Moving very heavy loads around whilst guaranteeing they cannot be dropped are attributes which are very attractive to the nuclear industry, so this is where we call on the utilitarian bogie to deliver. They are a cross between a railway flatrol and the bridge of an EOT crane (which we shall discuss in Chapter 9) so are perfect for moving heavy nuclear payloads from one location to another, and, just as useful, stopping at predetermined points along the way.

Their one operational constraint, if we can call it that, is that they work best when moving back and forward in a straight line, or a linear path as it is sometimes termed. And certainly their recovery, although possible, is more problematic if we try to route them around bends.

Apart from their load bearing capacity, which can reach over 100 tonnes, they are very compact so are adept at squeezing through tight spaces such as shield door openings, or traveling down long tunnels with only millimeters to spare on all sides. As if all this wasn't enough they can also be deployed over considerable distances, so with the right drives and recovery mechanisms can easily make journeys in excess of 300 m.

With so much going for them it is no surprise that we find bogies, or trolleys as they are often referred to in the United States, playing a crucial role in the day-to-day operation of so many nuclear facilities. Normally the only significant issues we need to address are how to propel them and, for those that operate in an inaccessible environment, how to recover them if they happen to fail whilst in mid-journey?

Drive mechanisms fall broadly into the two categories of on-board and remote, with those needing backup often using a combination of both. On-board electrical motors receive their power from a cable reeling drum, which can be carried on the bogie itself or alternatively located at the end of its tracks. Other bogies are a much simpler affair and comprise not much more than a robust chassis with four wheels.

These bogies are connected to a steel rope which loops around rollers located at both ends of their track, so can be moved back and forth by an electrical motor rotating one of the rollers. To keep the cable taut, at least one roller is fitted with an adjustable tensioner.

The alternative to electrical motors and steel ropes is to use a hydraulic ram or rigid chain, both of which simply push or pull a bogie along. They work very well, but unlike other systems, they are more constrained in the distances they can cover. Realistically they are limited to traveling around 5 m, maybe 10 m at most.

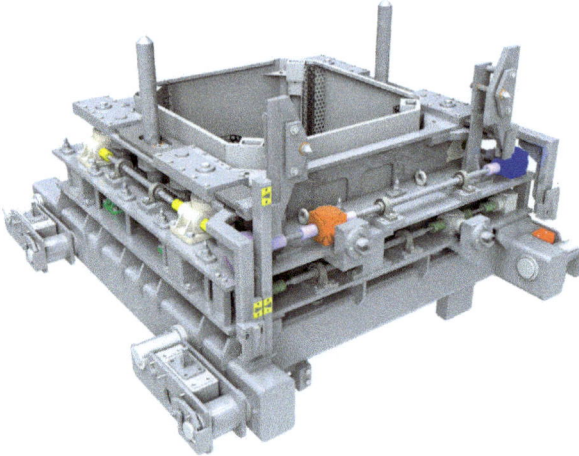

Fig. 7.24 Remotely operated bogie with raise and lower capability.
© Nuclear Decommissioning Authority.

Bogies that are only required to carry loads from one given point to another need some relatively simple additions before they are deployed, such as shoot bolts which lock them in place at their parking positions. And bogies that need an ability to raise and lower their payload will have some additional equipment installed on their chassis.

For bogies such as that shown in Fig. 7.24, which operate in a hostile radiological environment, there is the additional constraint of ensuring their on-board equipment will always be failsafe. Or alternatively that it is possible to remotely recover them. As mentioned earlier, bogies have a good deal in common with EOT cranes and this extends to considerable similarities in ways they are recovered, the provision of duplicate drives, how they *know where they are*, how their cables pass through shield doors, and so on. Such themes are covered in some detail in Chapter 9, so we can hold those thoughts for the moment and come back to them later.

7.6 Decontamination

One of the things that probably strikes you when first entering nuclear facilities is that there is an awful lot of stainless steel on show. In fact practically everything that comes into contact with radioactive materials is made from it. More often than not it is type 304L, which incidentally is the low carbon version of 304 stainless steel, the one widely used for manufacturing saucepans and kitchen sinks. For the nuclear industry, though, the low carbon properties of 304L are better suited to welding than the type we find in our homes.

As you might expect, when it comes to selecting materials this is not one from the bargain basement area of a store, so there had better be a good reason for choosing it. There is—actually, there are several reasons. When you deploy equipment in a nuclear environment, you do not really want to spend more time caring for it than is absolutely necessary. You want to send it on its way, use it when you feel like it and, apart from routine maintenance, not have to worry about it. Carbon steel does not meet that profile.

We do not need to be materials specialists to realize the silky smooth surface of stainless steel will be less prone than carbon steel to trapping minute particles of radioactive contamination. In addition exposed carbon steel tends to rust, which only enhances its contamination trapping abilities. It might seem that applying paint before it gets the chance to rust would be a good idea and indeed it would be fine for a short period, but unfortunately painting leads to a never ending cycle of touch-up and repair which is not great in a nuclear environment. All in all then, its less than ideal surface and inability of painted carbon steel to withstand repeated decontamination, make it unsuitable for use in a hostile radiological environment. So stainless steel it is.

There are three main reasons why we need decontamination processes. Firstly, to reduce contamination levels on items of equipment so that operators can safely handle them. Secondly, to meet the conditions for acceptance stipulated for transporting radioactive materials, or set by a facility for which they are destined. This includes surface contamination on packages being dispatched for long-term storage or disposal. And last but not least because we have an obligation to do so.

The first two are unambiguous enough, but the last is a little more multifaceted and is a theme we shall return to in Chapter 13 when we discuss waste management. Essentially, though, it is about the regulatory and indeed moral imperative, that we minimize volumes of high and intermediate waste as much as possible, not forgetting the financial burden of disposing of waste, which escalates more or less in line with how radioactive it is.

So there are lots of good reasons to decontaminate and happily there are a wide variety of processes to choose from. The thing is, deciding which is most appropriate for a particular situation can be quite a challenge, not least because plant operators will have to live with it for years, maybe decades to come. As a result trials are often instigated, possibly backed up by a research and development program, where simulated contaminants are tackled by an array of potential cleanup process until eventually a winning candidate emerges.

7.6.1 Swabbing

This particular technique is what we might term the entry-level decontamination process. In its simplest guise, swabbing is conducted by operators using small filter type papers or cloths, say around the lid of a transport flask which is being checked prior to dispatch. When done in this way specialists will be in attendance to monitor swabbed items and ensure they achieve an appropriate level of radiological cleanliness before leaving the area.

Alternatively, where items are too radioactive to approach, swabbing is performed by robots using what looks like a large cotton bud, with a tip is about the size of a golf ball. Once swabs have been taken, they are deposited into a detector and analyzed for traces of contamination. Depending on results, an item can be swabbed again until it reaches its specified level of external cleanliness. So although in this particular situation swabbing is strictly speaking part of the monitoring process, it can also double-up in performing some light decontamination duties.

7.6.2   Water

The most common decontamination processes utilize water, which can also be delivered as steam. Ordinarily it comes in liquid form, with its various guises depending mainly on the pressure at which it is delivered. Low pressure and deluge systems are used to remove loose contamination, say on an in-cave crane, whilst high or even ultra-high pressure is capable of dislodging more stubborn substances, such as the grout which is often used to immobilize intermediate-level waste. By the way, high pressure water can cut through steel so has no trouble at all eating into surfaces at which it is directed. For this reason we need to be quite selective in its use and be prepared to factor in an allowance for metal erosion whenever it is used.

The one problem inherent in using water as a decontamination medium, is that it creates a liquid effluent stream which itself must be collected and processed, no doubt using techniques such as ion exchange and ultra-filtration, which we discussed in the previous chapter. Where contamination levels are not too high, it may be possible to recirculate the water and top it up now and again, but eventually it will become loaded to a point where it too must be processed.

7.6.3   Submersion

One of the challenges faced by mechanical engineers in the nuclear industry is designing equipment in a way that minimizes any voidage, or nooks and crannies, where contamination may accumulate. However, in cases such as say a mechanical pump it is impossible to satisfy this objective entirely. Invariably then we find equipment which cannot be decontaminated by spraying water at it, no matter how powerfully it is delivered, so this is where we turn to submersion processes. The favorite medium here is nitric acid, which is very effective at dissolving most of the contaminants we come across in the nuclear industry.

As an alternative to acid, contaminated components may be placed in a bath of water and bombarded with ultrasonic waves. They send pulses through the water, creating an effect similar to the way in which Hi Fi speakers are made to tremble, and again can do a thorough job of dislodging contamination. Whether it be water or acid, not all nuclear sites have the infrastructure or spare capacity to deal with the liquid effluents that are created. Happily then, there are dry decontamination techniques available which are also very effective.

7.6.4 Dry pellets

Fig. 7.25 Waste drum after dry pellet cleaning process. © Nuclear Decommissioning Authority.

Pellets come in various forms, from plastic to sand, to ball bearings, and are most effective when a range of various sizes are blended together. Apart from being a dry process this technique is also extremely aggressive, so can obliterate the most obstinate of contaminants and transform a component back to its original pristine condition (Fig. 7.25). Whichever medium is used it is invariably recirculated several times, but eventually a combination of accumulating contamination and disintegration of the pellets themselves dictates that they must be replaced by a fresh batch.

7.6.5 Carbon dioxide

A variation on the pellet theme is delivered by liquid carbon dioxide (CO_2), which when forced through appropriately configured nozzles streams out in a form akin to the hailstones that occasionally sting our skin on bleak winter days. The beauty of CO_2 is that it vaporizes on contact with a hard surface, so can be routed through a facility's ventilation filtration systems. All of which means that if the contamination being removed is radiologically quite clean, say inactive grout on the surface of a disposal container, then CO_2 does not create too much of a waste stream. In the right situation therefore, it can have a lot going for it.

7.7 Cave arrangements

Now that we have seen something of the equipment which occupies caves and their environs, we can take a look at how caves themselves may be arranged. We shall assume here that our example cave accommodates both a crane and a power manipulator. And, in terms of its adjoining areas, will take the most onerous case, namely that a decontamination bay is needed to clean up in-cave equipment before it is maintained. This just leaves us with the dilemma which arises every time a crane and a power manipulator are deployed in the same cave.

Put simply the choice is this. Should a crane and power manipulator share the same set of crane rails and work side by side, or should we install two sets of rails, one above the other, then deploy the crane on one set and power manipulator on the other? Everyone who listens to arguments in support of both alternatives knows the answer straight away. The problem is they don't all agree, which tells us this is a tricky one to crack. Let's look at shared rails first.

7.7.1 Shared crane rails

7.7.1.1 Combined crane and power manipulator on a two girder bridge

There are a few variations on this particular theme, but they are so closely related as to form a single family. The common denominator is a single set of crane rails, just as you would see in any factory up and down the land, the difference being that they carry two items of equipment.

The entry-level option is where we find the crane and power manipulator mounted on a single crab unit and sharing a fairly standard two girder bridge. This particular set up is pretty limiting because where one item goes the other must ride along with it. The biggest constraint arises because the crane and power manipulator block each other's sideways travel. It means that ordinarily you can only get one or other of them to perform tasks on whichever side of the cave they happen to be located. However, this configuration does come with a fallback option which delivers some additional flexibility.

Once you get the bridge into a maintenance area, it is possible to switch its power manipulator to the other side of the bridge by lifting it over the crane. Power manipulators are easier to transfer because they are smaller and lighter than their crane companion. It takes a little time, but doing the switch does allow each item of equipment to cover a cave's entire footprint.

Understandably, there can be a tendency to look at such a basic arrangement, figure it is close to useless and wonder why anyone would bother installing it. However, in the right situation, with light duties and little need to keep switching equipment back and forth, it could be perfectly adequate. If it was our money and we could get along fine with this arrangement, we would go for it.

7.7.1.2   Combined crane and power manipulator on a four girder bridge

Clearly the main drawback with that basic configuration is having to switch over the crane and power manipulator to get full handling coverage within a cave. Fig. 7.26 shows an enhanced arrangement which utilizes a four girder bridge to overcome the problem. Although they still travel along a cave together, the crane and power manipulator now get complete independence when it comes to moving sideways.

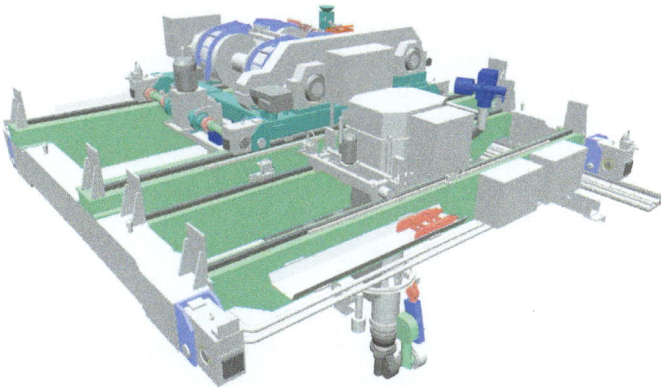

Fig. 7.26 Combined crane and power manipulator on a four girder bridge.
© Nuclear Decommissioning Authority.

Once again there is nothing wrong with this arrangement if you expect one item, probably the crane, to do most of the work, with the power manipulator being called upon only very occasionally, say, for maintenance tasks. However, if both items of equipment are going to share the workload, then as with the previous arrangement joining them together really only makes sense if their duties are very light indeed.

7.7.1.3   Independent crane and power manipulator

The configuration most commonly associated with a shared rail approach is to have an independent crane and power manipulator. With this arrangement both can go about their duties without dragging their companion along. From a safety perspective all that is needed is keep out of each other's way, which is something that interlocks and limit switches can comfortably deal with. However, an interesting debate begins when we consider how both items of equipment will be decontaminated and maintained.

7.7.1.4   Single decontamination and maintenance bays

Let's assume we have an independent crane and power manipulator on the same set of rails, with the power manipulator closest to the decontamination bay. Here's the quandary. When the time comes for a crane to be maintained it cannot bypass the power manipulator as it heads for its decontamination bay, so both items must leave the cave together. A perfectly good power manipulator that could be getting on with its duties whilst the crane is out of the way must instead stop work because it has become an obstruction. On top of that, since the power manipulator is entering the maintenance area it too must be decontaminated. The maintenance team will take advantage of the situation by checking out the power manipulator whilst it is there, but really they are having their maintenance regime dictated to them rather than choosing their moment, which is not ideal.

Recognizing that from time to time one of the units is going to be landlocked by the other, there is quite a complicated debate to be had in deciding which way round to place them on the rails. The default position assumes that because power manipulators are more complex they should be closest to the maintenance area, that way they can get some care and attention without affecting the crane's operation. However, some analysis needs to be none to understand which unit will be doing most of the work and crucially whether loss of the crane or power manipulator would be most detrimental to a cave's operation. As ever it is important to get it right, otherwise facility operators will find themselves frustrated by downtime that, with a little more foresight and planning, could have been avoided.

One of the postulated benefits of opting for this arrangement is that it saves the additional space that would be required if decontamination and maintenance facilities were duplicated at opposite ends of a cave. On the face of it, it sounds plausible enough but when you look a little closer it is not entirely accurate. It does stack up for the decontamination bay which, as the power manipulator and crane can pass through it one at a time, will remain the same size. However, the adjacent maintenance area more or less doubles in length, because it must now accommodate both items of equipment parked side by side. Typically then this arrangement takes less space than full duplication of decontamination and maintenance bays, but will nevertheless add to the overall length of a cave's support areas.

Of course there is nothing inherently wrong with the notion of a crane and power manipulator sharing common decontamination and maintenance facilities, but there is a crucial factor to consider. This option is only viable if throughput will not be compromised during periods when both the crane and power manipulator are out of action. If it can demonstrated that this arrangement will not affect productivity then it should be grabbed with both hands. But if there is any doubt we should not be tempted to go there.

7.7.1.5 Duplicate decontamination and maintenance bays

None of the options we are considering here could be proclaimed as the best solution for every set of circumstances; if it were that simple, we could all go home early on the day this decision was due. Each has pros and cons which must be carefully weighed for every individual case; having said that, there is another option that has lots to commend it.

Here the crane and power mulator enter a cave from opposite ends, so they both have their own decontamination bay and maintenance area. The beauty of this arrangement is that either unit can leave the cave without interrupting the other's work plan. The only significant problem, from an operational point of view, is that the crane and power manipulator cannot pass each other on their rails. So if a crane is busily occupied in the center of a cave and the power manipulator needs to get past, it just has to wait. Having said that, if you decide early on that this is the way you are going to configure the crane and power manipulator, then you can arrange equipment within a cave so as to minimize any possible disruption. Not for the first time we find that if a potential problem is recognized early enough we have a decent chance of doing something to work around it.

Of course there is a price to pay, literally, for so much flexibility. We must bear the cost of duplicating those maintenance and decontamination bays, not to mention the shield doors that go with them. Whether it is worth the additional capital outlay depends on what a cave's throughput modeling is telling us; if the numbers show that its crane and power manipulator need to busy themselves to keep up with production, then it is probably too risky to opt for anything less. This option might have the appearance of being over indulgent, but the thing to weigh here is repercussions of lost production against the cost of duplicating decontamination and maintenance facilities. The difficulty is figuring out the likelihood of one being surpassed by the other.

7.7.2   Twin crane rails

Finally there is one option that comes with all the bells and whistles and in pure engineering terms is a fabulous solution, namely two sets of crane rails with a single decontamination bay and an adjoining maintenance bay.

Standing inside a cave which is equipped in this way (before it goes active of course) you really are confronted by quite a magnificent scene. Its main claim to fame is that this arrangement manages to overcome the problem of a crane and power manipulator being unable to pass each other, by simply locating one above the other, which is fine when you say it quickly, but it does take a bit of doing.

As I mentioned earlier none of the alternatives we are considering here can claim the outright title as *Best Option*. Even this one, which seems to answer all prayers, has characteristics that might not always be too desirable. Most notably a power manipulator must always be on the lowest set of rails, which is all to do with the stability and effective reach that we touched on earlier in this chapter. In essence the longer a mast is the more it wobbles about when fully extended.

I suppose the most notable consequence of this particular option is the extra height it demands within a cave. Certainly the additional volume created by double stacking rails runs counter to the usual objective of keeping caves as small as possible—not too small, mind you, but fit for purpose. Since the additional volume is upwards and almost empty, it does not contribute at all to expanding the really precious space down on a cave's floor; it just creates more air space and a larger surface area of cave wall.

The thing to keep in mind is that sometimes in can be well worth creating that space just to squeeze some additional productivity from the equipment moving back and forward on those rails. The other drawback is that, although this arrangement makes it possible to locate decontamination and maintenance bays at the same end of a cave, it does require an increase in the height of their shield doors, which is not an insignificant challenge.

On the upside this double stacked arrangement has more flexibility than the option with duplicated decontamination and maintenance facilities, and its overall length is significantly less; in fact, the whole thing is about the same length as a traditional cave with single decontamination and maintenance bays. Yes there is some additional height, but the bonus of being able to create a facility which is significantly shorter than the closest alternative cannot be underestimated.

What this tells us is that the debate on which crane and power manipulator arrangement to go for is not simply confined to the internals of a cave with its decontamination and maintenance facilities. We must make our selection with an eye on the bigger picture, examining the impact of each alternative on the rest of a facility and on its surrounding site. In some locations, where available real estate is at a premium, the saving on footprint that this option delivers could be the difference between a project which is viable and one that is not. It is never too easy, is it?

7.7.3   Cave arrangement selection

In those situations where a crane and power manipulator will be heavily utilized there is minimal debate. One way or another the flexibility that is needed can only be achieved by installing independent equipment, whether it be on one or two pairs of rails. On the other hand, where their duties are relatively light one of the more restrictive options will be perfectly acceptable. However, in the real world there are plenty of cases which fall between the two easily identifiable extremes. Understandably when this occurs some considerable justification will be needed for anything other than a fairly basic arrangement. During such a debate there is no denying that some combination of two sets of crane rails, two decontamination bays, two crane maintenance areas and a rather magnificent shield door could look like quite a luxury.

In the end we all look to production modeling and equipment reliability data to highlight the right answer, but even this, at least in the early stages, will be based on assumptions that cannot be absolutely guaranteed. So secretly we all have our own preference. Personally where there is any doubt over the possibility of compromising throughput, I tend to favor a single set of rails with decontamination bay and a maintenance bays at both ends. It keeps the cave volume down and hence the overall height of a facility, it gets the crane closer to the action (rather than have it ride on rails above a power manipulator), makes for a simpler shield door arrangement and enables each item to be decontaminated or maintained without worrying where the other one is. It also makes for a good arrangement when it comes to recovering a crane or power manipulator which has broken down, as each can help to rescue the other, something we shall look at in Chapter 9. Yes it costs to double up on some of the support equipment, but where production is crucial it is a sure-fire investment.

I also believe this is one of those areas where clients should be very much involved in the decision making process, since it is they who will ultimately inherit what is delivered. The main difficulty we face is that very often there is no right or wrong answer to this particular conundrum, so it is seldom going to be an easy call. Having a customer's take, particularly on the subtleties of capital versus lifetime costs, can only help when it comes to selecting an appropriate arrangement for each unique situation.

7.8   Flasks

Flasks, also known as *shipping containers*, *casks* and *overpacks*, are used to transport radioactive waste and divide broadly into two distinct types: those that always remain within the confines of a nuclear site and those that venture far beyond, traveling on public highways, rail networks, and by sea. Apart from providing shielding, the main operational requirement for all flasks is that they must not be breached in any way and lose their containment, so the main differences between the two generic types arise primarily from speeds at which they travel.

On-site transports move very slowly indeed, with trains, for example, typically limited to around 15 km/h (10 mph), whereas flasks traveling in the public domain must move along at a much faster pace. Since just doubling the speed of an object results in four times the impact energy, it is no surprise that flasks traveling beyond a site must be considerably more robust than those which stick to its environs and move along at a genteel pace.

7.8.1   Top loading flasks

7.8.1.1   Cuboidal and cylindrical flasks

Fig. 7.27 50 tonne cuboidal flask on transport bogie.
© Nuclear Decommissioning Authority.

Both generic types come in either a cuboidal (Fig. 7.27) or cylindrical form and whilst most have their opening via a lid that bolts solidly in place, some are accessed through a sliding door in their base. I should say, though, that the sliding door type are inherently weaker than flasks with a more conventional bolted lid, so are invariably limited to on-site duties. Let's take a look at the more commonplace top loading flask first, then examine the bottom loading alternative.

Fig. 7.28 Cylindrical flask.
© Nuclear Decommissioning Authority.

The reason for two different shapes stems from geometry of the inventory being carried, with cuboidal flasks being used for a wide range of payloads, from baskets of spent AGR or Magnox fuel, to packages of decommissioning waste. Cylindrical flasks (Fig. 7.28) are often divided internally by a stainless steel framework and used to transport assemblies of irradiated fuel, such as PWR and BWR, or maybe hold several containers of vitrified waste stacked neatly one on top of the other.

The shielding performance demanded from flasks also varies depending on the inventory they are designed to carry, but invariably results in a substantial fabrication. You will recall from our discussions in Chapter 2 that one thickness of lead, say, 50 mm, can deliver the same shielding performance as two and a half of steel, or 125 mm; therefore, to keep dimensions manageable, flasks may be fabricated from a sandwich made up of steel outer faces with a denser lead filling at their core. Nevertheless wall thickness can easily reach 300 mm or more.

On top of their hefty construction cylindrical flasks have end shock absorbers fitted when they are being transported (Fig. 7.29), whilst cuboidal flasks have additional corner protection. In addition both have robust steel fins covering much of their external surface. The fins are there help to dissipate heat emanating from their payload, say irradiated fuel rods; however, just as importantly, they also supplement a flask's protection by deforming in the event of a collision and absorbing much of the impact force. When you put the whole thing together their construction can easily result in cuboidal flasks weighing 50 tonnes or more, and large cylindrical flasks tipping the scales at around 100 tonnes.

Fig. 7.29 Cylindrical flask with shock absorbers fitted.
Science Photo Library.

When we consider the kind of radiological inventory that flasks carry, it soon becomes clear that their lids, which can weigh over 5 tonnes, have an important role to play in maintaining containment. For this reason we can expect to see them secured by an awful lot of bolts—typically at least 16, but often many more.

Apart from keeping a lid firmly in place, the bolts compress a sophisticated sealing arrangement which stops even minute traces of contamination escaping into the surrounding environment. The seals, which always run in parallel pairs, are tested by pressurizing their interspace or alternatively creating a vacuum within it. Either way special instruments are used to confirm the double seal is performing as it should and that there are no leaks. In addition, before a loaded flask is allowed to leave a nuclear facility its outside surfaces are swabbed and monitored, with particular attention being paid to the area where a lid meets the main body. And of course it is not until all tests have been satisfactorily passed that a flask is declared good to go.

7.8.1.2 Impact testing

Ordinarily flasks just get on with doing a job of work without any fanfare. It is possible, though, that their normally uneventful existence could be eclipsed by an unexpected turn of events, in other words an accident. So as I mentioned a moment ago, it important that whatever happens they are not breached in any way. To achieve this all flasks are assigned a height above which they must not be lifted. Typically it ranges from less than 1 m, for a flask with a sliding door at its base, to maybe 10 m or more for some of those with a bolted lid. If a flask was to fall from above its designated safe height then its integrity may indeed be compromised. Many of the tests needed to determine what constitutes a safe drop height are performed by computers or dropping small scale models, neither of which is much of a spectacle.

Now and again, however, an actual full size flask is dropped from its predetermined safe height, or even higher, onto steel plate or a concrete slab then minutely examined to see how it has lived up to simulated expectations. To make the tests as challenging as possible flasks are normally dropped in various orientations, including onto their corners, rather than simply spreading the impact load across their entire base. Assuming all goes well then, after a loud bang, there should be a little scratched paint and some deformation of the impact absorbing features. That said, these tests play an important role in verifying engineers' number crunching and also provide reassurance to regulators and the public that flasks are every bit as robust as is claimed.

Fig. 7.30 Flask crash test. Science Photo Library.

The most famous UK test (Fig. 7.30) was conducted in 1984 when a cuboidal flask, along with its flatrol, was placed on a railway track and hit by an unmanned locomotive traveling at around 145 km/h (90 mph).

Fig. 7.31 Flask after crash test. © EDF Energy.

The locomotive was destroyed whilst, but for some superficial damage, the flask survived intact (Fig. 7.31). It was quite a stunt, one staged to reassure the public in a most spectacular fashion, and is certainly an event that lives long in the memory. No doubt engineers involved in the flask's design were feeling a little jittery as a 140 tonne train, towing four carriages, thundered towards their newly painted pride and joy, but once the dust had settled, they must have exuded blissful confidence for weeks, more likely years afterwards.

7.8.1.3 Lifting above assigned drop height

Having gone to so much trouble to ensure flasks can safely withstand the impact of falling from a particular height, there are no prizes for guessing that the first rule is never to lift them any higher than is absolutely necessary. When flasks are being traversed by a crane it is preferable to skim them just above the floor, ideally at a height of say 300 mm. Unfortunately, though, it is not possible to stick to that objective for all transfers, but it is a good place to start. If a flask must be hoisted then it is only under exceptional circumstances that it can be lifted any higher than its assigned drop height and then only with additional protection measures in place.

One of the most common (but by no means only) times that this happens is when flasks are lifted from their transport vehicle and into a facility. Process flows may demand that a transport flask must interface with operations high above the ground floor where it first enters a building. The sequence is that a flask is first lifted from its transport, staying within its designated lift height, and then moves sideways until it hovers above a crush pad. From that position it is safe to continue lifting the flask onto the floor above.

A crush pad is typically made up of a metal honeycomb or balsa wood and is specifically designed for its particular situation. If the worst should happen and a flask made a free fall decent, the honeycomb structure would cushion the fall and stop it from rupturing. Understandably, though, a crush pad is never a first choice solution, far from it, but is always held in reserve for those occasions when sticking to designated lifting heights becomes an impossibility. Personally I have been in many a nuclear facility but have only come across them a couple of times.

7.8.1.4 Fire testing

Apart from impact, the other notable risk that flask designers need to guard against is fire, particularly in an enclosed space such as a road or rail tunnel. To demonstrate that they are up to the challenge, flasks must withstand the effects of being heated to several hundred degrees centigrade for a lengthy period. Once again it is a realistic simulation of an accident scenario which flasks may have to contend with.

7.8.1.5 Lid removal

The one problem, if we can call it that, inherent in all top loading flasks is that their inventory needs to get in and out of them without creating a hazard. The difficulty arises because their payload will, by definition, be radioactive, so adequate shielding and containment must be provided around the area where material transfers take place.

In some facilities such operations only occur when flasks are totally submerged (Fig. 7.32) and therefore able to take advantage of deep water to provide the necessary shielding. For most, though, flask loading and unloading takes place on dry land, so this is where a lid lifting machine comes in.

Fig. 7.32 Flask unloading underwater.
Science Photo Library.

There are several ways of safely removing the lid from a flask containing radioactive materials, but essentially they are all variations on the sequence shown in Fig. 7.33. By the way, loading a flask is simply a reversal of the steps we are about to follow. We shall assume the flask in question has a cuboidal profile and weighs 50 tonnes; the package within it is highly radioactive, but its surface is radiologically clean so will not release any airborne contamination when it is handled. In a link back to our discussions in Chapter 3, I have shown radiological zones for each step. As we know, zoning denotes the highest levels that may be reached during normal operations.

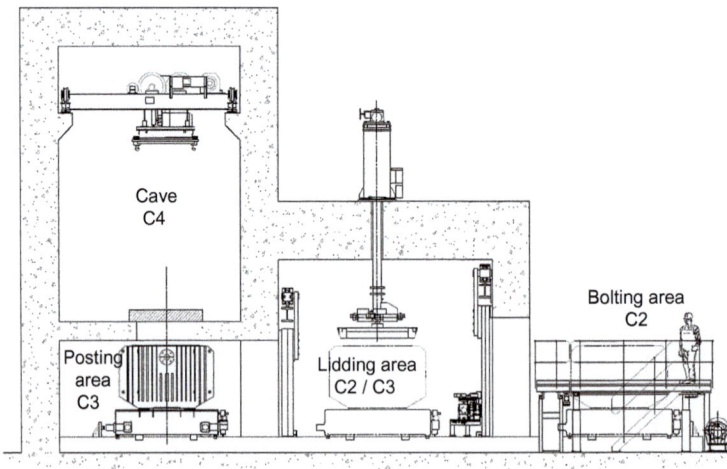

Cave
C4

Posting
area
C3

Lidding area
C2 / C3

Bolting area
C2

Fig. 7.33 Lid lifting and flask unloading sequence.
© Nuclear Decommissioning Authority.

An overhead building crane lowers the flask onto a bogie in the bolting area, where it is surrounded by platforming, sometimes mobile, on all four sides. From here, operators can comfortably access the flask lid, so can use pneumatic nutrunners to loosen its bolts. In some cases bolts are removed whilst in others they are held captive in the lid, so are already in position for retightening. On the return journey operators use torque wrenches to ensure bolts are tightened precisely in accordance with their specification. With a lid weighing around 5 tonnes, even with its bolts loosened, it will stay firmly in place, and so ensure radiation is contained within the confines of a flask.

With preliminaries completed, the adjacent shield door opens and a bogie moves the flask forward. Having previously lidded a departing flask, the lidding machine will already be lowered into position and ready to engage with a pintle on the incoming flask's lid.

Fig. 7.34 Pintle arrangement. © Nuclear Decommissioning Authority.

We shall discuss pintles in Chapter 9, but essentially they are lifting features on the lids of flasks and other containers that can be firmly engaged by a partner mechanical device. In one variation they may look like the example shown in Fig. 7.34, where we can see a heavy lid being held above its companion shielded container.

Once the shield door is closed again, the lidding machine's mast is raised vertically, taking a flask's lid just clear of its body. At this stage indirect radiation, or *scatter*, will emerge from the gap between a flask and its lid, but the surrounding concrete structure and outer shield door protects those in the vicinity.

Lidding machines, which are normally hydraulically driven, are designed to hold suspended lids more or less indefinitely, so once it has been raised the lid remains attached to it. The inner shield door now opens and a bogie moves the open flask forward until it stops below a shielded opening into the cave. All that remains now is to close the inner shield door and open the shielded *gate* above the flask, allowing a cave's crane to reach down and retrieve the radioactive package.

Although in this case the flask's package is radiologically clean, the lidding and posting areas are effectively a buffer between the cave's C4 contaminated environment and the plant's main C2 operational areas. For that reason, and in keeping with good radiological zoning practice, the posting area is assigned a C3 classification, with the lidding area acting as a C2/C3 buffer.

We have previously touched on the fact that air is always drawn from lower contamination zones towards higher ones, a subject we shall discuss in detail in the next chapter. For the moment we can note that air from the C3 posting area will be drawn into the C4 cave above it. To help maintain containment the top of a parked flask will be positioned just millimeters from the opening above it, forcing air to accelerate as it squeezes through on its way to the cave. As a result, air is moving with sufficient velocity to make it virtually impossible for any contamination within the cave to float down and settle on a flask. With its inner package dispatched the flask retraces its steps until it is hoisted away again by the building's overhead crane.

For completeness, it is worth a reminder that in situations where the packages being handled are contaminated, a departing flask's external surfaces must be checked to ensure they are radiologically clean. This may be done on the return journey to the bolting station, or at a similarly platformed area nearby. Either way, the flask monitoring area will be cordoned off and attract a radiological classification of at least *C3, surface contamination only*, more specifically the C3 (S) we discussed in Chapter 3.

7.8.2 Bottom loading flasks

Fig. 7.35 Bottom loading flask. © Nuclear Decommissioning Authority.

If, for operational reasons, we need to introduce or retrieve radioactive packages from above, say, a shielded cave or storage vault, then we can employ a bottom loading flask (Fig. 7.35). With this type of flask the radioactive payload may be carried in a steel skip, or basket, which is raised and lowered by an electrical hoist mounted externally on top of the flask. At the base of the flask is a shield door which slides open, allowing the contents to pass through.

When shield doors operate in this way, sliding horizontally, they are known as gamma gates. To be really precise, a single shield door of this type is only half a gamma gate, since their operation depends on them operating in pairs.

7.8.2.1 Permanently installed gamma gates

The arrangement in Fig. 7.36 shows a permanently installed gamma gate which remains in place throughout the entire operational life of a facility. It divides into two main areas, the gamma gate itself (the other half) and a box that contains the opened gate along with its drive mechanism. Below the gamma gate is a same-sized hole in the concrete roof of a shielded cave.

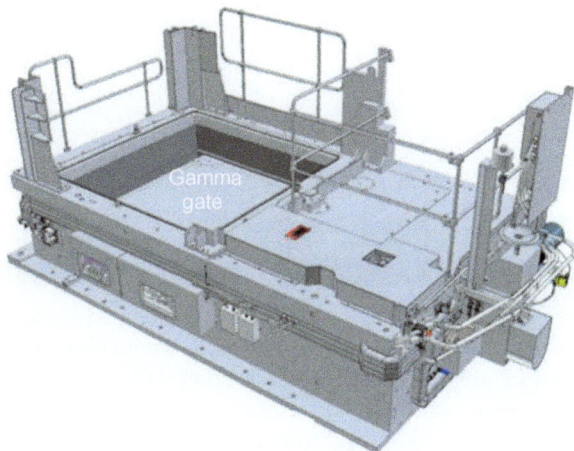

Fig. 7.36 Permanently installed gamma gate. © Nuclear Decommissioning Authority.

To initiate a transfer the flask is lifted by a crane and lowered onto a companion gamma gate such as the one shown. Crucially a steel peg in one gate slots into a corresponding aperture in the other, effectively melding the two into a single entity. With a flask in place, the stationary gamma gate's drive mechanism (typically a hydraulic or screw jack system) retracts its gate, and in so doing drags the flask's gate with it. With that, the flask's hoist can be triggered, lowering or raising its payload from the area below.

So it turns out that when we get right down to it, the business end of this particular operation is remarkably simple. Incidentally, to counteract the possibility of a stationary gamma gate being opened without a flask in place to maintain containment, they are fitted with a locking device which can only be released by the considerable weight of a bottom loading flask being placed on it.

7.8.2.2   Mobile gamma gates

In addition to permanently installed gamma gates, there is a variation which operates in a similar same way but is designed to be mobile. They are needed in situations such as that shown in Fig. 7.37, where an array of shield plugs covers a charge floor. Radioactive packages are stored in channels directly below the plugs, so a gamma gate is needed to gain access to them.

Fig. 7.37 Mobile gamma gates.
Science Photo Library.

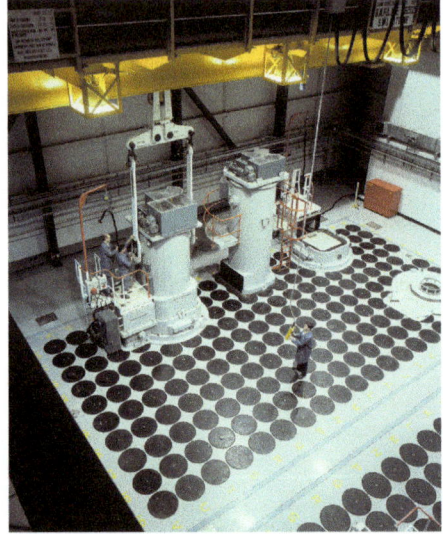

An overhead crane positions a mobile gamma gate above its target shield plug, but before transfers can take place, a specially designed bottom opening flask is used to extract the plug. With that done the gamma gate is slid back into its closed position and the plug-carrying flask parked to one side of the charge floor. Once the shield plug has been removed, a mobile gamma gate is used in exactly the same way as its stationary counterparts.

7.8.2.3   Nappy

When bottom loading flasks are transported across nuclear sites, they are fitted with what is known as a *nappy* (Fig. 7.38) which is effectively a hefty steel shroud that envelops the gamma gate.

To be honest I always feel that nappy, or the US equivalent of diaper is an unfortunate term, especially (as is so often the case) when a flask's inner skip is carrying some sort of liquid waste. The name suggests that its role is to capture slurries and the likes that may escape from a skip whilst in transit, but this is most definitely not the case.

Skips carried within bottom loading flasks have various designs, but if holding wet payloads have a heavy lid which sits in place and does a good job of containing their

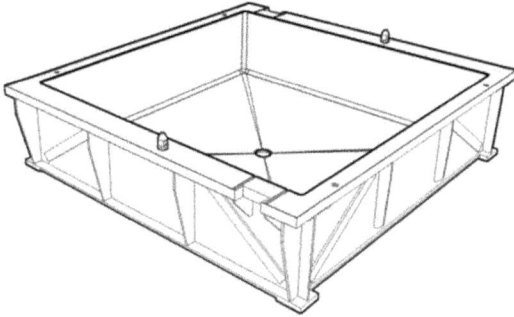

Fig. 7.38 Flask nappy.
© Nuclear Decommissioning
Authority.

inventory whilst being transported. In addition a flask's gamma gate has its own seals, so the possibility of a spillage occurring and making it as far as the nappy is quite remote. Realistically in normal operations a nappy may be called on to contain a little moisture now and again, mainly condensation, but certainly no more than that. If we think of a nappy as drip tray, then it gives more of a clue as to its real purpose.

When a bottom loading flask arrives at its destination, an overhead building crane deposits it on a nappy stand where the first job is to release clamps which hold the nappy in place. Bearing in mind that the nappy may have slight traces of contamination on its inner surfaces, this operation is normally carried out in an area with a radiological classification of C3(S). With a flask free of its nappy it can be transferred to a waiting companion gamma gate which, now that its own gate is exposed, should always be nearby.

7.8.3 Flask design durations

Once we get to see something of how flasks must perform and how surprisingly complex their fabrication is, it becomes clear that it must take a lot longer to design them than would be obvious from their simple external appearance. Here's a thought, however. If it was decided that a new flask was needed and that it was destined to operate outside the confines of nuclear sites, then the process of developing a new design, conducting the necessary analysis and tests and obtaining the all-important transport license, would take at least a couple of years and maybe a good deal longer. And if a flask will be called upon to cross international borders, then we could add another year or two to the whole process. It is a great example of the forward planning which is so important when dealing with nuclear facilities.

Although we have covered quite a bit of ground in this chapter, we have nevertheless confined ourselves to examining what might be considered the essential elements of a nuclear mechanical engineer's knowhow: shielding windows, manipulators, flasks, and so on. As mentioned earlier, this gives an insight into the design of more bespoke mechanical handling equipment, but beyond this their remit spreads even further. In particular there are two areas that demand chapters of their own; one is devoted to *cranes*, but first we will look at *ventilation*.

Ventilation

8

This discipline, which is part of the mechanical function, is normally referred to as the HVAC group. It stands for *heating ventilation and air conditioning*, which means there should be absolutely no doubt about what these particular engineers do all day. Indeed it is true of most industries, where HVAC engineers do exactly as their name implies; in the nuclear industry, though, they are responsible for something quite different. They still have their regular HVAC responsibilities, like ensuring we have fresh air to breathe, humidity is agreeable, and the temperature just right. But for these engineers, the greatest portion of their time is dedicated to maintaining radiological containment within nuclear facilities. In other words they protect those of us who venture inside, by ensuring airborne contamination is directed away from us and to areas where it can be contained or treated. There are of course physical barriers to the spread of radio-active contamination, such as walls and doors, but if it were not for their expertise it would waft around and gather pretty much wherever it pleased.

I must admit that if you have not worked too closely with a nuclear HVAC team, it would be all too easy to underestimate the significance of their contribution to the design process. On the face of it moving air around a building does not sound too complicated: a few fans and filters, some ductwork and off you go. In truth, this group of engineers play a crucial role in turning a facility's all important radiological zoning philosophy into reality, and for that they need a particularly thorough grasp of how nuclear plants function: all in all then quite a demanding job. In the nuclear world you will sometimes see HVAC used as an acronym for *heating ventilation and containment*, which does give a better clue as to what their role entails. Whatever we call them, it is what they do that matters. Let's take a look.

8.1 Role of nuclear ventilation system

At this point we would normally get straight into our subject, but in this particular case it is worth taking a moment to clear up a popular misconception, one that has a bearing on so much of what follows. It is quite often assumed that the role of a nuclear ventilation system is to keep contaminated environments clean, by sucking airborne particles out of the air and sending them hurtling along ductwork to places where they can be dealt with, but this is not the way it works.

At one time or another we have all come across extractor fans either in a domestic situation, in kitchens and bathrooms, or in restaurants, offices and the likes, so will have witnessed first-hand that even where fans are pretty big and powerful, they struggle to seize smoke or steam until it wafts right up to them. Exactly the same occurs with installed ventilation systems where effects, in terms of airflow around extract duct openings, are precisely the same. Fig. 8.1 shows what is happening.

If we take it that airflow at the mouth of a circular extract duct has a velocity of 100%, then we can see that at a distance equivalent to half a duct's diameter, its

Nuclear Facilities. http://dx.doi.org/10.1016/B978-0-08-101938-2.00008-8

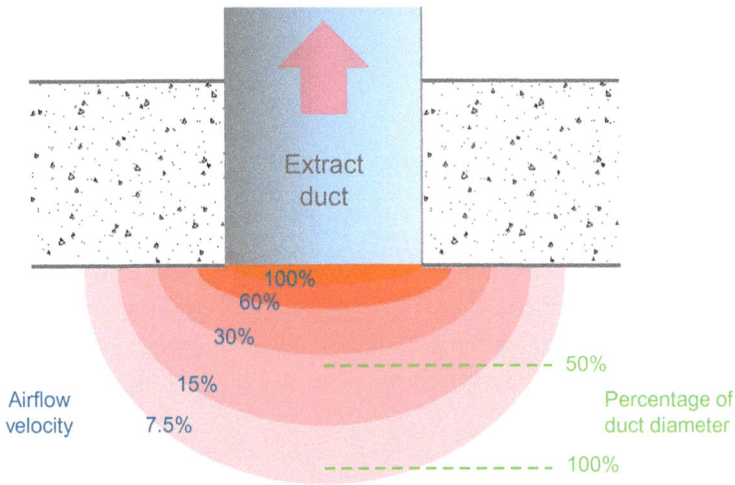

Fig. 8.1 Extract velocity contours.
© Bill Collum.

velocity drops below 30%. And by the time we reach a distance which is equal to a duct's diameter, airspeed has plummeted to less than 10% of that at its open mouth. So, for example, 5 m/s through a 1 m diameter duct opening would drop below 0.5 m/s at the same distance of 1 m.

If we visualize an extract duct in the ceiling of a heavily contaminated environment, say a shielded cave, then it is clear that its *sphere of influence*, a duct's ability to suck contamination from the air, is very limited indeed. So much so, that in practice there is precious little air movement within the body of a cave and certainly no prospect of significant contamination being plucked out of the area. Yes, a relatively small amount of airborne contamination will be captured and carried along until it reaches filtration units, but this is just a by-product of the system's primary purpose.

The crucial point to keep in mind is that nuclear ventilation systems are all about maintaining containment at *interfaces* between the radiological zones we discussed in Chapter 3. Put simply, if airborne contamination tries to drift from a high contamination zone towards a lower one, the system's airflows will be setup to drag it back to where it belongs. No matter how sophisticated a ventilation system may be, it cannot provide powerful suction into every nook and cranny to vacuum out contamination; the *extract velocity contours* illustrate just how impossible that is.

Restraining the spread of contamination is the responsibility of others. It starts with a design team developing equipment and processes which minimize the amount of contamination generated in the first place, then falls to plant operators to follow comprehensive maintenance and housekeeping regimens that will keep facilities as radiologically clean as possible. Now that we have cleared that up we are ready to move on.

8.2 Integration with radiological zoning

The guiding principle for when HVAC engineers enter the design process is very simple; as with so many other disciplines, they must be involved very early on and not merely invited to the table after a facility's fundamentals have been established. It is not just routine stuff, like the size and location of ventilation plant rooms that needs their input, more crucially they must be involved in the exercise of combining a facility's ventilation and radiological zoning philosophies; the two must dovetail together.

As we saw in Chapter 3, there are several variations in the way radiological zones are represented; fortunately they all boil down to the same thing. They all have a range of concentration levels for contamination, starting with regular clean air then going through various levels from virtually clean to very highly contaminated. In our case we are following the naming convention that starts with clean air being C1 (white), through C2 (green), to C3 (amber), with C4 (red) applied to zones with the highest levels of contamination. Radiation levels, for our purposes R1 to R4, do not fall directly within the jurisdiction of nuclear ventilation engineers, so we shall concentrate here on the contamination aspects of radiological zoning, particularly dealing with airborne contamination.

The thing to recognize about rad zoning in a ventilation context (you can call it "rad" zoning when you are in the business) is the importance of congregating like contamination zones together. So, for example, from the ventilation point of view all C3 (amber) zones should ideally be grouped together. Of course it is pretty much impossible to achieve this objective, but it should be recognized as an important goal. The further we stray from the ideal of congregating like zones together, the more complex and therefore costly a ventilation system will be to install, and, just as significant, to operate.

If you observe HVAC engineers studying evolving schemes for a new facility, even one at the very early concept phase, you will notice the first thing they do is take out their highlighter pens and color the building's different radiological zones. They are assessing the complexity of zoning relationships and counting sub changerooms and other interfaces between zones. In essence they are figuring out the most straightforward and economical way to move air around the building, while at the same time maintaining its radiological containment.

If they found that all C3 (amber) zones were congregated together, it would be a relatively straightforward matter to arrange airflows to and from that common location. If on the other hand those zones were dispersed all over a facility, then its air supply and extract network would clearly be more extensive, more convoluted. This in turn would add to the duty on a facility's extract fans, which in turn would add to the cost of electricity, which can be considerable, and so it goes.

Actually, as we shall see next, it is a lot more complicated than the relatively conventional challenge of routing ductwork to and from various radiological zones. Air in nuclear facilities does not necessarily travel via ductwork; instead much of it must be cajoled to drift, in a predetermined manner, from one radiological zone to another. This additional constraint is another good reason for endeavoring to gather like radiological zones together.

8.3   Cascade philosophy

Let's go back to the importance of combining a facility's approach to ventilation with its radiological zoning. Why does it matter? The reason for making so much effort to integrate the two is connected with the way air is handled inside nuclear plants. In the UK, it is referred to as the *cascade* philosophy. The United States, for example, sometimes adopts a slightly different approach, with a little more emphasis on using fans and ductwork to move air around. However, the essential element of all nuclear ventilation systems is the same, in that the higher an area's contamination zone is, the more emphasis there is on drawing air into it. This ensures that adjacent less contaminated air always gets sucked into a higher zone, and particles of airborne contamination are unable to waft in the wrong direction.

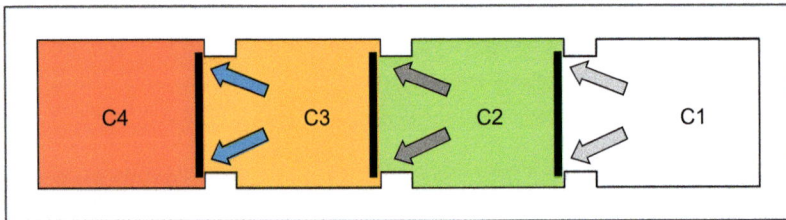

Fig. 8.2 Cascade philosophy—direction of airflow.
© Bill Collum.

Fig. 8.2 captures the basic principles of a cascade system, or indeed any nuclear ventilation system, where air always moves from one radiological zone to another and always towards an area that is potentially more contaminated than the one it is leaving. It cannot make its way back to areas with a lower contamination category and certainly not towards C2 (green) areas which should always be free of any airborne contamination.

8.3.1   Reliability

You may be thinking to yourself that this cascade business is all very well but surely it depends on two things. The suction between one area and another needs to be pretty good, otherwise particles might not flow in the right direction. And the system needs to be very reliable because if it failed contamination could take on a mind of its own and float wherever it pleased. Let's take reliability first.

If any aspect of a nuclear facility is deemed *important to its safety*, then it is not acceptable to rely on one safety mechanism or one system to keep that crucial function in operation. You cannot put all your eggs in one basket so to speak. For most nuclear plants ventilation, particularly of areas designated C3 and C4, falls within that safety category, so needs some form of backup. It is all about having spare capacity ready to step in just in case one of the extract fans should fail, or equally important if part of the system needs to be shut down for maintenance or repair.

When it comes to backup you will hear nuclear ventilation systems described as 3 times 50%, or 2 times 100%. In the case of a 3 times 50% version (Fig. 8.3), a system's extract fans are replicated three times with each unit able to deliver 50% of total

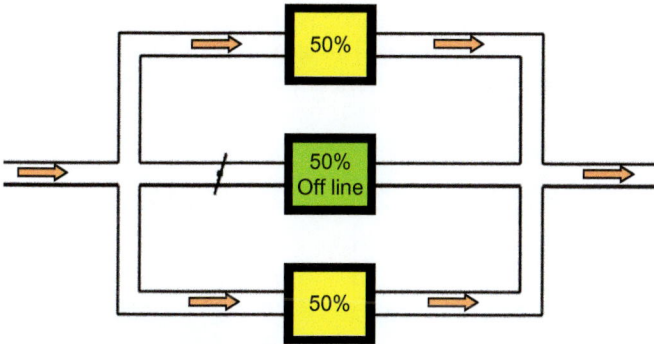

Fig. 8.3 Three times 50% ventilation system.
© Bill Collum.

demand, so only two of the available centrifugal fans need to be running to satisfy a facility's ventilation extract requirements.

If a serious malfunction caused two 50% fans to fail, we could reasonably expect the one remaining fan to deliver 50% of a facility's needs. However, with less air tumbling through ductwork and competing for space the resistance it experiences is significantly reduced, enabling a lone fan to meet around 65% of normal demand. Running on a single fan would be well outside a facility's operating parameters, but could nevertheless keep the more crucial elements of a ventilation system ticking over until full capacity was restored.

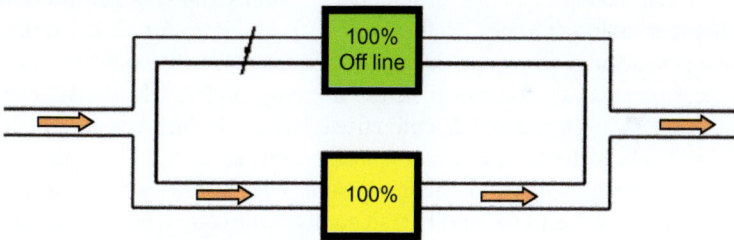

Fig. 8.4 Two times 100% ventilation system.
© Bill Collum.

With a 2 times 100% system (Fig. 8.4) there are two powerful fans, each capable of satisfying a facility's entire demand: in other words, complete duplication of the extract system. In Chapter 10 we shall look at how electricity is delivered to nuclear facilities and see that, when necessary, power supplies are also duplicated. So all in all, nuclear ventilation systems with their standby equipment and duplicated electricity supplies are very dependable indeed.

8.3.2 Air pressure

As for the intensity of suction, or negative air pressure, it can vary a little from one country to another, with each having national guidelines that must be followed by their HVAC community. Whatever their slight differences, the guiding principle is always the same, namely that a facility's depression is greatest within radiological zones that are potentially the most contaminated. This guarantees that where two zones meet, airborne contamination always flows from a lower zone towards a higher one.

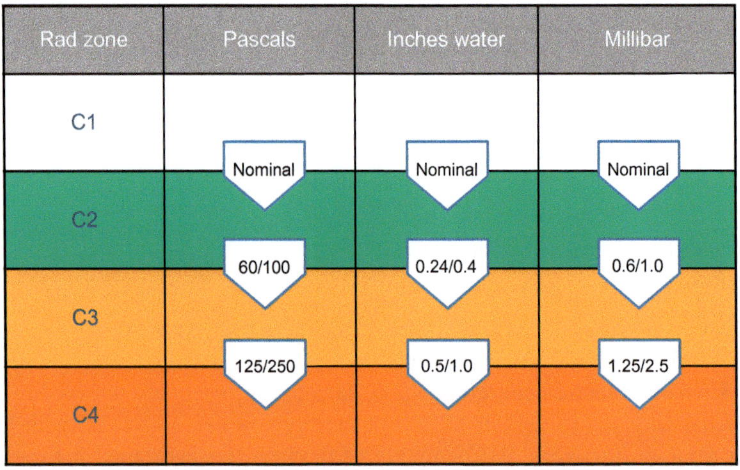

Rad zone	Pascals	Inches water	Millibar
C1			
C2	Nominal	Nominal	Nominal
C3	60/100	0.24/0.4	0.6/1.0
C4	125/250	0.5/1.0	1.25/2.5

Fig. 8.5 Differential pressure between radiological zones.
© Bill Collum.

Fig. 8.5 gives examples of the differential pressures (DPs) which are typically required between adjacent zones. Rather than a fixed figure for each interface, the requirement is to achieve a pressure that sits within the range shown. I have used three types of measurement, so you can pick the one you prefer. UK HVAC engineers ordinarily favor Pascals, while the US preference is liable to be inches water. Having said that, millibar tends to be the most recognizable unit, as it is the one that weather forecasters use when quoting atmospheric pressure. They also work in the hectopascal, which is identical in value to the millibar, so numbers turn out to be the same anyway, which just goes to show that you need to stay alert when dealing with meteorologists.

You will notice that only a nominal or minimal depression is required between C1 (white) and C2 (green) zones. The reasoning goes that a C2 environment should be free of airborne contamination and therefore on a par with fresh air passing by outside. This being the case, there is no imperative to micromanage their interface so a nominal depression works just fine.

8.3.3 Velocity

There is another factor that is very closely allied to the relative depression between various radiological zones, namely the velocity, or speed at which air is traveling. We know that when the wind drops an airborne leaf will fall to the ground, so exactly the same is true for particles of contamination. Having higher and higher negative air pressure between the more onerous radiological zones helps to move things along, but if we want to be sure of stopping contamination drifting in the wrong direction then we also need some airspeed at interfaces between zones.

To illustrate how velocity dovetails with differential pressure we can look at air traveling through a sub changeroom, but keep in mind that the same principles apply to any interface between radiological zones. We discussed changerooms and sub changerooms in Chapter 4 so know that although they are a single room, one airspace, they are divided by a boot barrier which has a different radiological classification on each side.

You will recall that those crossing a sub changeroom sit on the barrier and perform a particular routine, which ensures they do not carry contamination about their person as they move from a high zone to a lower one. The ventilation system then deals with airborne contamination. It adopts a working assumption that it is always present, and that the side of a barrier with the highest classification is more contaminated than its contiguous neighbor.

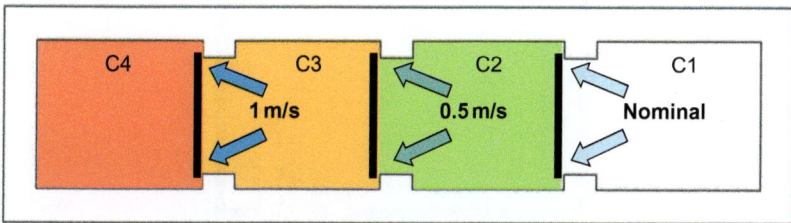

Fig. 8.6 Airspeed at interface between radiological zones.
© Bill Collum.

In the case of a C2/C3 (green/amber) sub changeroom or indeed any such interface between these zones, the ventilation system is designed to keep bulk air moving towards the C3 side at around 0.5 m/s (Fig. 8.6). As you might expect the velocity required at an interface between C3 and C4 zones is greater and so moves continually at around 1 m/s. In everyday discussions air is normally referred to as "crossing the boot barrier" at these speeds, but strictly speaking it applies to air passing through open doors.

With both doors of a sub changeroom closed, the ventilation system will be nicely balanced so have no trouble coaxing air in the right direction. However, once either door opens that balance is disturbed, so the system must be designed to counteract any possibility of air heading off through a door and in the wrong direction. Momentarily opening doors to both sides of a sub changeroom is not a problem, but longer durations may risk contamination absconding. For this reason plant personnel are trained in operational procedures that govern the correct routine for accessing sub changerooms.

So-called "standard" door dimensions vary a lot, so for our purposes we can assume the door, or *interface*, is 1 m wide × 2 m high. When setting up a ventilation system to achieve a particular airspeed, the crucial factor to establish is how much air must be moved in a given period to deliver that flow. The answer lies in

$$\text{Velocity (m/s)} \times \text{Area (m}^2） = \text{Quantity of air (m}^3/\text{s)}$$

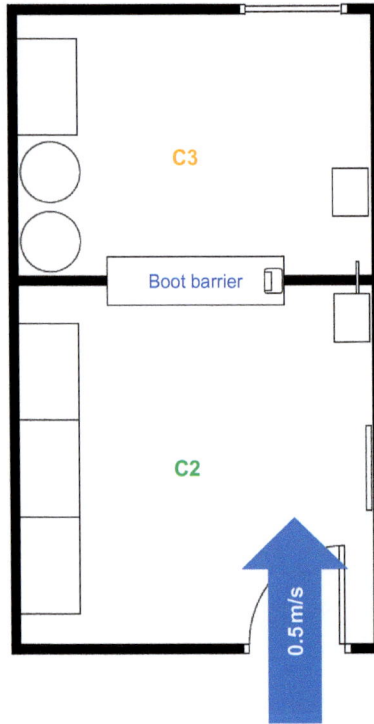

Fig. 8.7 Airflow across C2/C3 sub changeroom.
© Bill Collum.

Since we are dealing with just one opening at a very handy size of 2 m^2, the calculation is easy:

$$0.5\,\text{m/s} \times 2\,\text{m}^2 = 1\,\text{m}^3/\text{s}$$

So even if a door on say the C2 side of a sub changeroom is open (Fig. 8.7) as long as the cascade system is designed to pull 1 m^3 of air through it every second, then it will automatically travel at the required speed. It may be a simple illustration, no doubt far too simple for the liking of our HVAC friends, but for the rest of us it demonstrates the point pretty well.

8.4 Engineered gaps

Closely related to the subject of moving air through personnel doors which are on a boundary between radiological zones, is doing the same thing with hefty shield doors that are in a similar situation. More often than not, apart from their shielding function, they stand between C3 (amber) and C4 (red) areas, so need air to hurry through at a rate of around 1 m/s. That said, it is not always necessary, or indeed practical, to maintain that velocity when a shield door is open.

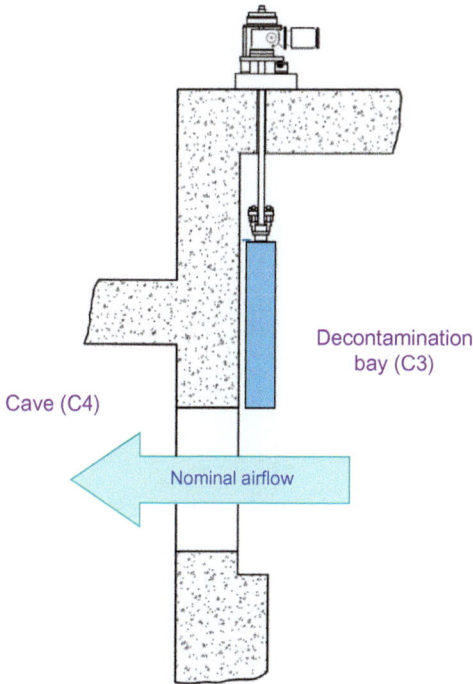

Fig. 8.8 Airflow through open shield door.
© Bill Collum.

Decontamination bay (C3)

Cave (C4)

Nominal airflow

A good example is a door such as that shown in Fig. 8.8, which stands between a highly radioactive cave and its adjacent crane decontamination bay. In a situation like this, when the bay is unoccupied, air need only drift through the open door at a nominal speed, just as long as it moves in the right direction. What matters is that when the door is closed air will squeeze around its edges and flow into a cave at the mandatory speed. I'll tell you why.

You may at some time or other find yourself inside a crane decontamination bay of the kind we are discussing. If you take a close look at the shield door that is protecting you from contamination and fierce radiation within the cave, it may appear to be badly fitted. It will overlap with its concrete opening alright so there is no danger from radiation shine, but the door needs to open and close now and again so will not be a perfectly tight fit and certainly not hermetically sealed. The resulting narrow space is known as an *engineered gap* and it has an important role to play.

We understand the cascade system, so realize that air is being drawn into the cave, but with potentially gross contamination to contend with we may reasonably wonder if some of it might find its way to where we are, just on the other side of the door. So this is where the 1 m/s comes in. With air hurrying through the gap at such a speed, there is precious little chance of contamination being able to flow in the opposite direction. It turns out then, that what may look like sloppy detailing around a shield door is a

carefully considered engineered gap, one which is helping to deliver containment control. In fact, in the nuclear world, sealing rooms up tightly is seldom a good idea.

Fig. 8.9 Adjustable
airspeed fin.
© Bill Collum.

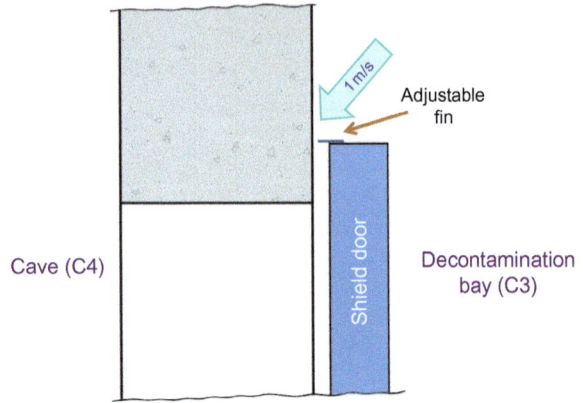

To help get air velocity just right the sides of shield doors have fins attached
(Fig. 8.9) which can slide back and forth so as to make fine adjustments to the width
of an engineered gap. The steel fins are adjusted during a facility's commissioning
phase, when a cave's ventilation system is being balanced and when airspeeds are just
right, locked in place.

8.5   Maintaining containment at truck bays

So we have seen something of how air moves along and contributes to a facility's
containment philosophy, but sometimes persuading it to do our bidding needs even
more of a helping hand. Picture a facility with a large truck bay at one end. It has
to have direct access to the road outside so the bay will be categorized as a C1 (white)
area—in other words, regular fresh air. If you looked up at the soffit of the truck bay
you would see the underside of a floor within the building, let's say at a level of 8 m.
Of course the whole point of this type of truck bay is that they are used to import and
export heavy flasks and other packages. So typically the soffit of a bay has a large
hoistwell within it to accommodate transfers in and out of the facility, usually via
a crane located within the body of the main building. To complete the picture the
hoistwell, up at 8 m level, can be opened and closed by sliding a horizontal door across
it. This type of opening forms a major interface between the nuclear side of a facility
and the world outside. So although the truck bay is categorized as C1 (white) the area
immediately above it is invariably categorized as C2 (green).

We have seen how the cascade philosophy is applied to air movements within a
nuclear plant, so will quite rightly assume air must be drawn into the building from
a truck bay and not the other way round, otherwise C2 air could drift down from a
hoistwell and out through the large industrial doors. The thing is that truck bay and
hoistwell doors are so large there is no way a ventilation system could possibly pull
sufficient depression if both doors were left open at the same time. Worse than that if a
strong breeze was blowing outside then C2 air would not just drift down into a truck

bay, it would be sucked out of the building with some force. Even more problematic, the ideal truck bay configuration has a large door at each end, so that vehicles can pass through without reversing. Imagine the drafts that would be generated if three large doors were left open at the same time. This is where interlocking comes in.

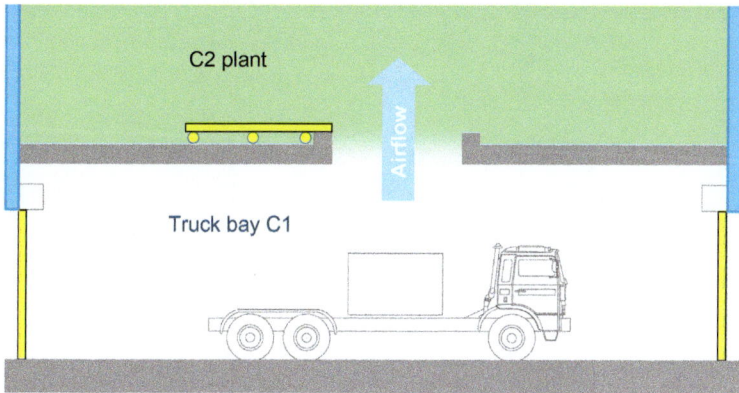

Fig. 8.10 Interlocked doors.
© Bill Collum.

In cases such as these, doors must be interlocked so that a hoistwell can only be opened when the large doors which give access to the street are closed (Fig. 8.10). If, as an example, we follow the import process, we shall see the sequence is straightforward enough. We just need to keep in mind that engine fumes are dealt with by a separate local ventilation system.

Interlocks see to it that the hoistwell door remains closed while a vehicle enters the truck bay through one of its large industrial doors. Once the vehicle is inside, both external doors are closed and a series of package import procedures carried out. With the external doors still closed, it is only then that interlocks will allow the hoistwell door to open and affect a package transfer. Sequencing door opening in this way ensures a ventilation system is not called upon to do battle with the elements outside, so will have no difficulty keeping C2 air from tumbling down through the open hoistwell and out of a building.

8.6 Maintaining containment on building perimeter

There is another important principle that sits alongside that of the cascade philosophy, namely that it is not always acceptable to locate an area classified higher than C2 (green) on the outside wall of a building. A constraint that is all to do with the risk of contamination escaping through a building's external shell. There is no problem with locating C1 (white) zones on the perimeter, since in radiological terms they are no different to air outside. Similarly there is no restriction on locating C2 zones behind a building's external skin, as these areas are not subject to airborne contamination. However, once we get to zone C3 (amber) and above it is not quite so straightforward, because now airborne contamination is not a theoretical possibility. It will almost certainly be there, particularly in C4 (red) areas where levels can be very high indeed.

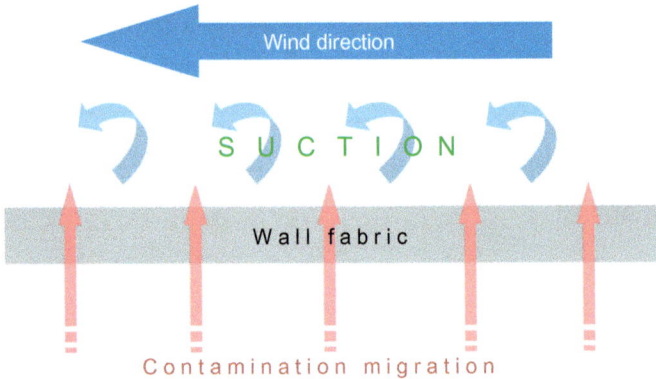

Fig. 8.11 Wind suction.
© Bill Collum.

Therein lies the basis for a principle of not straying above a C2 zone around a building's perimeter. The concern is that airborne contamination, present in higher zones, could migrate through external walls and spread outside. We have seen how high contamination zones are held at a constant depression, or negative pressure, so might reasonably assume that any seepage would be drawn *into* a building rather than the other way round. However, it is possible that in the right conditions, or maybe that should be wrong conditions, wind passing by a building (Fig. 8.11) could cause enough suction, enough negative pressure, to draw contaminated particles through imperceptible fissures in a building's external brickwork, cladding, and so on.

Any proposal to locate an area rated C3 on the perimeter of a building must therefore be assessed on a case by case basis. For example, on its own standard insulated building cladding is unlikely to provide sufficient containment for this type of environment. However, if an additional impermeable barrier is installed behind the cladding, then it is normally permissible to locate such a zone directly behind a facility's external wall.

So with adequate safeguards in place it can be acceptable to locate C3 zones on the perimeter of a nuclear plant, but what about C4 zones where highly radioactive materials are housed? Ordinarily we can expect these areas to be enclosed by hefty shielding, say a meter of concrete, so surely we do not expect contamination to make its way through such thick walls?

By now we are accustomed to the idea that when it comes to catering for extreme possibilities, the design of nuclear facilities is necessarily extremely cautious. So discovering that we must consider the possibility of contamination making its way through brickwork or building cladding does seem to fit the bill. However, you

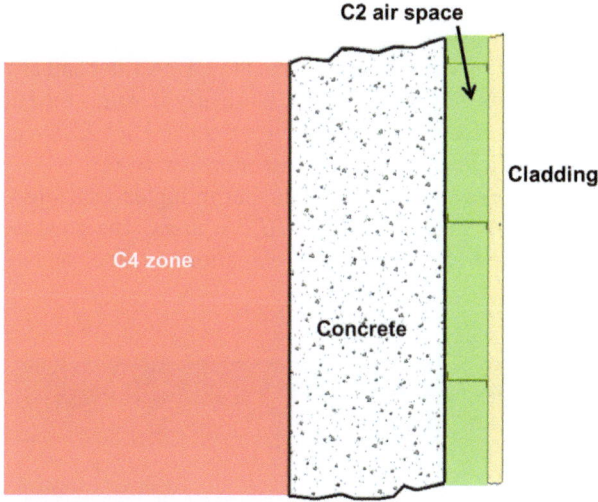

Fig. 8.12 Containment of a C4 zone.
© Bill Collum.

may be a little surprised to hear that designers must also consider the possibility, due to the same strong winds, of contamination escaping from C4 (red) areas by migrating through fine cracks in thick concrete walls.

In reality, the issue does not arise too often, as access requirements often dictate that such areas are located deep inside a building and surrounded by zones of a lower category. These lesser zones then act as a buffer between high levels of contamination at the heart of a plant and the environment outside. However, occasionally C4 zones do find themselves behind a building's external walls.

On those occasions where there is just a concrete shield wall between a C4 zone and the outside world, then ordinarily the wall's integrity must be enhanced with some additional protection. A typical solution is shown in Fig. 8.12 where building cladding, or siding as it is known in the United States, is fixed to sheeting rails which are fastened to the shield wall's surface. Apart from sheltering walls from the effects of passing winds the resulting space, which may be 200–300 mm deep, is categorized as a C2 radiological zone. A narrow cloak of C2 air might not sound like much but as with all such environments it is overseen by a facility's monitoring systems, so provides excellent insurance against contamination absconding from a C4 environment.

8.7   Filtration

We cannot go too far on the subject of ventilation, particularly the nuclear kind, before we get onto the crucial subject of how air is filtered. All of the air that flows through a nuclear facility, even its most highly contaminated areas, eventually ends up back outside where it started. Environmental regulations demand that aerial discharges from all nuclear facilities are thoroughly cleaned, which is exactly what we would insist on ourselves.

8.7.1   HEPA filters

The device used to clean up contaminated air is the HEPA filter and as is so often the case it is an acronym with several meanings. Normally it stands for High Efficiency Particulate Air filter, but also has other meanings, such as High Efficiency Particulate Arrestance and High Efficiency Particulate Arrestor. Just to keep us on our toes, they are also referred to as *absolute* or *true* filters and in higher echelons of the HVAC community you may even hear talk of *H14 HEPA filters*. The good news, for those of us who are not enthralled by interesting variations in filter names, is that they all mean the same thing.

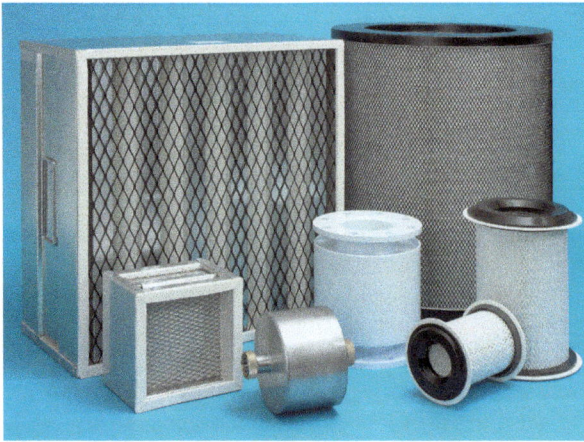

Fig. 8.13 HEPA filter types.
© M C Air Filtration.

HEPA filters, which have a pleated fiberglass core, are available in two generic shapes, rectangular and round (Fig. 8.13) with the round, or circular varieties tending to be the profile of choice for the UK nuclear industry and many others around the world. In the case of rectangular filters air is drawn through from one side to the other, which results in trapped particles accumulate on the surface exposed to incoming air. With circular filters, air enters via their central axis which is blocked at the opposite end, forcing air to push through the filter around their perimeter. The attraction with circular types is that even though used filters are contained and carefully handled, if contamination does come loose then, that which has not penetrated the filter will gather in the void at its core rather than fall away from a flat surface, as can happen with rectangular filters.

In addition, filters with a circular profile do not need to be aligned in a particular orientation when being inserted into their housing. This helps to speed along manual change operations and is even more useful when filters are being handled and changed remotely. Furthermore, sealing arrangements for circular filters are generally viewed as having the edge over those on their rectangular counterparts.

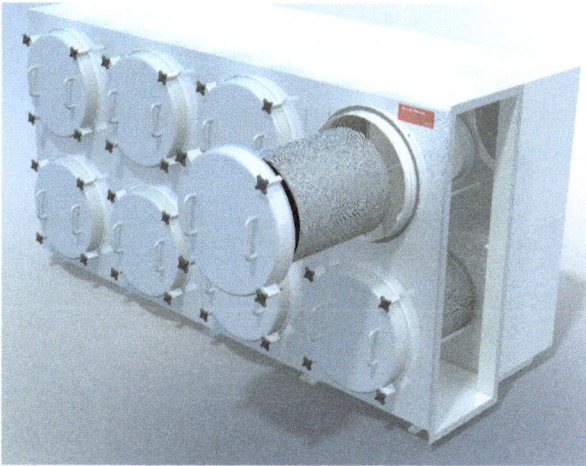

Fig. 8.14 HEPA filter bank.
© M C Air Filtration.

The maximum flowrate that can be achieved through a standard circular HEPA filter, of nuclear grade, is a little under 1 m^3/s, or 0.95 m^3 to be precise. So a requirement to handle say 7.5 m^3 of air per second will demand a bank of 8 HEPA filters (Fig. 8.14) or *inserts* as they are known. It is an important consideration, one that can figure quite prominently during the design optioneering phase. For example, thoughts of using high volumes of air to keep an area cooled may need to be discounted due to the sheer number of HEPA filters that would be required to make the system operable.

HEPA filters have an efficiency of 99.97% in capturing particles down to submicron size. To put that in perspective, 1 micron (μm) is equivalent to 1 millionth of a meter; human hairs are around 70 μm across and if a particle is less than 10–15 μm across, it will not be visible to the naked eye.

The science in this area is extremely complicated but in essence boils down to two things. Firstly, a HEPA filter's 99.97% efficiency is ascribed to the most penetrating particle size normally taken to be 0.3 μm, traveling at a filter's rated flow capacity (0.95 m^3/s for the filters we are considering here). Secondly, HEPA filters are even more efficient at trapping particles of a different size—not just those which are larger than 0.3 μm, but rather surprisingly smaller particles as well. Clearly HEPA filters can trap very small stuff but there is no getting away from the fact that 99.97% is not the same as 100%, so in some situations we need to do better.

8.7.2 HEPA filtration—Beta-gamma plants

The approach taken differs a little depending on the radiological inventory being handled within a facility. Ordinarily, for plants housing beta and gamma emitting materials, air circulating within C2 (green) areas is considered free of airborne contamination, so does not need any form of filtration before being discharged. That said, and as we shall see in Chapter 11, C2 environments are continually scrutinized by *alpha and beta particulate monitors* which double-check that all is as it should be. If traces of contamination are detected then extract systems can be shut down while an investigation is conducted and the situation recovered.

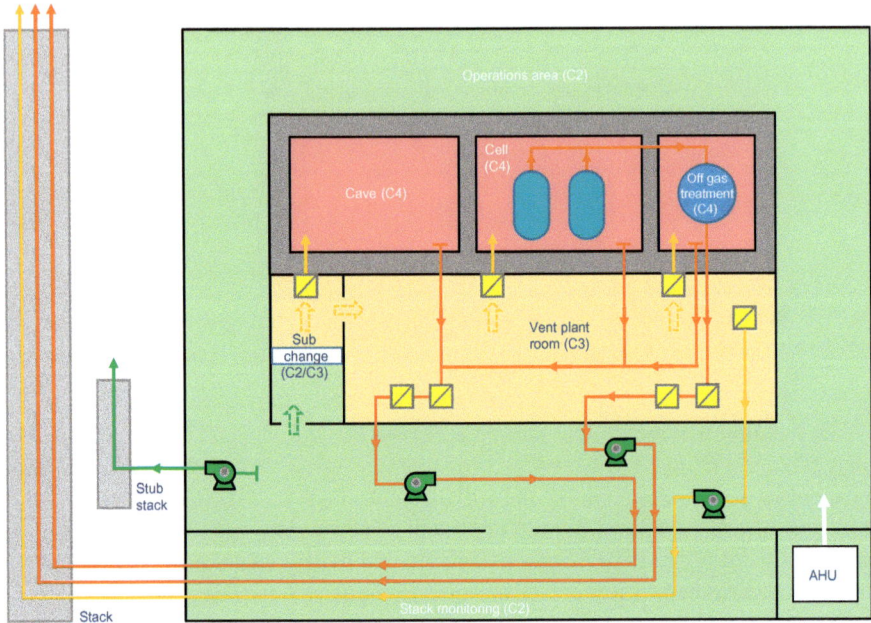

Fig. 8.15 HEPA filtration—beta-gamma plants.
© Bill Collum.

The level of HEPA filtration required for C3 and C4 zones can vary a little from one beta-gamma facility to another. So case by case is determined by calculations performed by a project's process engineers, along with some input from radiation and safety specialists. Ordinarily, the minimum expectation (Fig. 8.15) is that air which has passed through C3 areas will be routed through a single bank of HEPA filters (shown in yellow) before being discharged, while air that has occupied the higher C4 zones passes through two banks of HEPA filters before it is released back to the surrounding environment. In other words, two times 99.97% filtration of those submicron particles, which is pretty thorough However, even that is not good enough for an alpha plant.

8.7.3   HEPA filtration—Alpha plants

You will recall from Chapter 2 that alpha particles pose quite a different threat to those arising from gamma materials. So although they require comparatively little shielding, alpha substances can be particularly harmful if inhaled or ingested. For this reason, if following a UK type zoning philosophy, alpha plants have at least one additional layer of extract filtration (Fig. 8.16).

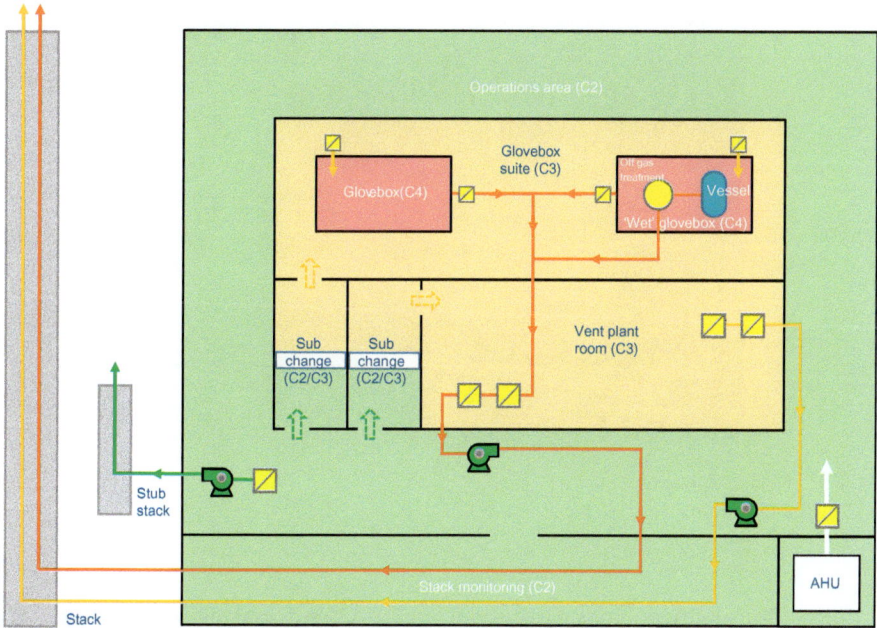

Fig. 8.16 HEPA filtration—alpha plants.
© Bill Collum.

Again, exact details are confirmed by the process engineering team, with some assistance from other specialists. Ordinarily, unlike a beta-gamma plant, incoming air passes through a bank of HEPA filters as it enters a building. Air being discharged from C2 areas is routed through a single bank of HEPA filters, and that which has occupied C3 areas through two banks. Any air which originates from C4 (red) zones, typically within gloveboxes, must pass through three banks of HEPA filtration before being discharged again.

8.7.4 Mobile filtration unit

Clearly ventilating a whole nuclear facility can only sensibly be achieved by a permanently installed system, an extensive one at that, but for ad hoc activities of a temporary or semipermanent nature, we can turn to the rather nifty mobile filtration unit (MFU). Invariably they are skid-mounted and in most cases compact enough to be wheeled through a standard-sized door opening.

The example shown in Fig. 8.17 houses two circular HEPA filters, or *inserts*. They can be assigned as primary and secondary filters and employed to filter air from C4 (red) areas, or alternatively filter higher volumes of air from C3 (amber) areas. In this arrangement contaminated air is drawn into the top flue through a flexible duct, passes through the HEPA filters and is expelled via the lower flue. Filtered air can be blown directly into a facility's C2 (green) environment, or by use of flexible ducting routed elsewhere.

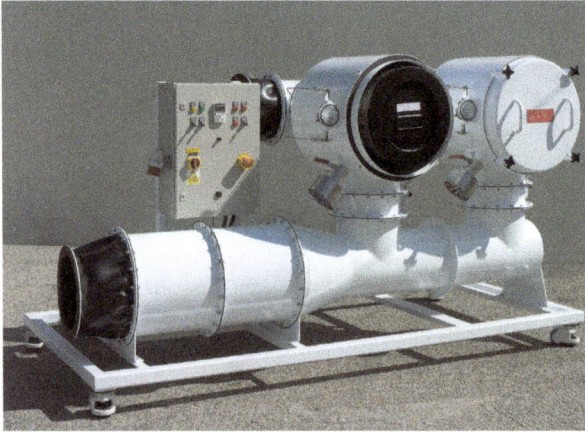

Fig. 8.17 Mobile filtration unit.
© M C Air Filtration.

Other MFU configurations carry a single HEPA filter, making them suitable for filtration of C3 (amber) areas. All in all then, MFUs are an invaluable piece of kit and a common sight around nuclear sites, particularly during decommissioning. Speaking of which....

One of their neatest applications occurs during decommissioning, when we quite often see air from MFUs routed directly into a facility's primary ventilation system. It is achieved by attaching flexible ducting to connection points, which are preinstalled in permanent ductwork as part of its original design. Of course this means the HVAC folks must plan, decades in advance, for scenarios where MFUs may be needed in the future. "All part of the service" as they are only too keen to remind us.

8.7.5   Air in-bleed filters

You will recall we discussed engineered gaps a little earlier and their contribution to cascading air into C4 (red) areas. However, on their own they are not a terribly scientific way of routing air into environments that demand a finely balanced air supply. Even more problematic, it is possible for a ventilation system to be momentarily overpowered and send air in the wrong direction, dragging contamination in its wake. The phenomenon is known as a *back pulse, or backflow* and can happen due to say a sudden pressure release from the ullage of a processing vessel. There are two ways of combating the possibility of C4 air heading the wrong way, the first of which makes use of a HEPA filter.

An air in-bleed filter can be used to supplement flows into C4 areas and at the same time provide a means of trapping contamination should a back pulse occur. In fact we could think of them as the HVAC equivalent of a pressure relief valve. They are quite a

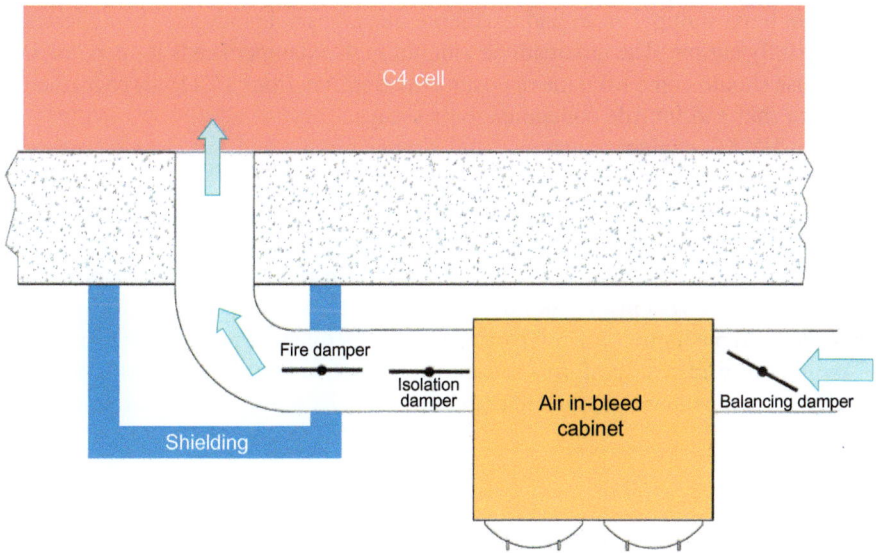

Fig. 8.18 Air in-bleed filter.
© Bill Collum.

simple affair really (Fig. 8.18) with HEPA filters housed in a cabinet which is connected to a cave, or cell via a short length of ductwork. Ductwork adjacent to the cabinet contains three dampers: one that closes in the event of a fire; an isolation damper which is closed during filter changeout operations; and a balancing damper which can be rotated so as to make fine adjustments to airflows and depression within the area being served. Apart from their important safety role, air in-bleed cabinets can also house conditioning plant that may be required to heat or chill air before it is routed into the shielded environment.

We can make sound arguments in support of locating in-bleed filters in C2 (green) areas rather than C3 (amber); after all, they may spend their entire operational life, as many do, waiting to capture a pulse of contamination but never be called upon to perform. However, in the event that it is needed, an in-bleed filter will become contaminated. With this in mind it may be prudent to invoke the principle of *defense in depth* and locate in-bleed filters in a C3 area, as by definition they are already setup to deal with any particles that might possibly escape during the filter change procedure.

Although such an escape is unlikely, many designers nevertheless consider it good practice to search out a C3 area that can accommodate their in-bleed filters. Where possible, the usual preference is to locate them either in a C3 workshop or on the C3 side of a sub changeroom. It does mean one of these areas needs to be stretched a little, but on the upside it saves having to provide dedicated containment that may never be called upon to fulfill its role.

As an alternative to the in-bleed filter we can instead install a pressure regulating damper. In their "single vane" version, they look like a sophisticated cat flap. Other variations have multiple vanes and look more like adjustable louvers. Their operation is beautifully simple. The airstream passing through a damper keeps it open, but if a back pulse should come from the opposite direction, say from a C4 (red) cave, then it will slam shut and impede contamination's escape.

Many HVAC engineers expound the simple virtues and low cost of pressure regulating dampers, while others are a little wary of them, expressing concerns that their vanes may jam at the very moment they may be called upon to close in a hurry. In the interests of fair-mindedness, such fears can be expunged by ensuring there is an appropriate maintenance and testing regimen in place. However, it has to be said that the two systems are not quite comparable. Whereas an in-bleed filter is designed to trap submicron sized particles, a damper cannot provide the same level of assurance so instead performs more as a restrictor. With that proviso, both are equally serious contenders, such that appropriate selection is dependent on specific circumstances.

8.7.6 Manual versus remote filter change

Circular HEPA filters, which are typically around 600 mm long and 500 mm in diameter, do not last forever. Gradually they become clogged up with tiny particles and need changing, or simply exceed their shelf life. When developing designs for new facilities and looking ahead to the time when spent filters will need to be replaced, quite often an interesting problem arises when considering how that change will be effected.

It all comes down to a debate on whether filters will be located in an easily accessible area and replaced manually, or located behind thick concrete shielding and changed remotely. Everything hinges on how much contamination may be airborne, and therefore drawn towards filter banks, along with what isotopes may be present, and from that determining the levels of radiation emitted by particles of contamination as they build up on filters.

At the extremes of the debate on whether to opt for manual or remote change, calculations are straightforward and the answer is obvious. For example, if particles trapped on filters originate from materials at the low end of the intermediate level waste (ILW) spectrum and from an environment where little airborne contamination is present, then even those that are quite heavily loaded can be managed to remain within low level waste (LLW) limits. So in this case HEPA filters can be approached and changed by hand. Having said that, safety considerations always dictate that any exposure to radiation is minimized, so a filter change team will not stand around discussing the previous evening's TV. The exchange will be carried out and they will be on their way.

On the other hand, if contamination originates from highly radioactive sources that also generate appreciable airborne contamination, then the solution is equally unambiguous; here even small accumulations will emit significant radiation, so these filters need to be located behind hefty shielding and out of harm's way. Let's look at manual change first.

Fig. 8.19 Manual filter change (testing).
© M C Air Filtration.

If you think of changing the air filter in your car you can unclip its cover, take it out, hold it up to the light and have a good look at it, then if you do decide to change it you can leave the old one on the driveway until you get around to tidying up. Changing HEPA filters which are loaded with contaminated particles is not such a casual affair. Happily, the technique used to change them, without breaking containment, is another good example of a simple low-tech solution to a tricky problem.

In essence, it is accomplished by withdrawing a spent filter into a plastic bag (Fig. 8.19) which is then heat sealed to ensure any loose contamination is contained. In addition, the procedure for fitting a new bag includes wrapping and heat sealing the remainder of the previous bag, so that it too can be safely disposed of.

We do not need a calculator to figure that the costs, both capital and operational, associated with remotely changing filters in a shielded cave (Fig. 8.20) will be exponentially higher than those of a simple hands-on approach. Sometimes costs can be minimized considerably by locating highly active filters within a cave which is needed for some other purpose, rather than building a dedicated filter cave. That way we get most of the filters remote handling equipment for free, so to speak. Of course the cave itself would need to be stretched somewhat to accommodate a bank of HEPA filters, but that is a relatively small price to pay.

Quite often, other factors come into play that scupper plans to locate HEPA filters within a convenient cave. There are obvious issues such as imposing potentially unwelcome changes to the way an operational cave is configured, and other more subtle ones such as diverting remote handling equipment from important production duties, something that is unlikely to go down too well. The upshot is that some very detailed analysis will be needed to evaluate the pros and cons of providing a dedicated filter cave, versus forcing an operational cave to take on a certain amount of multitasking.

Fig. 8.20 Remote filter change.
© Nuclear Decommissioning Authority.

So we have considered both extremes, where on the one hand it is crystal clear that filters can be changed manually and on the other there is no doubt they must be shielded and changed remotely, either in a dedicated cave or by using space in one provided for other operational purposes. Unsurprisingly, many cases sit in that awkward gray area between the two. The favored solution is always going to be manual change, so the challenge is striving to achieve it, while at the same time complying with the nuclear industry's myriad of regulations. In this particular case, the main objective is to ensure any exposure to operators is, *as low as reasonably practicable*, a subject we shall discuss in Chapter 14.

There are some fairly simple techniques that can make a contribution, such as constructing shield walls between banks of filters (Fig. 8.21) so that operators

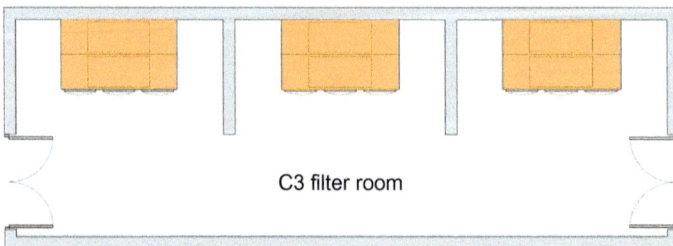

C3 filter room

Fig. 8.21 Shielding between filter banks.
© Bill Collum.

exchanging a filter in one bank are not needlessly subjected to radiation shining from others nearby. I must stress, however, that such shield walls are certainly not the primary form of radiation protection for a workforce, but are rather a prudent measure that can be utilized to keep operator exposure to the lowest possible level.

In striving to achieve manual change there is one design approach that, in the right circumstances, can work very well indeed. If we step back from the usual assumption that filters should be fully loaded or close to it before they are changed, then there can be some room for maneuver in avoiding potentially dangerous levels of radiation accumulating on them. With this approach filters are exchanged before high levels of contamination, and therefore radiation, get a chance to build up. The downside is easy enough to see; filters must be changed more often.

8.7.7 Filter life

The life of a HEPA filter is generally considered to be around 7 years from its date of manufacture, with maybe 2 years of shelf life and up to 5 years in operation. In practice their actual operational life can vary considerably, depending on the particular radiological burden and environmental conditions to which a filter is subjected. Typically they remain in service for say 12–18 months, but in favorable conditions may even exceed the normal expectation of 5 years.

However, if they are to be replaced on reaching a predetermined level of radiation, which allows them to be handled manually, then they may only be employed for around 3–6 months. With so many variables, it is not possible to make a generic comparison between the numbers of filters that may be needed in a manual versus remote change situation. What we can say is that over the operational life of a facility many more filters will be required if the manual option is selected.

But how do we know when it is time to change them?

8.7.8 Determining when to change filters

If you remember tea when it could only be bought loose, in little boxes, the technique used to determine when it is safe to manually change a filter is easy to follow. Then again if you have been brought up on teabags you may need someone older and wiser tell you how it was in bygone times, when tea was poured from a pot and irksome tealeaves came with it. To keep your brew free of these tiny pollutants you held a strainer in the flow of tea to capture the leaves. At first tea would pass straight through the strainer with no noticeable effect on the flow, but as more and more leaves gathered in its bowl the flow would reduce to a trickle.

It was most noticeable if you filled several cups without emptying the strainer, as by the time you reached the last one tea could barely get through the accumulation of leaves which were blocking its path. Exactly the same thing happens with HEPA filters; it is just that there is more science involved in ensuring they are changed before becoming impenetrable.

The mental image of a HEPA filter becoming more and more clogged up with contamination, often leads to an assumption that air passing through a "dirty" one will somehow be less clean when it exits. Interestingly enough the opposite is true. As pores within a filter begin trapping particles, their apertures narrow, which makes them more efficient than a perfectly clean one. The downside is that air trying to get through meets increasing resistance so needs more energy, delivered by adjusting the variable speed centrifugal fans, to move it along. However, our HVAC friends have managed to turn this minor irritation to their advantage.

Resistance faced by air trying to get through a filter can be quantified, which in turn can be used to determine just how clogged up it is, or *blinded* as the HVAC folks say. Instruments measuring the pressure drop across a HEPA filter; air rushing up on one side and trickling out on the other can warn plant operators when it is time to make a change. So we do not need to examine a filter to know how much contamination it is holding, or in HVAC speak, how *loaded* it is.

This same technique is used to keep a check on filters that are destined to be changed when they reach a predetermined activity level, one which allows them to be handled and changed manually. Sources of upstream radiation will be well known and quantifiable, so armed with that information it is not too difficult to predict how much contamination a filter will be holding when it reaches a level that would be too dangerous to approach, and from there we can extrapolate what the air resistance will be. Differential pressure (DP) instrumentation, set to a safe level, triggers an alarm in plenty of time to warn plant operators that a manual filter change is due. As a final check filters are also monitored to confirm that radiation levels are indeed as predicted.

8.7.9   Filter disposal

It may seem that changing filters more frequently is not too much of an issue, particularly as filters themselves are relatively inexpensive; unfortunately, the same cannot be said of their disposal. It is a factor that becomes more significant when we bear in mind that filters are housed in cabinets, with generally with between four and eight in each, and they are all changed at the same time. In addition, many areas are served by banks of several cabinets, so there can be a fair number of filters involved. To minimize costs, filters can be compacted before they are containerized and dispatched to their disposal destination, even so the associated costs will still be high. The tricky arithmetic comes when we tot up figures involved for the whole life of a facility, which might be say 20 or 30 years.

The relevant cost analysis comes down to weighing two extremes: the high capital cost of providing remote change facilities for a relatively small number of filters that will reach ILW levels, along with their high disposal costs. Versus minimal upfront expenditure on manual change equipment, along with a lifetime burden of purchasing considerably more filters, but paying less for their disposal as LLW.

Of course there is always pressure to minimize initial capital expenditure, as by its very nature it looks good and the plaudits are more or less instant. However, we must avoid any temptation to take the comfortable short-term route when deciding how HEPA filters will be deployed, and instead think long and hard about a plant's lifetime of operational costs. It is not easy is it, but if we analyze the pros and cons of manual versus remote change objectively, then we can come up with the right answer.

8.7.10 Push-through filters

Where we consider ventilation of the caves and cells that prevail in beta-gamma plants, then airflow rates, levels of contamination, or both, dictate use of the 600×500 mm diameter HEPA filters that we have discussed so far. However, where relatively small areas are being ventilated, which includes the gloveboxes we find in alpha plants, then it opens up the possibility of using a more compact HEPA filter.

As with their larger counterparts, push-through filters come in various sizes, but are typically in the order of 400 mm long and 150–200 mm in diameter. Being so small, the volume of air they can handle drops from 0.95 m^3/s (950 L) to around 150 L/s. Of course they can be deployed in multiple units so can cope with a reasonable volume of air, but once we approach double-digit numbers of these filters they lose their appeal and we are better sticking with the larger types.

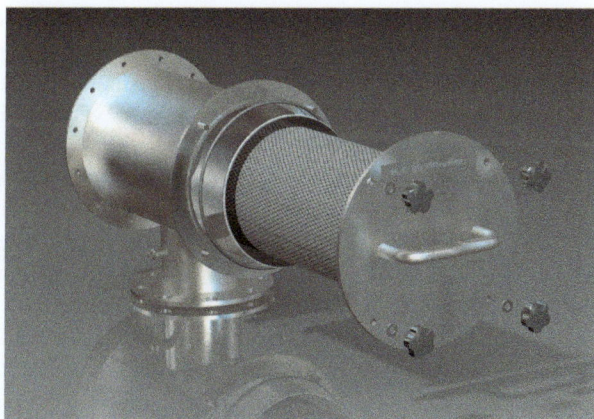

Fig. 8.22 Push-through filter partly inserted into housing. © M C Air Filtration.

Push-through filters derive their name from the way they are changed, which interestingly enough straddles the line between the purely manual or remote techniques which are employed for larger HEPA filters. In their operational state these filters are loaded into a stainless steel tube (Fig. 8.22) and, even in the normal vertical orientation, are held in place by a pair of silicone rubber O-rings that grip the tube's sides. Just as with larger HEPA filters, air enters via the filter's central axis which is blocked at the opposite end. This forces incoming air to push through the filter medium and leave its contamination behind, before exiting again via a smaller tube on the housing's side.

When the time comes to remove a spent filter, an operator opens the accessible end of its housing and inserts a new filter, which in turn pushes the spent filter out of its stainless steel housing. The now redundant filter either falls directly onto its collection point, or is directed there by a chute. Either way it is ready to be picked up, manually if in a glovebox or remotely in a cave, and dispatched via the facility's waste disposal routes.

There is no doubt that in the right situation, push-through filters do provide rather a neat and elegant solution, particularly when associated with the gloveboxes we find in alpha plants. For beta-gamma plants, opportunities to use them can be quite limited.

8.8   Air conditioning

If you get into a car on a very hot day, you have two options to cool it down and keep it cool: If a car is fitted with air conditioning you can keep the windows closed, open the air vents, and turn the air conditioning to maximum chill. In fact, and this is still part of option one, if the car has climate control you can just select a nice cool temperature, say 16°C and let the car's systems do the rest. Alternatively, option two, if a car is devoid of air conditioning you can set the temperature dial to the cool end, turn the vent fan up to full speed, drop the windows, and off you go. With our first approach the ambiance within a car remains quite serene, while with the other wind buffets your ears and you may not be able to hear the radio too well. However, both methods will get the temperature down to an acceptable level.

When it comes to cooling buildings, the options are more or less the same. On a warm day, just as with our well-specified car, we can chill air entering a building before directing it to its ultimate destination. But then it costs money to cool air, even in a car you may notice a slight dent in fuel economy. The alternative is often misleadingly referred to as *free air cooling*. Instead of chilling incoming air you draw in huge quantities and rely on its sheer volume to deliver the cooling instead. This approach needs more powerful fans, which must revolve at significantly higher speeds, and the increased volume of air can lead to a proliferation of very large ductwork. What's more, care is needed not to force too much air through a building, as it could compromise the finely balanced flows required of a cascade system.

The reality then, is that this apparently simple method of cooling a building has its challenges and certainly does not come for free, not even close. The ductwork itself, along with space needed to house it has a cost, and all that extra ventilation equipment will not come cheap either. But the costs that really grab our attention are associated with electricity usage and an increased number of HEPA filters.

As with any major industrial facility, the cost of cooling a nuclear plant through decades of operations could easily run to several million pounds, or dollars, a factor which, on fine days at least, is compounded by running beefed up fans. As we know standard circular HEPA filters, of nuclear grade, are rated for a flow of 0.95 m^3/s. So increased volumes of cooling air demands more filters, which results in more radiological waste, which is expensive to process and store, and so it goes. The upshot is that in those parts of the world that enjoy an agreeable climate for most of their year, the so called *free cooling* option is not even a contender. However, in the UK it is definitely worth a closer look.

You may have guessed that the debate on how to cool a building on a warm day is only half the story, particularly in a climate such as that experienced by the UK. On those days when basking in the Sun is a distant memory, no matter what system we use incoming cold air needs to be heated, normally with low pressure hot water (LPHW) or with electricity. If operating a *high air volume* option, then the massive quantity of air that had previously been used to cool a building on a fine day may now need to be heated. Then, cooling with lots of air, which may have had something going for it on a pleasantly warm day, does not look quite so appealing after all.

The reality therefore, is that most ventilation systems have some flexibility built into them and house a combination of equipment capable of delivering high volumes of cooling air, or alternatively chilling less air, or indeed heating it, when it becomes necessary. On the face of it selecting appropriate cooling and heating systems looks reasonably straightforward. Study the meteorological data, cost the plant and equipment required for different configurations and estimate a lifetime of running costs for the various contenders. Yes, HVAC engineers would not be too stretched if that exercise was sufficient to make an informed decision. Just to make it interesting, there is a twist in the tale—actually, a couple of twists.

8.9 Heat recovery

Some of the processes and materials handled inside nuclear facilities generate heat, very often considerable heat. For example, molten glass produced within vitrification or glassification facilities flows at temperatures of well over a thousand degrees centigrade. Likewise, significant is heat is produced due to the decay of other radioactive products. Once processed, such materials may be deposited in stainless steel containers and placed in large storage vaults, each holding anything from a few hundred, to in some cases several thousand containers of radioactive waste, all pumping out heat like a huge storage heater. Added to these major sources, significant heat is also generated by electrical equipment such as motors, pumps, and fans. The reasoning goes that if heat generated within a facility could only be harnessed, which it can, then when the temperature drops outside it could be put to use in reducing the need to heat incoming air.

So here we have another angle for the HVAC folks to think about. It might just be possible to get the best of both worlds. Make use of *free cooling* during summer months and still manage to be practically self-sufficient for the remainder of the year, by recovering heat that would otherwise be wasted, a theme we shall touch on later when we examine how ventilation plant rooms should be arranged. It all comes down to the specifics of heat generating activities within individual facilitates and of course their geographical location. There's something else.

8.10   Solar heat gain

I recall some years ago working on the design of a new facility, when the lead HVAC engineer dropped by to ask if I knew what color the building cladding, or siding would be. At the time we were in the very early stages of design evolution, just getting to grips with various options on how the main processes might be configured and how other areas of the building might be zoned. I thought it an odd question for a project that was just coming out of its starting blocks, so made some smart remark about architects being the only people to get excited about such things. I felt guilty straight away. It was nothing to do with ventilation, but because my architect friends quite rightly get infuriated by jibes about them picking colors. Fortunately there were none around, so there was no harm done. It was clear, though, that my HVAC colleague had a serious question and was genuinely interested in the building's color. It turned out he was working on the effects of a phenomenon that sheep have understood for a very long time.

You may have noticed that if you drive through countryside at the end of a sunny day, sheep, if there are any around, will be lying along the sides of the road. They may not know too much about road safety but they are familiar with the basic principles of solar gain, or *solar heat gain* as it is more properly known. Radiant energy from the Sun is absorbed by any surface exposed to direct sunlight and the darker the color the more energy, or heat, it soaks up.

The dark asphalt used to surface so many roadways is an excellent example of a heat absorbing surface. Sheep love it so much because the depth of road construction below the asphalt allows it to retain warmth, much as storage heaters do, then slowly release that energy long after the Sun has gone down. In the same way, a dark-colored building absorbs much more heat than an identical pallid one. And much of that heat penetrates the building's outer skin and warms up the area inside. It then falls to the HVAC team to deal with that heat and keep temperatures inside at an acceptable level.

In this particular case, the cladding color was an important consideration, because the building we were working on was destined to spend most of its days bathed in glorious sunshine—not in the UK, then. I think even the architects were a bit stunned when I raised the matter, but they made out like they dealt with such emergencies every day and quickly came back with an answer. They confirmed the facility's outer skin would be predominately silver, which although it was chosen because it looks pretty good also has the added benefit of reflecting some heat away from a building. All in all then, the cladding was exactly what the HVAC folks would have selected themselves.

So when it comes to cooling a building the HVAC team have a lot to wrestle with, from whether to chill incoming air or force more of it into a building, to balancing cascade airflows, factoring in prevailing weather conditions, the cost of electricity, waste disposal and even a building's color. Still, they enjoy a challenge and it keeps them quiet.

8.11   The ventilation sequence

Now that we have seen something of the role ventilation plays in maintaining radiological containment; along with examining HEPA filtration and challenges of keeping buildings at the right temperature, we are ready to examine how the theory is applied to real buildings.

Many large nuclear facilities have an administration center, or services annex attached to them. Typically, they house office accommodation, conference rooms, training facilities, dining areas and the like, and maybe some electrical equipment for the adjacent nuclear facility. When these C1 (white) blocks are present they will almost certainly have their own independent ventilation system, all of which is no different to that found within any other commercial premises. So we shall concentrate here on how air flows through the nuclear facility next door.

As you might expect, there is a logic to how ventilation equipment should be arranged within a nuclear plant. Put simply, air should enter at one point, make its way around the building then, depending on where it has been, exit again via a couple of alternative routes. I suppose we could leave it at that, but then my HVAC friends would be a bit disappointed at their contribution being so understated, so I had better say some more.

In terms of the scale of a nuclear ventilation system, it is worth mentioning that, for those facility's spread over several floor levels, practically the whole top floor can be occupied by ventilation equipment of one kind or another, such as air handling units, fans, filters, and much of the main ductwork distribution network. And remember we can be talking here about pretty big buildings, maybe the size of a soccer pitch, so clearly ventilation makes a major claim on real estate.

8.11.1   Generic arrangement of ventilation plant rooms

To get a feel for what a system might entail, let's assume that the entire top floor of a proposed new facility is indeed given over to ventilation equipment. We can start with a quick tour of how, in an ideal world, the area would be arranged (Fig. 8.23) then examine the system in a little more detail. We shall use a beta-gamma plant as our example and keep in mind that, as we discussed earlier, alpha plants need an additional stage of HEPA filtration for each of the radiological airstreams. In addition their incoming air is almost always filtered.

Air is drawn into a building via large air handling units, or AHUs as they are more commonly known, which sit behind a bank of louvers located high on an outside wall. Immediately behind the louvers we shall find a coalescing filter, which squeezes moisture from incoming fresh air before it begins its journey. Following that, the AHUs either heat or chill the air, as necessary, then send it on its way.

If we stick with the passage of C2 (green) air to begin with, it is routed around a building, without entering any areas that are subject to airborne contamination, then

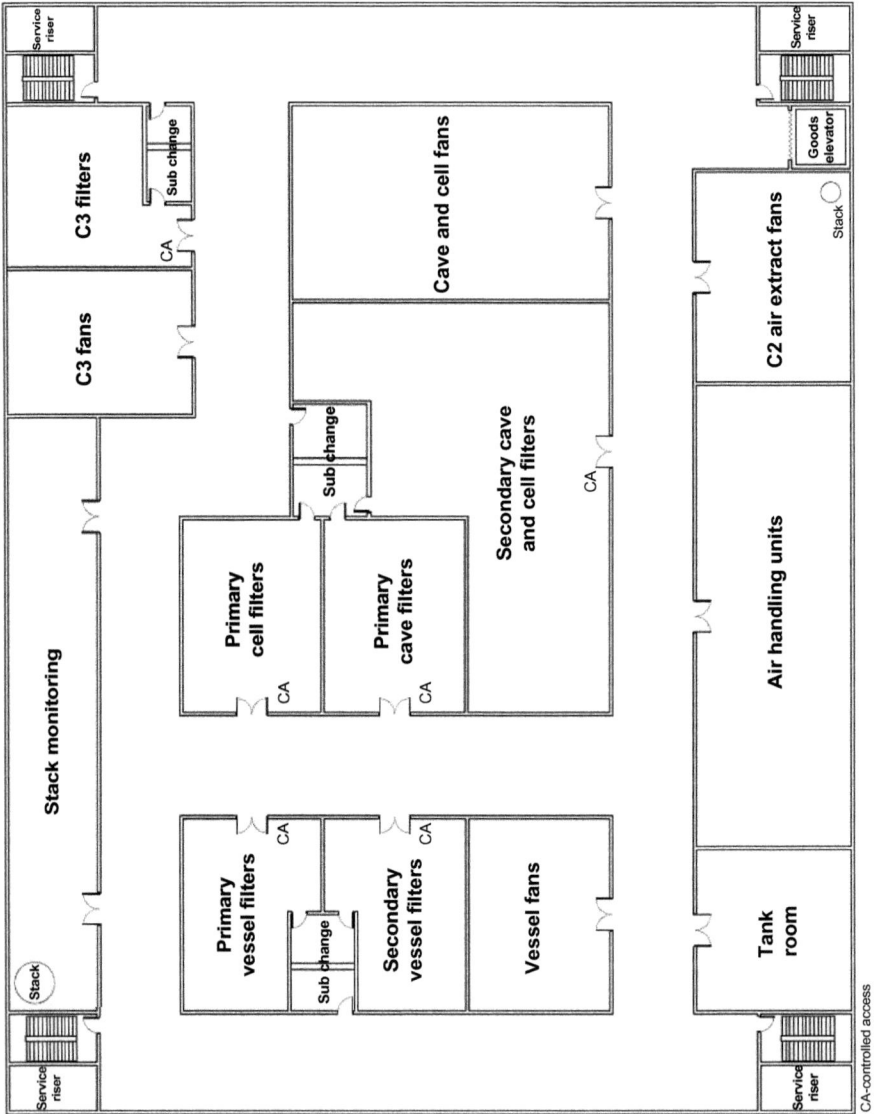

Fig. 8.23 Generic arrangement of ventilation plant rooms—beta–gamma plant.
© Bill Collum.

eventually drawn out again by fans in the C2 air extract room. When you think about it, air leaving a building is always going to have been warmed significantly; after all, it will have been conditioned on its way in so as to be at an agreeable temperature for those inside, and also have picked up heat from operating plant while on its travels.

For this reason, the ideal location for a C2 air extract room is adjacent to the AHUs, so that heat recovery systems can be used to draw heat from exiting air and use it to warm air heading through AHUs in the opposite direction. It is not a process that comes for free, as there is an outlay for the additional equipment plus its ongoing operation and maintenance costs. Nevertheless, heating incoming air in this way delivers significant cost savings over the operational life of a facility.

To complete this particular leg of the journey, air that has occupied C2 zones is routed back outside via a short, or stub stack, which need only be a little higher than the roof, or its parapet if it has one, so maybe 3 m at most. For beta-gamma plants C2 air is not filtered, whereas for alpha plants it passes through a single stage of HEPA filtration before being monitored and sent on its way.

As we have seen, air entering areas of a facility rated C3 (amber) and above does so via the cascade system. For beta-gamma facilities with the full array of radiological activities under their roof there are four such high category airstreams to deal with (1) air that has passed through C3 areas of the building; (2) C4 airstreams which originate from caves; (3) C4 airstreams which originate from cells; and (4) air from the vessel ventilation system. In every case, air from these zones is cleansed by various stages of HEPA filtration before reaching the centrifugal fans that are pulling it along.

We know from earlier discussions on mass and energy balance that process engineers determine a facility's precise filtration requirements, but in most cases it follows a familiar pattern. Namely, air that originates from C3 zones such as workshops, maintenance areas, and so on is the least onerous of the group, so passes through just one stage of HEPA filtration. Airstreams from C4 caves and C4 cells are the next most contaminated so are routed through two filtration stages.

There are two ways of configuring this, with the correct solution dependent on levels of contamination present within individual caves and cells. In the earlier Fig. 8.15, I have depicted the most common approach, whereby air from caves and cells is routed through a common filtration system. However, in Fig. 8.23, where we have more of an interested in potential demands on floor space, I have shown the most onerous situation. In this setup caves and cells each have a dedicated bank of primary HEPA filters, after which they are allowed to mingle and share a common secondary stage.

The final member of what we might call the *active airstream group* is the C4 vessel ventilation system. Initially this air is routed through moisture elimination and off gas treatment equipment which is part of the liquor processing system, after which it is picked up by the building ventilation that we are following here. Ordinarily this stream carries more contamination than arises within caves and cells so is always routed through a standalone system, one which follows the same pattern of primary and secondary HEPA filtration, followed by dedicated extract fans.

Ideally air leaving any C4 (red) environment, such as caves, cells and vessels should be filtered locally, otherwise ductwork carrying it to distant filters may become contaminated and almost certainly need to be shielded, something which only serves to complicate maintenance activities and later decommissioning, so is best avoided. With this in mind, Fig. 8.23 depicts an ideal arrangement, where primary and secondary HEPA filters are located directly above the C4 caves, cell and vessels they serve.

As for ventilation plant rooms, it is worth noting that those housing filters are themselves designated a C3 radiological zone. It makes sense, after all filters will be loaded with contamination which has originated from areas that are potentially extremely radioactive. So even though safe-change techniques are used to replace them it is, as we discussed earlier, a prudent precaution to zone these rooms in a way that guards against the possibility of contamination escaping during filter change operations. There is, though, one relaxation, in that it is not necessary to provide every filter room with a dedicated sub changeroom; instead it is permissible to access two or even three filter rooms from a common sub changeroom.

You will notice that large doors are shown on each of the ventilation plant rooms. They are there to accommodate the occasional transfer of equipment, all of which travel via the goods elevator, although some plants may opt to use a hoistwell instead. Where transfers are made between C3 plant rooms and C2 corridors a nominated specialist, such as a radiological protection advisor, must be in attendance to supervise the move. For this reason these doors are always designated as *controlled access* points.

8.11.2 Ductwork distribution network

Invariably huge ducts are needed at the start of a ventilation system to carry high volumes of air rushing through AHUs, so care is needed in planning the air distribution network (Fig. 8.24). Ventilation engineers often start by dividing the network into four approximately equal segments, all originating from a header duct that joins the AHUs together From here ductwork arteries, maybe a meter square, drop down into the building below via service risers located in its four corners. From there air continues its journey via progressively smaller ducts until finally reaching its destination. This makes for a far better arrangement than trying to spread the entire network from a single very large duct behind a bank of AHUs.

Ducts with a square cross section tend to be used in the initial stages to transport clean air; however, their angular corners are prone to accumulating airborne particles, so contaminated air which arises later in the system is best routed along ducts with a circular profile.

Whichever route it takes, whether along ductwork, cascading through a facility's various radiological zones, or from vessels deep within a process cell, air is pulled along by huge centrifugal fans, normally located on the top floor. Before it is allowed to leave the building, air that has visited contaminated areas is routed through a monitoring process, and from there to a stack, often quite a tall one, from which it is expelled. However, we shall save our discussion on those final steps for Chapter 17, where we look in more detail at what it takes to discharge previously contaminated air from a nuclear facility.

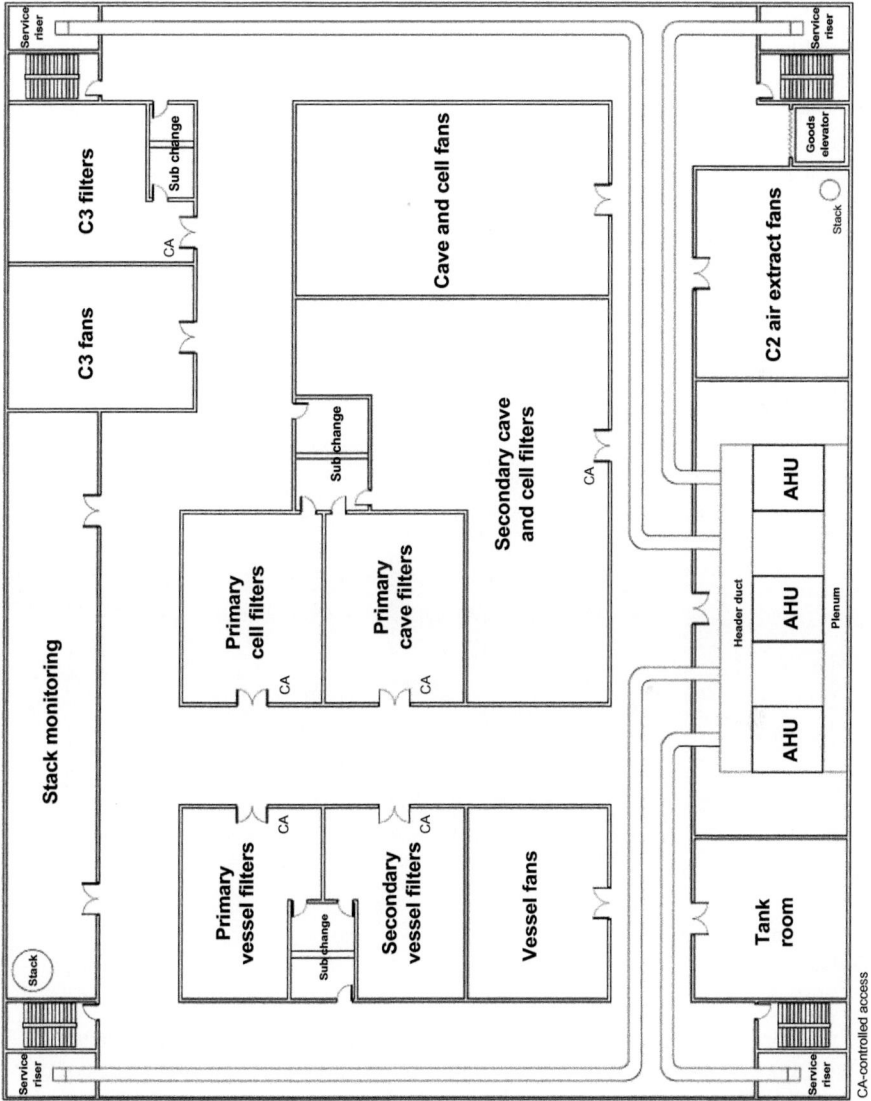

Fig. 8.24 Ductwork distribution network—quartering.
© Bill Collum.

The "top floor" arrangement we have just examined may suit the HVAC folks but not necessarily everyone else. As always these things are all about compromise, an ideal compromise, so realistically we are unlikely ever to encounter a layout identical to the one shown. For one thing, nuclear facilities sometimes find it difficult to accommodate the full array of ventilation equipment on a single floor, particularly as other equipment completely unrelated to ventilation often finds an excellent reason to elbow its way into the area, squeezing ventilation plant out of the way.

By way of an example, I have shown a tank room in among the ventilation plant, as they tend to head the list of contenders with an equally valid reason to be on the top floor, after all their tanks often hold liquids that are best released by gravity, so the higher they are located the better.

The good news, when it comes to locating ventilation plant rooms, is that there is a degree of flexibility. HVAC folks may not always admit it, but there is. Of course, we should always strive to get as close as possible to satisfying their requirements. Apart from keeping them happy it makes for a more efficient and cost effective ventilation system, so all in all is well worth the effort.

8.12   Air handling units

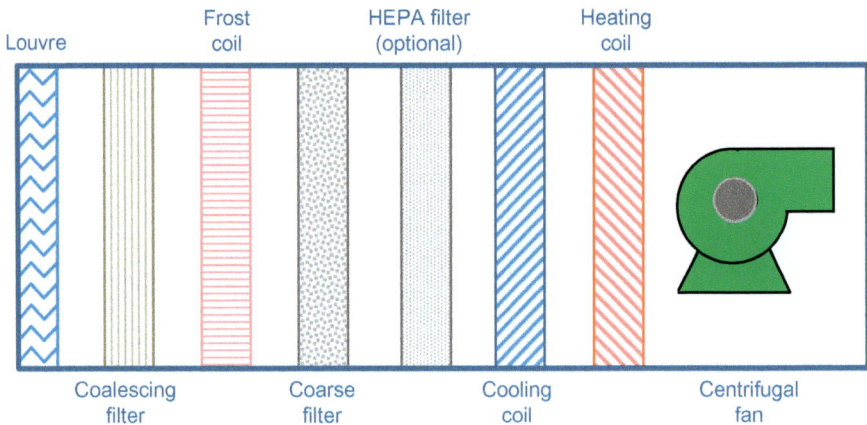

Fig. 8.25 Air handling unit.
© Bill Collum.

From their rather bland appearance it would seem that AHUs are nothing more than a big box with a powerful fan inside, but they are actually quite sophisticated pieces of equipment (Fig. 8.25). Their opening, which may face into a plenum, is protected by louvers that keep the elements at bay. In the UK where rain is almost omnipresent, we can expect to see triple bank louvers, whereas in other parts of the world a sand trap louver may be more appropriate. Next a coalescing filter removes excess moisture from incoming air and in coastal locations also captures salt, which, with its corrosion properties, is best segregated from air as soon as possible.

A frost coil, whose operation may be supplemented by the reclaimed heat we discussed earlier, kicks in when air temperatures fall below zero. Its job is to protect the coarse filter that follows by raising the temperature of incoming air to say 3–5°C. A coarse filter is not in the same league as a HEPA filter, but nevertheless does a good job of trapping larger particles and keeping them out of the system. For some facilities, such as alpha plants, the coarse filter is followed by a HEPA filter which as we know can halt particles down to a submicron sizes.

Where incoming air is warm enough to create a stifling atmosphere, its temperature is lowered by passing it through a cooling coil, supplied by either chilled water or a refrigerant. Conversely, if air is too cold the main heater, which is normally powered by either electricity or LPHW raises its temperature to a more agreeable 20°C or thereabouts. Finally, a powerful centrifugal fan sends the newly conditioned air hurtling into a ductwork distribution network.

8.12.1   AHU location

As with every major item of equipment there are a whole host of issues associated with choosing the right location for a facility's AHUs. They can be placed outside, say on a roof, but for major nuclear plants are normally located on a building's top floor, for several reasons. Being indoors affords them weather protection, so minimizes upkeep and extends their operational life. In addition, it keeps them above the dust, debris, and litter that may get sucked in down at street level, and as we have seen the top floor is a good place to start a ventilation distribution network.

Being inside a building, we need to ensure AHUs do not clash with structural steelwork or bracing, something which in itself can be a significant challenge, particularly when we consider how big these things can be. And so it goes; there is lots to think about. If we look at just two AHU location issues in a bit of detail, it will be enough to give us a flavor of some of the thought processes involved.

The first issue we need to consider is wind direction. Strictly speaking it is not the "first" as we really need to consider all factors at the same time. After all it would be a pity to get too far on resolving one issue in isolation, only to find an apparent winning solution undermined by the "next" problem on the list. Still, if we stick with wind then the best location for air intake louvers is one with minimal turbulence on the outside. So facing into the prevailing wind is not a good idea.

On the other hand, it can be even more challenging if the open mouths of air handlers are subjected to the considerable depressions generated by wind suction. In such situations Mother Nature will be doing her best to suck air out of AHUs, while fans struggle to pull it the other way. For that reason it is advisable to avoid walls that run parallel to the prevailing wind, the corners of a building, anywhere that is subject to a wind tunnel effect due to a nearby building and any location close to a step in a building's profile, which is quite a long list.

The ideal location then from the wind pressure perspective, is on the lee side of a building in the center of a long featureless wall. Having said all that, I do not want to give an impression that the ideal location for a bank of AHUs should determine the profile of a building, or dictate how its internal spaces are arranged. It is a significant factor but by no means a dominant one. It is important to recognize there are issues associated with their location and be ready to acknowledge that AHUs have their place in a facility's queue of competing priorities.

The second of our location issues centers on ensuring AHU intakes do not become blocked. Imagine a situation where the roof of a building is stepped between one roof deck at, say, 20 m level and an adjoining one at 30 m. We may weigh up all of the contributing factors and conclude that the 10 m high *wall*, created by the step, is an excellent location for a facility's AHUs. Unlike competing locations it even has the added benefit of easy access to louvers from the 20 m roof deck. You may remember that in Chapter 5 we looked at snow loadings and considered how deep it can get over a period of 10,000 years, something that many nuclear facilities must consider and be designed for.

We all know that the actual depth of falling snow is only half the story, since the deepest snows are caused by drifting and the very deepest accumulations occur where drifts meet vertical faces, like the side of a house, or as in this case a wall of louvers rising between two roof decks. What may appear to be an ideal location for AHUs is also the one most susceptible to blockage by drifting snow. We may amuse ourselves with thoughts of plant operators up on the roof in a blizzard, shoveling tonnes of snow over the side, but they are unlikely to crack a smile so we need to think again.

We can see that even for a seemingly nondescript task there is much to think about, from the relationship of AHUs to other ventilation equipment, to wind direction, to the risk of blocking air intakes, not forgetting the need to give a wide berth to structural steelwork. It does show that precious little is as straightforward as it seems. Of course selecting the right location for a few AHUs is never going to be the focus of attention—unless, that is, they end up in the wrong place.

8.13   Air quality

We know that the first destination for all air entering a nuclear facility is its C2 (green) areas, from where it cascades into other areas around the plant. In effect higher radiological zones receive second, or even third hand air from the lower ones. It is a concept that paints a picture of cascaded air being stale and suffocating, especially as it moves deeper into potentially more contaminated areas, so we need to take a closer look at what is going on.

I guess if these buildings were populated like stores in the January sales then the atmosphere would become very sticky indeed. But the thing is most nuclear facilities, even enormous ones, have relatively few people inside them. In fact some are manned by just a handful of operators, each with a cathedral full of air to themselves. Even those with scores of people working under the same roof can have a population best described as sparse. Crucially for those facilities that do employ large numbers, the vast majority occupy C2 (green) radiological zones. In other words, those areas supplied directly with fresh, nicely conditioned air. So far so good, but what about air quality in higher zones?

There is no denying that personnel occupying C3 (amber) areas are breathing air passed on from the C2 (green) zones adjacent to them. However, there are a few things mitigating the possibility of an unpleasant environment. Ordinarily the volume of a building's C3 areas is very small indeed when compared to that of C2, so only a comparatively small volume of air makes the transition to higher zones.

Supply	
C2	100%
Extract	
C2	50–80%
C3	10–30%
C4	10–20%

Fig. 8.26 Typical airflow by volume.
© Bill Collum.

Percentages vary from one facility to another but Fig. 8.26 gives an idea of how airflow quantities might divide up. These numbers reflect the *cascade* approach with 100% of air being supplied direct to C2 areas, and from there being split in various ways before it is finally discharged again.

Clearly if a ventilation system tried to force all of that C2 air into the much smaller space occupied by C3 and C4 zones, then the airspeed in those areas would be enough to knock us off our feet. Happily then, with the cascade approach at least half of a facility's C2 air tends to be routed back out of a building, having never entered a higher zone. And most important, in terms of second hand air, that which does get passed on will have had little or no direct contact with mankind. In addition the main routes between C2 and C3 zones will, where necessary, have conditioning plant attached, so that air quality can be improved before entering its new C3 neighborhood. So if we set aside the issue of potential airborne contamination in C3 areas, which is dealt with by respirators and the likes, then the air itself is perfectly breathable.

The question of breathing quality air in C4 areas does not arise, since these zones are potentially far too contaminated for anyone to even think about inhaling their atmosphere. In any event, personnel who do enter will almost certainly be fully covered by a protective suit and have their own air supply, either carried with them or supplied via a flexible pipeline. All in all then the relatively small number of occupants in nuclear plants, along with some supplementary conditioning, ensures air is always of an appropriate breathing quality.

8.14   Vessel ventilation

You will recall from Chapter 6 that air which has visited processing equipment is routed through a facility's off gas treatment plant, which removes aerosol and noxious gases such as iodine, radon and tritium. If you study a few process flow diagrams or more detailed drawings for many processing activities, you will notice arrows pointing off the page signifying that, following off gas treatment, this still contaminated air is someone else's problem. That "someone" is the HVAC team.

They deal with air resulting from off gas treatment in much the same way as air flowing through other areas of a facility. However, this particular airstream can have the feel of a drizzly day about it, or contain *free moisture* as the HVAC folks like to say, so the first step is to route it through a moisture eliminator which is placed within ductwork. As air hits its metal vanes, moisture coalesces and flows down to a collection point from where it is consigned to an appropriate effluent treatment stream.

At this stage air may have passed through two moisture elimination processes, the first within breakpots, evaporators, and so on. Even so, it could still be at a level of 100% relative humidity (RH) which is too high for HEPA filters to cope with. Ideally its RH needs to be reduced to no more than 80% otherwise the saturation process may eventually compromise a filter's physical strength.

There are two ways of reducing humidity: either blend moisture laden air with a dry airstream which has originated from an entirely different area, say, a C4 (red) cell, or heat moist air before it reaches HEPA filters by passing it through a pocket, or coaxial type heater. Interestingly heating air does not make its moisture disappear as it is still there, trapped within the ductwork. What does happen is that the moisture gets dissipated to the point that it will not condense out too readily.

Let's say, for example, that 1 m^3 of air with a RH of 100% is heated, as with all other substances the increase in temperature causes it to expand. In round numbers we can assume the 1 m^3 swells to become 2 m^3 which would give the air a RH of 50%, and so it goes. HVAC engineers run the calculations to ensure air is sufficiently dry so as not cause a problem when it passes through their HEPA filters.

8.15 Gloveboxes

Strictly speaking gloveboxes do not sit entirely neatly under the *ventilation* heading, but neither do they fall directly within the jurisdiction of any other single engineering discipline. In fact one way or another practically everyone can have at least some involvement in their design. That being said, this chapter is certainly an essential precursor to any discussion on gloveboxes, so this is a good time to examine them.

When, rather than dealing with beta-gamma materials and their inherent radiation, we are instead handling alpha or beta products, we need a very different type of containment to the heavily shielded structures found elsewhere. For these materials we turn instead to the glovebox. They may not look too complicated (Fig. 8.27) but their seemingly simple appearance belies some awfully sophisticated engineering.

We know from Chapter 2 that when handling alpha products such as plutonium-236 and uranium-238, or beta emitting isotopes, say strontium-90 or technetium-99, there is barely any hazard at all from radiation. Instead, the primary risk is from inhalation of microscopic particles which are exceptionally harmful to our health. Gloveboxes then are all about containment, as in nothing, no matter how infinitesimal can be allowed to

Fig. 8.27 Glovebox suite.
© Nuclear Decommissioning Authority.

escape. In addition some gloveboxes do indeed handle radioactive materials, albeit not at the levels found in beta-gamma plants.

So yes, they present quite a challenge, but what are they for?

8.15.1 Purpose

If we peel back the layers of a heavily shielded structure such as a cave or canyon, it is relatively straightforward to figure out what is going on inside, say, a mechanical handling process to segregate and package various radioactive materials. The goings-on inside gloveboxes are often less manifest, so we need to look a little harder.

The use of gloveboxes divides into three main activities. Some are used in a laboratory setting for experiments, research and analytical purposes. Others are deployed in nuclear facilities, in this context referred to as *alpha plants*, where they perform an operational role, say, converting plutonium nitrate solution into a powder, or packaging transuranic wastes. While a smaller number are employed to pilot a proposed new process which, if successful, may ultimately be scaled-up into a fully operational facility.

8.15.2 Ventilation

As I say, if we set aside their actual purpose, then the most important role of a glovebox is to provide unassailable containment. In essence the principles of containment are aligned to those applied to beta-gamma plants, with which we are now familiar; it is just that their inhalation hazards demand some additional layers of protection.

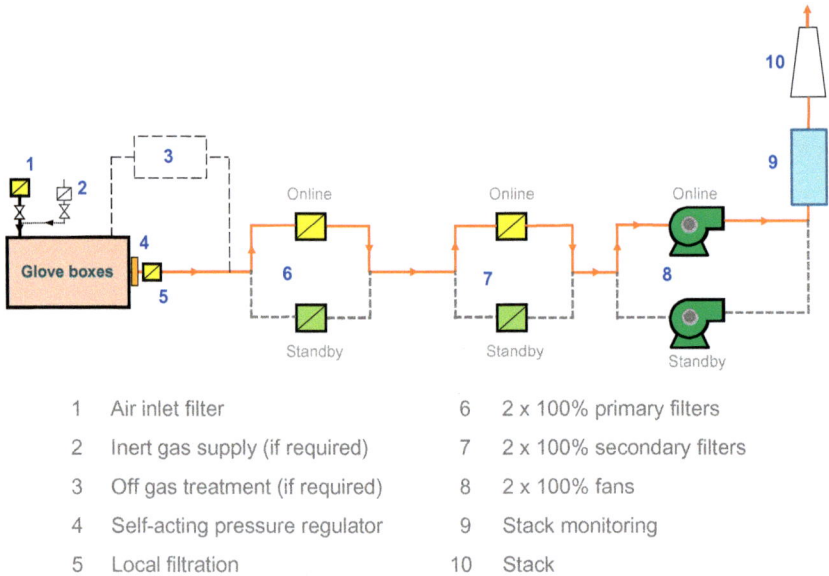

1	Air inlet filter	6	2 x 100% primary filters
2	Inert gas supply (if required)	7	2 x 100% secondary filters
3	Off gas treatment (if required)	8	2 x 100% fans
4	Self-acting pressure regulator	9	Stack monitoring
5	Local filtration	10	Stack

Fig. 8.28 Glovebox ventilation.
© Nuclear Decommissioning Authority.

One of the most notable differences is that gloveboxes have at least three stages of HEPA filtration (Fig. 8.28) rather than the two seen on beta-gamma plants. Each glovebox has a local filter directly attached, often the push through variety (Fig. 8.29) discussed earlier. This ensures any gross contamination is trapped immediately and does not get the opportunity to invade adjacent ductwork. Subsequent primary and secondary HEPA filters are shared between multiple gloveboxes and can be located in a plant room some distance away. And if there is any risk from ignition sources the atmosphere within a glovebox is inserted with a *protective gas*, typically either argon or nitrogen.

Where a glovebox houses wet process, then an off-gas scrubber system may be needed ahead of the primary HEPA filters. And if processes give rise to the possibility of condensation accumulating on the inside of glovebox screens, air must be changed at a rate of 10–15 times per hour.

If you stand beside a glovebox and consider how harmful its invisible environment is, you may well wonder how safe you might be if one of its gloves was torn or even became detached altogether. Could contamination escape? There are several systems available, known collectively as *self-acting pressure regulators*, which can immediately sense even a slight change in pressure within a glovebox. By the way, for *immediate* we are talking around two tenths of a second, so these systems are pretty quick off the mark.

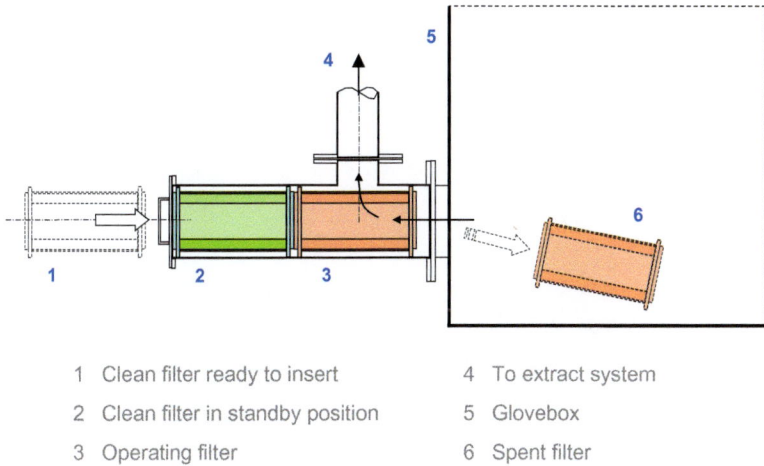

1 Clean filter ready to insert   4 To extract system
2 Clean filter in standby position   5 Glovebox
3 Operating filter   6 Spent filter

Fig. 8.29 Push-through filter.
© Nuclear Decommissioning Authority.

If a pressure change is detected, control systems instantly send extract fans into a frenzied acceleration. This creates an enormous depression within a glovebox, sufficient to draw air through a breached glove port with a velocity of at least 1 m/s. Faced with that kind of speed contamination is unable to drift out of a glovebox, so the surrounding environment remains radiologically clean while recovery procedures are implemented.

Under normal operating conditions most nuclear gloveboxes are held at a depression, or *negative pressure*, of between -250 and -500 pascals (Pa). So the obvious question is, "Why not hold them under a constant and much deeper depression, that way they are always poised to counter any breach of containment?" This sounds reasonable; however, the problem is that once negative pressure exceeds -500 Pa, gloves begin to balloon and pull away from an operator's fingers, stiffening up as they do, which of course impedes dexterity. In practice, then, gloveboxes are typically held at around -375 Pa.

8.15.3 Glove ports

Easily the most recognizable feature of a glovebox is (no surprise) its glove ports; although interestingly, they are not necessarily used too much. For gloveboxes in a laboratory setting the emphasis tends to be on ad hoc tasks performed on say chemical or biological materials. Here, there is significant reliance on hands-on activities so gloves are heavily utilized.

At the opposite end of the spectrum we find gloveboxes in operating nuclear facilities (Fig. 8.30) where tasks can be repetitive, often automated, and where radiation levels may creep into the R4 (red) band. In these conditions gloves are reserved for maintenance activities, with maybe the odd foray for an infrequent procedure. Apart from anything else this, quite rightly, minimizes the incidence of exposing operators to radiation, a theme we shall return to in Chapter 14.

Fig. 8.30 Glovebox housing automated processes.
© Nuclear Decommissioning Authority.

In the middle we find gloveboxes which are again located in nuclear facilities, although on this occasion employed for low, or nonradioactive operations, but of course still carrying the inhalation risks pertaining to alpha and beta products. In these cases gloves are widely used, occasionally coming close to the levels seen in a laboratory setting.

When the time comes to change a glove, it is not necessary to extract the existing one and risk losing containment. Rather there are proprietary systems available, whereby a new glove can be inserted while at the same time pushing a used glove into the glovebox, which is rather neat. Once detached, used gloves follow the route assigned to other contaminated wastes.

As for gloves themselves, they are formed from materials such as latex, neoprene and butyl, with selection being based on operating conditions and tasks being performed. And where radiation is present gloves are lead-loaded. In some situations multiple pairs are worn. So there is always a fine balance between selecting gloves, or combinations of gloves which are appropriate for prevailing conditions, while at the same time not compromising an operator's ability to perform their various tasks.

8.15.4   Shielding

It may sound like a contradiction to say that on the one hand gloveboxes are all about containment, while on the other raising the specter of sufficient radiation to bring shielding into the equation. In nuclear circles, it is not uncommon and as

I mentioned a moment ago can even reach into the R4 (red) spectrum. This may be due to the presence of americium-241 or other isotopes, with associated gamma emitters, which can accompany alpha and beta products.

Ordinarily glovebox screens are made from sheets of PVC, polycarbonate, acrylic materials, and so on. However, when radiation is present they may be formed instead form the leaded glass discussed in the previous chapter. In addition local shielding can be provided, as required, by neutron absorbers such as blocks of paraffin wax, or sheets of laminated beech wood, both of which are relatively thin and very effective.

We know from Chapter 2 that if fissile materials are incorrectly handled we run the risk of initiating a criticality. For gloveboxes with day-to-day variation in the way such materials are handled, the risk of a criticality is assuaged by a combination of operating procedures, how much material is present at any one time, the proximity of those materials and the geometric shape in which they are held. Where gloveboxes are designed to perform repetitive tasks, all of the criticality avoidance measures are built-in, so they are deemed *safe by design*.

8.15.5 Maintenance

When it comes to maintenance everything hinges on meticulous planning and attention to detail. For a start equipment within a glovebox is, as far as possible, entirely modularized. So rather than face the prospect of wholesale disassembly to access a single component, the exchange process is streamlined considerably. And of course inbound equipment is pretested before entering a glovebox's hostile radiological environment. In addition, every conceivable point within a glovebox that may need some form of operator intervention, no matter how infrequent, will be within reach of a designated glove port. True, some of these ports may be almost permanently blanked off, but they are there, just in case.

One way or another, whether it be for operations, inspection, testing, maintenance, or waste management, gloveboxes must be able to interact safely with the world beyond their confines. For this there are posting ports which can conduct transfers in and out of a glovebox without compromising their containment.

It turns out then, that there is very much more to ventilating nuclear facilities than meets the eye. When you think about it, the HVAC folks deliver a beautifully elegant solution to what is after all a pretty complex challenge. By concentrating on the areas that matter most, those interfaces between radiological zones, they keep air exchange volumes down to an absolute minimum. Apart from providing excellent containment control, deploying such an efficient ventilation philosophy also saves a fortune in running costs, so they have all of the bases covered, and then some.

Cranes

9

As we make our way through these pages, we are seeing time and time again that two simple words, "radiation" and "contamination," can keep nuclear engineers on their toes in ways undreamt of by colleagues in other industries. Having said that, I can just imagine engineers in the petrochemical, automotive, aeronautical, and other industries thinking, "Hang on a minute, we don't have it too easy either." I suppose the point is we all face challenges that are unique to our particular industry and it would be unwise for us to underestimate what colleagues in other spheres are up against.

Why do I mention this, and why now? Well, for the next few pages we are going to look at cranes, the ones that operate inside buildings; what could be simpler? The world is full of them, trundling up and down every day of the week without so much as a glance from those nearby. Surely this has got to be routine engineering, even for the nuclear industry? No, I'm afraid not. Cranes are an excellent example of what would normally be considered straightforward, workaday engineering, taking on a whole new dimension when it enters a nuclear world. What we shall see is that even cranes located in benign nuclear settings, such as C1 (white) or C2 (green) radiological environments, are significantly more sophisticated than their counterparts in say a warehouse or fabrication workshop. We shall also discover that cranes operating in extremely hostile radiological environments are barely related to what passes for a crane in the outside world.

We need to start with the elements that make up a crane, any crane, and take an overview of how they operate, that way we can get all the terminology straight. As we examine conventional cranes we shall also look at differences applied to the design of those operating in nuclear facilities, what we might call the basic nuclear crane, often referred to as *high integrity*, then later on look at the highly specialized cranes that operate in hostile radiological environments.

9.1 Conventional cranes and high integrity nuclear cranes

9.1.1 Crane components

Strictly speaking, cranes that operate on rails fixed some place high up in a building are known as *Overhead Traveling Cranes*, they are however commonly referred to as EOT cranes which stands for *Electric Overhead Traveling*. You can take your pick from either of those names and everyone will know what you are talking about. In addition they are sometimes referred to as *gantry* cranes. However, as we shall see in a moment, gantry is one of the names given to steelwork track on which they travel, so is not really suitable as a generic nametag.

There is, however, another family of cranes, those mounted on long legs, normally four of them and therefore able to straddle whatever it is they are carrying. They too go by several names, one of which is indeed *gantry*. So there we have another reason for

Nuclear Facilities. http://dx.doi.org/10.1016/B978-0-08-101938-2.00009-X

not applying it to the EOT type. That's it; now we have their name straight we can look at the cranes themselves.

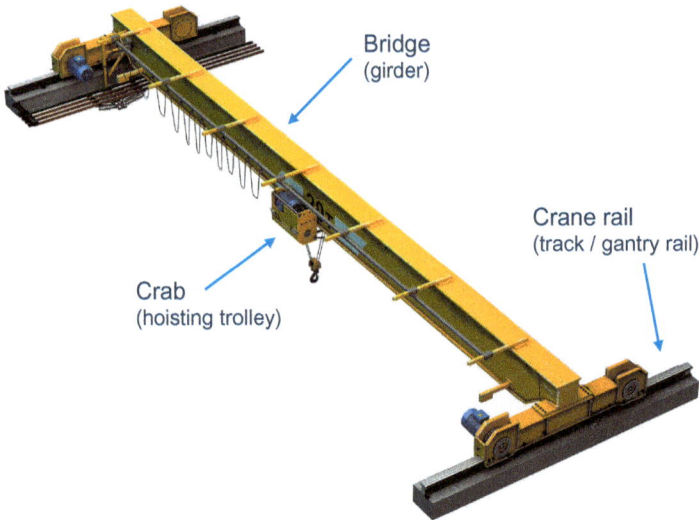

Fig. 9.1 Single girder EOT crane.
Istock.

When you get right down to it, an EOT crane does not have too many major elements. There are a myriad of components making the whole thing up, but when you stand back and take it in there are just three main items on show (Fig. 9.1). As usual each has a selection of names to choose from. The main structure is known as a *bridge*, or simply referred to as *girders*. The *crab* becomes a *hoisting trolley* and *crane rails* are often referred to as *tracks* or *gantry rails*.

There are several permutations for the main crane elements. Most notably some cranes have a single bridge while others have two. And whereas the majority of cranes, certainly the really heavy ones, run on top of their crane rails, there are plenty that run on the bottom flanges of the "I" section beams on which they travel (the same shape as stanchions we examined in Chapter 5). For our discussion here it is best to visualize a top running crane with a double bridge, as shown in Fig. 9.2. Although the nuclear industry deploys cranes of every configuration, this arrangement could justifiably claim to represent the typical model.

While we are on the subject of a crane's components, there is one other item that does not quite fit within that category but I need to mention it anyway. Most of us would use the term steel *cable* to describe the strands that wind up and down to do a crane's lifting, but the correct descriptor is *rope*. Of course in normal parlance *rope* is a heavy cord of some type, but in the world of crane design there is a clear differentiation. *Cable* is the word used for stuff that conducts electricity, while *rope* (Fig. 9.3) is the one that carries heavy loads. Describing something as cable in the

Fig. 9.2 Twin girder EOT crane.
Shutterstock.com.

Fig. 9.3 Rope drum.
Shutterstock.com.

wrong context might lead to confusion and the possibility of a discussion going round in circles for a while.

To complete a crane's overview, what about their color? I must admit that for a long time I assumed all cranes had to be painted bright yellow. I presumed it was some statutory requirement or other, based on the fact that brightly painted moving objects are hard to miss. But no, it turns out you can paint a crane any color you like and indeed some manufactures do opt for less striking colors. It turns out that the fact they are almost always bright yellow is a kind of tradition. There, don't you feel like a crane expert already?

9.1.2   Design standards

Rules governing the design of standard cranes are comprehensive to say the least. You will know by now that I spare you the minutia of such things, but I thought you might like a couple of examples just to show what you are missing.

I was amazed to discover how tight a crane's tolerances must be. For example, for spans up to 15 m, crane rails must be positioned to within ±3 mm along their entire length, which equates to about the thickness of a metal watch strap. Beyond that span the tolerance increases incrementally, so that by 25 m the permissible variation it is a touch over ±5 mm. And no matter how great the span, the deviation must not exceed ±15 mm.

If you looked along the top of a crane rail you would expect it to be pretty flat, otherwise cranes would twist and contort as they traveled across troughs in the rails. Flatness is important because unlike cars cranes do not have a suspension system to even out their ride. The rule is that if you surveyed a crane rail from one end to the other, it should not deviate more than 10 mm either side of its center line along the entire length. Those are very tight tolerances particularly when you consider that all cranes must be thoroughly examined, tested, and (apart from a few exemptions that we shall come to later) recertified every year.

If you think those tolerances are demanding and they are, think about this. Many of the cranes operating in nuclear facilities must be designed to withstand the worst earthquake that might possibly come along in a period of 10,000 years. Not only that, if they happen to be carrying a load at the time then they must not drop it. In addition, in many situations they must be capable of taking that load to a position where it can safely be lowered to the floor. In fact there are even cases where cranes must be capable of carrying on working, almost as if nothing had happened.

We shall return to this theme later when we look at how cranes cope with seismic events. In the meantime here is one more constraint for you. When a crane has its maximum working load suspended at its central position, deflection of the bridge must not exceed 1/750 of its span. If we stick with our example of a crane spanning 15 m, then even with a maximum load of say 100 t its bridge must not deflect more than 20 mm. That's enough of that; you get the idea, lots of tight tolerances and exceptionally demanding requirements made even more challenging in a nuclear environment.

9.1.3   Operation

EOT cranes are normally operated in one of three ways: remotely or *wirelessly* as it is also known; from a pendant; or by an on-board driver in a cabin. Remote operation can be carried out from either a control desk, with some assistance from cameras, or from a mobile control unit (Fig. 9.4) carried by an operator. However, in nuclear facilities proposals to use wireless systems demand a little more scrutiny than elsewhere.

The challenge for engineers is demonstrating that the system can satisfy a safety case to the same standard as a hard-wired control desk. Of course it can be done

Fig. 9.4 Wireless crane operation.
Istock.

and there are plenty of examples around that prove the case, it just takes some additional effort. It can be worth it, because the beauty of a wireless system is that it gives operators freedom to adjust their position in order to get the best view of a load, which is particularly important when it is being set down. In the nuclear industry, wireless systems have the added bonus of enabling operators to stay clear of areas where there may be increased levels of background radiation. So in certain circumstances they are a very attractive option.

Fig. 9.5 Pendant crane operation.
© Nuclear Decommissioning Authority.

Pendants hang from a crane's bridge on a strengthened control cable and are operated by a person walking with the crane (Fig. 9.5). The best pendants are suspended from cables that can be moved back and forward along the bridge, so enabling operators to negotiate their way around obstacles as they move through the crane area.

Fig. 9.6 On-board crane
operation.
Shutterstock.com.

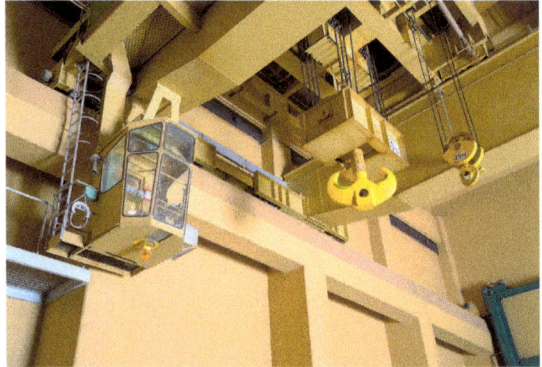

Some of the bigger cranes are operated by an on-board driver who sits in a cabin fixed to their underside. Cabins can be located in a static position on a crane's bridge (Fig. 9.6) but then as a crab moves away so does its suspended load, leaving the driver with a diminishing view of whatever is going on. Where drivers need to maintain a closer view of the action, their cab can instead be fixed to the underside of a crane's crab unit.

On the face of it, having an on-board driver floating above the action feels like an ideal arrangement, but in reality they cannot always see too much at the business end of a hoisting operation. Being in an enclosure they are also somewhat isolated so are unable to converse directly with their colleagues. To operate safely they often need an assistant, known as a *banksman*, stationed down below to guide them. Communication can be electronic but very often hand signals are used, a bit like the tic-tac system employed by racecourse bookies.

9.1.4   Lifting accessories

If out of the blue we were asked to describe what attachment cranes use to pick things up, most of us would respond immediately with, "A hook." If we were given a bit more time to think about it we would recall seeing things picked up with magnets, large buckets, slings, and grabs. The nuclear industry makes use of all of these but tends to use different equipment when handling a radiological payload. Such apparatus is not unique to the Industry, but is not quite so commonplace as hooks and the likes we are all familiar with.

Before we move on to look at actual equipment I have a bit more terminology to share with you. As usual it is important because innocent use of a wrong term can lead to all sorts of confusion. The term *device* should not be used to describe a mechanical grab (Fig. 9.7) or anything else that picks things up. In this context a device is the crane itself, the whole thing. The equipment we are about to examine, the kind of kit that grabs onto fair sized loads to pick them up, is referred to collectively as lifting *accessories*. And our old friend the hook (Fig. 9.8) is a lifting *feature* which is the term reserved for any item suspended directly from hoisting ropes.

Fig. 9.7 Lifting accessory—mechanical grab.
© Nuclear Decommissioning Authority

Fig. 9.8 Lifting feature—crane hook.
Shutterstock.com.

Ordinarily magnets, buckets, slings, and petal grabs are categorized as accessories; however, in the odd case where they are attached directly to the end of hoisting ropes, they become lifting features. As I say, the nuclear industry uses all manner of accessories, but there are a couple of particular favorites.

9.1.4.1   Spreader frame

Fig. 9.9 Spreader frame.
Shutterstock.com.

Spreader frames (Fig. 9.9), which you can see operating at any container handling port, can make a four-point lift by slotting into oblong shaped holes on top of the package being moved. The frames are always square or rectangular and have what is known as a twist-lock in each of their four corners. Twist-locks are the same oblong shape as corresponding slots on top of their target package, but slightly smaller.

Fig. 9.10 Twist-lock.
© Nuclear
Decommissioning
Authority.

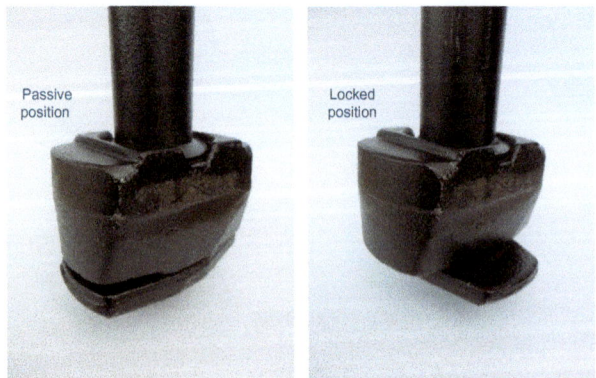

Operation of this device is beautifully simple (Fig. 9.10). As a crane lowers a spreader frame over its target package, the four twist-locks pass through corresponding slots on its top. Once the locks are through a motor at the center of the spreader frame is triggered, turning all four twist-locks through 90 degrees. That's it; it is impossible to disengage the spreader frame again until the twist-locks are turned and realigned with their slots.

Where spreader frames are used to handle radioactive packages, there is a risk that if their motor failed it would be difficult to activate twist-locks and release a load, primarily because any form of direct manual access would be ruled out because of the radiation. For this reason, nuclearized spreader frames incorporate backup systems that can be used if their primary turning mechanism fails. This may be in the form of a supplementary electrical motor, or manual recovery techniques (discussed later) which can be implemented from a safe distance.

Fig. 9.11 Four drum stillage.
© Nuclear Decommissioning Authority.

Twist-locks provide a solution which is both elegant and robust—always a winning combination and one particularly well suited to the nuclear industry. Applications are numerous but Fig. 9.11 shows a typical example, with drums of radioactive waste placed in a stillage where they can safely be handled four at a time.

9.1.4.2 Lifting beam

In its most basic form this kit comprises nothing more than a steel beam with a chain hanging from a shackle at each end. Attachments on the chains simply connect to whatever is being lifted and off it goes. However, the nuclear industry tends to employ a more sophisticated, more robust version, where chains are replaced by arms formed from steel plate (Fig. 9.12). In their lift position arms are at 90 degrees to the beam and effectively locked onto a pair of trunnions. The thing is, to get them into that position the arms need to widen out so they can get past the trunion heads.

This is often done by hand, but if gaining access is difficult it can be performed remotely. In one solution the operation is accomplished by energizing a motor attached to the main beam. It turns a threaded steel bar which passes through attachments on top

Fig. 9.12 Lifting beam.
© Nuclear Decommissioning Authority.

of both lifting arms. The slight twist to this arrangement comes from threading both ends of the bar in opposite directions. As a result, when the bar rotates a lifting beam's arms either both move in or both move out together, depending on which way the bar turns. All in all then this is a very simple, very sturdy piece of equipment. So if you need to make a rock solid connection for a two-point lift then this is the kit to go for.

9.1.5 Pintle

For completeness, having looked at ways of achieving both a four-point and a two-point lift, we must include a mention of the pintle. They are used to make a single-point lift and come in several shapes and sizes.

Typically they look like the example in Fig. 9.13, in this case attached to a flask lid, and are engaged by either a shrouded fork, which simply slots around the pintle, or by a grab designed to envelop the pintle before lifting commences. Other pintles are more of an oblong shape and are engaged by using the twist-lock technique we looked at earlier.

9.1.6 Performance

In this push-button hey presto age, we tend to assume just about everything can be made to happen more or less instantly. So when designing nuclear facilities it can be tempting to presume a crane can be here one moment and there the next, lifting here, lowering there, all to suit our bidding. It may be frustrating when we want to

Fig. 9.13 Pintle.
© Nuclear
Decommissioning
Authority.

increase a facility's production, but in reality designers can only squeeze so much performance, so much speed from an EOT crane.

For example, a 20 t crane might typically travel at no more than 20 m a minute; a 120 t crane (and there are plenty of those in the nuclear industry) probably travels at only half that speed. Then of course a crane must gather speed at the start of a journey and slow down again before it finally stops, otherwise whatever is dangling from it will swing around like an unwieldy pendulum. In large facilities it can easily take 5 minutes or more just to move from point "A" to point "B." And because cranes themselves are so heavy, the actual load they may be carrying does not make that much difference to their performance. So even when unladen, it is not possible to speed them up too much.

Then there are the hoists which are even slower. The lifting or lowering speed of a 20 t hoist is around 5 m a minute, while a 120 t hoist might operate at just 2.5 m a minute. If you think about the time it will take to connect a spreader frame or lifting beam, then hoist a load, transport it, lower it off, and disconnect the load from its lifting accessory, it is not difficult to imagine 15 or 20 minutes going by. It might not seem like a long time, but in a busy facility those are precious moments and can have a considerable impact on how much throughput can be achieved in a day. It is possible to improve performance but it can have a disproportionate effect on the size of gearboxes and electric motors, so needs to be carefully considered.

In many industries there is no problem with designing cranes to operate as quickly as possible, then working them right up to those limits. However, in the nuclear industry speed might not be such a good idea. Crane designers will tell you that collision damage, due to kinetic energy, is determined by the formula $\frac{1}{2}MV^2$ (half mass × velocity squared). In other words, doubling the speed results in four times the impact energy. So here is something else to think about: in situations where cranes are carrying a nuclear payload it can be prudent to throttle back on performance. It may slow productivity a little but is inherently so much safer.

One way to safely speed things along is to fit cranes with a smaller auxiliary hoist, one which handles less weight but can raise and lower a good deal faster. Usually they are fitted to the same crab unit but carry out lighter duties than the main, much slower hoist. These auxiliary hoists are often used for maintenance duties, but in the nuclear industry can have a much higher calling.

As we know, radioactive products coming and going from nuclear facilities are transported in heavily shielded flasks, also known as *shipping containers*, *casks* and *overpacks*. Oddly enough, the cranes used to move these monster loads around are far too powerful for some operations. Let's see why.

Fig. 9.14 Shielded transport flask.
© Nuclear Decommissioning Authority.

We know from Chapter 7 that there is no such thing as a standard shielded flask. That being said, the example shown in Fig. 9.14 is a common enough sight around nuclear facilities. Let's say the whole flask weighs 50 t with its lid contributing 5 t to that total, or the *all-up* weight as it is known. When a crane moves this type of flask around it is carried, as shown, by a lifting beam locked onto the trunnions on its sides. Whenever the time comes to remove a lid, say for maintenance purposes, its many bolts are loosened and either removed or held captive in readiness for retightening. With that done, the lid is picked up by its pintle and deposited on a parking stand. All of which should be entirely routine; however…

The chances are pretty slim that one or two bolts may not have been fully loosened, but a rigorous safety analysis will determine it is a possibility. If this were to happen we can confidently predict how things would unfold. With a 50 t hoist pulling on a pintle, the whole flask would be lifted from its parking position. Of course one or two bolts, even sizable ones, could not take that strain for too long. So eventually they would snap, releasing the flask to crash back to where it started, or someplace nearby: clearly a situation to be avoided.

This is where an auxiliary hoist comes in. If, as is the case, a pintle is designed so that it can only be engaged by a matching attachment on a second 5 t hoist, the possibility of accidentally lifting the whole flask goes away. Firstly, any attempt to raise a lid which is still secured to its flask will be detected by a load cell (discussed later) and halted, but even if this failed the 5 t hoist motor would stall. A stall occurs when the shaft or *rotor* of an electrical motor experiences so much torque that it is unable to turn. Typically this occurs when a motor faces 3.5 times its rated capacity, which in this case equates to 17.5 t. With a flask weighing 50 t, even when empty, there is no possibility that a 5 t auxiliary hoist could come close to getting it off the ground.

If a lifting operation were to stall in this way then everything would come to a halt until the problem was investigated. Rogue bolts would be identified as the cause and removed before operations safely recommenced. Designers and safety specialists love to proclaim they have made a potential problem go away, using the phrase "fault sequence terminated." And this is a good example of how they can do just that.

If you really want to improve performance it is possible to operate two cranes on the same set of rails. On the face of it this might sound a bit dangerous, but there are reliable protection systems which ensure they will not collide with each other. When installed as a pair, cranes employ various technologies such as laser, infrared, and radar to monitor distance between them and even their relative closing speed. Armed with this information cranes can, if necessary, be automatically slowed down and even brought to a complete halt.

Having said that, even though technology is available to operate them safely, placing two cranes on the same set of rails is not normally contemplated in situations where they may be vying for the same space and at the same time. It is, however, an option that makes perfect sense where a heavy duty crane, say 60 t capacity or more, is only used occasionally. In between times it can be parked up to one end of its crane rails, leaving the coast clear for a much more agile 5 or 10 t crane to get on with the day's routine. As I say, there are ways of keeping them apart but it is inherently much safer to use them one at a time.

9.1.7   Hook approach

So we might be able to squeeze a bit more performance from them, but there is one limitation on all overhead traveling cranes that we can do nothing about. They all have what we might term a no-go area. If you imagine a hook hanging down from a crane's hoist and picture the bridge moving from end to end along its rails, it is clear that the sheer size of a double bridge will stop its hook reaching as far as the end walls. Likewise when a crab moves from side to side on its bridge, a suspended hook will not be able to travel out as far as the crane rails on both sides (Fig. 9.15). It is simply a crane's own dimensions that make it impossible for a hook to provide full coverage of the floor area below it. This limitation on a crane's effective working area is known as its *hook approach*, often referred to as *end approach* for the bridge and *side approach* for the crab unit.

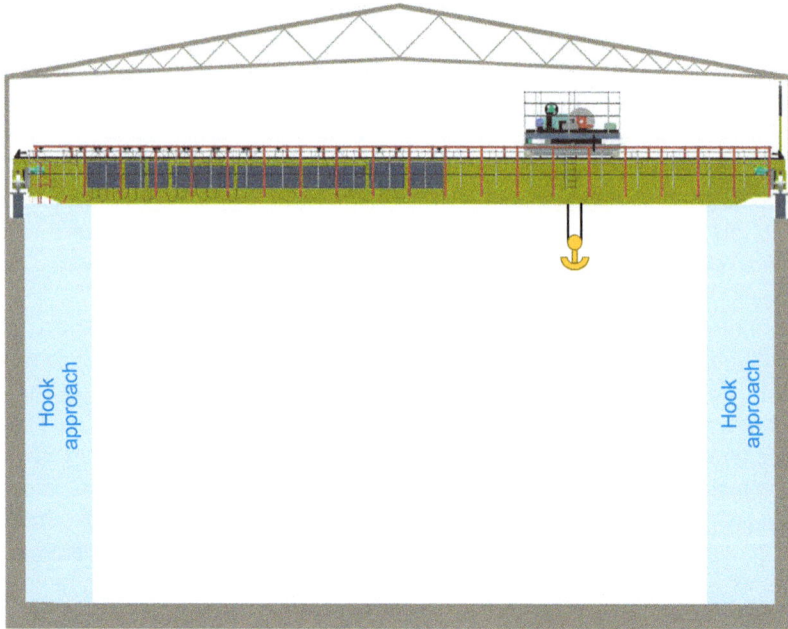

Fig. 9.15 Hook approach.
© Nuclear Decommissioning Authority.

You may wonder why I mention something so obvious, after all it is not too much or a revelation that EOT cranes cannot reach around their edges. The thing that is a surprise is how extensive that out of bounds area can be, especially in the nuclear industry where heavy duty cranes are so commonplace. There are so many variables that it is impossible to create a generic table showing various crane capacities, along with hook approach distances for their sides and ends; however, I can give you some ballpark numbers.

It would raise no eyebrows, for example, if a 120 t nuclear crane had an end approach of 5 m and a side approach of 3 m. Even a fairly modest 25 t crane could have an end approach of 3 m, with around 2 m effectively blanked out along the sides. For those planning operations below a crane it can be tempting to assume its hook will have more or less full floor coverage, so it is important to establish early on what the limitations will be, especially when your bear in mind that the distances for both sides and both ends will not necessarily be the same.

As I say, the issue of hook approach is more acute in the nuclear industry, because load for load cranes tend to be bigger than comparable equipment in other industries. There are quite a few reasons, but if you trace them back many stem from the space needed to accommodate additional safety features on-board a crane. For example, cranes carrying a radioactive payload are fitted with more safety interlocks than most. For instance if a crane is poised to lower a load through a hoistwell, interlocks will ensure the hoistwell door is fully open before allowing a decent to commence.

This negates the possibility of operator error and provides assurance that loads will always pass through a guaranteed clear opening.

All this electronics leads to more control panels which are usually looking for space on-board the crane itself. On top of that there are additional braking mechanisms, bigger and more robust gearboxes, and so on. A by-product of all this additional hardware is a need for more maintenance and hence additional personnel access platforms, which add even more bulk to a crane. You can see how it goes: a few more electronics here, some extra kit there and pretty soon an ordinarily hefty crane becomes a leviathan. Yes it does leave quite a large area beyond a hook's reach, but is the price we pay for the enhanced safety features these cranes deliver.

9.1.8 Safe working load

All cranes are designed to comfortably carry what is known as their *safe working load*, abbreviated to SWL. This figure excludes lifting accessories such as grapples and lifting beams, so their weight must be added to that of any load being carried. The SWL limit is so crucially important that you will see it spelt out in large letters on the bridge of every crane, in fact on both sides of the bridge. That way there is no excuse for anyone trying to hoist a load which is too heavy.

Apart from clearly advertising the SWL there are a couple of other things that help ensure a crane is not inadvertently subjected to demands beyond its lifting capacity. Firstly, as we would expect, engineers must establish during the design stage what a crane's maximum credible load will be when it comes into operation, including the weight of lifting beams, spreader frames and the likes. However, this is not the load a crane is designed to carry; to be on the safe side the SWL is set at 10% or so above that weight. But it does not end there. Design codes incorporate an additional safety margin, effectively an inbuilt overdesign that ensures EOT cranes are capable of lifting at least 25% more weight than the SWL they are authorized to carry.

I'll give you an example. Let's say engineers have done their homework and concluded the maximum load a crane will carry during its operational life is 90 t. If you add 10% it brings the SWL to 99 t, but we are dealing in round numbers, as sensible engineers do, so will set the actual SWL at 100 t. With the SWL established, crane engineers use statutory codes to develop a detailed design and one fine day a shiny new crane gets installed in a building.

With a crane designed to carry 10% more than its maximum credible load, you might reasonably assume its commissioning tests will be based on exactly the same margin of safety, in our case a weight of 100 t. But no, before an EOT crane can begin operating, it must be tested to ensure it complies with the additional 25% stipulated by design codes, in this example 125 t. A 90 t load carried by equipment with at least a 125 t capability is the kind of margin most of us would settle for. Even my ultracautious safety specialist friends will accept percentages like that.

Having taken in those numbers you might be thinking something along the lines of, "Even though a crane has 'SWL 100t' picked out in big letters on its bridge, we all know it could really lift 125 t if it had to. So maybe we can bypass the SWL now and again?" I was pretty sure the answer would be, "No" but checked it out, just in case. Sure enough the stated SWL is a crane's absolute working limit. If operators find their load demands are higher, then regulations dictate that a crane must be requalified and if necessary upgraded to meet a new SWL. It just goes to show how important it is to get that number right while a crane is nothing more than lines on paper, or an image on a computer screen.

9.1.9   Annual examination

It would be a pity to go to so much trouble to prove a new crane can comfortably carry its maximum load only to neglect its maintenance, maybe for years, to the point that lifting any load would pose a danger to those in the vicinity. Happily legislation steps in to guard against such a possibility. In the United Kingdom and generally throughout the world it is a statutory requirement that all EOT cranes must be subjected to an annual examination, without which they cannot legally continue to operate. Although I should mention there are a small number of exceptions to the rule, such as the in-cave cranes, we shall discuss later.

To be honest, I had assumed for a long time that the annual test entailed a crane being pushed to the limit, much as it is during its very first trial. I had visions of steel ropes straining and the whole crane trembling, until it passed through the ordeal and someone in a hard hat signed off a certificate to empower it for another year. It turns out however, that things are much less dramatic, and cranes are subjected to a very thorough examination rather than making them quiver until the paint falls off. I must admit I felt somewhat cheated by the news until I discovered the reason why.

As we might expect, a crane's design process incorporates projections for the metal fatigue cycles it will experience during its working life, and ensures it can cope with those stresses and strains over the decades that follow. Quite sensibly, experts consider that pushing a crane to its limit every year will simply induce unnecessary metal fatigue and speed up the aging process, without proving anything. Instead inspectors conduct a very thorough examination of all components, including ropes and crane rails, and from this determine a crane's fitness to continue operating safely to its SWL for another year.

9.1.10   Load limiting devices

Advertising a crane's SWL in big letters can be helpful, but is ineffective when it comes to countering human error. For that reason cranes are also fitted with load limiting devices, which cut off power if someone erroneously tries to make a lift beyond the SWL.

In its simplest form, a mechanical-type load limiting device is based on the action of a stiff spring. There are several ways of fitting them, but whatever the configuration they invariably compress slightly when a load is lifted. If a hoist tries to lift a load heavier than its SWL, the spring will compress beyond a predetermined setting and trigger a simple switch that cuts power to the hoist motor. This type of mechanism works just fine, but on its own is a little too basic for the needs of cranes operating in nuclear facilities. For one thing, because they are either fully on or fully off they give no indication of an impending problem and just trigger without warning. Limiting devices employed by the nuclear industry tend to be based on strain gauge technology.

A load cell, which utilizes strain gauge technology, is fitted close to a crane's lifting feature or lifting accessory and crucially must always be located within the load path so as to feel the strain, so to speak, when an object is being lifted. Typically a load cell comprises thin strips of metallic film or several snaked wires, around 0.025 mm in diameter, embedded within a silicon wafer. During a lifting sequence hoisting ropes tighten up, become taut and finally take the full strain of their load. If tension mounts beyond a predetermined level, the wafer will begin to deform ever so slightly.

To become a strain gauge expert you need to picture what happens when a rubber band is stretched: it thins out. Exactly the same thing happens when the metallic film or wire of a strain gauge starts to bend: it stretches and thins. It may be indiscernible to you and me but the more it bends the thinner it gets. Now, hold that thought for a moment and then think about this.

We shall see in Chapter 11 that the way electricity flows along a wire is quite analogous to how water behaves; so, for example, the bigger the pipe, the more water (or *current*) that can flow through it. Equally then, a heavy copper wire can carry more current than a thinner one.

Strain gauges take advantage of this behavior by passing an electric current through the film or wires embedded within them. Control systems are then set to detect any change in electrical resistance, and warn of too heavy a load *before* motor stall torque gets the chance to cause an unwanted drama. It is a wonderfully elegant solution, particularly for such an industrial scale problem.

9.1.11 Protection against dropped loads

I suppose that if we were asked to speculate on what the main risk of using a crane might be, most of us would have to mention the possibility of dropping a load, where we would envision some heavy object breaking free of, say, its hook, or lifting beam and plummeting to the ground. Fortunately, there are a couple of things that mitigate against such a calamity. Firstly, there are the annual examinations I mentioned a moment ago, so actual equipment, including a crane's steel ropes should be in good condition and unlikely to fail. Added to that, lifting beams, shackles, and other accessories are fitted with safety catches which lock them onto a crane's hook and also to the load being lifted. So although it is possible to lose a load in mid-air the chances of it happening are slender to say the least.

Having said that, there is a more credible cause of a free fall event, known as *double blocking*, a situation we might not visualize quite so readily as a load breaking free of its lifting accessory. It happens when a crane's hook block rises too far and catches the underside of the crab unit above it. The initial impact, along with the power of a winch mechanism, that may still be turning, can rapidly overload ropes and result in a suspended load, along with its lifting accessory and hook block, breaking away.

To guard against this possibility conventional EOT cranes are fitted with a limit switch, which triggers if a hook block gets too close to its crab unit and cuts power to the crane's motor. For the nuclear industry, however, the specter of such a sudden stop is not too welcome, so nuclear cranes are fitted with a rotary switch which counts revolutions of a rope barrel and stops it turning well before an impending clash. However, just to be doubly cautious a limit switch is also fitted as a backup.

So although free fall events are rare, even for conventional EOT cranes, there is a different kind of *dropped load*, one which is more credible and does happen occasionally—very occasionally, you will be pleased to hear. It may not be quite as dramatic as a full-blown free fall, but is still serious enough to have engineers doing all they can to avert it.

Crane designers refer to it as an *uncontrolled descent*. The main difference between this and an all-out free fall is that the lifting accessory and its load stay connected to the crane. It is an important distinction because a crane's drive system and pulleys create considerable inertia, which limits the acceleration of a falling load significantly. Crucially with this kind of event a crane's apparatus holds a falling load back. As I say it is not quite as dramatic or destructive as a plummeting load, but no doubt still an unsettling experience for those in the area.

Standard EOT cranes can conservatively expect to experience a dropped load, or uncontrolled descent once in every 10,000 lifts, which sounds pretty impressive until you consider it equates to the operation of a crane doing less than six lifts a day over a period of 5 years. The additional safety features built into nuclear cranes seek to reduce that statistical probability to less than one in a million lifts, which is much better, but the risk still remains. As a result, the all-pervading safety case quite rightly demands an assumption that a nuclear payload could make an uncontrolled descent at any time, thus putting the onus on engineers to develop a fallback position for that eventuality. You will recall we examined this theme in Chapter 7 when we discussed how shielded transport flasks are designed and tested.

If then, as is normally the case, packages are handled within the bounds of their safety parameters and can therefore tolerate being dropped without releasing their contents, the question sometimes follows as to why bother going to all the trouble and expense of installing a high integrity nuclear crane? Unsurprisingly the answer comes back to safety, and more specifically in this context, risk reduction.

Even though packages are tested and certified to withstand falling from a particular height, it would still be irresponsible to proceed on the basis that a nuclear payload may make a free fall, or even uncontrolled descent now and again, without risking some harm. Apart from the conventional dangers of a falling load, it is just possible that a package may not live up to its test standards, which would risk releasing radio-active materials. A combination of prudence and common sense therefore dictate that all reasonable measures must be taken to ensure nuclear materials are always moved with minimal possibility of creating a drama.

So what else can crane designers do take to stop an uncontrolled decent from happening? The brakes fitted to your car are only engaged if you press the brake pedal or apply the parking brake, so unless you deliberately intervene the brakes are not applied. The philosophy adopted for brakes on any crane, whether for its bridge, crab, or hoist, is the other way round.

There are various braking mechanisms available but they all have one thing in common; in their normal position the brakes are on so that everything is effectively locked in place. Before a hoisting operation, either up or down, can commence powerful springs must be forced back to get a hoist's brakes to disengage. If there is an electrical power failure or some other control system malfunction, springs will revert to their unrestrained position, forcing them back onto the drum or disk. In other words hoisting mechanisms always fail-safe by applying brakes if the need arises.

I had thought such a system must be about as good as it gets and would easily satisfy even the most demanding of safety cases. But it turns out nuclear cranes are often fitted with two braking systems: one is responsible for day-to-day operations while the other, which is not subject to any wear and tear, is permanently on standby just in case the primary system should fail. In addition, rope barrels may be fitted with braking systems which are triggered by an overspeed condition, essentially a mismatch between input and output speeds. So high integrity cranes incorporate a fairly comprehensive suite of safety features, but what else can be done to obviate risks.

9.1.12 Zoning

We have seen there are limitations on how high nuclear transport flasks may be raised, but in addition there are also constraints on how and where they can travel. Even though cranes can quite happily traverse their loads diagonally, the normal practice in nuclear facilities is to move either parallel to crane rails or at right angles to them, as if moving around a grid pattern. It is not that there is anything inherently dangerous about diagonal transfers; it is more to do with restrictions on transfer routes, and the safety implications of compelling operators to carry out maneuvers that are more complex than necessary.

In addition, it is imperative that a suspended load must never pass above equipment which is crucial to the safe operation or controlled shut down of a nuclear plant; this equipment is normally referred to as safety-critical equipment. In practice, the rule is expanded to include virtually any item of plant and equipment. In addition the protocol also extends to a ban on lifting one transport flask over another, even flasks designed to withstand collision with a runaway locomotive. It is just not done.

These restrictions can make it appear there are large areas of wasted space beneath a crane and that the whole area could have been better configured. However, the open spaces are transfer zones, a bit like air corridors assigned for use by aircraft. Apart from keeping equipment out of harm's way, these unobstructed routes also maximize opportunities for loads to "skim" the floor, something that is always a good idea.

There is another restriction on lifting and transferring nuclear payloads which is simple enough to spot, but has repercussions that we may not deduce quite so easily. It is not advisable to transfer, or suspend loads above personnel, including crane operators themselves. No surprise there. Even though we understand the reliability figures and know how unlikely a calamity is, none of us would want to look up at the underside of 100 t load, or even one weighing just a few kilograms.

The less obvious situation occurs where loads are being hoisted across a high level floor and therefore out of sight to those on the floor below. In such a situation you could be blissfully unaware that something weighing umpteen tonnes was dangling above your head, which would be a dangerous circumstance unless you were protected in some way. In practice then, where there is a potential risk of a floor being impacted, it is reinforced to eliminate any possibility of a load plunging through it. If you happened to be beneath a falling load you might hear a thud and wonder if someone had left a door open, but you would come to no harm. It's amazing isn't it, how the tentacles of a comprehensive safety analysis grow and grow until they expose ramifications we could barely foresee, something we shall examine more fully in Chapter 14.

Having gone to so much trouble to minimize lifting heights, establish safe transfer routes and ensure equipment and people are kept out of harm's way, how do we ensure crane operators will observe the relevant predetermined rules all day, every day? It comes down to those interlocks I mentioned earlier, but also to preprogramming and crucially to the way a crane is zoned.

Zoning, which dictates where cranes can go and what they can do when they get there, is built into the control and protection systems that restrict what operators can make them do. Employing such systems prevents a crane obeying any command which is outside its preordained safe operating regimen. If required, it is possible to change those parameters, but only if a rigorous assessment concludes it is safe to do so. From a safety point of view it is a pretty good arrangement, since we can be sure a crane will stick to the script unless it is reprogrammed with a new one.

9.1.13   Seismic performance

By now we are getting the feel that some serious engineering goes into designing and manufacturing any crane that operates in a nuclear facility, so it would be a pity if one fell off its rails or dropped its load during an earthquake. It is worth noting that the load itself may only account for one third of the problem, since cranes themselves can often weigh more than twice their SWL. So, for example, a load of 120 t may be carried by a crane weighing say 250 t, or an all-up weight of 370 t, which is quite a big number to be hanging up in the air. Furthermore, we know from Chapter 5 that nuclear structures are often designed to cope with seismic accelerations of, say, $1g$, but for cranes, spanning between supports which are some distance apart, accelerations can rise to three times that, or even more. Clearly this is an area that keeps crane designers and structural engineers very busy indeed.

If a crane happens to be carrying a nuclear payload when an earthquake rides through, we really do not want that load to be left dangling there for weeks or months after the event. The prerequisite therefore, for most nuclear cranes, is that they must be sufficiently operable after a severe earthquake to get their load to a safe place and set it down. Bearing in mind that power supplies may have been interrupted, it is permissible to employ manual means to recover a suspended load, say by releasing brakes on a hoist, the priority being to restore a safe environment as quickly as possible.

If an earthquake does strike, cranes tend to take on a mind of their own and move back and forward along their rails, which I suppose is just what we would expect. However, a particularly fierce earthquake can also force a crane, even an exceptionally heavy one, to move from side to side and even jump up and down, a display of raw power that we can do without. Let's take a look at how conventional cranes roll along their tracks and then compare the different approach taken by nuclear engineers.

If you have ever studied the wheels on a train you will have noticed they have a flange around their inside edge. It is these flanges, on opposite wheels, which stop a train from wandering off its tracks. Conventional crane wheels have a flange on both sides, which is even better at keeping wheels in place. And just as with a train there is some breathing space, or float, between tracks and the flanges that kiss their sides as they roll along.

This is an excellent arrangement for a regular crane; it is safe, robust, and unlikely ever to give any trouble. However, for nuclear cranes which must be capable of coping with severe earthquakes, maybe the worst that could come along in a period of 10,000 years, this kind of arrangement does not work terribly well, and it is not too difficult to see why. As an earthquake hits its stride, steel stanchions supporting crane rails will start to twist and turn and inevitably the crane rails will flex in unison with them.

Although there is some float between the flanges of conventional crane wheels and rails they ride on, it would be insufficient for the turmoil created by a ferocious earthquake. As crane rails are pulled this way and that they will collide with flanges on a crane's wheels, and once that starts the consequences are unpredictable. For one thing more force will be transferred into the crane itself which only exacerbates the situation. In such extreme circumstances these enormous forces could even compel the heaviest of crane's to jump clean off their rails. In any event there would certainly be enough energy to cause considerable damage to the double flanged wheels along with their bearings and mountings, turning an expensive crane into an immovable quandary. Incidentally rails supported on concrete structures are not immune from these same consequences; movement may not be quite so pronounced, but a crane will still get a rough ride.

Occasionally a facility's seismic analysis may confirm that a conventional wheel design will work just fine, even in a nuclear facility. It is all down to specifics of the crane and its supporting structure, so this particular detail needs to be synchronized with the analysis. In cases where a conventional design can satisfy seismic demands there is no need to provide anything more sophisticated. In most cases, the challenge is to come up with a design that allows a building's structure and EOT crane to move independently during an earthquake.

The way it is achieved is beautifully simple. Rather than employing flanges both sets of wheels are ordinarily perfectly flat, with one set captured onto their crane rail by means of guide rollers. During an earthquake the guided wheels stay centered on their rail, while wheels on the opposite side of a bridge are free to slide around without causing any damage. I say "ordinarily" because wheels on the guided side may be flanged but in most situations a plain profile prevails. As an added precaution, steel brackets are fitted which will hold a crane in place if its guide rollers were to fail. This guided wheel arrangement works well with any type of building structure, but really comes into its own when different construction materials are used to support the two crane rails.

Many nuclear facilities have a shielded concrete structure at their core, surrounded by a steelwork structure which houses services and support activities. As a result, many EOT cranes find themselves supported by concrete on one side and steelwork on the other. In situations where cranes are supported only by concrete or only by steel, there is a fair chance that during an earthquake their supports will dance with some kind of harmony. So crane rails sway together and to some extent a crane can go with the flow.

All that poetry in motion goes out the window when a crane spans between concrete and steel. Now when an earthquake hits, concrete will stand almost motionless while by comparison steel becomes much more agile. In these situations guide rollers are fitted to the side which seismic specialists deem will experience the lowest accelerations, normally concrete; that way they can hold a crane steady until the movements calm down.

9.1.14 Polar crane

You would think EOT cranes are always going to be confined to running up and down rails and simply moving from left to right but there is a version, the polar crane (Fig. 9.16) which rotates on a circular track.

Fig. 9.16 Polar crane.
Getty Images.

The first time I came across one in the flesh, so to speak, I was visiting an aging reactor facility to discuss its decommissioning plans. As we entered the reactor hall my guide set off in full flow, pointing at this and that and no doubt sharing lots of crucial information, but no one was listening. I stood routed to the spot, staring up at the magnificent crane and marveling at how splendid it looked. As I regained my composure and contemplated some suitable words of apology, the crane began turning and my head shot upwards again; this was too good to miss. From the corner of my eye I glimpsed my host, arms folded but clearly prepared to leave me alone until the show was over. I watched the crane carry out a few maneuvers and eventually dragged myself back to the business at hand. It was a good day.

Faced with the challenge of providing crane coverage within a circular building the polar crane now looks like such an obvious solution, but we can just imagine how eyebrows must have been raised the first time someone suggested making an EOT crane turn like a carousel. In every other regard a polar crane behaves in exactly the same way as its linear counterpart, but its circular tracks do make it perfect for a small number of uncommon situations. Without it, for example, operations within many a reactor building would be, at best, problematic.

As I indicated at the beginning of this chapter, cranes operating in benign nuclear surroundings (the kind we have been discussing so far) may have their heritage in conventional cranes, but when you get up close they do differ quite markedly. We have seen, for example, that nuclear cranes incorporate safety interlock systems, carry auxiliary hoists, are fitted with dual braking systems and can cope with severe earthquakes. Armed with this broad understanding of the workings of a basic *high integrity* nuclear crane, it is time to move into the world of a harsh nuclear environment. The cranes we shall come across in here are a bit special to say the least.

9.2 In-cave cranes

Outside the nuclear industry you are unlikely come across the type of crane we shall be discussing here. Others may have to contend with hostile chemical and environmental conditions, but it is only nuclear facilities that carry the added burden of radiation and the contamination which often accompanies it. The bottom line here is that these cranes operate in places which are completely out of bounds to us humans, so C4/R4 (red) radiological environments.

The whole emphasis therefore is on making sure that whatever happens, which covers an awful lot of ground, an in-cave crane will not, for example, get marooned in its working environment. They must be robust, exceptionally reliable and most important of all, it is unthinkable that we could find ourselves gazing at a stranded crane, through a heavily shielded window, and be powerless to retrieve it. If they do break down then it must be possible to recover them to a safe area where they can be repaired or maintained. The reliability issue is particularly challenging since these cranes do not benefit from dependability that flows from mass production; they are almost always a one-off, a bespoke design for a unique set of circumstances.

9.2.1 Cartesian crane

If you are well up on your mathematical terminology, you will be familiar with the term "Cartesian coordinates" from which these cranes derive their name. It is all about identifying the location of a point in space, by way of its distance from intersecting lines and planes that are at right angles to each other, no doubt a fascinating subject for those keen enough to delve into it. Mind you having said that, knowing what Cartesian means just highlights what an excellent name it is for this particular type of crane. After all their life is spent moving up, down, and sideways, all at right angles and in a bid to locate some point in space.

The position of any item of interest within a cave, such as a lifting beam, is always referenced by its X, Y, and Z coordinates, with X equating to distance along routes parallel to crane rails, Y at right angles to them, and Z denoting distance up and down. If you have used any 3D drafting system you will know that points within a computer model are referenced in exactly the same way. It is true that the whole family of EOT cranes operate in the same three directions. It is just that a higher degree of positional accuracy is demanded of in-cave cranes, and in this regard the Cartesian crane excels.

9.2.2   Polar jib crane

If there was a competition for impressive looking machines, the polar jib crane would be right up there with the best of them. Even those with zero interest in all things mechanical would have to look again. They are highly complex yet still manage to be exceptionally robust, and even though they look quite purposeful somehow exude elegance. As if combining such diverse attributes were not enough, the whole package can perform in a setting as hostile as any manmade environment on earth. And by the way, they operate remotely and can be recovered without the need of a human entering their world. Pretty impressive.

Fig. 9.17 Polar jib crane handling vitrified waste container.
© Nuclear Decommissioning Authority.

The major difference between polar jib cranes (Fig. 9.17) and EOT cranes is that they do not have a bridge, instead their crab unit travels on rails which are close together and normally attached the underside of a cave's roof. As you might expect with something so unlike its competitors, in some ways they outshine EOT cranes while in others they are not quite so useful. Occasionally, their unique characteristics do make them the ideal choice.

There are two areas where polar jib cranes score way ahead of the opposition; firstly they can fit through a much smaller shield door, which is important because it brings down the initial price tag of a door and helps reduce ongoing maintenance costs. Not only that, a smaller door reduces the containment boundary between one radiological zone and another, which pleases just about everyone. The best bit however, is that polar cranes do not suffer too much from hook approach restrictions which curb the reach of every other crane. So much so that the end of their jib arm could touch the sidewalls and ends of a cave, if indeed they were allowed to. So very little is out of bounds.

As with the Cartesian crane, height is referenced by position on the Z, or vertical axis, but because they rotate about a center point their horizontal positioning is determined by a combination of angle and radius, or more specifically Y movement, by travel, and angular position on a fixed circle. As for their tooling, it is very similar to that deployed on the manipulators we discussed in Chapter 7. As I say, quite a package.

9.2.3 Modularization

We also looked at a cave's decontamination and maintenance areas in Chapter 7, so need not dwell on them here. It is worth noting that in-cave cranes rely heavily on modular design to speed along their maintenance and repair. It is an approach which enables self-contained assemblies to be quickly exchanged, allowing faulty modules to be refurbished and tested in a workshop without the time pressure of holding up a crane that needs to get back to work.

One example is a crane wheel module which can be bolted into place in 1–2 hours, rather than the whole day or more that it can take to do the job in a more traditional piecemeal way. Another favorite module is an electric motor and gearbox assembly, but I suppose the biggest crane module we can expect to find is an entire hoist unit. It comprises the motor and gearbox assembly, along with a barrel already wound with its rope and can even have a lifting frame attached. I should say that generally a module of this magnitude is rarely contemplated within the confines of an in-cave crane maintenance area (CMA). It is possible, which is reassuring, but there would need to be a compelling case for taking on such a challenge.

9.2.4 Written scheme of examination

Over the preceding pages, I have mentioned that ordinarily there is a statutory requirement to examine cranes every a year. However, in the nuclear industry, it is deemed appropriate that some cranes skip their annual checkup and are instead examined every 2 or 3 years. Sounds a bit odd doesn't it? If anything we might expect things to be the other way round, with nuclear cranes being examined more frequently than others, so I need to explain.

The option exists to invoke what is known as a *written scheme of examination*, which is ideal for cranes operating in hostile areas such as C4/R4 (red) radiological environments within heavily shielded caves. Other cranes operating in the nuclear industry, including the high integrity type we have just discussed, are subject to exactly the same examination regimen as everywhere else.

To appreciate the logic behind applying this apparent relaxation we need to understand why annual examinations are conducted in the first place. There are two reasons: firstly to verify that a crane is safe to operate for another year without risk of harming anyone that happens to be in the vicinity, and secondly (but linked to the first reason), normally cranes can only legally continue to operate if they pass their annual check. The situation has many parallels with the United Kingdom's Ministry of Transport (MOT) test, to which vehicles are subjected. Think of it like this. If a car was entirely remotely operated and spent its entire life performing tasks in an enclosed and inaccessible environment, there would be little point in subjecting it to an MOT test. That same reasoning applies to the kind of crane we are discussing here.

If there is one thing we know for sure about cranes operating in caves, it is that no matter what they do, no matter how they perform, no one is likely to come to any direct harm as a result of their failure. The demands of a facility's safety case will see to it that whatever happens within it, a cave's containment will not be breached. Even if an entire crane fell off its rails, which is quite improbable, no one would get so much as a scratch. Now I am not suggesting such an event would be inconsequential—far from it—but what I am saying is that in itself it would not be unsafe.

Even so we might still surmise that, while from a safety perspective these cranes do not need to be tested quite so often, they do after all operate in a pretty challenging nuclear environment. So maybe we should just test them annually anyway? Indeed it does sound reasonable until you factor in how these cranes are maintained, and also the potential exposure to radiation for those involved in their examinations.

Normally when a crane is being inspected, its guards are removed and it is very carefully scrutinized. However, taking that same approach when examining in-cave cranes carries a risk of unnecessarily exposing surveyors and those assisting them to radiation. We have seen that unlike conventional cranes, modularization is central to the design of these particular cranes. Happily then, when it comes to examinations this unique characteristic can be made to work in everyone's favor.

Rather than strip modules down and risk the possibility of exposure to radiation for those involved in examinations, it can sometimes make more sense to dispose of a whole module and replace it with a brand new one. When this happens the new module will already have been tested and come with appropriate certification in place, so a surveyor can be confident everything is in good order. On the other hand throwing modules away during inspections would be quite expensive; therefore, in practice operators synchronize their maintenance program with that of official examinations.

In reality then, surveyors find themselves dealing with cranes which are constantly being renewed and operating in areas where they are incapable of harming anyone. When you add the whole thing together, common sense dictates there is little justification for slavishly adhering to a practice which, although right and proper elsewhere, is not really appropriate in these specific circumstances.

9.2.5 Recovery

If one thing occupies the minds of engineers engaged in designing in-cave equipment then this is it: recovery. I suppose it could also be described as retrieval, but in a nuclear world the term retrieval tends to be associated with the decommissioning task of gathering radioactive waste before it is processed and stored. Recovery is the term normally used to describe getting failed equipment out of a radioactive environment so that it can be repaired and set to work again. Apart from a couple of notable exceptions, such as deep-sea or space exploration, this is not an issue most engineers need concern themselves with. If, for example, equipment on a car assembly line breaks down you can prod and fiddle with it until you get it working again—not so easy if it is inside a highly radioactive cave.

As we have seen, before equipment that resides in a hostile radiological environment can be handled, it probably needs to be decontaminated, which is a relatively straightforward task. Before we get to that stage, equipment needs to be safely extracted from wherever it happens to be.

The first principle of effective in-cave recovery is to minimize the likelihood of having to do it in the first place. The objective is to locate as much as possible of what might conceivably break down on the outside a cave, that way if it does give trouble you can simply walk up to it and get to work. A good example of putting equipment in a convenient, easy to get at location, is associated with the control systems for a crane or indeed a power manipulator. It is another of those ever so simple ideas that works a treat: simply locate control equipment, fuses, electrical relays, and so on the outside of a cave, rather than on-board a crane or any other apparatus which might be difficult to get at. Putting them somewhere that is easily accessible means a fuse can be changed in 2 minutes rather than say a couple of days.

The second line of defense against gazing helplessly at motionless equipment is not to put all your eggs in one basket. If you have ever dreamt of having a spare engine tucked inside your car, ready to surge into action if the primary one should splutter, then this will appeal. The type of crane we are discussing here never relies on a single electrical motor to drive them along their rails, to operate their hoists, or to activate twist locks on a spreader frame; in fact they have dual drives for everything.

It is comforting to know that if an electrical motor fails to respond to a command, or even worse stops in the middle of an operation, operators can fire up another one capable of executing exactly the same task. But then you will already have guessed what the response to that level of duplication is, "What happens if a backup fails to respond or only manages to operate for a few moments?" The answer is, "We shift from defending against equipment failure and move into recovery mode."

When it comes to designing for recovery rather than against it there is only one objective, albeit a multifaceted one. Identify everything that might conceivably break down and every permutation of how and where it might happen. Then ensure you can recover from all those situations and get back to normal operations as safely and quickly as possible. A layperson could spend hours drawing up that kind of, "What if?" list and of course the experts are even better at it, so this is a tall order.

One of the things to keep in mind when you look inside a cave, any cave, is that it is not an open space to be utilized pretty much how individual operators see fit. No, there is order and meticulous planning, everything with its designated role and in its proper place. Some caves may change usage during different phases of their operational life, and if well planned can even embark on entirely different tasks during a facility's decommissioning. Ideally all of this should be figured out early on and built into a cave's original design. It is possible to amend a cave's configuration after it becomes operational, but unless preplanned the task can be substantial enough to become a project in its own right, so not something to be tackled lightly.

Recovery planning assumes that the motors on a crane's bridge, its crab, hoist and twist locks could malfunction at any point on a journey. If any of the primary motors and their backup should fail at the same time, which would be mildly irritating, then it is always possible to continue an operation by instigating recovery from outside a cave.

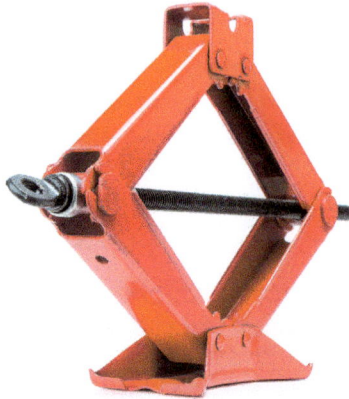

Fig. 9.18 Scissor jack.
Shutterstock.com.

The way it works is not too dissimilar to the operation of some car jacks, the scissor type that have a long threaded bar running through the middle (Fig. 9.18). As its bar is turned the jack's struts slowly pull themselves along it, forcing the jack to rise vertically and lift a car. To operate this type of jack you simply connect a handheld rod to an attachment on one end of the threaded bar and start to turn. If you have ever used this type of jack the thing that strikes you immediately is how easy it is to lift a car. Okay there are a couple of tires touching the ground so you are not lifting a car's entire weight, but nevertheless you could still be lifting half a tonne or more and doing it effortlessly. Unfortunately many of us have had this roadside experience, so have discovered that with the aid of a small and simple mechanism we can move exceptionally heavy loads. The secret to easily moving so much weight is in the very high ratio of turns we make just to raise the load a few centimeters. Going back to recovery of our in-cave crane, the same principles apply: it is slow but not too difficult.

Faced with using a lance for recovery it is always preferable to minimize the amount of effort required to use it. If dealing with a suspended load, the lance should ideally be used to lower it back to where it started, not to lift it even higher. Once it is safely down, twist locks can be disengaged from a load and recovery moved to the next phase. By the way if twist locks also fail to respond, even to their backup recovery drives, then they in turn can be activated by another lance, or better still by a nearby manipulator. It is possible to raise a suspended load higher and move it to another set-down location, but then turning a lance to raise a heavy load which is several meters away is always going to be more difficult. It can be done, but ideally should only figure as a fallback plan when developing recovery scenarios.

Clearly the big difference between doing this sort of exercise at the roadside and doing it from outside a radioactive cave is physically getting at whatever it is you want to twist. If you examine the walls and roof of a typical cave closely, you will notice there are lots of small plugs, removable shield plugs to be precise, typically around 170 mm in diameter. Each plug lines up with the business end of an in-cave mechanism that can manually lift or lower something, or move it from side to side. It is vital here to ensure that operators maneuvering a lance will not be subjected to unacceptable radiation exposure while they are doing it. Clearly this cannot be left to chance, so is one of the issues factored in at the design stage when locations of plugs and recovery mechanisms are synchronized.

Corralling radiation may appear to be quite a daunting task, but there is one characteristic of a typical cave configuration that simplifies it somewhat: distance. Sources of radiation, in other words the materials being handled and processed, tend to reside close to the floor of a cave, while recovery plugs are generally located either in the roof of a cave or high up on its walls.

When we discussed radiation in Chapter 2, we saw how its intensity declines as the distance from a source increases, and drops by its inverse square each time the distance is doubled. Added to that, radiation shine is not unlike torchlight, in that apart from some scatter it tends to travel in straight lines. It is possible then, to arrange recovery ports in locations that minimize potential for direct shine from sources within a cave. And as you will have deduced the recovery preparation phase will, where possible, involve moving radioactive materials away from the scene.

So when effecting crane recovery there are ways to deal with radiation, but having removed a shield plug it might seem that stopping airborne contamination escaping from a cave presents an even greater challenge. However, we know from Chapter 8 that the atmosphere within a cave is always held at a depression relative to the area which surrounds it. In practice then, when a small shield plug is removed air is continually sucked through the hole it creates and into the cave. In fact the ventilation system will be set up so as to draw air through this type of penetration at a speed of around 1 m per second, so contamination will find it difficult to drift through an opening and get to where operators are working. Nevertheless, as a precaution they may still wear protective clothing and breathing apparatus, and also tent-out their workspace so as to segregate it from adjacent areas.

To get a feel for how recovery might be instigated we can use a simple example. Our imaginary task is to lower a stainless steel box to the floor of a cave. The box is a 1 m cube and is lifted by means of a simple spreader frame. It is just routine workaday stuff, the idea is to keep it routine when equipment malfunctions. In our recovery example the spreader frame is functioning perfectly but both hoist motors have failed, so we need to lower the box manually.

With preparations complete, a local shield plug is removed and replaced by a temporary plug which can accommodate a recovery lance. Once inserted, the lance is attached to a connector on the stalled hoist and slowly turned until the stranded box is set down again. With that done the spreader frame is disengaged. Once the box is safely down, a standard shield plug is reinserted, the out-cave vicinity monitored to ensure it complies with normal operating conditions and the area returned to its regular duties.

It is easy to see that before getting to a stage where recovery operations can be smoothly executed, even for our simple example, there is a fair bit of upfront coordination required to ensure everything is in the right place and that a myriad of interfaces have been adequately addressed. This explains the need for all that careful in-cave planning I mentioned earlier. For example, crane designers need to identify every position where an object will routinely be picked up or put down, so that they and their engineering colleagues can synchronize a cave's layout and equipment design with corresponding shield plug locations.

In recovery terms the situation I have just described would be ranked in the straightforward category. Simply lower a load back down, de-latch it from its still-operating spreader frame, get the crane back to its maintenance area and repair the hoist. It becomes a bit more challenging if a crane itself stops while halfway between transferring a load from point A to point B, particularly as the load may be left dangling over some precious equipment. Even though a crane's control systems are always zoned to stop loads being lifted over safety-critical equipment, there will still be plenty of other apparatus that we want to safeguard for straightforward commercial reasons, a design activity normally referred to as investment protection.

For occasions when recovery is not quite so straightforward, caves invariably have an emergency set-down position. It is a grand title for some floor space where any troublesome load can be lowered off and parked for a while. The beauty of an emergency set-down position is that it is surrounded by shield plugs, which are strategically arranged to effect all manner of preplanned recovery scenarios. This even includes several plugs in a cave's roof, directly above the set-down position, which give flexibility to access recovery points from above.

Those charged with developing a cave's configuration, would have a much easier life if they could select any spot they wished for emergency recovery. Unfortunately, the location is pretty much determined for them. By definition, a primary emergency set-down position must be capable of serving every conceivable eventuality, so must therefore be located at the end of a cave which is closest to its maintenance bay. It is all to do with the way a crane itself is recovered, which we shall get to in a moment. Caves may also have one or two auxiliary set-down positions which are located to accommodate localized recovery, but generally the primary set-down position is the focus of most recovery scenarios.

If a crane does break down, there are no prizes for guessing that the main priority is to get it back to its maintenance area so that it can be repaired. As a consequence of that sensible objective, recovery systems are biased in favor of moving, or dragging a crane towards the safety of its maintenance bay. It is possible to use heavy steel rods to push a crane in the opposite direction, but this technique is normally only employed if it becomes absolutely necessary.

This is a good time to mention the golden rule of in-cave crane recovery. Having picked up a load, never ever move a crane unless the load has a route to an emergency set-down position. Once a crane moves away from one of its designated pickup or set-down points, its crab and hoist are no longer backed up by recovery shield plugs assigned to that particular location. In effect a crane has moved away from its comfort zone, which is why it is so important to avoid situations where a suspended load gets trapped behind an obstacle that it cannot maneuver around. Having said that, the array of recovery mechanisms and their associated shield plugs do make it possible to work around obstructions. However, if at all possible, it is always prudent to keep the coast clear for direct access to one of a cave's emergency set-down positions.

To be confident of getting a crane back to the security of its maintenance bay we really must depend on something that is both simple and robust. There are two main ways of doing it. The first employs a simple bridge, commonly known as a "rabbit," which is permanently parked at the far end of a cave. In itself, this type of recovery bridge cannot be used as a crane, since apart from its wheels it has no moving parts. Steel wires, connected to each end of the rabbit, run alongside both crane rails and back to the maintenance area. If a crane loses both its drive motors these wires can be connected to winches in the maintenance area, which then drag the recovery rabbit back down the cave. En route it will meet up with and capture the stranded crane, taking it back to the maintenance bay.

Once a crane is repaired its first job is to push the rabbit back to its parking position at the far end of a cave, where it patiently awaits another adventure. As an alternative, it is possible to dispense with the bridge and opt for an arrangement which simply deploys a separate "rabbit" at the end of each crane rail, again with steel wires attached.

The other popular recovery method involves connecting strainer wires, as they are known, directly to a crane's bridge. However, unlike the rabbit technique these wires are not static, since they must move with a crane and in fact are normally bundled with other cables, something we shall discuss more fully in the next section when we examine how cranes receive their power supply.

Apart from the rabbit and strainer wire systems, there is another recovery technique available, but only to caves that have opted to install two cranes or an independent crane and power manipulator on the same tracks, something we discussed in Chapter 7. Although it must be said this one comes along as a bonus, and should not be seen as replacing those primary recovery systems, one of which will still be needed. There is nothing to it really. If either a crane or power manipulator should become stranded its partner can be used to either push it, or tow it out of a cave and into a maintenance bay. The ability to push or pull is pretty much what we would expect from two items of equipment working alongside each other, but the relationship brings with it another means of recovery, one which is not quite so obvious.

It is possible that a crane or power manipulator may become stranded through loss of its power supply, due to say damaged electrical wiring. If this should happen, either device can plug an auxiliary power supply into its neighbor; then once any suspended load is lowered they can travel in tandem back to their maintenance bay.

Another aid to some of the more complex recovery scenarios involves placing shield plugs in a line along the entire length of a cave. This is over and above those used for localized recovery operations that we discussed a moment ago. These plugs are spaced at appropriate intervals, maybe a couple of meters or so apart, and positioned in a wall just a little distance above one of the crane rails, or alternatively in a cave's roof and directly above a crane rail. A lance, inserted through one of these ports, can move a crab along its bridge, which, coupled with the ability of a rabbit or strainer wires to drag a stranded crane along its rails, gives operators the flexibility to move a crane around pretty much as it would when fully operational. This is particularly important when recovering a crane with a suspended load attached, as it can maneuver around obstacles if it has to. It might be a slow process but it will get there in the end.

As we can see the basic principles of crane recovery are all about keeping it as simple as possible, a philosophy which even extends to recovery scenarios that are beyond the abilities of a simple lance. For example, a situation where a crane's hoisting ropes become hopelessly entangled may appear to be one which can only be overcome by an extraordinary feat of ingenuity. However, the simple answer might to be to use a manipulator to cut the hoisting ropes then get the crane back to its maintenance area, re-reeve the rope barrels and send a crane or power manipulator back into the cave to tidy up. There is no doubt that recovering equipment from a hostile radiological environment is always going to present difficulties, but with sufficient forethought and careful planning it need not be the intractable challenge it first appears.

9.2.6   Power supply

9.2.6.1   Catenary

If you examine most conventional overhead traveling cranes, you will notice their power is delivered by an electrical cable which loops up and down alongside one of the crane rails (Fig. 9.19). This type of cable arrangement is known as either a *catenary* or *festoon*, with both names being equally valid.

Fig. 9.19 Catenary cable.
Shutterstock.com.

As a crane moves off down its tracks the cable stretches out and the depth of its loops gets shallower. When it comes back the loops sag back down and the cable bunches up again. This is a good example of a fit-for-purpose solution—nothing fancy, it just does its job. However, even though it performs perfectly elsewhere it is not suitable for cranes that operate inside a shielded cave.

The main problem area for a catenary type cable occurs where it meets the hefty shield doors that segregate radioactive caves from their decontamination and maintenance bays. We know from Chapter 7 that these doors fit fairly snugly around a cave's crane rails, so as a result there is no space for a drooping electrical cable to get through. In fact if a suitably sized gap was provided for this type of cable it would also allow radiation to shine through the same opening. So some other solution is required. What is needed is an arrangement that keeps the cable taut, like a tightrope, so that it can slide through a very small gap between a shield door and crane rail.

Fig. 9.20 Cable reeling system.
© Metreel Ltd.

9.2.6.2 Cable reeling

This is where a cable reeling system comes in (Fig. 9.20). It pays out electrical cable from a large wheel which is located close to the CMA, typically either on a platform below the CMA or in a room at the far end of it. The wheel operates in a fashion not unlike that of a garden hose reel, in that once the hose is connected to a tap we can unwind it and water will continue to flow. In the case of a cable reel, *slip rings* are used to keep electricity flowing and unlike a garden hose the reel is motorized so as to keep tension in the cable.

Because a taut cable can easily fit through a small hole beside a shield door, it neatly solves the problem encountered by looped catenary cables. In addition, as

I mentioned when we discussed crane recovery, strainer wires can be bundled with a crane's power and instrumentation cables so that, should the need arise, they are always poised and ready to drag a crane back to its maintenance area. The one downside to recovering a crane in this way is that the steel wire becomes so tense it cuts into sheathing on cables as it is drawn onto a reeling wheel. So the whole bundle needs to be replaced if it is used for to retrieve a stranded crane: having said that it is an excellent system and has a lot going for it.

Unsurprisingly with such credentials, cable reeling is the favored system when it comes to providing power to cranes operating in hostile radiological environments, which brings me onto an issue which is linked to their use, one that may have crossed your mind already.

Here we have a crane with its power and instrumentation cables in tow and operating in a highly contaminated environment, so on the face of it, it seems anomalous that while the crane operates behind a hefty door, its accompanying cable moves in and out of an accessible maintenance bay. Admittedly access to this C3 (amber) area is restricted, but nevertheless there would seem to be a risk of contamination on the cable migrating into an area occupied by maintenance personnel. To counter this problem a system is installed which continually wipes cables clean as they exit a cave. It is a simple solution which barely appears to warrant mentioning, but I raise it to highlight the attention to detail which is required when dealing with contamination, particularly the high levels found in many caves.

One of the tasks designers face when putting a nuclear facility together, is identifying any equipment that might possibly need replacing during its operational life, which is quite a long list. With that done they must figure out how to get failed, or redundant equipment out of a building and its replacement back in. It might even be that two routes are required, with exiting contaminated equipment being cut up and going in one direction and a shiny new substitute arriving from another. To do this exercise well, it is imperative that the largest and heaviest potential items are identified; that way we guarantee all equipment can be changed without getting jammed in a tight spot on route.

It you are ever involved in the exercise of ensuring there is space to replace equipment and are examining an area in the vicinity of a cave's maintenance bay, then I can save you some time. The largest potential candidate for replacement will almost certainly be a cable reeling wheel. Their size varies according to the length of cave they serve, but a diameter of 2 m is not unusual. Even though the chances of ever needing to change one are extremely remote, there is a slim possibility. It comes under the heading of, "You never know," so these wheels always feature in the list of candidates for replacement. It may be possible to break the old one up to get rid of it but ideally the new one should arrive in one piece, so handling them does demand a fair bit of space.

Incidentally, I can tell you that although the chances are a cable reeling wheel will last more or less forever, it is not too uncommon to find that cables themselves do need replacing. If you have ever unraveled plastic piping you will have discovered that it insists on twisting and curling up a most unhelpful way, so have some idea of what a cable changing exercise is like. So even if we dismiss the possibility of needing to change a cable reeling wheel, we must in any event cater for re-reeving one, an exercise which takes about as much space as changing an entire wheel. For one reason or another then, designers cannot afford to skimp on space in a CMA; one day plant operators will need it.

It might have crossed your mind that 2 m seems like a disproportionately large diameter for sufficient cable to cover the length of a cave, even a particularly long one. After all, we have all seen never ending lengths of heavy-duty electrical cable wrapped around very much smaller barrels than that. Even allowing for the inclusion of a strainer wire, 2 m does seem excessive. Ordinarily it would be but as we can see in Fig. 9.20 this cable is wound in what is known as a mono-spiral. It takes up space but stacking cable in this way rather than bunching in on a wide drum is an excellent way of minimizing twists and turns, something which would be difficult to deal with inside a shielded cave.

9.2.6.3 Busbars

A busbar system adopts the same principle as a live rail for electrically powered trains. If you look down from the platform of just about any underground railway or subway system, you will notice an extra rail running parallel to the regular tracks. This additional rail delivers power to a train's electrical motors and is picked up by a connector, or *shoe*, which protrudes from beneath the train. To ensure that the all-important power flows without a break, connectors are sprung so as to push against a track and maintain contact at all times.

Cranes can employ similar systems, although their three-phase electricity supply is carried by three bars rather than the single rail used to deliver direct current down on the subways. In fact as we can see in Fig. 9.21, the red busbars are excellent

Fig. 9.21 Busbar power supply.
© Metreel Ltd.

in situations where bends need to be negotiated. However, use of busbars for cranes operating in a radioactive environment tends to be pretty limited, partly because of their inability to help with recovery, but mainly because over a long period they are prone to gathering dust. It is not an issue of worrying what visitors might think, but because dust creates the possibility of losing power due to a poor connection. Busbars are used occasionally, but not nearly so widely as a cable reeling system.

9.2.6.4 Drag chain

If you took an open-ended length of bicycle chain, cleaned the oil off it and laid it on the worktop in your kitchen you would be ready to simulate a drag chain. First off, nail one end of the chain firmly to the worktop (don't try this at home) then pick up the other end and place it on top of that which is nailed down, so that the chain loops back on itself from the middle. The loose end can now be moved back and forward, sliding over the chain below and extending almost full length in one direction and then the other. If you were to embed a small electrical cable within the chain it could power a moving object and follow it up and down the worktop.

Fig. 9.22 Drag chain.
© Metreel Ltd.

The production version of a drag chain (Fig. 9.22) is similar to the do-it-yourself version, essentially a long jointed tray carrying cables within it. The kind we are talking of here can be connected to a power supply within the CMA and laid halfway down a cave, then loop back on themselves and connect to the crane itself. It is quite a neat solution with cable furling and unfurling as a crane moves along, and is more reliable than a cable reeling system. Once again, though, it has limited use within a radioactive cave environment, mainly because drag chains are unable to contribute to a crane's recovery.

So there are several reliable ways of getting electrical power to a crane. However, when it comes to nuclear caves, choosing the right approach is always a twofold challenge; select a system which is most appropriate for the unique operational demands of a particular cave, but then also keep an eye on how recovery will be instigated.

9.2.7 Guidance systems

With in-cave cranes being so inaccessible, it is vitally important that they go exactly where they are instructed with considerable accuracy and tedious regularity. Being in roughly the right spot is not good enough in this environment. There are two main methods employed and both are wonderfully unsophisticated.

Fig. 9.23 Rack and pinion.
Shutterstock.com.

The first system utilizes a rack and pinion (Fig. 9.23) with a rack laid alongside one of the crane rails, and does nothing more than count revolutions of a pinion wheel. The wheels are well engineered, have a known circumference and do not skid of slide when they turn. Linking a computer to a pinion's rotation is all it takes to accurately calculate where a crane is at any given moment. As we would expect the same system can also be used to measure cross travel of a crane's crab unit.

The other favored method of positioning a crane is visual, so feels a little more intuitive. White disks, about 100 mm in diameter are marked with a crosshair and fixed to exposed surfaces inside a cave. If you were able to look up from inside a cave you might see them attached to the underside of its roof, but they can also be fixed to walls or even to immovable objects within a cave.

As operators maneuver a crane into one of its many predetermined positions, they align crosshairs on its on-board cameras with one or more of the fixed targets around a cave, just like overlaying the image from one telescopic sight on another. Targets and cameras will have been synchronized during the commissioning phase and are regularly revalidated during operations, so we can be confident that when crosshairs are superimposed a crane is exactly where we want it to be.

For completeness, I should also mention there is another popular method of guiding cranes to a given spot, one that is widely used in nonnuclear settings. Quite often a spotlight with a narrow beam, or even a laser, is mounted on a crane's crab unit and aligned with a simple target on a floor or wall. Lining a crane up in this way is very much faster than tweaking it into position every time it gets close to its destination. It might not be quite as hi tech as monitoring pinion revolutions or viewing crosshairs on a monitor, but it is utterly dependable and in the right environment perfectly fit for purpose.

At the beginning of this chapter, I mentioned there is more to nuclear cranes than meets the eye. Since then we have discovered they must be able to ride out a violent earthquake and still be operable, or at least function well enough to lower off a load and be recovered to a safe area. They are designed to stick to a predetermined safe operating regimen, one which demands sophisticated control systems and interlocks. Auxiliary hoists are used not just to speed along operations but also as a safety feature. In-cave cranes must be designed to withstand prolonged exposure to high levels of radiation; their recovery systems, such as dual drives and cable reeling, must be robust and reliable enough to retrieve a crane without need of human intervention. Their design must embrace the very best of modularization techniques, and so on. In addition, we saw in Chapter 7 that decontamination techniques can be quite aggressive, so can add that crane materials and components must be capable of withstanding a thorough soaking from water and attack from acid. The list of demands is very long indeed.

When you visit a nuclear facility either to work there or as a visitor, it is worth taking a look at one or two of its cranes, both in regular operational areas and those only visible through heavily shielded windows. Reflecting on their hidden depths for a moment, reminds us that things in this environment are seldom as they seem. The apparently humble crane, demonstrates quite nicely that in a nuclear world peering beyond our first impressions often reveals quite an unexpected story.

Electrical

10

The one thing that differentiates this group from other engineering disciplines is how very compartmentalized each of its subgroups can be. Engineers in other disciplines may major in a particular area, such as a structural seismic specialist or mechanical manipulator expert, but with a little encouragement can wax lyrical about a wide gamut of their group's capabilities, albeit not to the same level of competence. These folks, though, tend to stick a little more rigidly to their own field, such as delivering electricity to the point of need or developing control systems that oversee a facility's operations. Being so specialized is not a shortcoming in their proficiency, but rather an indication of how very diverse the responsibilities of an electrical engineer can be.

Unsurprisingly for a discipline that covers so much ground, they are known by a wide selection of names. Sometimes the word *Systems* creeps into their title but ordinarily they are referred to as either, *Electrical and Instrumentation* (E&I) or *Electrical, Controls, and Instrumentation* (EC&I). On its own *controls* does not explain too much of what they do, but it is actually a short form for control systems, which gives a bit more of a clue. As I say you can expect to come across several variations.

If you have had little or no involvement with the EC&I function, you might reasonably assume the scope of their involvement in a nuclear project is quite narrow. They will run cables around a building, install a few monitors and wire up panels in a control room. Oh yes, and make sure lights are bright enough and in the right locations. However, you may be surprised to hear their remit covers an awful lot of ground. To give an indication of how far their tentacles spread, think of this for a moment. In some nuclear facilities, particularly those with an emphasis on mechanical handling operations, electrical equipment of one type or another can occupy 15% or even 20% of a building's entire floor area, which is a pretty big number. We need to look closer.

10.1 Electricity supply

At its very highest level, electricity supply and distribution on nuclear sites is no different to that encountered on any other large industrial site. Beyond that, especially when it comes to backup supplies, nuclear facilities demand more than most. To set the scene we shall begin with a look at regular electricity supply and distribution, the kind of setup you come across at most power-hungry establishments, then move on to see how things ramp up in a nuclear environment.

10.1.1 Site ring main

The electricity supply entering an industrial site originates from either a nationwide network, such as the UK's National Grid, or from regional power companies. We shall assume an incoming supply of 132,000 V (132 kV) but keep in mind that it can be a

Nuclear Facilities. http://dx.doi.org/10.1016/B978-0-08-101938-2.00010-6

good deal higher. This sort of voltage is fine for a whole site, but needs to be reduced to a level more appropriate for the demands of individual facilities.

Fig. 10.1 132 kV–11 kV Substation.
© Nuclear Decommissioning Authority.

The first point of call therefore, for our lively 132 kV, is a substation (Fig. 10.1) more specifically a *step-down* substation, which reduces it to a still considerable 11 kV. Again this number can vary, but 11 kV suits most industrial situations. Having said that, electrical equipment in the United Kingdom and many other countries generally operates on a supply of between 220 and 400 V, so we need additional, or *satellite* substations to step the voltage down even further. Countries operating on supplies of 110–220 V follow the same principles.

It is possible to run individual 11 kV supplies directly from the main 132 kV substation to satellite substations on a site, in what is known as a *radial system*, but the widely adopted practice is to install a ring main or *ring circuit* (Fig. 10.2). With this approach 11 kV–400 V substations are connected to the ring in locations which are close to facilities they serve, and rather conveniently more substations can be added as the need arises. The major advantage of this arrangement is that individual substations can be taken offline, say for maintenance, whilst power continues flowing to others on the ring main.

10.1.2 Transformer

As we can see the role of a substation is to transform voltage, either upwards or downwards, so understandably the apparatus at its heart is known as a *transformer*. Fig. 10.3 shows a fairly typical configuration with our example 11 kV entering the primary side of a transformer via its delta, or *mesh* windings and, through a process

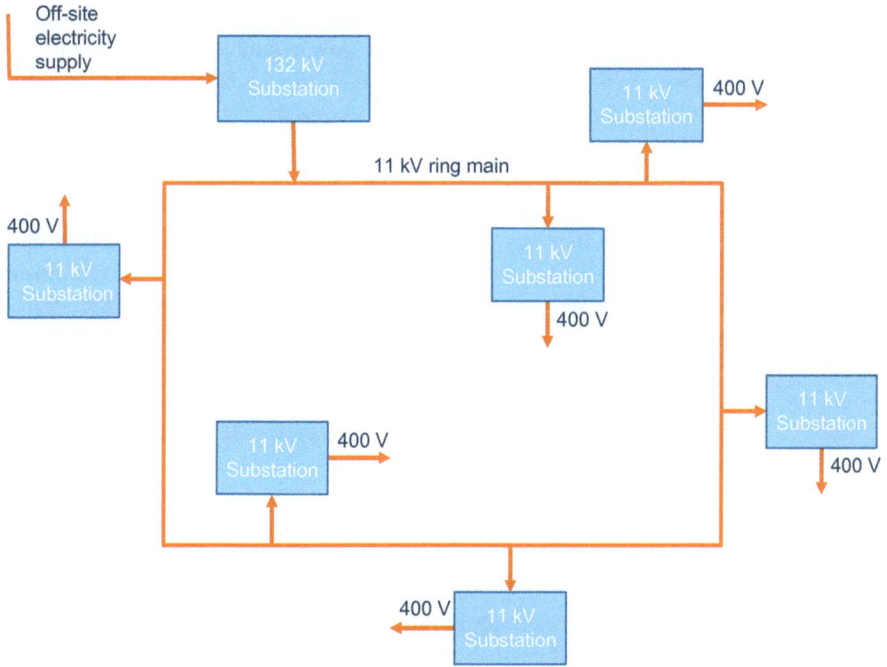

Fig. 10.2 Ring main.
© Bill Collum.

known as electromagnetic induction, 400 V exiting from the star, or *wye* windings. The diagram also highlights the very different roles played by a circuit's neutral leg and its earth, or *grounding* cable.

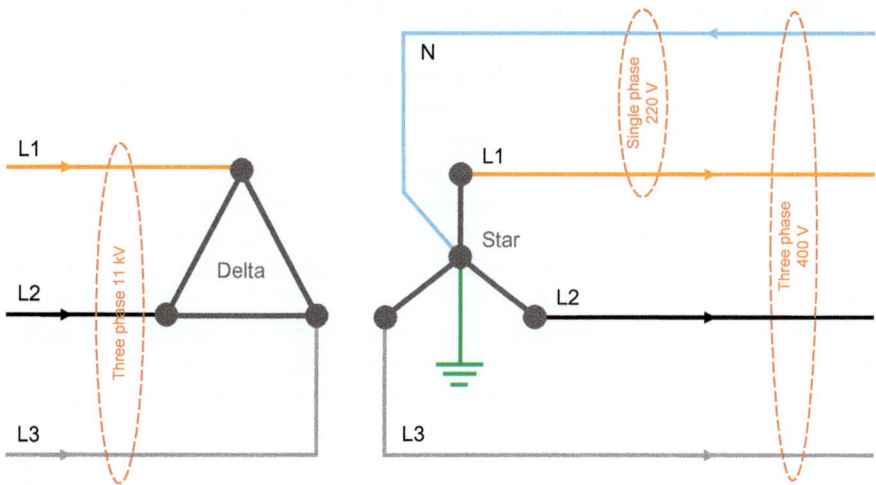

Fig. 10.3 Transformer.
© Bill Collum.

The role of a neutral cable varies depending on whether it is on a single phase or three-phase circuit. For a single phase, current must flow in a loop always returning to where it started. So we can imagine a live cable, in this case L1, delivering electricity, with the neutral leg playing more of a passive role in that it simply picks up the flow for its return journey.

For a three-phase supply a neutral cable is not always required, with specifics depending on whether the electrical system in question is balanced or unbalanced. For a balanced system, say where three phases are powering a delta-wound electrical motor, a neutral cable is not needed at all. Simply put there is no surplus current for a return journey, so its role becomes redundant. In a more representative situation, where three phases are distributed across an industrial electrical network, it is impossible to balance the system perfectly. In these cases a neutral leg picks up the surplus current and returns it to its originating transformer.

The role of an earth or *grounding* cable is one of protection. If a fault should occur in an electrical circuit, say due to a loose connection, it could cause electrical equipment to become live, resulting in an electric shock for anyone unfortunate enough to come into contact with it. An earth cable provides the path of least resistance for such a surge of electricity and literally directs it into the ground. Of course electrical circuits are protected by fuses but they may take a second, maybe a good deal longer, to "blow." In normal circumstances such durations would be quite inconsequential, but where electricity is involved those same moments could be fatal. An earth on the other hand is always poised and ready for immediate action.

10.1.3   Distribution board

Most heavy-duty industrial equipment is powered by three phases, L1, L2, and L3 which, as we have just seen, originate from the supplying substation. Typically then, a 400 V three-phase supply from a substation enters a switchroom, which is invariably located on a facility's perimeter and is effectively the hub from which all electrical cables fan out across a building.

The incoming supply is connected to a busbar which runs across the switchroom, and from there branches off to multiple distribution boards. And whereas the 400 V supply enters and exits a distribution board (Fig. 10.4) in pretty much the way we would expect, namely via a like-for-like connection, 220 V is delivered by connecting a live cable to any of the three-phase supplies. The same principle is reflected in the transformer diagram, where I have shown a single phase originating from the L1 cable, plus its neutral return leg. In reality the same 220 V supply could also be derived from either of the other live cables, L2 and L3.

Whatever the number of phases, distribution boards route all live cables through a fuse or miniature circuit breaker whilst neutral cables normally make a direct connection. For illustration purposes I have shown just one 400 V connection and one 220 V, but as we can see this particular board could accommodate a total of eight connections.

From distribution boards, cables snake-off to the four corners of a facility, with 400 V of three-phase supply heading for the likes of powerful ventilation fans, electric overhead travelling (EOT) cranes and all manner of electrical pumps, whilst single phase 220 V cables will be destined for less demanding lighting circuits along with sockets for power tools, computers, and the likes.

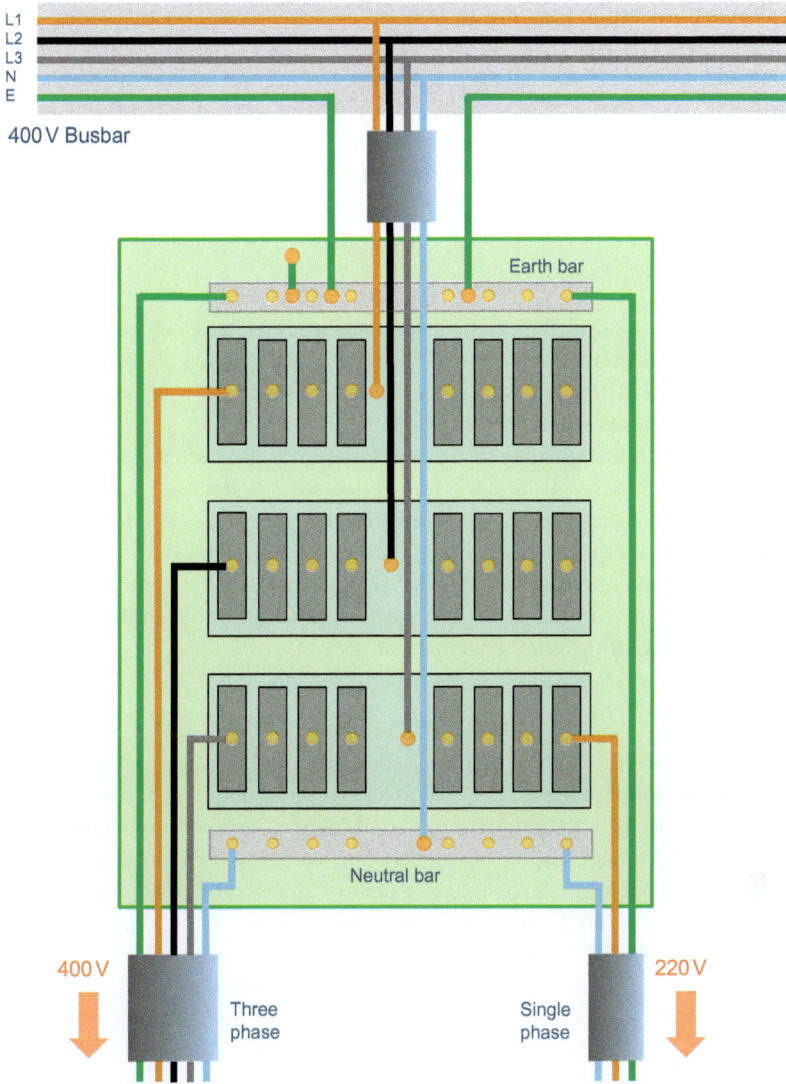

Fig. 10.4 Distribution board.
© Bill Collum.

10.1.4   Cable color coding

You will have noticed the different-colored cables in Fig. 10.4, so before moving on I must mention color coding. Whenever I come across an electrician pulling strands from a spaghetti bunch of cables, I always slow my step to marvel at how they could possibly know what goes where. So my admiration was compounded when I discovered the convention for color coding cables is not standardized across the world; in fact there are quite a few variations.

Of course most of the time this does not pose a problem, although connecting up three-phase electrical equipment which has been manufactured on another continent is bound to add nicely to the challenge. For our purposes, for a three-phase supply we shall adopt L1, brown; L2, black; and L3, grey. Whilst for a single phase, its one live cable will be brown. And in both cases we shall take neutral to be blue. Thankfully everyone agrees that either green, or green and yellow striped sheathing is appropriate for earth, or *grounding* cables.

10.1.5   Backup electricity

In our home or place of work it can be inconvenient if the electricity supply is lost, but ordinarily when the drama passes there is no harm done. Nuclear facilities, along with hospitals, financial institutions, air traffic control centers, and so on need to be better prepared. So how do we keep power flowing?

The first thing to recognize is that nuclear plants do not need a guaranteed round-the-clock supply of their full-blown, everyday electricity demand, so can get by very nicely on reduced power for quite some time. In fact, should the need arise, there is not that much electrical equipment needed to keep a plant ticking over, or shut it down safely. It is all a matter of establishing what, if anything, needs continual uninterrupted power, what needs power in a hurry and what can do without. With that done we can select from a range of different backup strategies.

10.1.5.1   Non-firm power supply

With a non-firm or *basic* option there is just one electricity supply entering a facility, so if it is lost everything goes off (Fig. 10.5) and stays off until power is restored again. It is comparable to the arrangement most of us have at home. Apart from a non-firm supply's obvious drawback of having no backup, there is also a maintenance issue. If it is necessary to conduct maintenance on a facility's upstream power supply, say at a substation, then power to the facility itself must be shut off and operations suspended until maintenance is concluded.

As I say, this is the bare minimum in terms of electricity supply, but if a facility can live with its limitations then there is no need to spend money on something more elaborate. Having said that, it is not really suitable for most nuclear facilities but, as with our homes, is perfectly adequate for many of the support facilities we find dotted around any nuclear site.

10.1.5.2   Firm power supply

If you eavesdrop on a group of nuclear electrical engineers for a while, you are almost bound to hear mention of "A" and "B" electrical supplies. It simply means that a facility has two supplies, usually described as being *independent and diverse*. Some industries use the term "redundant systems" to describe this same type of duplication. The principle is a good one and akin to that adopted for crucial aircraft control systems.

If an accident or malfunction on a nuclear site stopped crucial electricity flowing down a group of "A" cables, power would instead be routed along a duplicate set of "B" cables (Fig. 10.6). As an alternative, the system can be configured so that in

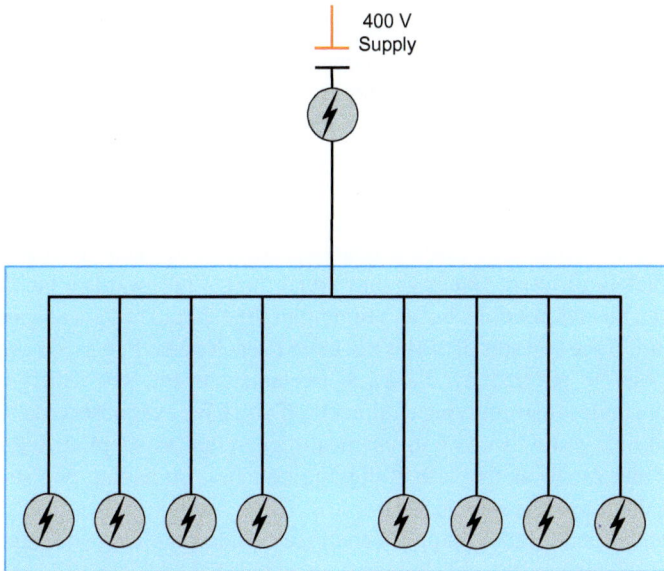

Fig. 10.5 Non-firm power supply.
© Bill Collum.

normal operating conditions a facility's electricity supply is shared, 50% each, between its "A" and "B" networks. The advantage being that if one arm should fail the switchover to full reliance on either "A" or "B" is instantaneous.

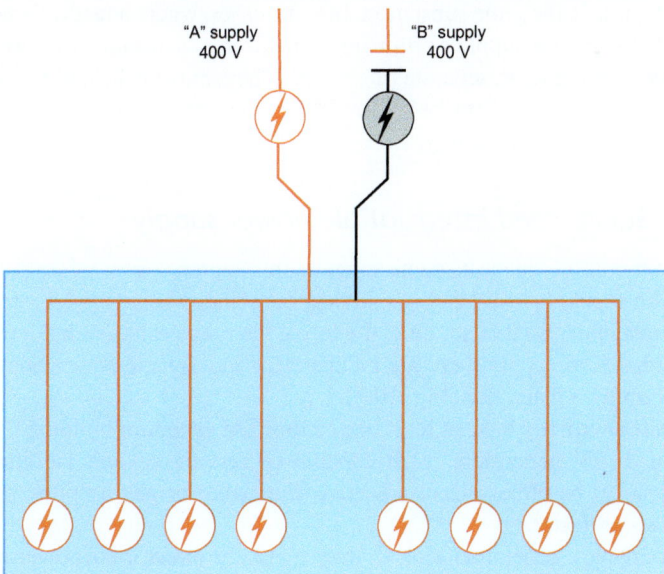

Fig. 10.6 Firm power supply.
© Bill Collum.

Clearly there are costs involved in doubling up on a power distribution network, but in the quest to keep electricity flowing it can turn out to be a worthwhile investment. As a bonus there is another benefit to duplicating systems, in that electricity can be kept flowing whilst upstream substations and the like are being maintained. It is just a fairly simple matter of shutting down one arm of the delivery network and routing power through the other.

Independence in this context refers to physical segregation of electricity supply lines, and begins before they reach a building. Ideally "A" and "B" cables should follow completely different routes as they wend their way towards a facility, which of course is not always possible, but as a minimum they should be kept at least a couple of meters apart. If the two sets of cables run close to each other there is always a danger that both could be severed by the same unplanned event, say damage from an earthquake or careless earthmoving equipment. If you have ever observed excavations being carried out, you will know mechanical diggers are as adept at finding power cables and water mains as they are at digging trenches. So having two supply lines and keeping them apart is a good idea.

The concept of independent supplies continues inside a building and right up to the point of need. However, it is not absolutely essential that "A" and "B" cables follow totally different paths within a building. They can often take the same course, say parallel to each other, provided there is again a couple of meters or so between them.

In terms of diversity the actual source of electricity for "A" and "B" supplies should be quite different. Of course if they were traced back far enough both will originate from the same local or national network, so if it is disrupted both supplies will be lost. However, as we shall see in a moment, there are ways around this. In practice then, "A" and "B" supplies for an individual facility generally originate from different distribution boards in the same substation. In some cases, where a facility's safety case demands it, these diverse supplies originate from different substations and even, say in the case of a nuclear reactor, separate ring mains. There may be the odd exception, but for most nuclear facilities a firm, or "A" and "B" supply represents the bare minimum it terms of a backup power supply.

10.1.5.3 Guaranteed interruptible power supply

It is a bit of a mouthful, but as its name suggests this approach guarantees power, even if a site's main supply is lost. Having said that, the standby supply it delivers may not be instantly available. So it is not quite the top of the range solution, but it does come close. As with a firm supply there are independent and diverse power lines, but this time backed up by a generator (Fig. 10.7).

Nuclear sites normally have at least one centralized generator building that houses a number of 11 kV generators, each capable of serving several facilities. These generators tend to be either diesel powered (Fig. 10.8) or gas turbines, which can to get up to speed faster.

When emergency generators kick in they normally direct their power to a site's substations and from there to facilities in need. Alternatively, if deemed necessary, electricity can be routed directly to safety-critical plant and equipment. We may

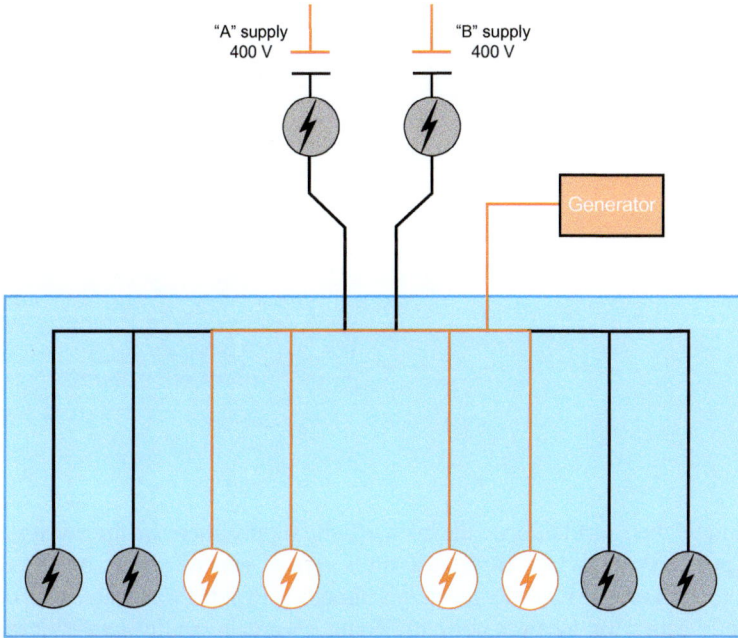

Fig. 10.7 Guaranteed interruptible power supply.
© Bill Collum.

Fig. 10.8 11 kV Generator.
Shutterstock.com.

reasonably assume that having several centrally located generators, mighty ones at that, would be insurance enough against loss of power, but no, this is the nuclear industry, so in some cases we need to do better. If a facility's safety case dictates that it must have a guaranteed interruptible supply then, rather than relying solely on a central backup system, it may also have its own emergency generator. Not only that, the chances are it will also have a connection point for a mobile generator (Fig. 10.9).

Fig. 10.9 Mobile generator.
Shutterstock.com.

Potentially then, in addition to the main off-site supply, there are three ways to keep backup power flowing to a facility: a site's central generators, its own dedicated generator, and a mobile unit. However, it is important to note that the ability to connect to a mobile generator does not normally count towards a facility's safety case, as it may sensibly be reasoned there is a high risk of roads being impassable at the very time it is needed. For this reason, the formal role of mobile generators tends to be restricted to backup during routine maintenance activities, but with some recognition that they may be able to help out in an emergency.

You may be thinking to yourself that this business of generators taking over from an off-site power supply does not quite add up, and you would be right. Even the mightiest of centralized generator suites could not meet the everyday demands of a nuclear site working at full tilt. Likewise for local generators, unless we throw a small fortune at it they cannot possibly satisfy the totality of demand for an operating nuclear facility.

Watts (W)	Equals
Thousand	Kilowatt (kW)
Million	Megawatt (MW)
Billion	Gigawatt (GW)

Fig. 10.10 Wattage units.
© Bill Collum.

Industrial power consumption is normally expressed in watts, with the usual high-value multiples shown in Fig. 10.10. Incidentally watts are calculated by multiplying a facility's supply voltage by its demand, measured in amps, hence the big numbers.

There is no such thing as a typical nuclear facility, but if we assume a power demand of 20–40 MW or indeed more, we shall not be too far wide of the mark. A site's main generators may pump out say 5–10 MW, and local generators serving individual facilities may deliver 1–2 MW. So whilst the accumulation of backup power is impressive, it is clearly incapable of replacing the potential shortfall for a power-hungry nuclear site. Fortunately electrical engineers have come up with an elegant solution to this particular quandary.

Fig. 10.11 EPD panels.
© Nuclear Decommissioning Authority.

If a nuclear site's main power supply is lost, its centralized emergency generators will automatically kick in to save the day. As their new power surges out across a site, circuit breakers at the various substations trip into Essential Power Distribution, or *EPD mode*, so that in a predetermined sequence power to nonessential services is shut off and directed to where it is needed most (Fig. 10.11). When you get right down to it, the power required to safely shut a nuclear facility down or keep it ticking over is not that great, so generators can cope admirably for quite a long time.

But what happens to plant and equipment that loses its power whilst in mid-operation? Will things start falling over or bumping into each other? Standard practice in the nuclear industry is to design equipment to failsafe, such that if power is lost it will adopt a safe quiescent state and remain there until it is reenergized, a good example being the braking mechanism on a crane's hoist which we discussed in the previous chapter. Essentially, if power goes off the brakes come on.

Guaranteed interruptible power can be an expensive option. Apart from the obvious capital costs of generators, along with additional cabling and equipment, there is another price to pay. There is little point in having generators sitting around if they let you down at the very moment they need to ride to the rescue. They must be dependable, which in practice means generators are subjected to a constant campaign of maintenance and testing. It certainly validates their reliability, but spread over a facility's lifetime the cost is significant. However, paying the price for such reliability is infinitely preferable to paying the potentially horrendous cost of finding oneself with no power at all.

10.1.5.4 Guaranteed uninterruptible power supply (UPS)

The backbone of a UPS (Fig. 10.12) is batteries, lots of them. If generators let us down, batteries will kick in and instantly provide essential power to the likes of computer systems, environmental monitors, level detection instruments and other equipment which is important to the safe and orderly shutdown of a nuclear facility's operations, and following that to keeping it in a quiescent state.

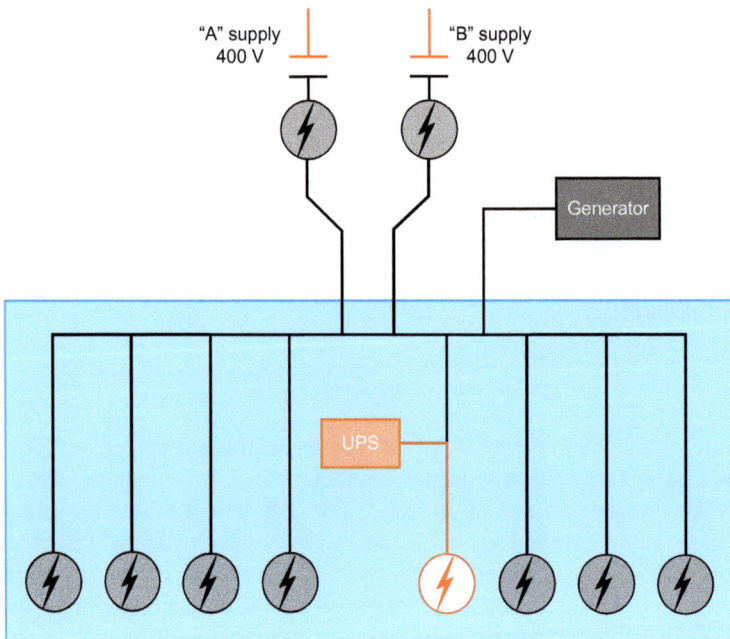

Fig. 10.12 Guaranteed uninterruptible power supply (UPS).
© Bill Collum.

Of course batteries do not last too long, so in the situation we are considering here could be expected to deliver power for around 4 hours. Not a fixed number that, because there are too many variables, but it gives a rough idea. We certainly cannot depend on them to keep going for a whole day. So the challenge is to get generators fired up before batteries give out. Ordinarily emergency generators should be pumping out electricity in less than an hour, two at the most, so that gives a 2 hour cushion before batteries start to fade. It may not sound too long, but in an emergency situation things will happen in a hurry and there are plenty of rehearsals.

So batteries are permanently poised and ready for action, but the trouble is they deliver the wrong type of current, and for many applications the wrong voltage. If ever there seemed to be a mismatch of requirement and solution, it looks like this could be it. So how do the Electrical folks get out of this one?

Fig. 10.13 Battery room.
Shutterstock.com.

Battery backup is produced by 2 V cells connected together in series, maybe 100 at a time (Fig. 10.13), so between them delivering 200 V of direct current (DC). There may be a small amount of equipment in a building that operates on DC, such as some of its instrumentation, but as with our domestic supplies practically everything else demands a supply of alternating current (AC).

Unlike DC, an AC current, is continually switching polarity. In other words, the current flows first in one direction, say, negative to positive, and then the other. Polarity of UK domestic supplies changes 100 times every second, which equates to 50 complete cycles. And the number of cycles per second determines a current's frequency, which is expressed as hertz (Hz).

If you look on a hidden side of any UK domestic electrical appliance you will find a label which, among other things, confirms the appliance is compatible with a power supply of 220 V and a frequency of 50 Hz. The same is true in many countries but others, such as the United States conform instead to 110 V and 60 Hz. In addition, as we have seen, much of the equipment in United Kingdom industrial facilities operates on a 400 V, three-phase supply, which is still at the same 50 Hz frequency. Other countries, primarily those with a 110 V domestic network, opt for a three-phase industrial supply of 220 V, again on a 60 Hz cycle.

So here we are with a battery backup system delivering, say, 200 V of DC, when we know that most of a facility's equipment is powered by 110–400 V of AC. This is where an inverter comes in. It takes a DC supply, in our case 200 V, chops it into separate pulses, then switches the polarity of those pulses back and forward, creating an alternating current. Not only that, but one of the reasons power companies like to supply us with AC is the ease with which it can be transformed from one voltage to another. So once a DC supply has been chopped and turned into AC, transformers have no trouble increasing or decreasing its voltage. Hey presto, 200 V of DC supply coming from lots of 2 V battery cells, suddenly becomes anything up to 400 V of AC.

10.1.6   Cable handling

On the subject of electricity supply it is worth mentioning that the cables we are discussing here bear little resemblance to the wires we see around our homes. In this high-power industrial league even low voltage (LV) cables carry between 600 and 1000 V and can have a physical diameter of 45 mm. I say *physical* diameter, because cable dimensions are quoted in terms of the cumulative cross sectional area of individual cables within them. So a LV cable with a "physical" diameter of 45 mm may be specified as 120 mm^2. An 11 kV cable, which is classed as medium voltage can have a physical diameter of over 65 mm. Interestingly enough, it may still be specified as 120 mm^2, with its greater physical diameter being due to increased insulation.

Fig. 10.14 Cable installation.
Shutterstock.com.

Of course with dimensions on this scale these cables, not to mention the drums carrying them, take a bit of handling, so their routing and final installation (Fig. 10.14) must be carefully planned. For example, the minimum bending radius for an 11 kV cable is around 700 mm. With such constraints it is no surprise that electrical engineers can be exasperated by jibes that they, "just need to run a few wires around the building." Very often it best to think "rainwater pipes" very heavy rainwater pipes.

10.1.7   Fire resistant cable

Nuclear facilities have an awful lot of electrical cable running around inside them (Fig. 10.15) in fact upwards of 100 km would not raise any eyebrows. Clearly if even a short length of it were involved in a fire, the situation would be a serious one. And if the cable concerned were sheathed in polyvinyl chloride (PVC), which is a common enough covering, the hazards would increase significantly.

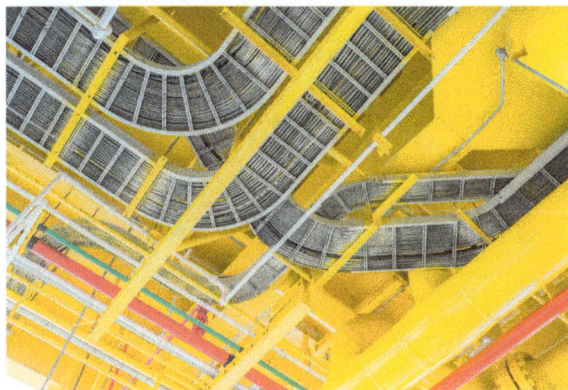

Fig. 10.15 Cable trays.
Shutterstock.com.

The reasons are twofold. PVC likes to burn, particularly in vertical runs. Once it starts it is self-sustaining, so is quite happy to spread and cause more destruction. More disquieting, when PVC burns it gives off hydrogen chloride fumes which are highly corrosive and harmful if ingested. Happily then, the cables we see snaking around nuclear facilities are a fire resistant type, such as *low smoke zero halogen* (LSZH). If a fire is initiated, the big advantage of LSZH, apart from an absence of halogen, is that it does not burn as readily as PVC and if it does the accompanying toxic fumes are greatly reduced.

10.2   Control systems

Nuclear plants are strewn with all manner of kit: everything from shield doors and cranes, to pumps, valves, conveyor systems and a whole lot more. All of this is fine, but how do we ensure they do exactly what they are supposed to do, when they are supposed to do it and crucially all operate in perfect harmony? For example it wouldn't be great if a pump dispatched radioactive liquor to a vessel, only to find a valve was asleep on the job and blocking the pipe earmarked to carry liquor to its destination. We need a system, some way of controlling an often vast array of plant and equipment.

10.2.1   Control hierarchy

As we shall see, this is not a task that can be dealt with by a one-size-fits-all approach, so sure enough it operates on multiple levels, all of which can be integrated by an all-seeing eye which resides at the top of the system. Let's start with the entry-level.

10.2.2 Basic control

With a label of *basic* this kind of control panel is well named. In its simplest form it comprises nothing more than a switch to turn on the power, a light to show it is on, a start button and a stop button, often supplemented by a hard-to-miss red button which initiates an emergency stop. And that's it. We often hear the term *intelligent system*, but in this particular case the system itself is devoid of such sophistication, so instead intelligence is dispensed by the index finger of a watchful operator.

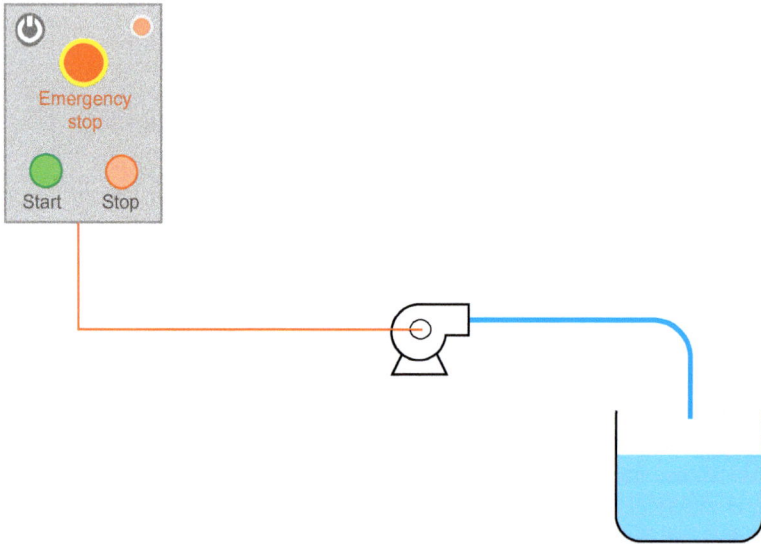

Fig. 10.16 Basic control.
© Bill Collum.

Fig. 10.16 shows how this type of control panel may be used. In this example, pressing the start button fires up a pump which dispenses liquid to a waiting tank. When an operator judges the tank is sufficiently full, they press a stop button and the pump shuts down.

As I say, all pretty basic. It is a bit like switching on a domestic food mixer: it will keep spinning until you intervene to make it stop. That being said, if we do not need anything fancy, then there is no justification for specifying something more hi-tech. For nuclear plants, there is not too much of a place for such basic technology so entry-level control systems tend to start at the next tier.

10.2.3 Sequence control

Although this is just one step up in the control hierarchy, in terms of technology it is already significantly more advanced than the simple on/off of a basic system. At its heart is a programmable logic controller (PLC) in effect a computer, one which is teamed with a motor control center (MCC). The big leap forward—well, two leaps—is that a PLC provides intelligence and an MCC the ability to switch electrical equipment on and off, all from within a control panel. With this setup the operator's push of a button delivers much more action.

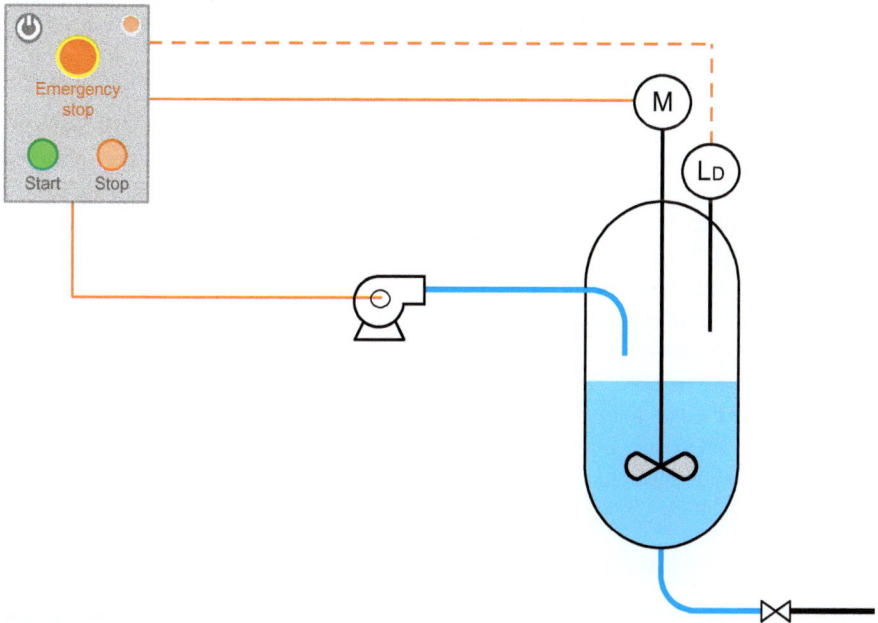

Fig. 10.17 Sequence control.
© Bill Collum.

Fig. 10.17 shows a simple sequence whereby pressing a button starts up a pump which sends liquid to a waiting vessel. For clarity I have used a solid red line to denote electrical power and a dotted red line for cables carrying instrument signals. This time a detector (discussed later) senses when the liquid reaches a predetermined level and informs the PLC which, through the MCC, stops the pump. At the same time the PLC instructs a mixer to start up, again via the MCC. We shall assume the mixer is on a timer so stops again after ten minutes. In this example further operations, such as emptying the vessel and sending its liquid on for further processing, are controlled by additional systems.

10.2.4 Motor control center (MCC)

Next in the control hierarchy is automation, but before we get to that we need to take a look at the MCCs that make it possible. Essentially they are devices that switch electrical power on and off, but of course it is not quite that simple. If we zoom in we find they divide into three distinct types, each serving quite a different purpose.

10.2.4.1 Direct on line (DOL)

First is the simple on/off function, known as DOL. It is useful and certainly has a place, but in many situations is far too crude, too sudden for the processes we are considering here.

Take a mixing paddle, one suspended in a vessel filled with some type of liquid, say a radioactive sludge, and visualize what would happen if its motor drive was simply switched on. From being stationary, the paddle would accelerate to full speed in an instant. Apart from subjecting the sludge to a rapidly unfolding agitation, it would place tremendous torque on the paddle's driveshaft. We need something more subtle.

10.2.4.2   Soft start

Next up is *soft start*, which sounds altogether far too genteel for an industrial setting. However, this technology has the welcome attribute of being able to start a motor slowly and over a short period, say 5–10 seconds, gradually gather pace until it hits full speed. Similarly *a soft stop* can also be implemented, whereby a motor slowly decelerates when it is switched off again. This is a much better way to start a mixer and equally useful for conveyors, pumps, fans and other applications. The one thing to note about a soft start is that once a motor hits full speed it will remain there until it is switched off and gradually decelerates. So although it offers significant advances over the all-or-nothing of DOL, it does have limitations.

Incidentally the technology that sits behind a slow-start MCC module is a thyristor, which operates on similar principles to the transistors we shall be discussing in the next chapter. Their main difference, in this context, is that thyristors can handle much larger currents.

10.2.4.3   Variable frequency drive (VFD)

Also known as an *inverter drive*, this system is delivered via a VFD controller back in the MCC. It offers total flexibility, not just during start up and stop, as it can also adjust a motor's speed and torque whilst it is busily whirling away.

The most impressive feature of a VFD is that they can be configured to sense a motor's torque in real time, say due to changes in a liquid's viscosity, and regulate rotation speed so as to maintain the most favorable performance. On the downside, VFDs are guilty of causing electromagnetic noise which, if unchecked, can interfere with nearby electrical equipment.

Now we have looked into the MCCs that make automation possible, we can move onto the subject itself.

10.2.5   Automation

We might reasonably describe the previous *sequence control* as performing activities in a *loop*, so to break out of that we need a system capable of overseeing a much wider vista, essentially multiple sequences, all of which are coordinated. For this we need a suite of MCCs, of the type just discussed, an ability to monitor several instruments, and finally a PLC which can examine incoming data from motors and instruments, then instantly issue instructions in the opposite direction.

Fig. 10.18 shows how the control panel for automated processes divides into three main zones: the PLC, MCC, and instruments marshalling area. Of course in reality there can be a whole plethora of MCCs serving multiple pumps and mixers, so in addition to the equipment shown we can add fans, compressors, heaters and more. Likewise for instruments, where we could add those measuring density, weight, viscosity, and so on.

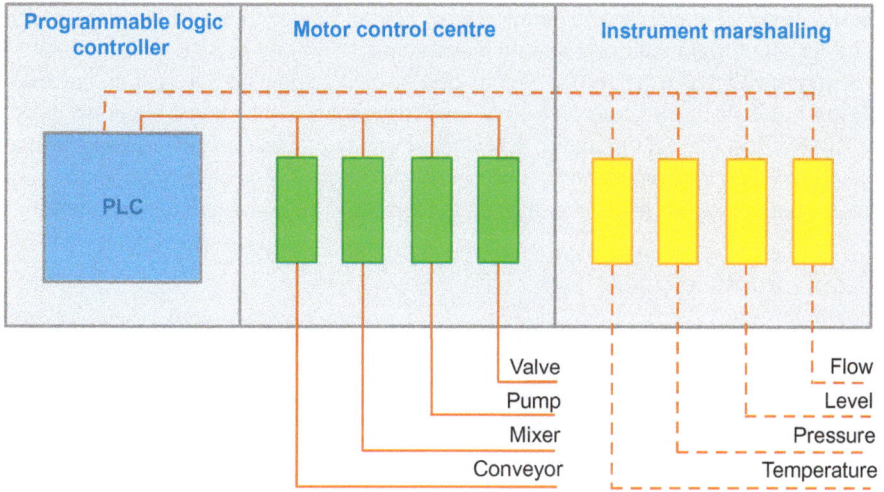

Fig. 10.18 Automation panel.
© Bill Collum.

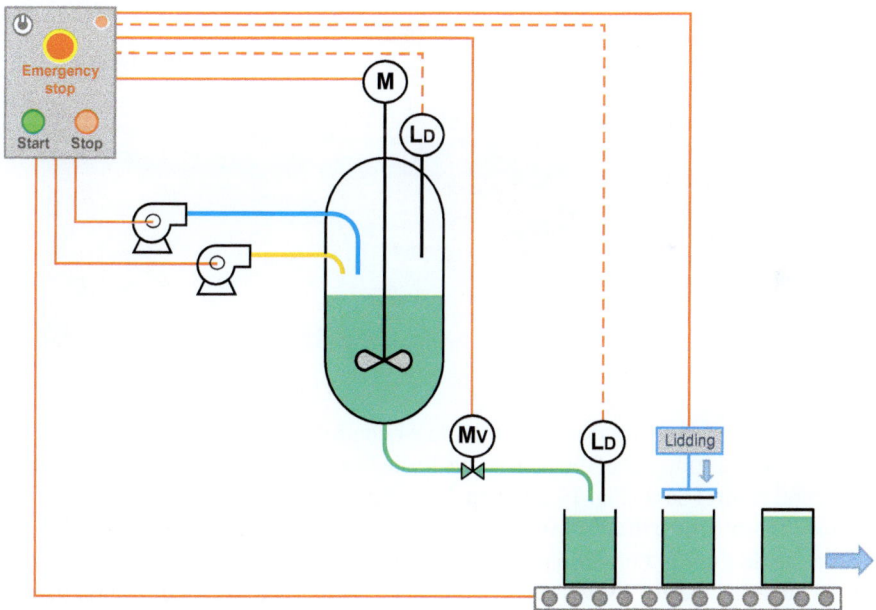

Fig. 10.19 Automated sequence.
© Bill Collum.

Fig. 10.19 shows an example automated process where liquids are blended together and transferred to a line of waiting drums. It begins with two liquids being pumped into a vessel, where they are monitored by a level detector. As the newly conjoined liquid reaches a predetermined level the mixer starts to rotate slowly; once the liquid reaches its maximum level, the pumps switch off. Mixing continues at varying speeds and after 10 min gradually slows down and stops.

A motorized valve below the vessel opens, dispatching the blended liquid to a waiting drum. It too is monitored by a level detector which sends a signal to close the valve at just the right moment. A conveyor gently moves forward, pausing whilst a lid is placed on the drum. And so it goes with one batch after another being processed through the vessel, all made possible by the workings of a nearby control panel.

10.2.6 Remote operation

For a self-contained production line the level of automation I have just described is excellent, but many industries must operate on several fronts at the same time. For instance my example process may well be preceded by receipt, batching and blending of the two liquids that it begins with. Automated sampling would no doubt be required to ensure product quality is maintained; ventilation may be needed to draw away harmful fumes, and so on. With so much going on, and spread over a wide area, we need some way of controlling events from a single location.

Fig. 10.20 Remote operation. Shutterstock.com.

Even though *remote* figures in its title, local control panels still feature out on the factory floor. This time, all of their data is channeled into a single network and duplicated at a central control desk (Fig. 10.20), which can be some distance away. From here an operator can watch over the whole production process and intervene should the need arise.

Once again this kind of setup is perfectly adequate for many production processes. For the nuclear industry, as with many others, we must move to a much higher plane, one where staggering complexity can be handled with relative ease.

10.2.7 Supervisory control and data acquisition (SCADA)

Its full title does not exactly trip off the tongue, so happily everyone adopts the acronym SCADA. Occasionally you may see *supervisory* replaced by *system*, but either way both are used to describe the same technology. On the face of it SCADA

(Fig. 10.21) provides a computerized overview which is similar to that employed by remote operations, just more of the same. But there are a couple of essential differences.

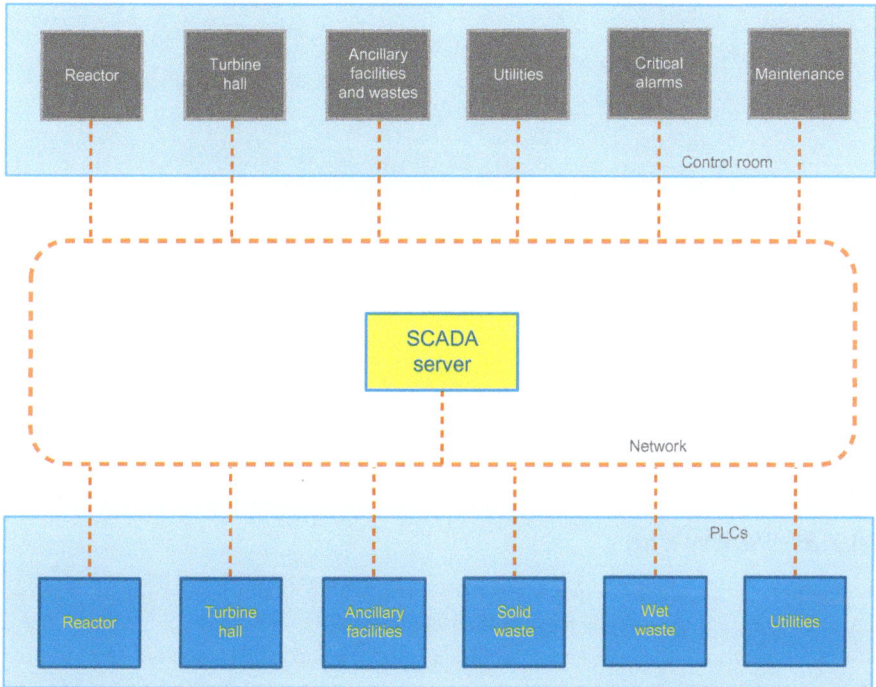

Fig. 10.21 SCADA system.
© Bill Collum.

Firstly, the PLCs are no longer in charge. True, they perform a similar role to that in other systems, but this time they take their instructions from SCADAs central computer and report back on what is happening within their sphere of influence. Secondly, SCADA systems can operate across multiple networks, spread if necessary across several geographical locations, so distance is not a limiting factor. And of course SCADA can coordinate all it sees. On the whole then this is multitasking on a grand scale.

The example shown reflects the usual philosophy of presenting low-level alarms on individual area display screens, whilst gathering all critical alarms onto a single screen at a dedicated workstation. In addition, there is also a maintenance screen which harvests data from all of a facility's intelligent instruments.

It may seem that with so much control technology at their disposal, nuclear facilities could simply fire up a SCADA system and leave it to get on with the daily routine but, no surprise, such a laissez-faire approach is out of the question. In practice the whole system, from local control panels out on a plant to a central control room (Fig. 10.22) are subject to continual scrutiny from a team of personnel who can intervene pretty much whenever they please. A SCADA system is good, very good indeed; even so, it is not capable of taking sole responsibility for the control of a nuclear facility.

Fig. 10.22 Central control room.
© EDF Energy.

10.2.8 Hardwiring

In Chapter 14 we shall discuss the *safety case*, which is developed for all nuclear facilities and without which they cannot operate. What we shall find is that where an item of equipment, say a valve, fulfills a safety function of some sort, then it must satisfy particular safety performance requirements. Simply put, the more we rely on an item of equipment to contribute to a facility's safety, the more we demand with respect to its reliability.

The demands of reliability are expressed it terms such as "ten to the minus six" (10^{-6}), 10^{-9}, 10^{-4}, and so on. In the case of say 10^{-6}, an item of equipment may only fail once for every million times it is called upon to perform, known as *failure on demand*. For 10^{-9} it would be one in a billion, so we can see how it goes.

For electrical equipment, reliability is defined by its assigned *safety integrity level* (SIL) which incorporates the possibility of failure on demand. There are four levels with one being the lowest and four the highest, and crucially it is not possible to claim a particular SIL by depending entirely on techniques such as computer modeling or calculation. Rather, a SIL is derived from a quantitative assessment of real data, such as exhaustive tests and historical evidence, so we know it is underpinned by substantive evidence.

What is of particular relevance here, is that to achieve a rating of SIL 4 an item of equipment, say a level detector, must be hardwired from its location all the way back to a control room. SCADA signals may travel in parallel, via a network, but on their own cannot ordinarily be relied on to control an instrument's role in safety-critical activities. I say *ordinarily* because in some system-specific circumstances the normal rules may not apply. However, for the most part, when it comes to safety it is sensible to assume hardwiring will have an essential role to play.

In addition to hardwiring, safety-critical instruments are also designed to fail-safe. So, for example, if the "safe" state for a valve is in its closed position, it will be fitted with a spring that automatically forces it to close if electrical power is lost.

10.3   Instrumentation

To say the term *instruments* covers a lot of ground would be quite an understatement. It really is a huge topic and a multifaceted one at that. To make sense of the subject we shall group instruments into four categories of level, temperature, pressure, and flow, and examine how each goes about taking their various measurements. In addition to these, what we might call "standard" types, the nuclear industry also makes use of a group known as radiometric instruments, a topic which is so specialized it needs a chapter of its own (coming next). As I say, it is a big subject. As if to emphasize the point, it turns out that instruments divide broadly into two distinct families, so we need to go there first.

10.3.1   Intelligent instruments

In earlier times, a maintenance team was continually deployed to examine instruments in accordance with a predetermined schedule. In other words they were forever tinkering with all manner of kit, whether it needed it or not, just in case, but not anymore. As we saw in our earlier discussion on how plants are controlled, PLCs are constantly reporting data such as temperature, flow, and pressure back to a SCADA system, so to make the most of this capability, intelligent instruments are employed.

It is directly analogous to the systems found in many cars, where the engine, cooling system, brakes, and so on feed information through to an on-board computer. As a result, rather than routinely booking in for an annual service or after a given number of miles, the timing of a car's maintenance along with details of what that maintenance entails, are dictated by its management system.

Intelligent instruments continually monitor their own performance and general wear and tear. At the same time the information they gather is routed to SCADA where it is displayed at a dedicated maintenance desk, usually in a central control room. With this approach, maintenance and replacement of components dovetails perfectly with real need, rather than diligently following a regimen that may not be entirely appropriate. Having said that, in situations where a standard instrument is entirely fit for purpose, there is no need to specify an intelligent counterpart.

10.3.2   Level measurement

If you have ever clambered around inside an attic space, you may have spotted an empty pipe that starts near the top of a cold water head tank, runs along the roof space and eventually sticks through to the outside of the house, maybe just below its roof guttering. If the shutoff valve on a head tank's supply pipeline were to stick open, water would continue gushing into the tank but instead of overflowing and pouring down through the house, floodwaters would be carried away by the pipe running across the attic. It may mess up the driveway but it is better than the deluged alternative.

Domestic plumbing systems normally make use of a ballcock or similar floating device to shut off valves that control the flow of water to head tanks and cisterns, but as many of us know from first-hand experience they do sometimes fail. Yes, there is nothing quite like the sound of indoor rain for hurrying along our plumbing skills.

The nuclear industry may occasionally make use of similar technology, say where innocuous liquids are involved, but it also deploys a selection of more hi-tech devices, all of which can be configured to keep their maintainable components out of harm's way. So whilst an instrument's business end may be exposed to intense radiation or some other hostile environment, operators can easily gain access to the components they need to tinker with. Common among these are bubblers, also referred to as *pneumercators*, so we shall start there.

10.3.2.1 Bubbler system

In its most basic form, air is forced at constant and known pressure down a bubbler tube, or *dip pipe*, which terminates close to the bottom of the vessel in which liquor level is being measured. The deeper a liquor the greater its hydrostatic head and therefore the more resistance air will experience as it exits the tube. By comparing a reference pressure with back pressure at a tube's opening, it is a simple matter for an instrument to extrapolate the variation and determine a liquor level.

This system works fine where the liquid in question is unchanging, say, a particular acid, and held at a constant temperature. However, if its characteristics are subject to variation then density comes into play, which has a bearing on hydrostatic head, so we need to enhance the system. The main difference (Fig. 10.23) involves introduction of a second bubbler tube. Granted it may not look like much, but it transforms the system into a much more sophisticated affair.

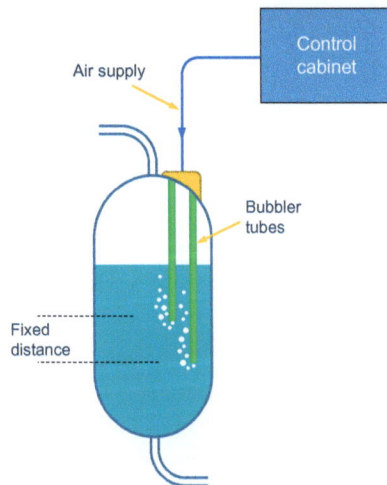

Fig. 10.23 Bubbler system.
© Bill Collum.

Both tubes are fed with air or, where there is risk of explosion, an inert gas such as nitrogen. The key feature is that both tubes terminate a fixed distance apart, so that at any given moment their differential pressure will mirror fluctuations in a liquor's density. In addition the longest bubbler tube, or sometimes a third one, is used to ascertain the overall hydrostatic head. By combining both figures, density and hydrostatic head, instruments can accurately determine the height of liquor within a vessel, no matter how often its characteristics may change. It really is a beautifully elegant solution.

10.3.2.2   Vibrating forks

You can imagine what would happen if you took a regular tuning fork, tapped it on a hard surface to get it vibrating and then slowly lowered it, fork first, into a bowl of water. The moment it touched the water its pitch would change. In essence vibrating fork instruments capitalize on this trait to determine when rising liquid has reached a particular level.

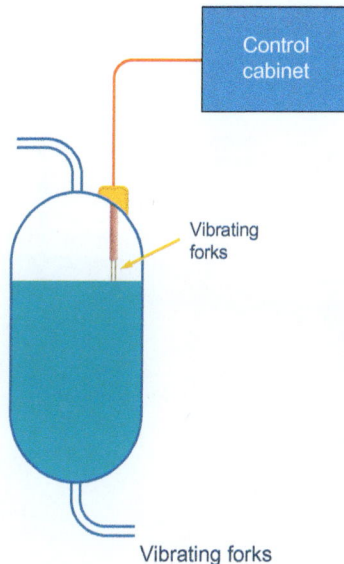

Fig. 10.24 Vibrating fork.
© Bill Collum.

A two-pronged fork is installed within a vessel (Fig. 10.24) and electronically vibrated at a constant frequency. Once liquid touches the fork its change in resonance is detected by an instrument which, if need be, can be set to terminate any inward flow. Even better these devices can cope with practically any type of liquor, no matter how dense or viscous, and rather handily the fork's vibrations help to make them somewhat self-cleaning. However, if deposits do build up, the instrument can be reset to its prevailing vibration and continue to function. Its main constraint relates to length, which for effective operations should not exceed 6 m. So all in all it is a very simple and dependable piece of kit.

10.3.2.3 Radar systems

We are all familiar with the acronym "radar," which stands for *RAdio Detection And Ranging* and its applications in tracking aircraft, radio astronomy, and so on. At a more micro level this same technology can also be harnessed to track liquor levels in a vessel. An antenna located at the top of a vessel beams radar waves, more specifically microwave pulses, down onto a liquid's surface. Their reflected echo bounces back to the antenna, which is connected to instruments that can extrapolate the time taken for its round trip into an actual level. Bearing in mind that microwave pulses travel at the speed of light, or around 300,000 km (186,000 miles) a second, it is impressive to say the least, that their short journey can be analyzed so accurately.

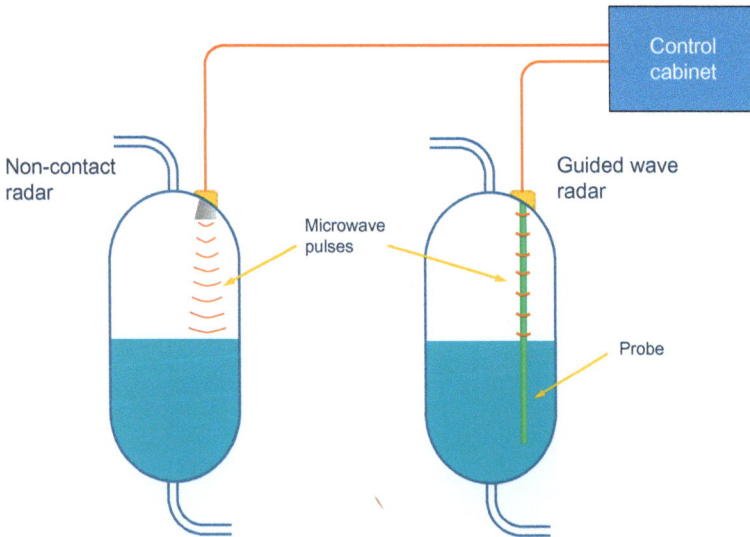

Fig. 10.25 Radar sensors.
© Bill Collum.

Radar detectors come in two guises (Fig. 10.25): the non-contact, or *through-air* type, where beams fan out in the traditional way, and a more constrained version, the *guided wave*, where microwave pulses are directed down a tube and can even negotiate 90 degree bends. The non-contact version is susceptible to signal scatter from vessel internals, such as agitators and pipework, which can generate false echoes, so ideally needs an unimpeded view of its target area. The probe, on the other hand, is unaffected by its surroundings, so functions perfectly well no matter what is going on around it.

A factor to bear in mind when considering use of radar is that not all liquids, or powders for that matter, share equal prowess when it comes to reflecting its signals. Differences arise due to the dielectric constant, or *permittivity* of individual liquids. If the dielectric constant is too low then much of a radar's energy will pass into a fluid, leaving little signal to be reflected back to the antenna, with consequent reduction in measurement accuracy. Water, for example, has a high permittivity, so reflects radar's microwave pulses very well indeed.

When processing liquors with a low dielectric constant, guided wave is the radar of choice as it is around twenty times more efficient at generating an echo from their surface. In addition this probe-based system is particularly adept at identifying interfaces between dissimilar liquids which have stratified into layers, so can also provide level measurements for individual strata.

10.3.2.4 Conductive sensors

This system works on the principle of using liquid to complete an open electrical circuit. There are two variations (Fig. 10.26). Where vessels are formed from non-metallic materials, such plastic or fiberglass, two electrodes are deployed a short distance apart. Whereas in situations where vessels are electrically conductive, stainless steel, and so on, there is the option to deploy a single electrode and use the vessel wall as its counterpart. In both cases an alternating voltage is fed into an electrode, which produces a current in liquid that comes into contact with it. Instruments sense that the conductive path has been completed and can trigger a response, such as halting any further ingress of liquor until its level subsides again.

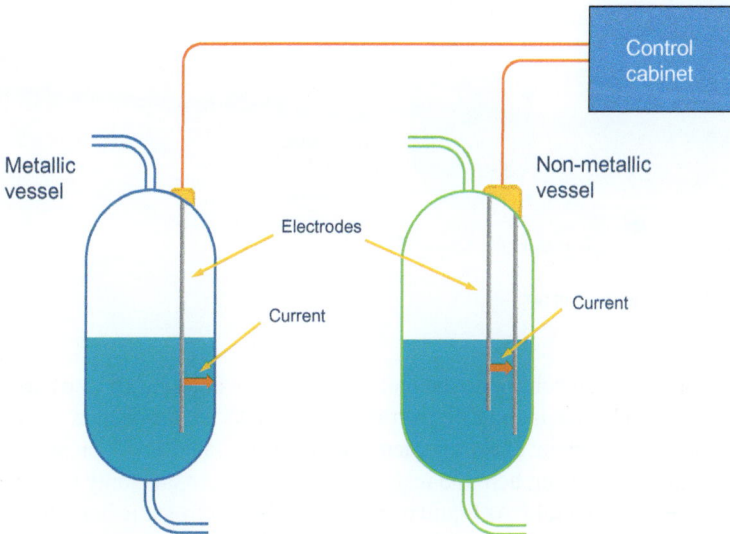

Fig. 10.26 Conductive sensors.
© Bill Collum.

On the face of it pumping electricity into a liquid sounds quite foolhardy, but *low current* levels are deemed inherently safe. Incidentally this system is not suitable for all liquids since some, such as demineralized water, do not have the conductive properties which allow a current to flow through them.

10.3.2.5 Capacitance sensors

Although this system shares some similarities with conductive level measurement, it operates on quite a different principle. In its simplest guise a capacitor is formed by introducing an alternating current (AC) to two conductive plates, or *electrodes*, which are spaced a little distance apart. The plates are separated by an insulating material within which energy is stored in the form of an electrostatic field. And it is this energy, measured in Farad (F) units, which is expressed as *capacitance*. Incidentally the AC is not of the type we discussed earlier, where it is associated with domestic and industrial electricity supplies and has a frequency of between 50 and 60 Hz. Capacitance sensors operate in the radio frequency (RF) spectrum which is measured in megahertz (MHz), where 1 MHz is equal to one million hertz. In this application then, measurements typically range between 0.1 and 100 MHz.

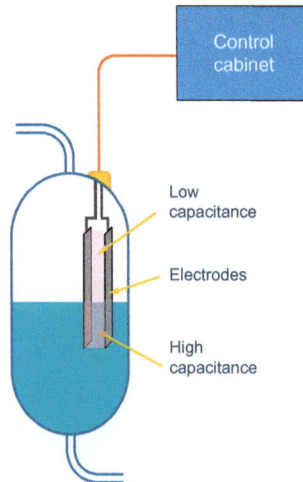

Fig. 10.27 Capacitance sensors.
© Bill Collum.

Capacitance varies according to the dielectric constant or *permittivity* of the insulator between electrodes and is quantified on a scale of 1–100. Air has a value of one whilst liquids, depending on their exact characteristics can be rated at 50 or more. In the situation we are considering here the insulator is either air, in an empty or part empty vessel, or some form of liquid. As liquid rises through the electrostatic field (Fig. 10.27), capacitance rises proportionately, so it is a relatively simple matter for RF instruments to convert this fluctuating capacitance into an incremental read-out of changing levels.

10.3.2.6 Ultrasonic

This system has some parallels with non-contact radar; its most notable difference is the use of high frequency acoustic waves in place of microwave pulses. A transducer, mounted above the vessel, transmits the waves down onto a liquid's surface (Fig. 10.28) or indeed onto the surface of a powder or other "solid" material. Their echo is reflected back, and once again time taken for the round trip is extrapolated into a level.

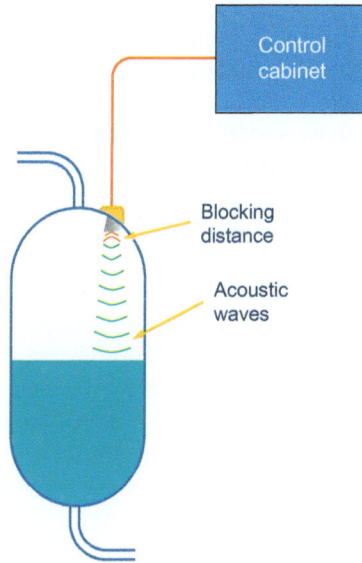

Fig. 10.28 Ultrasonic.
© Bill Collum.

As echoes return they collide with outbound pulses, creating some local buffeting and rendering a small zone ineffective for measurement purposes. It is known as the *blocking distance*, and although it varies from case to case tends to be around 150 mm so rarely presents a problem. As for acoustic waves themselves, the speed of sound varies in accordance with temperature, so, for example, at 15°C will zip along at 340 m/s. With this in mind instruments are calibrated to factor in the effects of temperature and so maintain accurate level readings.

There are several other methods available for measuring liquor levels, such as hydrostatic pressure and lasers, so engineers have plenty of nonintrusive systems to choose from.

10.3.3 Temperature measurement

When it comes to detecting heat, say the temperature of a simmering solution, or where steam is being used to drive liquors along, there are several contenders such as thermistors, bimetallic thermometers and infrared systems. We shall take a look at two of the most common, a *resistance temperature detector* (RTD) and a thermocouple.

Both are too fragile to be exposed directly to a hostile environment, so are invariably protected by a sheath or *thermowell*. And whilst both can deal with quite an array of tasks, the more accurate RTD is generally preferred for measurements up to around 600°C, whilst a thermocouple responds faster and is good for temperatures of well over 2000°C.

10.3.3.1   Resistance temperature detector (RTD)

Ohm's law demonstrates that a combination of voltage and current through a particular conductor determines its resistance. That resistance changes with variations in
temperature, and crucially those variations are predictable. There we have the basis
for a RTD also known as a resistance thermometer. The default setting for most RTDs
is that 100 ohms (Ω) of resistance is equal to 0°C. They come in various guises but all
need a wire to act as their resistor; most popular is platinum, with nickel and copper
among other favorites.

There are several variations in how RTD sensors are configured, with three of the
most common shown in Fig. 10.29. The thin-film version has its sensor sandwiched
between a ceramic base and a protective glass covering, whilst helical wire variations
are either wrapped around a ceramic spindle, or have their elements inserted into small
shafts within a ceramic tube. During operation a small current is passed through the
platinum wire and its fluctuating resistance converted into a digital temperature
readout. As is so often the case, an awful lot of nifty science is distilled into a remarkably
simple device.

10.3.3.2   Thermocouple

This device is founded on the fact that if two wires, made from different metals, are
welded or twisted together they will produce an electric current when heated or
cooled. The amount of current depends on materials used for the wires and crucially
the temperatures to which they are subjected. Capitalizing of this, instruments can be
set to detect a particular level of current, and therefore a particular temperature, say
close to the top of the breakpots we discussed in Chapter 6. If a predetermined level is
reached, then in the case of a breakpot, its steam supply can be terminated and an
alarm triggered to notify plant operators of the problem.

Of course thermocouples have plenty of other applications; in fact you will find one
inside many of the gas boilers that run domestic heating systems. They are there to
detect heat given off by a boiler's pilot light, so that if for some reason the small flame
is extinguished, maybe due to a loss of supply, the thermocouple will detect a drop in
temperature and shutdown the gas supply. Once again they stand guard as a safety
mechanism.

10.3.4   Pressure measurement

We are all familiar with the many ways in which pressure exerts force and how that
force can be translated into movement. Take the simple example of blowing into a
balloon; it will steadily inflate until reaching its fully expanded shape—no surprise
whatsoever. Likewise, if we were to blow very gently into the bottom end of a vertical
drinking straw, one half filled with water, the water would gradually rise up the tube as
the pressure increased.

Wire-wound RTD element

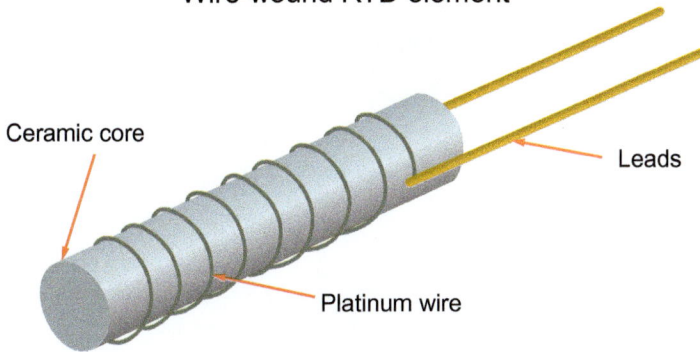

Ceramic core

Leads

Platinum wire

Thin-film RTD element

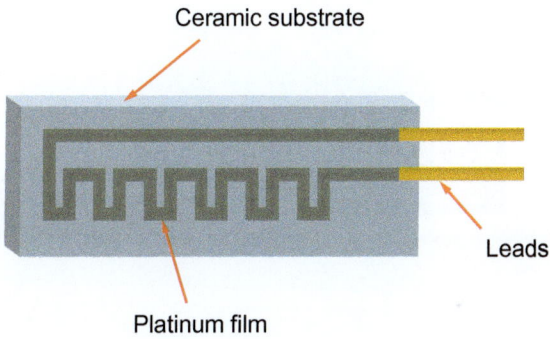

Ceramic substrate

Leads

Platinum film

Coiled elements

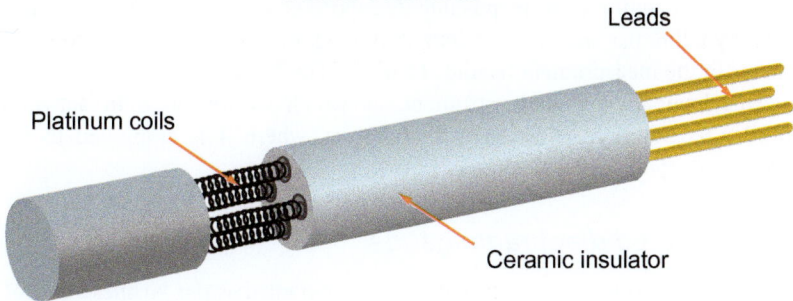

Leads

Platinum coils

Ceramic insulator

Fig. 10.29 Resistance temperature detectors.
© Elsevier.

10.3.4.1 Bourdon gauge

Instrument engineers have been paying close attention to such predictable behavior and made it work to their advantage. The Bourdon gauge (Fig. 10.30) is a good example.

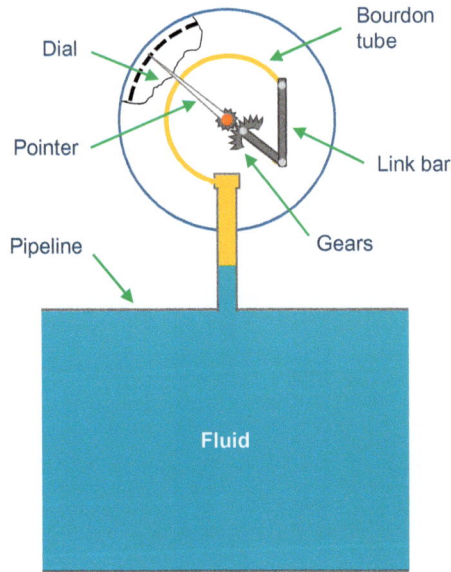

Fig. 10.30 Bourdon tube.
© Bill Collum.

It consists of a hollow tube which curves within itself. The end at the extremity of the curve is closed off, whilst the other end is open and therefore free to experience pressure exerted upon it, in this case from fluid flowing through a pipeline. As pressure increases, the tube unfurls a little, pushing its capped end outwards. This movement is picked up by a link bar and directed into a pivoting arm, which is geared to magnify movements in the measurement needle. That's it. The Bourdon gauge is a simple yet effective mechanically operated instrument, one which is widely used. In fact you will find similar technology inside many barometers, where it is employed to detect changes in atmospheric pressure.

10.3.4.2 Isolated diaphragm gauge

For many industries, including nuclear, it is important that the business end of an instrument does not come into contact with whatever is being measured. This may be due to properties of the fluids in question or because pressures involved are too high for sensitive instruments. The way it is achieved in this application is beautifully simple.

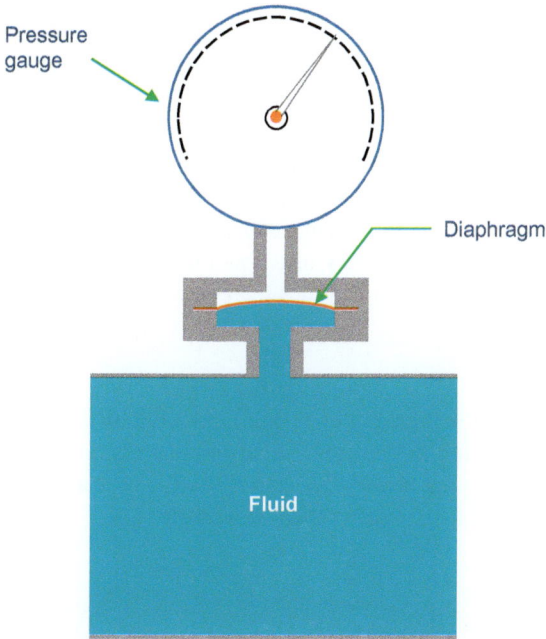

Fig. 10.31 Isolated diaphragm gauge.
© Bill Collum.

A diaphragm is captured between to plates, with the fluid whose pressure is being measured on one side and the instrument itself on the other (Fig. 10.31). In general, and certainly for our purposes here, we can assume that liquids are incompressible. So as pressure mounts within a pipeline the diaphragm flexes more and more, effectively transferring changes in pressure to its "clean" side where fluctuations are detected and displayed by an instrument. I have shown a Bourdon gauge but there are other alternatives.

One, for example, makes use of technology we discussed in the previous chapter, but manages to manipulate its purpose in quite a different way. You will recall that strain gauges can be formed by embedding metallic film within a thin silicon wafer. As the wafer flexes electrical resistance across the film changes and is detected by an instrument. In the application we are considering here a similar wafer is bonded to the "clean" side of a diaphragm, but apart from that it operates in exactly the same way. As the diaphragm flexes, electrical resistance across the film changes, but this time is converted into a pressure reading.

The diaphragm within these gauges is often formed from various types of plastic or rubber, which is absolutely fine where fluids being measured are chemically compatible and pressures involved are not too high. For radioactive substances, though, and where pressures and temperatures are high, the nuclear industry employs metal diaphragms such as stainless steel, hastelloy, and others. Plates capturing the diaphragm can be welded together, as in the example, or flanged and bolted.

10.3.5 Flow measurement

When it comes to measuring flowrate, the range of technologies on offer is almost bewildering; it includes ultrasonic, magnetic, and temperature differential systems, along with vortex meters, rotating vanes, and many others. To get a feel for this vast subject we shall take a look at two of the more commonplace types, an orifice meter and the pitot tube. Interestingly enough, many of the instruments which measure flowrate cannot be discussed without first having some understanding of ways in which pressure is measured, for it turns out that very often both depend on similar principles for their operation. Our first example illustrates the point.

10.3.5.1 Orifice meter

There are quite a few variations in how this technique is configured but all of them place a known obstruction within a pipeline, or duct, and take a pressure reading from both sides of it. Since the difference in pressure or *differential head* is central to its operation, this approach is also referred to as a differential flow meter.

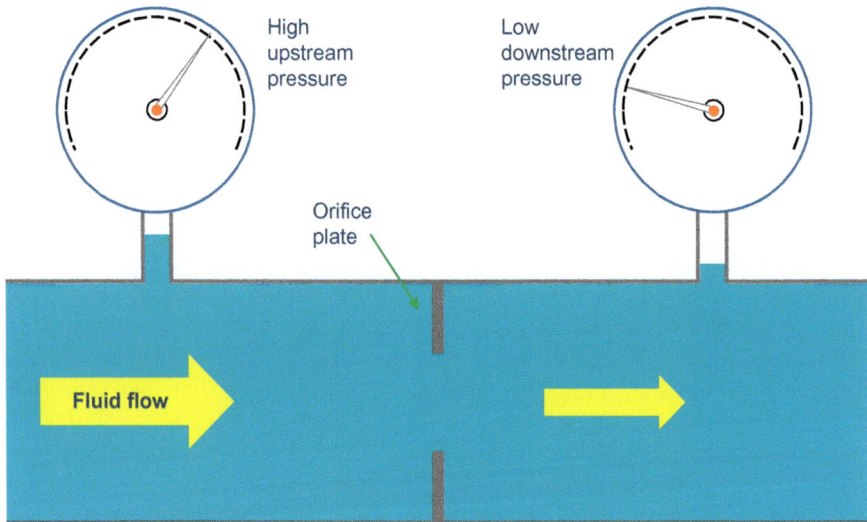

Fig. 10.32 Orifice meter.
© Bill Collum.

When measuring pressure and flow most technologies depend principally on Bernoulli's equation. Of relevance here it shows that the pressure drop across a constriction is proportional to flow rate. So we can see what is coming here. Fig. 10.32 shows the setup, with an orifice plate trapped within a pipeline and a pressure gauge on either side. Clearly with a plate impeding a fluid's progress, the pressure before or upstream of it will be greater than that experienced on the other side. Armed with two readings, giving us a pressure drop, and since the cross-sectional area of the orifice plate's opening is known, it is a simple matter of applying Bernoulli's equation to calculate the fluid flow.

10.3.5.2 Pitot tube

If we imagine a fluid flowing along a pipeline, we can visualize that the pressure exerted at an opening in the pipe's wall, its circumference, will be much less than that experienced within the flow, say at a pipe's center. Simply put, an opening in the pipe's wall experiences a sideways glance, whilst a point at mid-flow is subjected to the full-on force of a flow. Furthermore, and as demonstrated by Bernoulli's equation, the higher the velocity the greater the differential pressure will be.

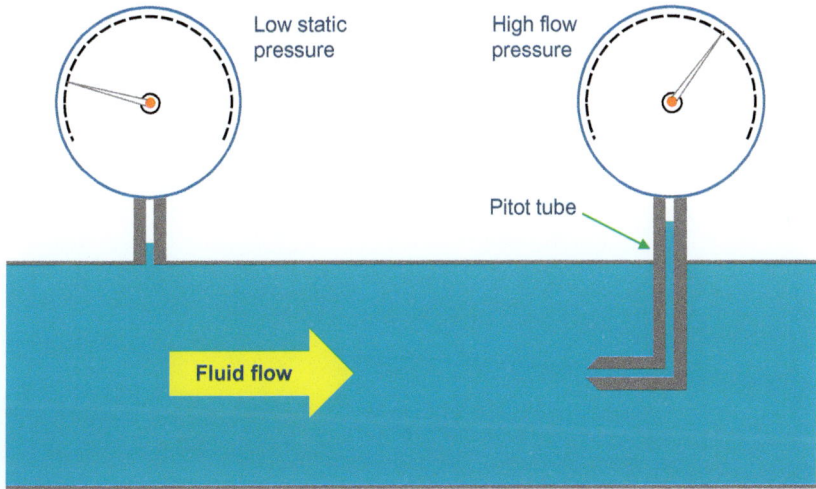

Fig. 10.33 Pitot tube.
© Bill Collum.

Once again this is a technology that can be configured in many ways. Fig. 10.33 shows a basic pitot tube within a pipeline, whilst a separate pressure reading is taken from the pipe's wall. One of the favorite alternatives dispenses with the pipe wall reading and deploys a more sophisticated pitot tube. As before it has an opening at its forward point but sideways flow, or static pressure, is measured from an opening in the circumference of the tube itself.

One of the limitations of pitot tubes is that they only measure flow at a specific point, so cannot provide data on the more representative average flowrate along a pipeline. To counter this pitot tubes can, if necessary, be configured to take multiple readings across the diameter of a pipeline, or indeed a duct. These can then be combined to provide an average flowrate.

If you take a good look around the next time you visit an airport, you may spot a pitot tube protruding from an aircraft's fuselage or beneath one of its wings. They operate in exactly the same way as those deployed in an industrial setting, but in this application are very effective at measuring airspeed.

At the beginning of this chapter, I mentioned how very diverse and therefore compartmentalized the role of an electrical engineer can be, evidenced by the very different challenges of distributing power, developing control systems, and deploying instruments. As if to underscore the point, there is an additional facet to consider. When it comes to instruments, there are what we might call the traditional types we have just discussed: those measuring temperature, level, pressure, and so on. But in the nuclear world there is another branch of electrical engineering, another specialism, one dedicated to measuring radiation and contamination, so this is where we shall go next.

Radiometric instruments

We have no trouble whatsoever in understanding what it is that most branches of engineering are all about. So, for example, any teenager, more or less, could have a go at describing the different attributes of mechanical, chemical and structural engineering. It is not quite so easy when we contemplate the life of a radiometrics engineer. Their role stems from a need to know what the nuclear industry is dealing with, whether it be products themselves or the environment in and around nuclear facilities. See what I mean—not a subject that conjures up a ready image. Undaunted, let's delve in.

Unsurprisingly there is a considerable volume of legislation attached to the control of discharges from nuclear sites, along with minimizing radiation exposure to nuclear operators and the general public. In addition, there are predetermined specifications for materials moving from one facility to another. Known as *conditions for acceptance* (CFA), they set limits, of say so many parts per million, on this or that radionuclide constituent and are tied to procedures which invoke an avalanche of penalties on any facility that steps out of line. And of course there are safety implications in ensuring hazardous or incompatible materials are always suitably segregated.

All of these constraints, whether they be regulatory, operational, safety related, or indeed straightforward good practice are fine, but some sophisticated equipment is needed to help set meaningful limitations in the first place, and then to verify they are being complied with. Enter the radiometrics engineer, or to give them their full title *radiometrics and special instruments* engineer. Let's look first at what they need to measure and then examine how they do it.

11.1 Monitoring requirements

Discharges from nuclear facilities may go to the air, to rivers and lakes or, if they are on the coast, out to sea. The thing to recognize here is that discharge authorizations, set by nuclear regulators, are for a site as a whole and rarely for individual facilities. So it is not permissible for a site's operators to dream up the need for a new facility, then simply ask regulatory authorities to grant say an additional 10% on top of their existing discharge limits. No, discharges from a new facility must be absorbed into those already authorized for the site as a whole. Limits for all sites are reviewed from time to time and as we might expect the impetus from regulators is always to drive them downwards.

Monitoring a nuclear site's various discharges, along with keeping a close eye on the environment inside its facilities, is pretty much what we would anticipate. Less obvious perhaps, when we zoom into individual facilities we find the analysis of product characteristics is every bit as stringent. It all stems from those CFA and has some parallels to facilities operating in a non-nuclear world. If, for example, you are expected to dispatch olive green paint to an associate factory, sending sky blue will not go down too well. Nor would it go unnoticed if a colleague expecting superheated

Nuclear Facilities. http://dx.doi.org/10.1016/B978-0-08-101938-2.00011-8

steam, instead received warm water. So setting ground rules and sticking to them is important, particularly in a nuclear context.

Nuclear facilities must know and crucially be able to prove the exact makeup, often down to individual isotopes, of materials they are processing, exporting or receiving. This applies not only to products transferred between facilities on the same site, but also those dispatched off-site, including to long-term storage facilities. Whatever their origin or destination, it is imperative that we have the means to analyze products at predetermined stages along the way.

The brief for any kind of radiation detection instrument is straightforward enough. We know our senses are totally incapable of detecting radiation, so all we want is an instrument that can do the job for us. Not only that, we also want to know what type of radiation it is and how much is out there. Not much to ask really.

11.2   Detection technologies

Although there are all manner of radiometric instruments, when you get right down to it their operations are based on one of four generic technologies. So we shall visit them first and then see how these fundamentals manifest themselves in various items of kit.

11.2.1   Gas-filled detectors

These detectors come in two generic types, both of which specialize in detecting ionization within a gas-filled chamber. In other words they pick up tell-tale signs of radiation as it interacts with a gas. The most prevalent technology is based on a Geiger-Müller (G-M) tube with the alternative being a *proportional counter*, which also comes in a slightly different variation known as the *gas flow* proportional counter. In truth there isn't an awful lot of difference between the three alternatives, so we shall start with a tour of their underlying technology and then see how they differ.

The main visible component of these detectors (Fig. 11.1) is a sealed metal chamber which is filled with gas. The exact gas varies so we shall come back to that in a moment. Running along the chamber's axis is a positively charged wire carrying a high voltage DC supply.

Radiation, whether it emanates from particles of contamination or pure energy, enters the chamber via a "window" which, for alpha and beta particles must be extremely thin so as not to impede even the weakest sources. One of the favorite window materials is *mica*, which is formed into sheets with a thickness of around 0.05 mm. For the more penetrating gamma and neutron radiation, windows are much thicker, say 2 mm, and made from more robust materials such as chrome steel.

Before moving on we need to remind ourselves of the fun we had with magnets during our schooldays. You will recall that the ends of a magnet are named as north and south, or positive and negative, and a pair of magnets only pull towards each other when unlike poles are aligned. Conversely, magnets push apart if identical poles face each other. The maxim of "opposites attract" applies equally to the situation that we are considering here.

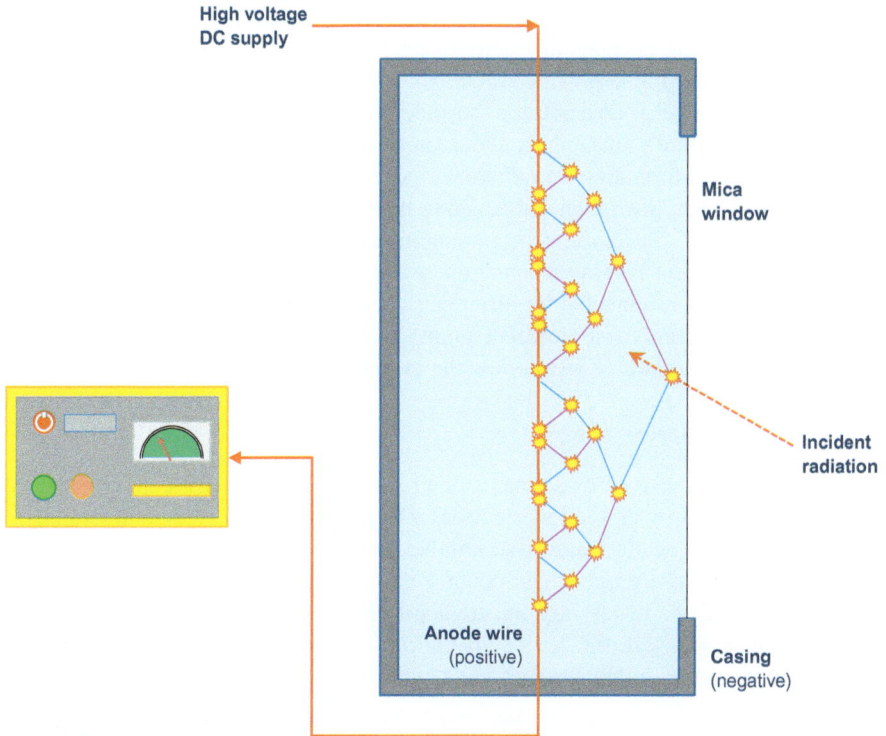

Fig. 11.1 Gas-filled detector.
© Bill Collum.

11.2.1.1   Detection process

As radiation enters a detection chamber, it triggers ionization of a small number of atoms within its gas, so to follow how things unfold we shall concentrate on just one of them. We shall also assume, for the moment, that the anode wire has just a handful of volts running through it.

Radiation striking a gas atom causes a negatively charged electron to break away, which is a good start because now we have the beginnings of something which can be detected. Once free, and because opposites attract, our newly liberated electron heads straight for the positively charged anode wire, creating a pulse as the two collide. Unfortunately, detecting the pulse from one lonely electron would take an eye watering investment, so for all practical purposes nothing happens. However, if we turn up the anode wire's DC flow, in the order of several hundred volts, it creates an electric field within the chamber and now things unfold very differently indeed.

Once again incident radiation strikes an atom liberating an electron exactly as before, but this time our lone electron, enlivened by the electric field, gains much more energy. As it heads towards the anode wire the lively electron hits another molecule so hard that it too is forced to release an electron. And so the process goes, with a rush of electrons colliding with unsuspecting molecules, triggering the avalanche we see in Fig. 11.1 and sending millions of electrons hurtling towards the anode wire.

So, radiation entering a gas-filled chamber produces a pro-rata number of ionizations, which in turn generate a pulse in the anode wire. The trouble is, fascinating as it may be, hardly anyone wants to hear about ionizations. What we need from radiometric instruments is information on radiation, and in real time.

Sure enough then, instruments attached to detectors convert electrical pulses into intelligible readings. The level of detail contained in these readings is influenced by the strength of current passing through an anode wire, so it is no surprise that this parameter is largely responsible for differences between the two main detector types. As I say, G-M tubes are most prevalent among gas-filled detectors, but that said their operation is easier to fathom if we view the proportional counter first.

11.2.1.2   Proportional counter

With an anode being struck by so many electrons their cumulative electrical pulse is readily detected. However, too few electrons will generate insufficient data for meaningful interpretation and too many will saturate sensitive downstream instruments. To get the balance right a proportional counter is *tuned*, so to speak, by adjusting the anode's voltage. Exact levels vary and are dependent on specifics such as a detector's chamber size, along with the type of radiation being detected and so on, but will certainly be measured in hundreds of volts and in many cases well over a thousand.

11.2.1.3   Gas flow proportional counter

As we have seen, when detecting alpha and beta radiation the mica window of a gas-filled chamber is necessarily extremely thin, so in some situations can be prone to picking up minute punctures. Even though they may be invisible to the naked eye, such punctures would allow the fill gas to escape, gradually degrading an instrument's performance. To counteract this possibility there is the option of directing a continual flow of gas through a detector, usually P-10 which is a mixture of 90% argon and 10% methane. The only limitation on this variation stems from the practicalities of maintaining connection to a gas bottle, which effectively rules out its use with portable instruments.

With both of these counters the number of ionizations generated, along with resulting strength of their pulse, is directly "proportional" to radiation energy which entered the chamber, so they are well named. Not only that, but the ability to fine-tune these instruments enables them to translate initiating radiation into dose, which can be displayed in Sieverts (Sv), or activity, shown in becquerels (Bq).

11.2.1.4   Geiger-Müller (G-M) tube

Operation of G-M tubes is very similar to that of proportional counters, with main differences arising from their operating voltage. G-M tubes still use a high voltage DC supply, but unlike proportional counters cannot be fine-tuned to extract maximum detail on the initiating radiation energy. Instead, instruments attached to a G-M tube simply inform that a radiation "pulse" or series of "pulses" has been detected. Clearly then, proportional counters are a more sophisticated affair, certainly in terms of the data they can yield, which is bound to make us wonder why G-M tubes are so popular.

There are two reasons. Firstly, the radiological environment within operating nuclear facilities is well understood, to the extent that isotopes present in various areas can be anticipated with quite a high degree of confidence. This being the case where a G-M tube detects radiation, its characteristics are predictable and easily translated into meaningful data on the source. And secondly, the diagnostic electronics downstream of a G-M tube are a good deal simpler than those attached to a proportional counter, so there are obvious advantages in terms of reliability and cost. Of course in situations where highly calibrated data is required a proportional counter, or gas flow proportional counter will always win out, otherwise a G-M tube is perfectly acceptable.

11.2.1.5 Gas types

Gas-filled detectors can choose from a variety of gases, depending on specifics of the radiation they are searching for. There are quite a few variables involved in teaming potential gases with available detectors, including the level of current applied to an anode wire, so we shall concentrate on the more familiar selections.

G-M tubes are quite versatile so can be targeted to searching for alpha, beta, and gamma radiation, and for this are charged with a noble gas such as krypton (Kr) or xenon (Xe). A gas flow proportional counter has similar prowess in detecting alpha, beta and gamma, but instead uses the P-10 gas I mentioned earlier.

Of the family of gas-filled detectors, a proportional counter generally has the narrowest focus, but that said it fulfills the important role of searching for bursts of neutron radiation. Again they can be charged with a noble gas such as helium-3 (He-3) or alternatively use my favorite-sounding gas, boron trifluoride (BF_3).

11.2.2 Scintillation detectors

Also known as *scintillation counters*, these devices capitalize on the fact that when ionizing radiation interacts with atoms in certain materials it generates a pulse of light. Ordinarily such a pulse is indiscernible, certainly to you and me, but these instruments have an ability to magnify the event considerably until it yields up meaningful data about what initiated it. They can detect all forms of radiation, but are primarily used to seek out the more penetrating gammas such as caesium-137, cobalt-60 and iodine-131. There are two main components: a scintillator, where light is engendered and a connected device which amplifies its characteristics.

11.2.2.1 Scintillator materials

Scintillators themselves come it two generic types, inorganic and organic, with each having traits that recommend them to different applications. Inorganic versions are formed from manmade crystals including bismuth germanate and caesium iodide, but the most popular choice is sodium iodide. Manufacturing processes and associated costs ordinarily limit the size of inorganic crystals to around 75 mm × 75 mm × 25 mm.

Organic versions are formed from plastic, or polyester such as polyethylene naphthalate. They have advantages of being low cost and easily formed into practically any shape or size, which in some situations can be very useful indeed. To give a feel for dimensions, it would not be unusual to come across organic scintillators of say 1000 mm × 500 mm × 100 mm, and there are plenty of larger examples around. On the downside, their signal does not quite match the quality of that from inorganic types, but it is still first-class and more than adequate for many applications. Whichever crystal is used, when ionizing radiation enters they all produce a pulse of light which can be translated into a well-defined reading.

11.2.2.2   Scintillation process

Radiation, whether it emanates from particles of contamination or pure energy, enters a scintillation block and strikes an atom, but unlike a Geiger-Müller tube its ionization does not result in an electron breaking away. Instead an electron is pushed into a higher orbit around its nucleus.

Almost instantly, nanosecond territory, the electron drops back into its regular orbit, or *orbital* to be precise. And it is the energy released by this rapid maneuver which results in a flash of light in the form of a photon. The light pulse may be within the visible range but could equally be ultraviolet or infrared. In other words, even if the "flash" was extremely intense our visual senses may not necessarily be able to detect it. Of course, as we shall see shortly, this is not a problem for the radiometric folks.

Scintillation blocks are enclosed within a highly reflective material such as magnesium oxide, aluminum oxide or titanium dioxide. Its purpose is to ensure a photon cannot escape. Instead the photon it is forced to bounce around, like 3D billiards, until finally it zips through a window which is its only exit and from there into the apparatus, a photomultiplier tube (PMT), which is waiting to amplify it.

11.2.2.3   Photomultiplier tube

Unsurprisingly, a solitary photon will not yield up too much information about the radiation which initiated it, so this is where a PMT comes in (Fig. 11.2). Actually a *photodiode* and *silicon photomultiplier* work on similar principles, but PMTs prevail so are most pertinent to our discussion. A photon's entry to the PMT is blocked by a solid photocathode plate, essentially a negatively charged electrode with a photosensitive coating. However, as a photon strikes the plate its energy is converted into an electron which is instantly drawn into the tube's core. And with that, things unfold very quickly indeed.

As with a Geiger-Müller tube, electrons are accelerated by the electric field which is created by a high voltage. However, this time it occurs several times and within a vacuum. Once an electron enters a PMT it is drawn towards a positively charged high voltage electrode, which in this application is known as a *dynode*.

As an electron strikes the first dynode it causes more electrons to be released which, looking for somewhere to go, head for the next dynode because it carries a higher

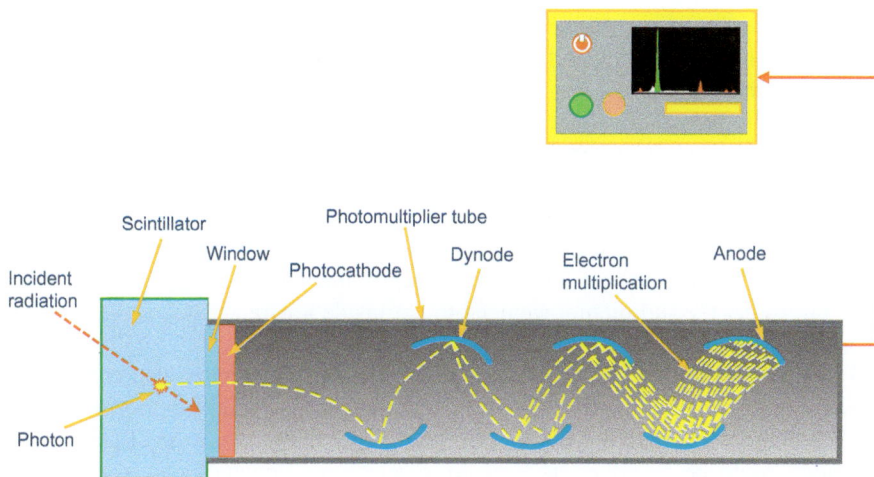

Fig. 11.2 Scintillation detector.
© Bill Collum.

voltage and therefore has a more powerful electric field. This cluster of electrons produces a rush of even more electrons which speed on to the next dynode, where the multiplication process is repeated.

Subsequent dynodes each carry a voltage which is around 100 V higher than the one which precedes it, so the avalanche continues until ultimately millions of electrons are ejected by the final dynode. Once again they head off en masse, all in search of a more powerful electric field. This time they find it in an anode, which the electrons bombard with so much energy that an electrical pulse is created within it. As a mark of how enthusiastic the growing swarm of electrons is, the whole photo-multiplication process takes about a nanosecond, or one billionth of a second.

11.2.2.4 Measurement

So we can see how a photon is initiated and cajoled into creating an electrical pulse in an anode, which is pretty impressive, but then it gets even better. Crucially such photons do not just emit a random flash of light, but have specific characteristics which are linked directly to the particular isotope which triggered them. Armed with that information and a photon's brightness, it is possible to determine what type of radiation has been detected and how intense it is. All that remains then, is to take the electrical pulse from a PMT's anode and direct it to an instrument which, as with a proportional counter, translates that data into sieverts, millisieverts, becquerels, and so on.

11.2.3 Semiconductor detectors

11.2.3.1 Conductors

Before we get into our discussion on semiconductors, we need to set the scene by examining regular conductors and how electricity moves through them. In the context we are considering here a conductor could be any metal, but invariably the choice is copper as it is the best performing of all non-precious metals. It is drawn into thin wires which may be used singularly or, for thicker cables, the copper is braded so as to maintain flexibility.

At its highest level the way electricity moves along a wire is analogous to how water behaves. Just as, when left to its own devices, water flows from a high point to a low one, electricity always flows from positive or plus, to negative or minus. In addition the flow, or volume of water passing down a river is measured by its current and exactly the same term is used to describe the amount of electricity flowing along a wire. Unfortunately the water analogy can only take us so far; to get the full picture we need to get inside a wire and see what is going on at a molecular level.

To get current flowing through a wire we need to apply a voltage, also known as a *potential difference*, for example, 100 V at one end and 0 V at the other. The current itself results from electrons jumping from one atom to another, so in simple terms electricity could be described as a stream of electrons flowing along a wire. The electron flow rate, or *current* is measured in amperes (amps) and voltage is the difference in all of this electrical energy between two points on a circuit.

Of particular interest to us, with regard to semiconductors, is that when an electron jumps from one atom to another it leaves a space behind which is filled immediately by another electron. Ordinarily we could expect this kind of space to be given a rather splendid if abstruse title, but sadly the opportunity has been lost and it is simply known as a *hole*. So, in the case of semiconductors electricity can still be described in terms of electrons zipping along, but then it can equally well be visualized as a stream of holes. On that curious note, we can take a closer look at semiconductors themselves.

11.2.3.2   Semiconductors

When it comes to conducting electricity, there are two extremes. At one end we have copper which reigns supreme, whilst the other is occupied by the likes of rubber, plastic, glass, wood and ceramics, none of which permit the passage of electricity. Between these extremities we have materials such as silicon and germanium which have quite remarkable properties and are classed as semiconductors. Silicon is by far the most prevalent, so we shall stick with it for most of our discussion.

In its ultrapure state silicon is a useless conductor, on a par with glass and all the rest. However, if a touch of impurity is added it "can" conduct electricity, but only do it effectively under particular circumstances. The process of adding impurities, which are often measured by just one part in several million, is known as *doping*, and is achieved by introducing elements such as boron or phosphorus to silicon crystals as they are being grown.

Doping produces two types of semiconductor, with phosphorus resulting in the negative charge or *N-type*, which has a surplus of electrons, and boron producing the positive charge or *P-type*, which has an excess of holes. These are used to manufacture transistors, so to follow this particular theme we need to touch on that subject for a moment.

11.2.3.3 Transistors

In its simplest form a transistor (Fig. 11.3) is nothing more than a switch. In this example current, depicted here as electrons, enters an N-type semiconductor, which is known in this configuration as a *collector*. However, its conductive properties are so poor that the current goes no further: the *off* condition.

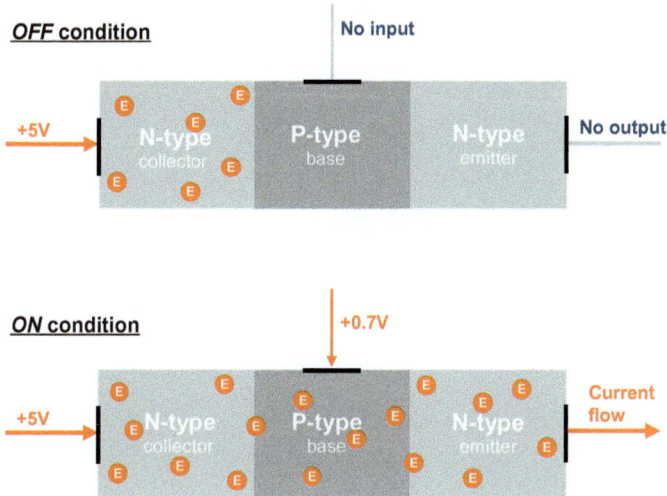

Fig. 11.3 Transistor.
© Bill Collum.

However, if just 0.7 V is routed into the central P-type semiconductor, or *base*, the additional electrons provide enough stimulus to get current flowing into the adjacent N-type semiconductor, or *emitter*: the *on* condition.

In addition to providing a switching function, transistors can also be used more subtly as amplifiers. The main difference to their setup revolves around base voltage which, rather than being at a constant level when it is on, can be adjusted. In this type of configuration very small variations in base voltage can result in large changes to output from an emitter, with these controlled fluctuations being described as either *current gain* or *voltage gain*.

11.2.3.4 Semiconductor detection process

Now that we are familiar with semiconductors and their role in the operation of transistors, we can complete the picture by examining how semiconductors are used to detect radiation. As we might expect the principles have much in common with transistors, albeit that with just a few twists the radiometric folks have managed to find a way to exploit them. Fig. 11.4 shows the setup.

The main differences in configuration are that rather than a transistor type sandwich of N, P and N-type semiconductors, there is just an N-type and P-type, with

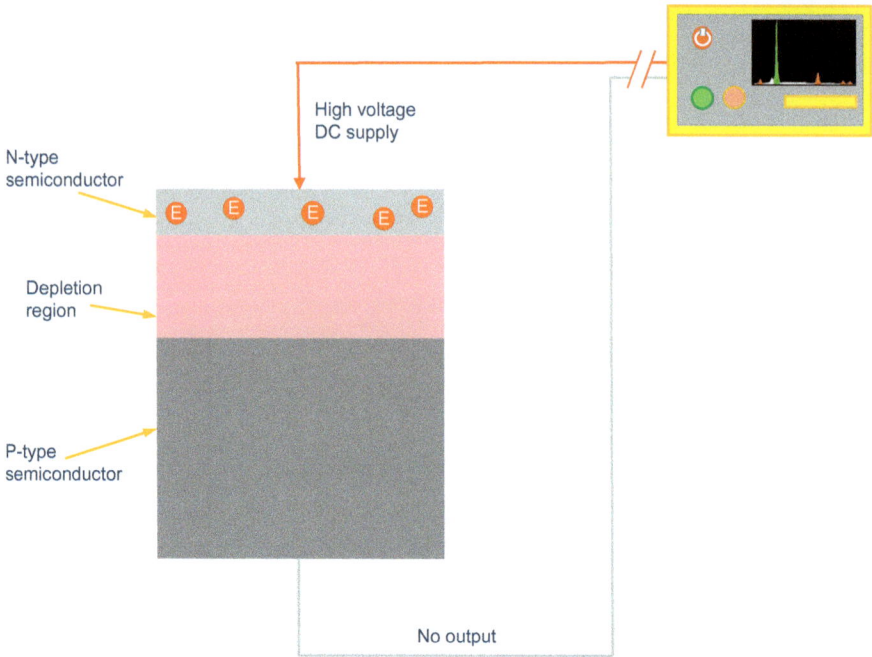

Fig. 11.4 Semiconductor detector—passive state.
© Bill Collum.

the area where the two meet known as a *depletion region*. In addition instead of a low, single-digit incoming voltage, detectors are fed by at least a few hundred, more likely well over a thousand volts. And finally there is no base supply. So, on the face of it, in the absence of a base to stimulate electrons, nothing is going to happen. But you have probably guessed, and you would be right, this time it is radiation which initiates a response, just as it does with proportional counters and scintillation detectors.

As incident radiation enters the depletion region (Fig. 11.5) it knocks electrons from their host atoms. This process is so sensitive that even weak radiation will dislodge thousands of electrons, and a more powerful source can liberate millions. Importantly it is not just electrons that are released, but what are known as *electron-hole pairs*. And as we know once electrons, and holes for that matter, are on the move a current is created. As with other detectors, electrons are accelerated by the electric field which accompanies a high voltage, so zip across the P-type semiconductor and out via the carrier wire. However, this time there is more to it than that.

Electrons continue with their normal behavior of following a circuit, in this case in a clockwise direction, so no change there. But electron-hole pairs, formed in the depletion region, tilt the usual equilibrium. They have a preference to flow in opposite directions, with slightly more holes than electrons being drawn towards the P-type semiconductor, and marginally more electrons heading to the N-type, traveling

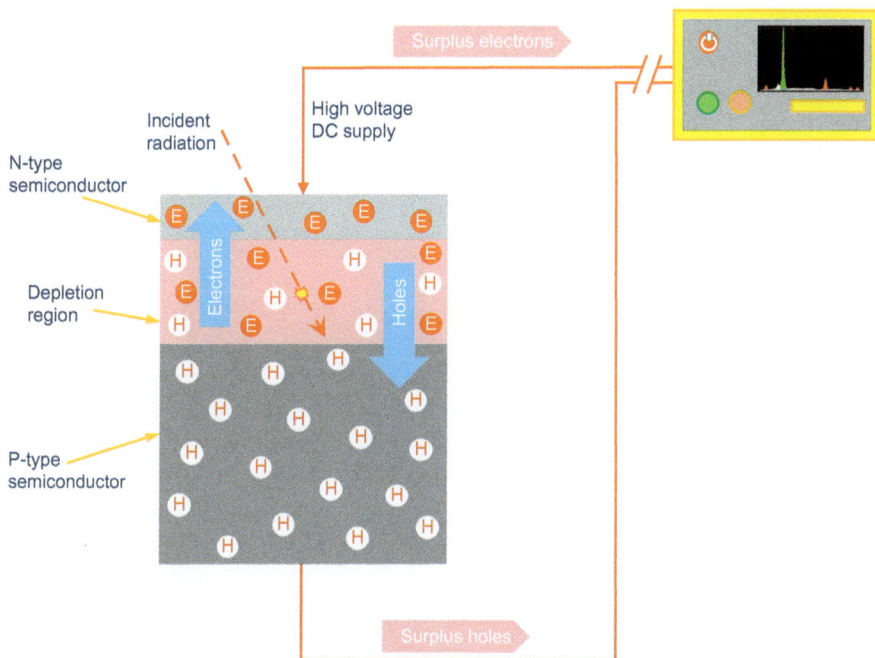

Fig. 11.5 Semiconductor detector—active state.
© Bill Collum.

upstream so to speak. These surplus electrons and holes are drawn into the high voltage carrier wire and from there to a measuring instrument, where they yield up accurate data on characteristics of radiation which triggered the whole sequence.

In terms of dimensions, detectors which are designed to seek out alpha particles are around 2 mm deep and 10 mm in diameter, like a small coin, whilst those searching for more penetrating gamma radiation might be say 50 mm deep and 75 mm in diameter.

11.2.3.5 Germanium detectors

As their name suggests, high-purity germanium detectors employ germanium semiconductors rather than silicon. They are often referred to as high resolution gamma spectroscopy (HRGS) detectors, but strictly speaking this is more of a reference to what they do, rather than what they are. Having said that, HRGS does summarize this instrument's claim to fame rather well, in that it yields the highest resolution of all detectors. However, this is specifically with regard to gamma radiation. In truth then, either nametag works just fine.

Unfortunately this kind of accuracy does come with some drawbacks. To begin with they are very expensive, think of a couple of family cars. What's more, to function properly these detectors need to be cooled with liquid nitrogen. Their operating temperature is normally described as 77K, which is on the *Kelvin* scale, but in more familiar units equates to minus 196°C. Above that temperature electrons have no difficulty crossing germanium semiconductors and forming a circuit, so liquid nitrogen gets them to behave and only zip across when stimulated by radiation.

Clearly the "plumbing" for all this cooling makes germanium detectors quite cumbersome, so ordinarily they are limited to being fixed in one location. For all of their shortcomings, where gamma radiation needs to be analyzed with supreme accuracy this is the detector to go for. This brings us onto the wider issue of matching detector technology to actual need.

11.3   Technology selection

With the technologies on offer being so diverse, from those based on gas, to plastics, crystals and semiconductors, it comes as no surprise that selecting the right one for any given situation is anything but straightforward. The good news is that each has particular characteristics which form the basis of a good place to start.

Technology	Cost	Resolution	Efficiency		
			Alpha & beta	Gamma	Neutron
G-M Tube Krypton (Kr) or Xenon (Xe)	Low	Low	Medium	Low	N/a
Proportional counter Helium-3 (He-3) or boron trifluoride (BF$_3$)	Low	Low	N/a	N/a	Low
Gas flow proportional counter P-10 Gas	Low	Low	Medium	Low	N/a
Scintillation detector organic	Medium	Medium	N/a	High	High
Scintillation detector inorganic	Medium	Medium	N/a	Medium	Low
Semiconductor detector	Low	Medium	High	Medium	Low
Germanium detector	High	High	N/a	Medium	N/a

Fig. 11.6 Technology capabilities.
© Bill Collum.

Fig. 11.6 presents an overview of the aptitudes for each technology, set against selection criteria of cost, resolution and efficiency, but as we might expect, there are layers of more subtle issues behind these headlines. So, for example, the G-M tube and gas flow proportional counter both achieve a "low" rating for gamma efficiency. However, if we were to get down in the detail we would find that a G-M tube is the best of the two, but still not quite efficient enough to achieve a "medium" rating. With a proviso on such fine detail, there is more than enough here to provide an insight into how a technology selection debate goes. Let's unwrap the criteria one at a time.

11.3.1 Cost

It could easily be assumed that high cost always equates to the "best," to the extent that equipment with an appealing price tag must be somehow second-rate, maybe not quite up to the job. However, this can be akin to suggesting that a microwave oven is inferior to a full-blown freestanding cooker, with its twin ovens, several hotplates, a grill and all the rest. If all you need to do is reheat or defrost food, for example, then a microwave oven is excellent and significantly less expensive. So as always, cost is important, but it must be factored in with the required functionality.

This is particularly relevant when we consider the cost of radiometric instruments, which can vary widely from one technology type to another. Fig. 11.6 depicts cost on an order of magnitude basis, so whether it be sterling, dollars, euros, and so on, "low" will be measured in hundreds of units, "medium" in thousands and "high" in tens of thousands. Clearly if low cost equipment can generate measurements which are entirely acceptable in a given situation, then it would make no sense to purchase something more expensive, especially when a top-end alternative might cost a hundred times more.

11.3.2 Resolution

One way to grasp what resolution is all about is to visualize the images provided by X-rays, magnetic resonance imaging and computed tomography scans. Each has particular viewing prowess across a range from solid bone, to dense tissue, to soft tissue, along with differing abilities to discern fine detail. Radiometric instruments follow a similar pattern.

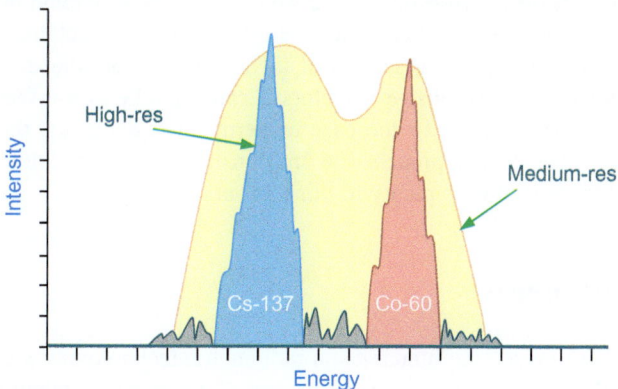

Fig. 11.7 Resolution comparison.
© Bill Collum.

Fig. 11.7 contrasts the output from high and medium resolution (res) instruments, where both are "viewing" the same mixed material. We can see that high resolution detectors have an ability to reveal fine detail, including individual isotopes, in this case caesium-137 and cobalt-60. Medium resolution detectors, on the other hand, give a good indication of radiation intensity, but can only distinguish between isotopes if they are not too closely related in terms of their gamma energy.

In addition both detectors can determine weight, down to the gram, of individual isotopes, again with high resolution instruments having the edge on performance. They can readily differentiate and calculate weights across a range of mixed isotopes, whereas medium resolution types are best suited to gaging weight when materials are dominated by a single isotope. So, for example, medium resolution instruments are effective at picking out the cobalt-60 which predominates in activated metals.

Low resolution detectors are more limited, but do have useful abilities in detecting and differentiating between alpha and beta particles. So whether it be high, medium or low resolution, each instrument has its place.

11.3.3 Efficiency

Ordinarily there could be some debate about the exact meaning of *efficiency*, but in radiometric circles there is no such room for maneuver. In this context it has a very specific meaning, which relates to how wide an area an instrument can "see" and how much data it can capture from that region.

I'm sure this interpretation is entirely predictable for the radiometric folks, but the rest of us can visualize it as though an instrument was shining a beam of light through the material or area being examined, with anything outside its beam not registering. A good example can be seen in the germanium, or HRGS detector, which as we know gets top marks for gamma resolution, so raises the expectation that it must also be super-efficient. However, its line-of-sight is relatively limited, so in terms of efficiency it only achieves a medium rating.

In addition to its prescriptive definition, we must also bear in mind that efficiency relates to a single detector, so can always be improved by say rotating materials in front of a detector or deploying several of them, but then of course cost escalates as well. For relatively inexpensive detectors it can often be cost-effective to array several of them together, but in the case of germanium detectors, their high cost may weigh against such an approach. The one constant in every technology selection debate, is that there is always plenty to think about.

11.4 Instruments

Now that we have visited the fundamental radiometric technologies, we are ready to examine how they are employed in various items of equipment. At their highest level radiometric instruments can be divided into three distinct categories: *health physics*, for those used directly by site-based personnel; *environmental*, which check aerial and effluent arisings; and *assay* instruments, which measure the radionuclide composition of materials.

11.4.1 Health physics

The majority of these instruments reside in changerooms and sub changerooms, so for those of us who routinely enter nuclear facilities it is the kit we are most familiar with.

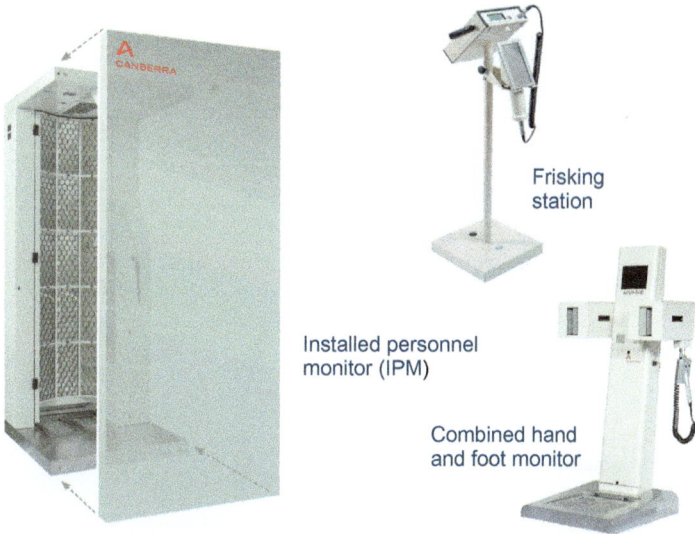

Frisking station

Installed personnel monitor (IPM)

Combined hand and foot monitor

Fig. 11.8 Health physics instruments.
© CANBERRA.

We discussed their purpose and how these instruments are used in Chapter 4, so will just remind ourselves here that it includes (Fig. 11.8) frisking stations, hand monitors, foot monitors and the installed personnel monitors that we step inside to scan our whole body before leaving a nuclear plant. All of these are searching for tell-tail signs of contamination about our person.

11.4.2 Environmental

These instruments, which operate in real-time, stand guard, permanently on the lookout for any hint of radiation or contamination which is above permissible levels. In other words they double check that everything is as it should be, whether it be inside a nuclear plant or monitoring aerial and effluent streams before they are discharged to the environment. Frequently then, this equipment (Fig. 11.9) operates in a passive state and only triggers, say by sounding an alarm, if it detects the consequences of a mal-operation or accident.

Alpha and beta particulate monitors, also known by their short form of *alpha in air* monitors are a common site around nuclear facilities. They operate by drawing air into an open-ended pipe and passing it through a filter which is continually monitored for signs of contamination. The mobile version, as shown, can be mounted on a trolley and is particularly useful in a decommissioning setting. For new facilities they are generally wall-mounted and present in sufficient numbers to obtain representative air samples.

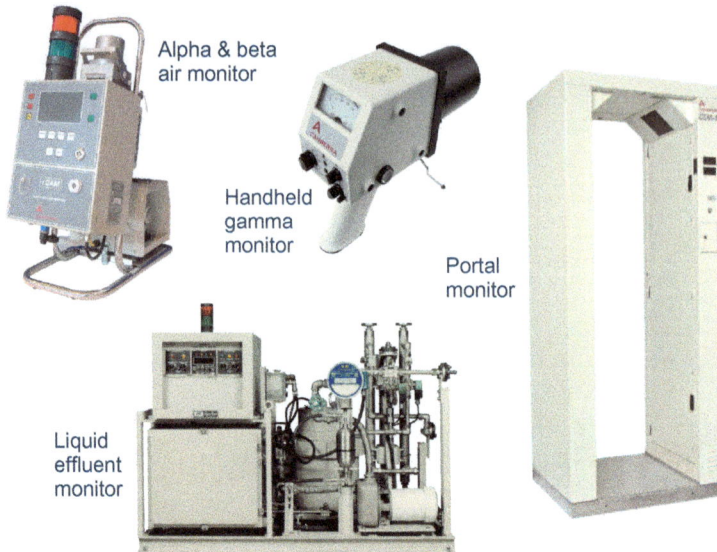

Fig. 11.9 Environmental monitoring instruments.
© CANBERRA.

Similar technology is used in stack monitoring systems which check the radiological quality of air before it is discharged to the environment. However, in that context instruments may be targeted towards searching for particular isotopes. We shall discuss stack monitoring more fully in Chapter 17.

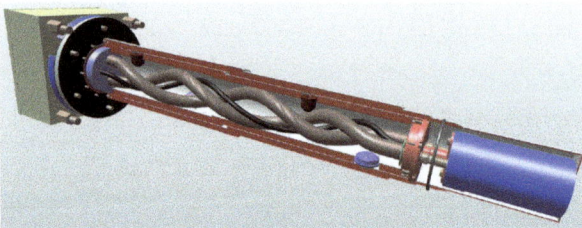

Fig. 11.10 Through-wall gamma monitor.
© Nuclear Decommissioning Authority.

Gamma monitors come in several guises, but all are used to police their environment in search of radiation. Wall-mounted monitors scan a predefined area, whereas handheld versions can be used to check an area before entering, or set up to monitor radiation levels whilst it is occupied. Portal monitors, depending on their dimensions, can scan personnel or vehicles as they pass through. The through-wall type shown in Fig. 11.10 can be inserted through a hefty concrete wall (omitted for clarity). The beauty of this setup is that whilst the detector, shown here in blue, may be subject to high levels of radiation on one side of the wall, it can safely be withdrawn by an operator standing on the other. As I say, there are lots of variations.

Liquid effluent monitoring is variously used to either check small samples, or installed online to continually monitor flows through particular lines of pipework. In both cases it may play an integral role in liquor processing operations, or alternatively be used to check liquids which are destined to be discharged to local waterways or the sea, say after undergoing an ion exchange process.

Although instruments are routinely divided into the health physics and environmental categories I have described, in truth there can be some overlap. So, for example, a handheld gamma monitor, being suitable for personal use, could equally well be tagged as a health physics instrument. As a general rule, we can think of health physics instruments as performing personal checks, whilst environmental instruments are more concerned with monitoring the goings-on within nuclear plants and demonstrating that discharges comply with regulatory authorizations.

11.4.3 Assay

Just as with gold, silver, and other metals, this is all about examining the constituents of a material, except that in this context, the processes employed are entirely non-destructive. Instead, radiometric instruments peer into a material to discern its radionuclide composition.

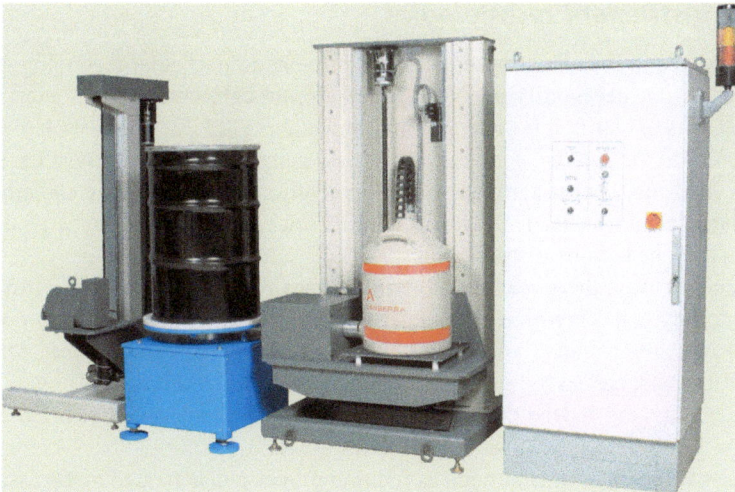

Fig. 11.11 Assay equipment.
© CANBERRA.

Assaying (Fig. 11.11) may be conducted whilst materials are being treated, as part of finalizing next steps in their processing, but its primary purpose stems from a need to demonstrate compliance with conditions for acceptance (CFA), a subject we discussed in Chapter 6. Assaying could be conducted to satisfy CFA for a receiving processing plant, or demonstrate compliance with transport regulations, or maybe to confirm that packaged waste is suitable for long-term storage. Whatever the reason, there will be prescriptive rules about the radionuclide makeup of a material, sometimes down to individual isotopes, so this is where assaying swings into action.

Assaying is performed on a wide variety of materials and across the full range of radiation levels, from low to medium to high level, among them, plutonium contaminated materials (PCM), activated metal, and irradiated fuel. In addition there are materials classed as *soft wastes*, which is a rather wide heading that divides into two categories. The first covers items which have been worn by personnel in either "suspect active" C2 (green) areas, or in contaminated areas governed by the radiological classifications of C3 (amber) and C4 (red). These articles are classed as personal protective equipment (PPE) and respiratory protective equipment (RPE), and include PVC suits, gloves, masks, respirators, and so on. The second category pertains to items of a more operational nature such as spent filters, swabs and the PVC sheeting which is used for temporary tented enclosures. The common trait of all soft wastes is that they can be compacted to save space and keep costs down.

Prior to assay, materials are normally packaged within carbon steel or stainless steel containers, depending on their activity levels, and in sizes ranging in size from 200 to 500 L and upwards to several cubic meters.

11.4.4 Instrument technologies

When it comes to matching radiometric instruments to particular technologies their partnering is not necessarily a given. There are, though, common associations, so in the interests of clarity it is these that are shown in Fig. 11.12. To illustrate the point of *exception to the rule* we can take a handheld gamma monitor. In most cases they utilize a G-M tube, but if particularly high resolution is required they can employ a germanium detector instead. However, with its price tag jumping maybe a hundred-fold the case needs to be a strong one.

In a similar way, there are several permutations for potential uses of instruments and ways they are categorized, so the listing depicts what we may term common practice. We are already familiar with most headings, frisking probe, activated metal, soft waste, and so on, so I will just highlight that *irradiated fuel* covers whole bars, fragments and indeed corroded fuel which is suspended within a radioactive sludge.

In cases where instruments, such as frisking probes and activated metal assay, can routinely choose from two technologies, the decision comes down to characteristics of whatever is due to be analyzed, the efficiency required and of course cost. However in the case of PCM, both of the available technologies must be used together, namely a He-3 or BF_3 tube and a germanium detector. The rationale goes that plutonium isotopes Pu-238, Pu-240 and Pu-242 all spontaneously emit neutrons, so need a He-3 or BF_3 tube to detect them, whilst isotopes Pu-239 and Pu-241 are predominately gamma emitters, so require the very different analytical prowess of a germanium detector.

			Technologies					
			Proportional counters			Scintillators		Semi-conductors
			G-M Tube (Kr or Xe)	He-3 or BF$_3$ tube	Gas flow tube (P-10)	Organic	Inorganic	Silicon	Germanium
Instruments	Health physics	Frisking probe			✓	✓			
		Hand & foot monitors			✓	✓			
		Installed personnel monitor (IPM)			✓	✓			
	Environmental	Handheld gamma monitor	✓						
		Alpha & beta particulate monitor						✓	
		Portal monitor				✓			
		Liquid effluent monitor							✓
	Assay	Plutonium contaminated materials (PCM)		✓					✓
		Activated metal					✓		✓
		Irradiated fuel (solid/corroded)					✓		✓
		Soft waste (PPE, filters etc)			✓	✓			

Fig. 11.12 Radiometric instrument technologies.
© Bill Collum.

11.5 Safeguards

As we know fissile materials, such as enriched uranium or plutonium, can be used in the manufacture of nuclear weapons, a subject which is outside the scope of these pages, where we are confining our discussion to civil nuclear activities. However, whilst we are on the subject of radiometric instruments it is appropriate to mention that they are used to track fissile materials within civil nuclear facilities, including storage vaults. In particular they identify specific isotopes, their amount and precise location. This comes under the umbrella of an activity known as *safeguards* which has many parallels with tracking money in financial institutions, balancing the books so to speak, so much so that the term used to describe many of its day-to-day processes is *accountancy*.

Of particular relevance, it is worth noting that the International Atomic Energy Agency (IAEA) take an active role in scrutinizing fissile materials around the world. They do not just ask for sight of relevant documentation but have their own measurement and tracking instruments, including cameras, in facilities which house such materials. Furthermore, they go to extraordinary lengths to ensure their equipment is secure, tamperproof and only accessible to nominated IAEA personnel. So in addition to their everyday regulatory, operational and safety related functions, radiometric instruments also make an important contribution to international safeguards.

Project planning

12

Imagine the situation where a client needs a new nuclear facility, they have an outline of what they want and funds available to turn their vision into a reality. The client holds some form of tendering competition, selects a competent organization to carry out the task and begins handing over the cash. What next?

We have seen something of what goes into a facility's engineering, but how do those charged with delivering a shiny new building coordinate the whole exercise? How do they ensure hundreds, maybe thousands of personnel, including engineers, manufacturers, constructors, and other specialists operate as a single entity? How do they demonstrate to a customer, month after month, that all is going swimmingly and their money is being wisely spent? It sounds tricky, and it is. Let's start at the beginning.

12.1 Client specification

When a project first becomes a reality, whether it be a relatively simple task such as remodeling a workshop, or a vast new nuclear facility that will take years to design and build, a document is produced that sets out what the task is. For short duration, comparatively straightforward tasks, it may be described in a note or letter from the customer, whereas for large projects it takes the form of a more comprehensive client specification, or *client spec*.

Bearing in mind the size, complexity and cost of major nuclear facilities, we might reasonably expect client specifications to be sizable tomes, bursting with useful information and packed with all manner of guidance and instruction. It is true that invitations to tender for these projects are hefty to say the least. But an actual client spec, the document that gets to the nub of what a customer wants, is quite a slim affair, normally numbering between just a handful and a couple of dozen pages. If a client spec were to go beyond describing "what" is required and specify "how" a task should be executed, it would become what is referred to as *preferential engineering* and stray into the optioneering territory that comes later on.

In the case of a new nuclear facility, a client specification will summarize its purpose and, depending on the facility's function, spell out the required throughput or generating capacity. With other more generic themes covering when it must become operational and for how many years, along with the importance of integrating a robust safety case, expectations in terms of environmental considerations, plans for waste management, and so on.

At a slightly more detailed level, there may be clauses that stipulate whether procedures used by a project team will be those of the client or contracting organization, along with direction on a customer's expectations in terms of reviews, approvals, and interface management. There may also be some indication of constrains such as a

Nuclear Facilities. http://dx.doi.org/10.1016/B978-0-08-101938-2.00012-X

nearby river or railway, the proximity of in-ground utilities, limitations on transport routes. Finally there may be some comment on processes for an orderly transition, probably during the commissioning phase, from a contractor to the facility's operations team. All in all, a mixture of a customer's general requirements and expectations, along with more specific guidance on the actual job at hand.

When you think about it, this is not the place for too much prescriptive direction; after all, a client is paying their supplier, or contracting organization, to figure out the details and crucially to shoulder as much risk and responsibility as possible. Equally, suppliers have no wish to be constrained into following a predefined plan; they just want clear guidance on what is required and freedom to use their expertise to best advantage, both for themselves and their customer. Of course there are plenty of hefty documents to cover the commercial and contractual stuff, but they are beyond this particular discussion.

So we have a client specification, we know what a customer wants, but how do we ensure their requirements are embedded into the heart of a project and always maintain an appropriately high profile?

12.2   Project controls

The term *project controls* is often used to describe management activities such as planning, cost estimating, resource forecasting, and so on, which are continually evolving and reported on a weekly or monthly basis. In its widest sense, it applies to the full gamut of activities that set the direction for a project and then, week by week, keep it on track. On the surface it can sometimes appear a little remote from duties of those involved in design activities. However, the reality is that one way or another, and to varying degrees, the whole design community has a role to play. Project controls is an important area, one that sets the framework within which a whole project team operates and in large part dictates success or failure of the whole endeavor.

12.2.1   Project control network

Every project manager secretly dreams of having unlimited funding, no particular deadlines to worry about and the pick of resources, always on standby and ready to swing into action at a moment's notice. Back in the real world, all three of those factors are tightly constrained and, just to make it interesting, very closely related. So when it comes to distilling what a client wants and figuring out how to deliver it, those *constraints* force their hard reality into every aspect of a project's planning.

With so much to think about, the priority in these early days is to establish the most comprehensive possible project control network, one that cascades out from the all-important client specification. Otherwise it will always be a matter of trying to catch up on being organized, which on major projects is practically impossible.

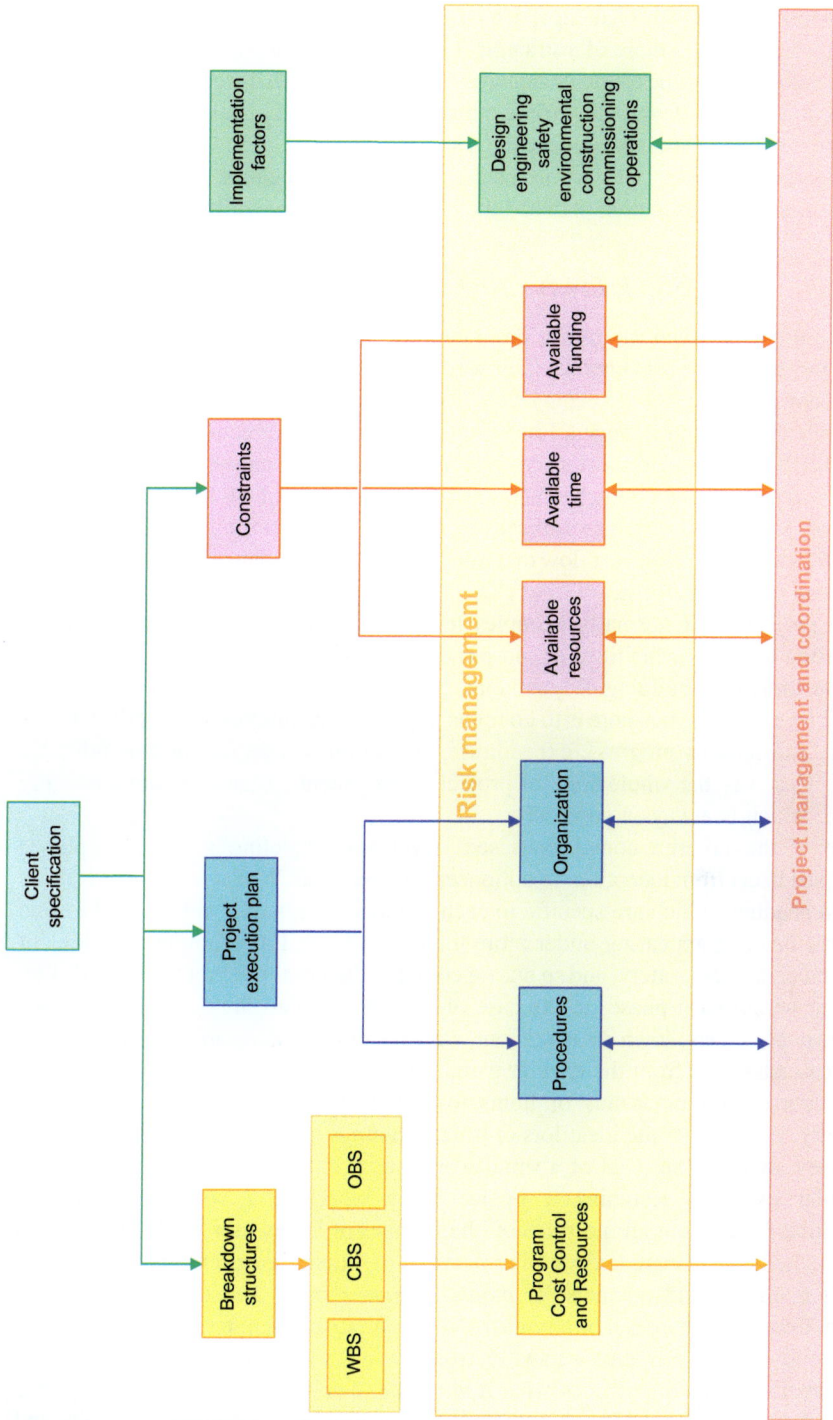

Fig. 12.1 Project control network.
© Bill Collum.

To help visualize the challenge, I have shown the network diagrammatically in Fig. 12.1. In truth, it is more of a strategic undercurrent which weaves its way through a project, albeit with some elements captured in its suite of formal documents. So although it exists at a strategic level, the project control network is too nebulous to figure as an actual deliverable. With that proviso, the network's headings of constraints and implementation factors are straightforward enough, so let's take a closer look at breakdown structures, a project execution plan and the all-pervading theme of risk management.

12.2.1.1   Work breakdown structure (WBS)

We shall see later, when we discuss programming, that to deliver an operating nuclear plant a project team must negotiate their way through thousands of individual tasks. So to have any chance at all of keeping track there must be some discipline in how those tasks are identified and arranged. Enter the work breakdown structure, widely referred to as a WBS. It divides a project's activities into elements, which are further subdivided until we arrive at tasks which are more readily managed that a single sprawling endeavor, such as "design mechanical equipment." In its purest form the hierarchy of a WBS can be depicted as a flow diagram (Fig. 12.2) with each layer revealing more detail than the one above it.

The project itself normally occupies level one, although for sizeable projects, comprising a group of facilities or sub-projects, it may be appropriate to assign level one to an individual building. Beyond that there are several ways of apportioning subsequent levels. So the key here is to go for an approach that mirrors a project's delivery strategy and how its progress, expenditure, and so on will be reported month after month. That way the whole suite of project management systems are automatically aligned, which is always a good idea.

One of the favorite conventions sees level two depicting a project's phases (discussed later) from inception, to optioneering, and so on. Whilst level three follows up with headings which are specific to each of those phases. It is not unusual to find the same heading appearing under more than one parent. For example, activities of mechanical, process, safety, and so on, are common themes throughout a project's life, so occur within each phase. In the case of a fairly standard WBS, such as the one shown, once it gets down to level four we find more manageable design activity items cascading out from their wider parent themes.

Of course with thousands of items to wrestle with this is a very condensed summary. But you get the idea: lots of broad headings gradually being broken down until they arrive at the level of a small work package, the *terminal element*, in this particular example designing a bogie during a project's optioneering phase. I should add that there are no rules on the number of levels in a WBS, so it comes down to the size and complexity of a project and to some extent preferences of clients, project managers and planners, or *planning engineers* as they are more properly known. Generally, there can be anywhere from four, to maybe three or four times that.

Beyond terminal elements we switch from the WBS to a program, discussed later, which spells out the minutiae of what it will take to deliver each element. Programs burrow down to the level of work packages such as drawings, documents, reviews, and

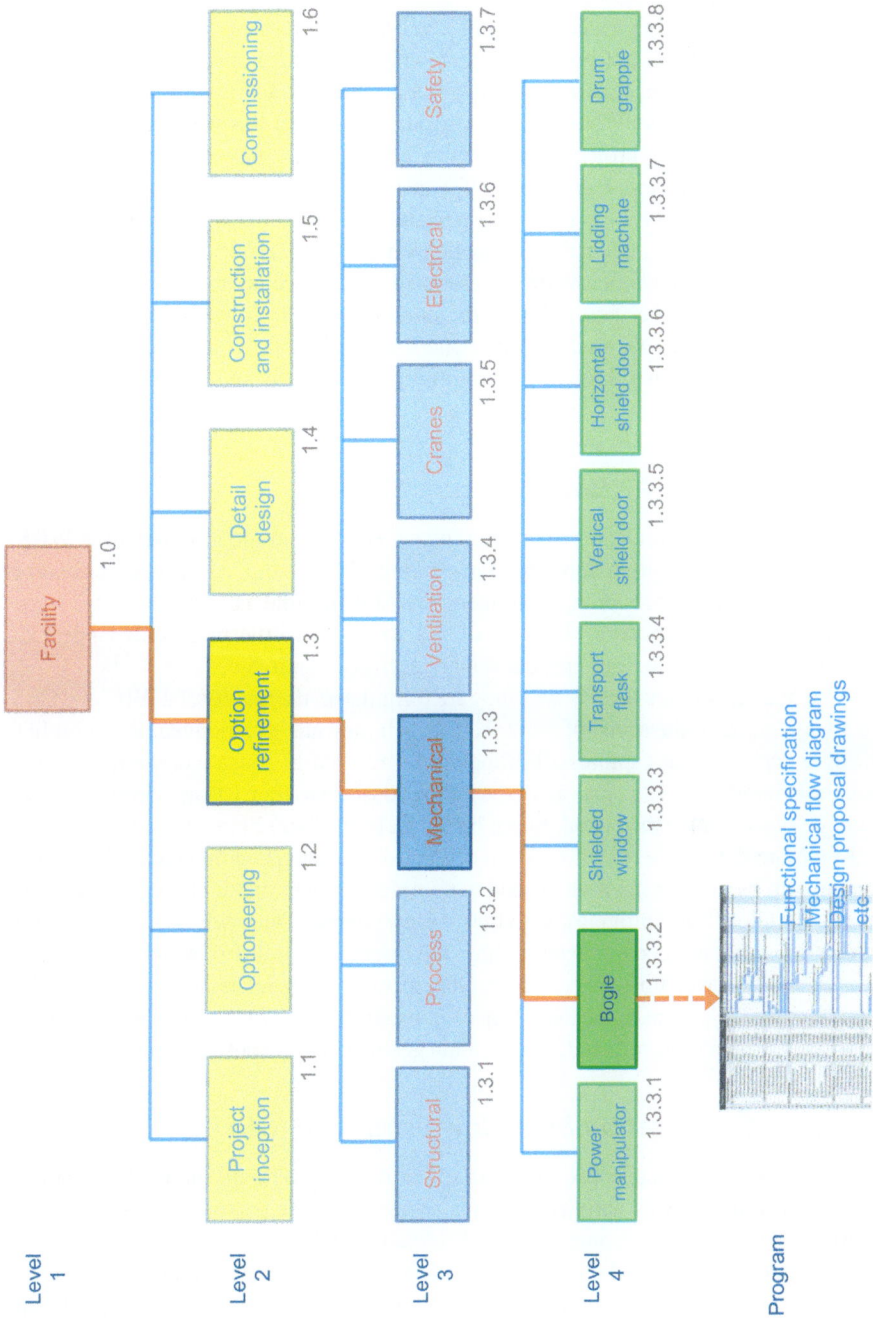

Fig. 12.2 Work breakdown structure.
© Bill Collum.

so on, show how much time each package will take and exactly when, in real time, it is planned to address and deliver them.

One of the golden rules when creating a WBS is that it must capture the *scope* of what needs to be done, not how (strategy), when (program) or for how much (cost estimate). What's more, it should always be possible to trace the origin of an item, or *child*, from its root upwards through various *parents*. In this way we can be confident that every task a project team is engaged in is legitimate and contributing to predetermined objectives. Conversely, if a task cannot be shown to have this kind of lineage, then we know for sure it is spurious and can expunge it. This brings me to the numbers we see attached to every box on a WBS.

Headings at each level of a WBS are assigned a reference number, or *code* with the first part of each number being that of the parent above it. So whatever the task, or *element* to be precise, we always know where it originated. What's more, the same numbering system flows through programming activities which cascade from each terminal element.

12.2.1.2   Cost breakdown structure (CBS)

One of the favored approaches to developing a CBS is to mimic its companion WBS. So, for example, they often adopt the same numbering system, a trait that goes some way to explaining those strange numbers seen on timesheets. Alternatively, some organizations may choose to develop a CBS which is a little more independent of its WBS, and of course both approaches are entirely legitimate.

What matters is that whatever way they are formulated, the totality of a WBS and CBS are always equal to the scope of work from which they are both derived. It is a bit like having a database that captures all of the work involved in constructing a new house, along with another one that lists all of the various costs involved. Clearly neither of them could be mistaken for the actual house, but in their own way both equate to it.

At a more detailed level, when a project's cost estimate is being prepared, much of it is built up by costing the time needed to complete each terminal element, essentially totting-up all of the program activities below them. Then when timesheets are completed, hours are booked against either the reference number of a terminal element or its subsidiary program activities, depending on how much detail the cost control folks need. This weekly reporting enables project managers to keep track of actual expenditure against that which was predicted, and no doubt get busy if they spot a looming overspend.

12.2.1.3   Organization breakdown structure (OBS)

To complete this particular trio we must add in those who do all the work. Although we tend to think of this activity in terms of resources or personnel, it is normally referred to in the wider context of OBS. Again it takes the same scope of work, but this time the WBS and CBS inform an analysis of the people required to deliver a project. On the face of it then, an OBS could simply be used to forecast future resource requirements, but as you might expect it is much more sophisticated than that.

12.2.1.4   Integrated breakdown structures

The trick here, the big payoff in developing a WBS, CBS, and OBS is that a project manager can balance them to smooth out their delivery strategy. For example, the trio will show where, if left unchecked, expenditure would rise one month and plummet the next, where during the course of a year resources would double, half, then double again, which is clearly unsustainable. Delivering vast nuclear facilities needs much more order than that. As with any other project they must follow a predictable pattern: steadily gathering pace at the beginning, running at a steady state, and then gradually tailing off again as they near completion. It is the three breakdown structures, working in concert, that help to make this possible.

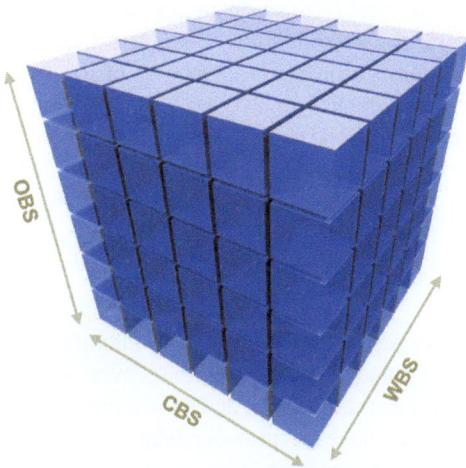

Fig. 12.3 Integrated breakdown structures.
© Bill Collum.

To appreciate how intertwined they are, we can view the WBS, CBS, and OBS as three facets of the same cube (Fig. 12.3). The message being that if they are not perfectly balanced the "cube" would become very contorted indeed, which, representing a whole project, is not where you want to be.

12.2.2   Project execution plan

In parallel to the exercise of creating breakdown structures, this is also the time to develop a *Project Execution Plan (PEP)*, equally well known as a *Project Implementation Plan*. This document takes the same scope and deliverables and spells out how they will be achieved. It covers everything from the organizational structure and responsibilities of key personnel, to reporting procedures, quality assurance systems, approaches to change control and reviews. Not forgetting the suite of design procedures that will be used, a description of the all-important document control systems, and so on. Essentially then, a PEP articulates the strategy for managing a project, who is responsible for its various elements and how they will go about their duties.

12.2.3 Risk management

Unfortunately, figuring out what a project needs to achieve and then setting it up to deliver is the easy bit, so there is barely time for a little self-congratulation before the real work begins. From this point one of the primary challenges can be summed up by the term, *risk management*. This is not risk in the sense of a catastrophe waiting around a corner, but risks to a project's ability to deliver on its objectives whilst at the same time sticking to its allotted time and budget.

It is akin a domestic project of having a new kitchen fitted. If we were to project manage the job ourselves, then we would order the materials and line up a joiner, electrician, plumber and plasterer, not forgetting a skip for the exiting kitchen. We might even draw up a simple program so that everyone involved knows exactly what they are doing and when they are due to arrive. One of the risks (and you may have experienced this) is that after weeks of careful planning, some of the new kitchen materials are not delivered on time. A mitigation strategy might be to hire a van and collect all the materials yourself, well in advance of the start date. It would certainly add to costs, but this has to be weighed against risks of the whole job being thrown into disarray because worktops are languishing in a warehouse on the other side of town.

12.2.3.1 Risk register

For the kind of major project we are considering here it is essential to identify and prioritize risks, calculate how much time and money it would take to recover if they became a reality, assess their potential knock-on consequences and have a mitigation strategy ready to address them. The tool used to assuage this project predator is the *risk register*. It can stretch to hundreds of items and cover everything from an essential regulatory license not being granted on time to borehole tests proving ground conditions are more challenging than originally envisaged, or even particular specialist resources not being available at the time they are needed.

The trick here is to avoid any temptation to log risks on a register, draft a mitigation plan, fill in the boxes and feel a sense of relief that the situation is under control. It isn't. The problem with some risks is that a few words on a register can belie the true nature of their reality. If we go back to our example of the kitchen, a risk may be that one of the tradespeople does not show up. We could faithfully log this on our risk register, assume there will be a day's disruption and assign a nominal cost to the delay. However, if, for example, a plumber was missing, other members of the team may be unable to continue and so head off to work elsewhere. In such circumstances it could be difficult to get them back and the whole project may be delayed for several weeks.

On a domestic scale it is unlikely we would plan for all such eventualities and therefore have to accept the consequences, but on major projects there is too much at stake, too many interfaces that could unravel if something goes awry. What is needed is active management, hyperactive actually, to constantly monitor potential risks and stay one step ahead of their disruptive power. Not least because they permeate practically every project activity and weave their way into knock-on consequences that could barely be foreseen. If a project is to be successfully delivered, then risk management must reside at its core and be tackled with renewed enthusiasm at the start of every new day.

12.3 Project management

At some time or other every member of a project team will have a smart observation to make about their project manager. It is not something that is absolutely mandatory, but is nevertheless a tradition which is generally embraced with some enthusiasm. The most popular theme centers on speculation about what exactly they do all day, a subject on which opinions can vary considerably.

When you get right down to it, they are directly responsible for ensuring a project is delivered on time and within budget, and equally important guaranteeing that the finished article performs in accordance with whatever it was that the client specified in the first place. However, that element of their responsibilities normally gets delegated to an engineering manager, leaving the project manager to concentrate on time and money. So this is what occupies them all day—apart from keeping an eye on their engineering manager and the rest of the team.

The tool that enables a project manager to keep track and ultimately to deliver is the program, also known as a *schedule*. So they will own it with a vengeance, monitor it continually and hold lots of meetings to ensure everyone involved understands their role and is on track to complete its various deliverables by the assigned deadlines.

12.4 Programming

Essentially a program is a rather comprehensive "to do" list, one which in this context flows from the WBS we discussed a moment ago. For major nuclear facilities they can easily run to 20,000 activities and I have seen some with more than 100,000. So their preparation is quite an exercise. In fact, I never cease to be amazed at the capacity of planning engineers to create these lists in the first place and then control their evolution as a project's timeline marches on.

Of course with thousands of interrelated activities all jostling for position it is never going to go smoothly, so certainly needs a watchful eye from the project manager. To be honest it is a task that requires considerable skill, but this is not something that is generally acknowledged too freely.

12.4.1 Nuclear factors

It would seem reasonable to assume that the programming exercise for nuclear plants must follow pretty much the same script as that adopted for any large scale commercial or industrial development. The thing is, when nuclear connotations enter the equation they take programming to a place far removed from that visited by any other type of development. It is worth being aware of what lies behind this disparity, as it has an influence on how these immense programs are assembled. Two examples will illustrate the point.

12.4.1.1   Division of costs

We saw in Chapter 5 that nuclear building construction is robust to say the least, so might sensibly reason that this element must account for a much higher percentage of a project's costs than would be encountered elsewhere. It turns out, however, that the opposite is the case.

There is no such thing as a standard spend profile, but broadly speaking the cost, including installation, of concrete, reinforcing bar, steelwork, cladding, roofing, and so on, can account for just 25% of the cost of a nuclear facility. Whereas comparable costs for more conventional structures might amount to 50% of a project's funding. Furthermore, design costs for a regular industrial facility might run to say 10%, but the complex nature of nuclear facilities can push this number up to a third, or even more of the entire budget.

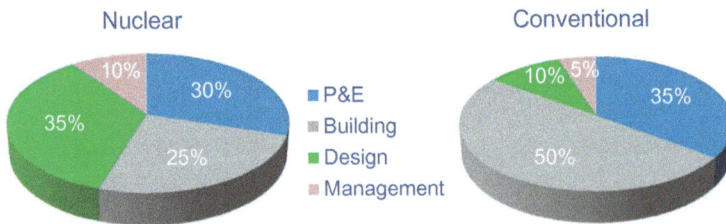

Fig. 12.4 Nuclear versus conventional expenditure.
© Bill Collum.

This leaves around 30% for often bespoke plant and equipment (P&E) along with its installation, and maybe 10% for the not insignificant management effort that such complex projects demand. Of course the figures for both nuclear and non-nuclear build can vary considerably, but the underlying trend (Fig. 12.4) of nuclear projects expending a comparatively low percentage on structures, even though they are significantly more robust than most, along with a higher than average spend on design holds true.

With say two thirds of a budget being expended on design, along with plant, equipment and its installation, this is clearly not a situation that can be tackled with what we might call a *generic* programming template. Nuclear planning engineers need quite a different mind-set, where it can be more appropriate to think of nuclear facilities as multiple complex processes, often unique and all carefully configured to create a single operating entity, all of which just happen to reside within a building.

Now that is not to make light of the building and civil engineering element. As we have seen, it is a monumental challenge in its own right, with facilities often having to withstand ravages of extreme environmental events, such as earthquakes and ferocious storms, which might come along just once in a period of 10,000 years. Ordinarily that would be enough for any project team to contend with, but in a nuclear world it *only* accounts for a quarter of the story.

12.4.1.2   Regulatory constraints

Our second example relates to the intense level of scrutiny that nuclear facilities are necessarily subjected to, both during the design period and beyond. For this we need to picture a plethora of regulators monitoring the design, construction and commissioning teams, all with requirements that must be satisfied in a particular sequence as a project progresses. Their remits range from health and safety, to environmental considerations, to specialist nuclear regulators, all with power to grant formal approvals at predetermined points in a project's evolution. Failure to satisfy their legislative demands can result in approvals being withheld, which could delay a project or even stop it altogether.

In programming terms, it is not simply a matter of allowing a week or two for preparations to meet with regulators in order to seek their endorsement. As we shall see, in Chapter 14 in particular, the level of detail they require is comprehensive to say the least and can take months or even years to prepare. This then must be woven into a project's timeline to ensure submissions to regulators are comprehensive, fully substantiated and crucially, delivered on time.

12.4.2   Programming logic

Clearly a project's program is fundamental to its successful delivery, so we need to be aware of the primary ground rule that must be observed in their execution. At its highest level, a program must have a clear logic, something we shall look at shortly. It sounds glaringly obvious doesn't it, but with thousands of tasks to deal with it would be all too easy to become engrossed in the details and lose sight of the overarching game plan. If this were to happen, if engineers from a particular discipline were to race ahead of their colleagues, they would in effect be forcing those who must be integrated with them to accept a fait accompli when they catch up.

An extreme example (just to illustrate the point) would be structural engineers freezing the positions of hefty concrete walls, along with penetrations through them, so that they could finalize their seismic analysis, whilst the rest of the team were still engaged in design development and needing more time to determine if proposed wall locations, not to mention penetrations, were even in the right place.

Project managers and planning engineers take the lead in coordinating this endeavor, but with design teams often numbering several hundred personnel it is incumbent on everyone to play their part. So, for example, if an engineer is working on the detailed design of a mechanical pump, before the processing equipment it serves has been adequately specified, then they will know for sure that something is misaligned and the pump's design will almost certainly have to be revised at a later date.

As we might expect, project managers are keenly aware that if activities are not continually monitored, adjusted and resynchronized, a project's program could unravel very quickly indeed. In such circumstances a project team would no doubt still be very busy, but I'm afraid actual progress would be nothing more than an illusion and considerable rework would lay ahead. It is no wonder therefore that project managers are continually questioning their teams, herding them along and generally harrying everyone in sight, which I guess explains why they are greeted by so much badinage in return for their efforts.

12.4.3 Activity links

Programs are normally displayed in the form of a Gantt chart (Fig. 12.5). If you study even a fairly basic one, you will notice a plethora of vertical lines linking the activity timelines that run from left to right across its pages. These lines show linkages, more formally known as *dependencies*. They determine when a particular activity must begin and, based on required duration, when it will be completed. Or if not completed, then at least developed to a predefined level of maturity before another related activity can even get started. In other words the output from one task, the *predecessor* is needed to feed into another one, the *successor*.

Fig. 12.5 Gantt chart.
Shutterstock.com.

So, for example, it is not possible to conduct a shielding assessment until results are available from analysis of a facility's radioactive inventory, either as it currently exists or projected into the future, whichever is relevant. Furthermore, without a shielding assessment it is not possible to determine the thickness of concrete shielding walls, which delays designing items embedded within them, such as shielding windows. And so it goes, thousands of dependencies and knock-on consequences all queuing up and waiting for a chance to run amok. It's not easy is it.

12.4.4   Critical path

It is worth mentioning that apart from a myriad of what we might term *local links* between activities, there is also a major line that connects one crucial step to another. It flows relentlessly through individual program items and across a project's phases, touching, as it progresses, on points such as when a tranche of funding is due to be released, or when regulators must issue approval to proceed beyond a given point. In essence it concentrates on any activity which, if not realized, could be a *showstopper* in terms of its ability to derail a project.

It is called the *critical path* and as its dramatic name suggests is fundamental to a project's success. If the due date for any item on a critical path is missed the ramifications could result in a project slipping on its programmed end date, so there would be an urgent requirement to conduct an investigation, develop recovery plans and get things back on track. Clearly the critical path is the most important line on any program, so it is monitored in microscopic detail, with any hint of delay being tackled straightaway.

12.4.5   Project phasing

We shall take a closer look at project phases in a moment, so in the meantime we just need the highlights. This will be enough to put our upcoming examination of *hold points* into context and conclude our discussion on programming.

Faced with a blank sheet of paper and thousands of activities to consider, where does a planning engineer begin? At their highest level programs for nuclear facilities must follow a series of predetermined steps, starting with the time a project is first suggested as a possibility, through to the time it becomes operational. With this overarching timeline in place they can concentrate on a myriad of activities which must occur, in the right sequence, within each of those major phases.

Project inception	Launch project, test the business case and prepare a client specification
Optioneering	Develop potential options, evaluate and select the preferred option
Option refinement	Refine the chosen option and prepare a close tolerance cost estimate
Detail design	Develop design in readiness for construction and equipment manufacture
Construction and installation	Build facility and install plant and equipment
Commissioning	Carry out active and inactive commissioning and handover to operations team

Fig. 12.6 Primary project phases.
© Bill Collum.

As is usually the case, different organizations have their own interpretation of the main steps a project must pass through before it becomes operational. Happily, logic is logic, so they all adopt similar themes and generally have around six phases to navigate along the way. Fig. 12.6 shows the phases along with a summary of what each entails, essentially starting with little more than a notion of what might be needed and ending with an operating plant. As I say, the details are coming up.

12.4.6   Hold points

We can see how a program is configured to manage the evolution of a project from the incubation of an idea through to a fully functioning facility. However, it can never be a matter of launching a project and unquestioningly working through each phase until it is completed. Rather it is imperative that controls are in place to ensure a project can literally justify its existence, before being allowed to continue its progression from one phase to the next.

After all things can change, to the extent that what may appear to be a good idea at the outset can, with improved definition, turn out to be anything but. What is needed is a *safety net*, a process to verify that everything is in good order before taking the next step. Interestingly, many organizations refer to this process as a *gate*, which gives a clue as to what happens when those points are reached.

As a project moves from one stage to the next its overall maturity and level of definition steadily improves. Factors such as how much it will cost, whether appropriate technology is available, whether the original brief still looks viable, and so on are all accorded more definitive answers with the passage of time. Hold points, which are an essential element of corporate governance, are effectively gates, where seasoned personnel decide whether or not a project is ready to pass through to the next phase.

Clearly if firmed up costs are turning out to be unaffordable, the necessary technology is still science fiction and the original brief has been sidelined by external events, there is little point in progressing any further. In fact stopping or redirecting a project can be every bit as successful an outcome as allowing it to proceed, although admittedly it may not feel like it at the time.

The timing and number of hold points varies from one organization to another, but for major projects there are usually three of them (Fig. 12.7). The first comes at the end of a Project's *Inception* phase, where a decision must be made on whether to release the considerable funding which is needed for *Optioneering*, followed by *Refinement* of the preferred option.

Project inception	Hold point	Optioneering	Option refinement	Hold point	Detail design	Hold point	Construction and installation	Commissioning

Fig. 12.7 Project hold points.
© Bill Collum.

At the end of those two phases, a second hold point review determines if the business case is still valid, the design robust, proposed technology adequately proven and the outturn cost estimate within budget. If these tests are passed then further funding is released to execute the *Detail Design* phase.

Notably, in addition to the cost of detail design activities, funds are almost always required for advance procurement. It applies to what are known as *long-lead* items, typically bespoke equipment that takes an awfully long time to manufacture, such as a high integrity nuclear crane. Unsurprisingly, no one wants to rush into signing contracts for expensive kit before a facility's construction has even been sanctioned. So those conducting this hold point review, and releasing the cash, must be convinced a project is destined to go the distance. Otherwise they face the prospect of needlessly committing to several contracts and being left with a pile of alarmingly expensive scrap metal.

Once the detail design stage is complete, there comes a final hold point to ensure previous evaluation parameters are still holding good, that any remaining project uncertainties are manageable, and that a close tolerance cost estimate confirms the project is still financially viable. With that gate successfully negotiated all of the funds necessary to complete a project, to turn it into reality, are released.

It might seem odd to dispense with further hold point reviews at a time when maybe more than half of a project's funding is yet to be expended. The thing is, once a project moves onto construction and beyond there really is no going back, so the post detail design review is about as serious as it gets. Of course for major projects there will be scores of other reviews along the way, but it terms of sanctioning a project to continue in existence, of releasing the money to keep going, it comes down to just a small number of crucial hold points.

12.5   Project phase activities

Now that the scene is set, we can look in a little more detail at what goes on as a project evolves across its various phases, from inception onwards. You will be pleased to hear we do not need to explore the tens of thousands of activities I mentioned earlier, but can stick instead to the headline challenges embraced by most nuclear projects.

For our purposes we shall assume the task at hand is to develop a program for a major new facility, but keep in mind that similar approaches apply to decommissioning existing facilities, and at a much smaller but not insignificant scale, to maintaining a pump or replacing a valve. I guess the underlying theme is that no matter what the task, little can be achieved by sending in a team and hoping to figure things out when they arrive.

12.5.1   Project inception

Project inception	Optioneering	Option refinement	Detail design	Construction and installation	Commissioning

Operating a nuclear site is a complicated business, so there is always a queue of potential new projects, all maneuvering themselves into position and eagerly looking for recognition. Some may establish their credibility quite quickly, say a store intended to house predictable future arisings of radioactive waste. Others, maybe a proposed new build replacement for an aging facility, will need more time to prove themselves. In such cases careful analysis is needed to establish if it might be more cost effective to upgrade an existing facility rather than start from scratch with a new one. Whatever the project some initial investigation is conducted to figure how it might be tackled, and substantiate the notion that it would make a good investment.

12.5.1.1 Investment justification

The *project inception* phase generally takes from a few weeks to a handful of months, so in the great scheme of things does not cost a great deal. Quite often it is referred to as the *business case*, in essence because this step entails gathering enough information to make an informed decision on whether it is sensible to invest some serious money in going any further.

Strictly speaking there should be no need for a great deal of engineering at this stage, so we should not expect to see too much by way of drawings and certainly no complex 3D building modeling. However, I should say that very often, even at this early stage, time is already pressing, so it may be legitimate to conduct some high level optioneering in order to discount options which are demonstrably unfeasible and clear the decks for development of serious contenders later on.

Ideally key members of the project team in waiting should be appointed, maybe on a part-time basis. It gives them the opportunity to become familiar with a potentially upcoming project, and just as importantly to work with their customer in developing the *client specification* we discussed earlier.

12.5.1.2 Technology identification

As with any activity that involves a fair bit of science, the nuclear industry never stands still. New, potentially better ways of achieving an improved outcome are always either under investigation or newly available. Whether it be improved product quality, increased productivity, enhanced safety, or an emerging technology about which little is yet known, a project team must examine latest developments before heading off down a path that may be obsolete before they reach the first junction. Technology identification is therefore an important element of this first stage.

12.5.1.3 Research and development (R&D)

Apart from a pressing requirement to know what technology might be available, it may also turn out that specific R&D is required. This may, for example, be to test the appropriateness of a potentially useful technology, or even develop a technology further so that a project team may make use of it. Where this is the case, an R&D program must be developed and carefully integrated within a project's overarching schedule.

Crucially, it must be feasible to realize the results of R&D in time to be amalgamated within a project's optioneering phase. Clearly if optioneering is due to be concluded in a year's time but the necessary R&D will take three years to complete, there is little point in a project team financing it. However, if pending results of R&D are deemed confirmatory rather than uncertain, a project may decide to proceed at risk by incorporating a particular technology, and allowing time to modify their design when the finer points of an R&D program do become available.

12.5.2   Optioneering

Project inception	Optioneering	Option refinement	Detail design	Construction and installation	Commissioning

There is absolutely no doubt that this is far and away the most significant design phase for any major project. When you think about it, the stage begins with precious little by way of engineering but ends with selection of the preferred option from which all else will follow. So although it is normally referred to simply as *optioneering*, I do feel *optioneering and selection* would give a better feel for the gravity of what it really means.

12.5.2.1   Preferred option selection

To illustrate how much ground this phase must cover, it is worth noting what is meant by a *preferred option*. It is certainly not a design proposal that looks "fairly close" and is deemed "near enough" to be knocked into shape at a later date. It will have taken a project team many months, possibly several years to arrive at the conclusion of this phase, and along the way a not insignificant portion of a customer's budget will have been expended. So the result had better be good.

If you picture yourself stepping up to the microphone, so to speak, to announce that the preferred option is ready, you are effectively saying to your customer that all other options have been considered and discounted and you have excellent explanations for why they were deemed unsuitable. Not only that, you can demonstrate that the chosen design proposal precisely satisfies requirements of the client specification, it will have no trouble achieving its throughput, is perfectly safe, can be built and operational on time, and you are confident that the cost estimate for delivery of the entire project is an accurate one. Accuracy at this stage can vary from one organization to another, but is generally expected to be within around +/−25% of the final outcome.

To focus the mind, I always think it is a good idea to visualize going back to your customer some months later, and having to explain that the *preferred option* has not quite lived up to expectations. They are unlikely to be impressed, or too enamored by requests for additional funding to revisit the optioneering stage they have already paid for, or your pleas for more time so that you can come back again with a new improved preferred option at a later date.

The main thing to keep in mind here is that once a project passes the point of selecting its preferred option, everything that follows should be either refinement or detail and to get to that level of confidence demands an enormous effort from everyone involved. There is something else, the significance of which can barely be overstated.

As I say, this phase begins with what is effectively a blank sheet, or at least very close to it. So in terms of design it is a fluid time, with concepts being thrown into the mix, some being discounted and others gradually coming to the fore. Crucially, in the great scheme of things, making changes at this stage costs very little indeed, which is certainly not the case when concrete has been poured and equipment is part way through its manufacture. In fact, as a project moves further and further from its optioneering days, the cost and knock-on consequences of implementing change rises exponentially (Fig. 12.8) to the extent that what might have cost a few tens of pounds or dollars at the outset, now devours hundreds of thousands, or even millions.

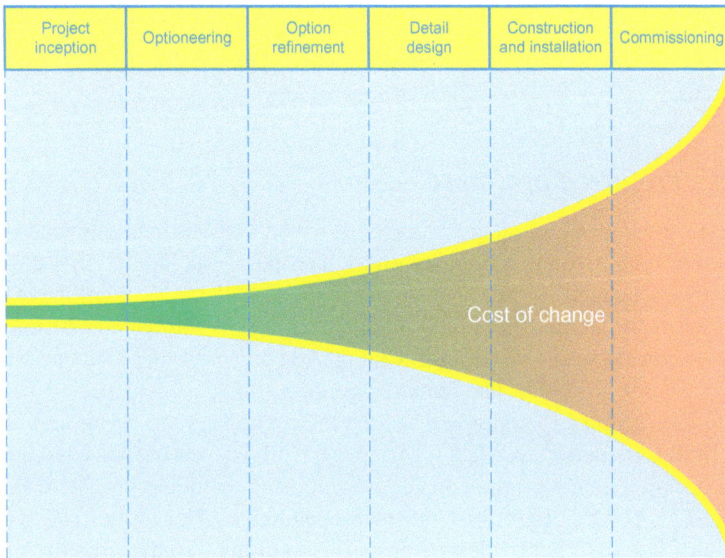

Fig. 12.8 Escalating cost of change over time.
© Bill Collum.

The situation is potentially even more problematic than one of *simple* cost and delay. For every project there comes a point where it is no longer even possible to carry out anything other than minor changes. In other words the cost, delay and general upheaval is just too much to countenance. If this were to happen, then operators would be forced to live with a facility that, at least in part, was less than perfect, which would be disappointing to say the least.

With commencement of a facility's operations being some years off, it is easy to be cajoled into believing that selection of a preferred option is no more than a stepping stone on the way to delivering that objective; in truth, few decisions come close to matching its far-reaching consequences.

12.5.2.2   Design philosophy documents

For an engineering community, the foundation of their designs is captured in philosophy documents for process, mechanical, electrical, structural, commissioning, operating and maintenance (O&M), and so on. This essential suite of documents sets the scene for everything from technologies which will be employed, to the maintenance regime that must be followed when a plant becomes operable.

Without the clear direction these documents provide, it would be difficult for an engineering team to make meaningful progress, so they are a priority which must be tackled early in a project's life. From them will flow a suite of basis of design documents, which add detail to the high level philosophy documents and provide guidance on how their objectives will be turned into reality.

12.5.2.3   Mass and energy balance

As we saw in Chapter 6, a facility's mass and energy balance is the cornerstone on which so much engineering is based. This in turn gives rise to the process flow sheet and from there more detailed flow diagrams for process (PFD), mechanical (MFD), ventilation (VFD), engineering (EFD) also known as a P&ID, which stands for *piping and instrumentation diagram*, and so on. Of course all of these take time to evolve and settle down, but as they do a clear picture begins to emerge on requirements for plant and equipment (P&E) and how the whole ensemble must interact.

Clearly this is a fast moving and iterative period, which is why it is so important to have a program in place that foresees all of the necessary activities and ensures their interactions are carefully coordinated.

12.5.2.4   Layout development

With the mass balance and suite of supporting documents all beginning to take shape, it is time to commence the layout development process. We shall cover this in more detail in Chapter 17, so for the moment I will just say it is important that layouts evolve in a structured manner, and absolutely essential that information fed into them from all members of the project team is at a comparable level of maturity.

In the early stages of this phase there will probably be a need to develop several layout concepts in parallel, after all there may be a range of potential technologies to investigate and even alternative facility locations, all of which will almost certainly entail a different building configuration. The crucial factor is that all options must be pursued with equal enthusiasm, and certainly none treated to special attention due to familiarity or an early hunch of a future preferred option.

12.5.3   Option refinement

Project inception	Optioneering	Option refinement	Detail design	Construction and installation	Commissioning

12.5.3.1   Refine preferred option

The easiest way to know if a project has succeeded in this phase is to compare the facility layout and supporting documents inherited at the beginning with those delivered at its conclusion. Bearing in mind that this stage is focused on pure "refinement" then all new or revised drawings, and indeed the accompanying raft of project documentation, should look and feel pretty much the same, the only notable difference being an additional layer of definition and detail. The exercise is comparable to taking an existing grainy photograph and putting it through a process that enhances it to one with excellent definition; the fundamentals do not change.

If there is a significant difference between the start and end points, then either the preferred design that resulted from optioneering was ill founded, or the option refinement phase has missed the point and launched into a redesign exercise which was unnecessary. Either way, both are situations to be monitored and strenuously avoided.

12.5.3.2   Review criteria

To underscore how crucial this phase is, it always ends with the hold point review we discussed earlier. Reviewers must be presented with evidence that gives them an unshakeable belief in a project's ability to achieve all of its objectives, which is quite a long list. It covers everything from demonstrating the business case and client specification are still relevant and can be satisfied, to proving the proposed design solution complies with its safety case and is not harboring unpleasant surprises that may come to light in the years ahead.

Crucially, there must be evidence that predicted costs of becoming operational are underpinned and demonstrably accurate: typically to within +/−5%. And equally important, decommissioning costs must also be accurately estimated, a subject we shall examine in Chapter 16. Clearly there is plenty to discuss. I have been both the reviewed and the reviewer, so can vouch that no matter what role you play it is an intense experience to say the least.

12.5.4   Detail design

Project inception	Optioneering	Option refinement	Detail design	Construction and installation	Commissioning

Detailed design is just that, where any hint of optioneering, other than at a micro level, should be a thing of the past. It is sometimes referred to as "handle turning," which is an unfortunate expression as it makes this phase sound quite routine, a notion that could hardly be further from the reality of what it entails.

12.5.4.1   Maintaining design intent

True, there will be an influx of personnel with skills more attuned to detailing than has been the case so far. But having said that, those who operate at the more detailed end of the design spectrum will cheerfully proclaim it is they who turn concepts, no matter how robust, into engineering reality, and of course the truth is that everyone has an important role to play.

The crucial factor, in terms of design skills, is that the early phases can only be successfully negotiated when each discipline has detailing expertise within it. And just as importantly, when a design is being detailed there must be ongoing engagement with those involved in its earlier development. In this way we can ensure design intent is recognized, clearly understood and smoothly integrated into the hundreds, or even thousands of details that flow from this stage.

12.5.4.2   Specification, design, and manufacturing

If we take the example of an item of mechanical handling equipment, say a bogie, then Fig. 12.9 puts the detailed design stage into context. Starting top right, we have seen that during the optioneering phase, *mechanical philosophy* and *basis of design* documents are developed which together set the framework for a facility's mechanical handling equipment.

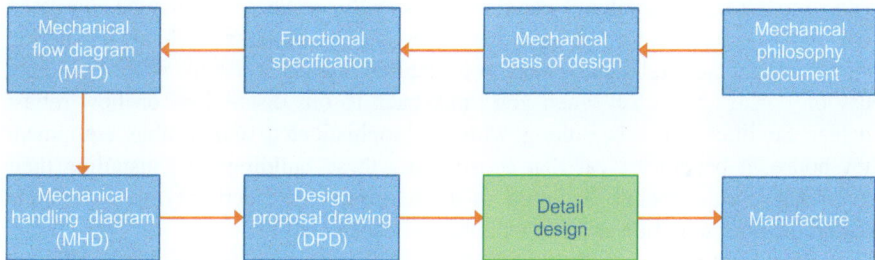

Fig. 12.9 Evolution of mechanical equipment design.
© Bill Collum.

Building on those documents, a functional specification is prepared which spells out the operational and performance requirements for our example bogie. With those documents in place, the design proper begins with a mechanical flow diagram (MFD) which shows how the bogie will interface with equipment and processes around it. And this is followed by a mechanical handling diagram (MHD) which articulates the bogie's functionality in more detail. Up to this stage the nearest we get to a recognizable design is an MHD's pictograms, which will give a high level, almost cartoon like view of the bogie's operations.

Finally then, with the preliminaries in good shape, design proposal drawings (DPDs) are produced, which show the bogie it all its dimensionally correct glory. However, even DPDs, which are prepared during a project's option refinement stage, do not contain sufficient information to manufacture a bogie. For that we need *detail design* drawings and a manufacturing specification, which may be prepared by experts within the project team but more likely by the manufacturer themselves.

These specialists not only have an intimate knowledge of how mechanical equipment operates, but also a comprehensive grasp of the regulations and engineering standards which apply to, for example, a bespoke bogie. Their remit ranges from specifying appropriate machining tolerances and welds, to minimizing sharp edges. They will also see to it that components handled by operators are within prescribed weight restrictions, and that lifting lugs are at the center of gravity for heavier modules which must be handled by cranes or other lifting devices. In addition, manufacturing specialists work closely with project teams to define quality assurance arrangements and develop an appropriate regimen for inspection and testing. And so it goes, for one item of equipment after another: words, and then lines on a drawing, steadily evolving until they are translated into a tangible reality.

12.5.5 Construction

Project inception	Optioneering	Option refinement	Detail design	Construction and installation	Commissioning

With so much emphasis on nuclear challenges, with a project team maybe engrossed in developing concepts for novel processing techniques, bespoke mechanical handling equipment, and so on, it can be easy to forget that one day someone will need to get busy on construction. Yet when you think back to our discussions on how robust nuclear facilities must be, along with the sophisticated and sizable equipment they house, it becomes clear that constructing these buildings and installing their equipment is an immense challenge in its own right. And certainly not something we can turn our attention to just as the heavy trucks arrive.

Whilst sitting in a comfortable office and studying design images on a computer monitor, it is a good idea to picture a construction team out on the proposed site, surrounded by concrete and steel, and all togged out in their hard hats, high-vis jackets, stout boots, and so on. To complete the scene, depending on a facility's location, I like to think of them being buffeted by strong winds and horizontal rain, or baking under a merciless Sun. Cruel images maybe, but it reminds us that construction is a tough job and design proposals that appear simplicity itself whilst sipping coffee in an agreeable environment, may be anything but for those out in the field.

What this tells us, is that design teams must work closely with construction specialists from the outset, otherwise their carefully crafted pretty pictures will literally never get out of the ground. I will give you an example of how important it is, and to illustrate the point this one originates far away from a facility's actual location.

Major construction projects can demand thousands of heavy truck movements, spread over a period of several years, each with an ability to cause upset to those living nearby or using local highways. Invariably transport regulators will impose constraints on where and when construction vehicles can take to the roads. For example, truck movements may not be permitted at all during busy periods of the early morning and evening rush hours, and some roads may even be ruled out of bounds to construction traffic at any time of day.

Such matters may appear remote to those charged with designing P&E. However, in some cases significant restrictions on vehicle movements can contribute to adopting a strategy of maximizing off-site modularization, so as to reduce the volume of construction traffic. Of course modular design is very different to in situ assembly, so designers would need to know it was a requirement before progressing too far with their creations.

See what I mean? Construction folks have a very different mind-set to designers. They must, for they occupy a totally different world, one where everything is solid and modifications take more than a few clicks and a new reference number. This forces them down a route of meticulous planning, which is described in a construction strategy document and reflected in a detailed program, both of which demand close liaison with the design community.

12.5.5.1   Construction programming

In programming terms the construction phase brings an interesting development: for it is at this stage that a project team's original schedule is overtaken by an even more detailed version, one from the contractor responsible for constructing a facility. In fact, depending on a project's contract strategy the construction contractor, which could be a consortium or joint venture, may also take responsibility for installing a facility's P&E and indeed commissioning it, which has even more ramifications for the program. Alternatively, an engineering, procurement and construction contract, also known as a *turnkey* contract, may be let, where a client hands over pretty much all responsibility to a single delivery organization. Whatever the contract strategy, the significant factor here is that when the construction phase begins much of the detailed design which is necessary for a facility's completion will still be incomplete, or maybe not even started.

It may sound like an odd concept, to launch into the construction of a particularly complicated facility without having finalized the details. For many projects, the build stage can easily take several years, so it makes sense to get started as soon as possible rather than wait until every full stop and comma of the detail is available. It is a bit like beginning to build a house without knowing *exactly* where all electrical sockets will be located, or how its kitchen will be configured. As long as the broad principles have been established, then details can be worked out as a house is being built. Of course, whether it be on a domestic or industrial scale, timing for the release of detailed information is crucial, as even a slight delay in one activity can have knock-on consequences for others. And they can accumulate very quickly.

12.5.5.2   Phased release of information

The way it works in practice is that final details are issued by design teams in batches, starting with the lowest levels of a facility and gradually working through successively higher levels, and even zones within levels, until finally all of the necessary detail has been released in an orderly manner. This includes everything from the size and configuration of steel reinforcing bars in concrete walls, to fixing details for external wall cladding, to floor finishes and paint colors.

The original project program will have been developed around a philosophy of releasing details in this way, but it is not until the actual construction contractor becomes involved that these aspects can be refined into the finished article. For a design team there are almost always surprises, where the contractor's program dictates an early release of information which was not anticipated, say the detailed design package for a facility's external wall cladding. The contractor will have good reason, maybe connected to procurement, for their seemingly odd request for details of something that is not needed for several years, so the sooner they are integrated into the design process the better.

12.5.5.3   Equipment installation

In the United Kingdom the tendency is towards erecting a building's structure first, but leaving large construction openings through which P&E is gingerly routed when the time comes for its installation. Once they have served their purpose these openings, which are often through hefty concrete walls, are closed up and blend invisibly with their surroundings. Whereas in the United States, the preference is more heavily biased in favor in minimizing construction openings, or *block outs* and using heavy lifting gear to lower equipment in from above.

The UK approach facilitates programming flexibility, by segregating construction as much as possible from the delivery and installation of P&E, whilst the US technique can make construction and installation less problematic. However, it does demand tighter programming control in terms of equipment procurement and delivery, as late arrival could delay construction on levels above areas where the overdue equipment should have been installed. Of course neither community sticks exclusively to these preferences, but they do tend to prevail. As usual both have inherent pros and cons and ultimately both work just fine, as long as we plan ahead.

12.5.6   Commissioning

Project inception	Optioneering	Option refinement	Detail design	Construction and installation	Commissioning

Commissioning divides into two distinct phases of *inactive* followed by *active*, which is when radioactive materials are first introduced, and for major projects it can easily take a year, or even considerably longer. It is not just about testing equipment and systems to demonstrate they function in accordance with their design, which is challenging enough. It is also the time to ensure a plant operates as safely and efficiently as possible. Although a year or more might seem like a long time, there is an awful lot to do, so commissioning is certainly an activity that must be woven into a project's program from its very earliest days.

12.5.6.1 Phased commissioning

For many nuclear facilities there is a programming constraint which adds additional layers of complexity to an already demanding challenge. It stems from the fact that no one rushes headlong into spending the considerable sums required to finance a nuclear plant, so by the time construction starts there is unlikely to be the luxury of proceeding at a relaxed pace. As a result, planning engineers must work closely with construction experts and other members of a project team, to search out ways of shaving as much time as possible from a program's timeline.

It can take many years to build a new nuclear facility, then kit it out and finally commission it. So ideally the objective is to find some way of overlapping all three activities. The principle appears simple enough, in that once construction folks have finished one area of a facility, say its lower levels, other colleagues move in to begin fitting out those areas and installing equipment. And with that done other specialists can get busy with commissioning, or *phased commissioning* to be precise. Essentially this means that various items of equipment, or suites of equipment, must be capable, initially at least, of operating on a standalone basis.

The trouble is there are two major difficulties to overcome. First is straightforward conventional safety, with the dangers of equipment installation teams and commissioners accessing a full-blown construction site where tons of concrete and steel are on the move every day. Addressing this concern is difficult enough, but nevertheless is the easier of the two and centers on ensuring a facility's various construction, installation and commissioning teams are segregated as much as possible.

Second is the considerable challenge of providing services—electricity, water, gas, compressed air, steam, ventilation, and so on—to pockets of a building before other areas are even built. Developing programs to deliver operable services to isolated pockets of a building is challenging enough, but is further compounded by presence of the construction openings I mentioned a moment ago. The problem here is that we cannot route services past construction openings until they have served their purpose and been sealed up. Flexibility can be gained by using temporary services; however, it is always going to be ad hoc and no substitute for having access to a fully operational suite of permanent services.

For phased commissioning to succeed well, its preparations must begin years in advance. Crucially, the engineering community must work with commissioning specialists in formulating its implementation strategy. And along with that, designs developed by all disciplines must be tailored to accommodate a phased, or *systemized* approach to commissioning. When we put the whole thing together, it becomes clear that precise programming of construction, installation, and discrete commissioning tasks is absolutely crucial to delivering a philosophy of phased commissioning and, most importantly, to getting a facility up and running on time.

12.6 Feedback

I must admit I always look forward to visiting a nuclear site and witnessing the birth of a new facility, particularly the vast structures often demanded by the nuclear industry. Even though I may already be familiar, on paper, with what is going on around me, I am always struck by the magnitude of what it takes to turn drawings and diagrams into a functioning facility.

Whenever I can, and all designers should do this, I like to ask the commissioning folks and others how their particular job is going. Unsurprisingly, they are always keen to share their views with anyone who might be remotely connected with whatever it is they are engaged in. Apart from picking up valuable gems which we can carry to our next project, it is a salutary reminder that there is always more to learn. But then standing still would be no fun at all.

Waste management

13

We know from our domestic situation how easily it is to accumulate waste, so can readily envisage how it will grow exponentially when encountered on an industrial scale. For nuclear facilities, conventional waste gives rise to the same segregation and disposal issues encountered elsewhere, so this is where we shall begin. With that done, we shall examine the approach taken to managing radiological wastes, and finally see how the strategic approach to waste management evolves as a project progresses across its timeline.

13.1 Waste management hierarchy—conventional

Predicting conventional waste arisings and figuring out how they can best be managed is a complicated business. In the nuclear industry, as with many others, the collective term used to describe this multifaceted activity is *waste minimization*. And the starting point for addressing it is establishing a hierarchy, where the most desirable solutions sit closest to the top of an inverted pyramid (Fig. 13.1). Even though this hierarchy is adopted by all types of industry, when applied in the nuclear world it can double up to deliver reductions in radiological wastes.

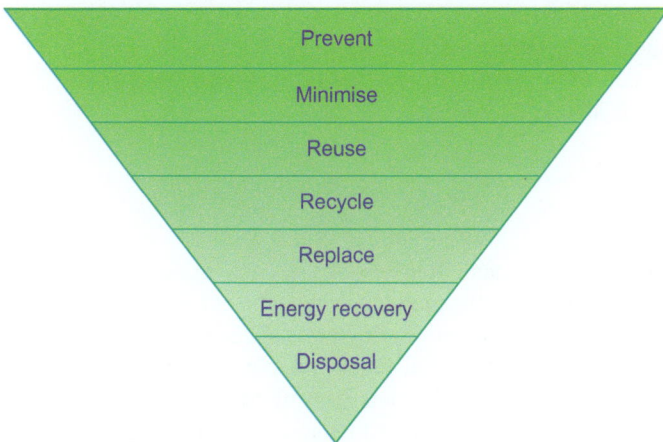

Fig. 13.1 Conventional waste management hierarchy.
© Bill Collum.

Nuclear Facilities. http://dx.doi.org/10.1016/B978-0-08-101938-2.00013-1
© 2017 Bill Collum. Published by Elsevier Ltd. All rights reserved.

13.1.1 Prevent

Also known as avoid, this is the ideal outcome for any waste management plan. If it can be achieved then it means some way has been found of circumventing an activity, or indeed an entire facility, which was destined to generate waste. It is a tall order but occasionally can be made to happen. In our case however, it is taken as read that we are considering the design and full lifecycle of an operating nuclear plant. So although it may be possible to design out discrete waste generating activities, in its widest context *prevent* is not really an option. We can however take steps to minimize any waste that does arise, so ordinarily this is our starting point.

13.1.2 Minimize

In many ways this is the simplest of themes to address, and yet it can make a significant contribution to keeping waste volumes in check. As we know C2 (green) zones within nuclear plants are not subject to airborne contamination but may have limited surface contamination in some areas. Nevertheless, any material entering these areas is viewed as *suspect contaminated* so may need to be disposed of as either very low level waste (VLLW) or low level waste (LLW). With this in mind, facilities should be configured and operated so as to minimize the potential for unnecessarily introducing materials into a radiologically controlled environment.

Obvious examples include removing packaging from consumables before they enter a nuclear plant, and wherever possible locating electrical switchgear in areas that can be accessed off-the-street, so to speak, rather than in a C2 area. In addition, and a favorite theme of mine, configuring any facility that conducts grouting operations in C4 (red) areas, to ensure grout preparation is conducted in an adjacent C1 (white) area. This has twofold benefits of allowing unused grout to be easily disposed of as non-radiological or *exempt* waste and keeping powders, along with dust hovering about them, out of a controlled environment, which, in terms of radiological housekeeping, is always a good idea.

13.1.3 Reuse

When developing complex and often bespoke designs for nuclear facilities, it is quite easy to become engrossed in the immediate challenge of making a process work, and indeed work as smoothly as possible. However, if initial concepts result in materials being used only once or just several times and then discarded, the challenge which automatically follows is to, ideally, find ways of continually reusing them.

Just as we would all be horrified at thoughts of using a saucepan once and then throwing it away, the same sort of thinking applies to industrial situations. For example, finding ways to continually reuse a waste transfer drum rather than dispose of it after only one outing, is infinitely preferable to disposing of the accumulation of drums that would otherwise arise.

13.1.4   Recycle

At first glance, reuse and recycle may appear to be quite indistinguishable or at least have a sizeable overlap, but when we look a little closer there is a clear differentiation. As we have just seen, *reuse* signifies that the same item is used over and over again, ideally indefinitely. *Recycle*, on the other hand, applies to materials which need some form of treatment, one that requires energy, either every time they are used or after a series of operations.

In the context we are discussing here, one of the ultimate examples of recycling has to be reprocessing spent nuclear fuel. Happily, there are many other opportunities to recycle which, although not on such a grand scale, can nevertheless make a significant contribution to a facility's financial performance and to minimizing its environmental impact.

A favorite approach to this theme is cleaning up wash-water from decontamination operations, so that it can be used again for the same or even different tasks. To achieve this, wash-water is routed through a process such as settling, ion exchange or evaporation. However, we need to be mindful that concentrating radioactivity in this way can generate an intermediate level waste (ILW) effluent stream. It may be small in comparison to the quantity of liquid LLW that might otherwise arise, but must still be assessed to ensure it synchronizes with site-wide waste management plans.

13.1.5   Replace

Also known as *substitute*, we participate in this particular theme every time we use a hot air hand dryer rather than paper towels, or opt for rechargeable batteries instead of disposable, or *primary* batteries as they are more properly known. In the nuclear world such opportunities do not always present themselves too readily, so waste management specialists keep us alert to the possibilities. For example, ordinarily when faced with a process which will generate contamination, the first reaction may be to clad the surrounding area in stainless steel. Of course it does a good job, but the material and installation costs are high and its disposal, as active waste, does not come cheap either.

The challenge here is to conduct timely analysis of projected activity levels and quantify the contamination that will arise during operations. In areas subject to gross contamination stainless steel is almost certainly the right answer, but as we slide down the contamination scale there may be opportunities to opt for epoxy finishes and even strippable coatings. Both perform perfectly well in the right environment and deliver significant reductions on the waste burden that would otherwise be created by using stainless steel.

13.1.6   Energy recovery

As with practically every other industry, nuclear sites are not ordinarily involved in the practice of converting conventional waste materials into energy. However, the scale of many nuclear operations does place a responsibility on sites to be scrupulous in sorting and segregating their wastes, so that others can realize the full potential of energy bound up within them. Some of the favored waste-to-energy technologies include incineration, gasification, pyrolysis and anaerobic digestion. All of which can provide heat or generate electricity from the by-products they create.

13.1.7   Disposal

Unfortunately no matter how hard we try, no matter how inventive we become, it is impossible to avoid generating at least some waste, so disposal must always figure as a potential outcome. Of course it is a last resort, but when faced with no other option we must dispatch conventional waste to a disposal site.

13.2   Waste management hierarchy—radiological

Cleary the conventional hierarchy of waste management has an important role to play, but in a nuclear world it is dwarfed by the much more challenging activity of managing radiological wastes. Again there is a hierarchy of preferences, with the ideal solution sitting on top of an inverted pyramid (Fig. 13.2).

Primary and secondary wastes

Before we look at the hierarchy itself, it is worth noting that all radiological wastes divide broadly into two categories: *primary waste* such as ion exchange resin and HEPA filters, which result directly from a facility's main operations, and importantly are foreseen by the mass balance we discussed in Chapter 6; and *secondary waste* such as redundant equipment components and protective clothing, which are a by-product of a facility's main operations and often referred to as *plant waste*.

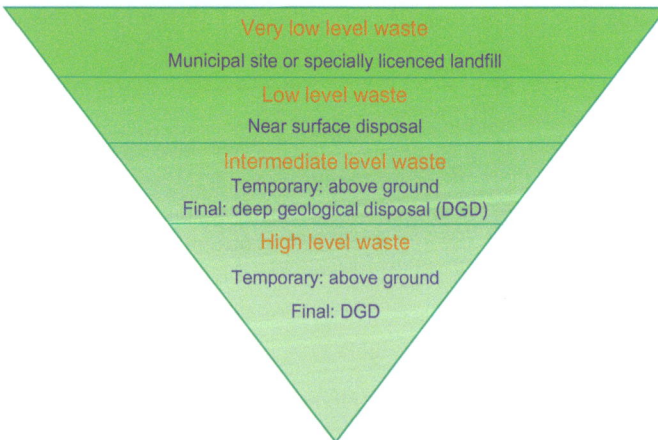

Very low level waste
Municipal site or specially licenced landfill

Low level waste
Near surface disposal

Intermediate level waste
Temporary: above ground
Final: deep geological disposal (DGD)

High level waste
Temporary: above ground
Final: DGD

Fig. 13.2 Radiological waste management hierarchy.
© Bill Collum.

13.2.1   Very low level waste (VLLW)

As its name suggests this is the lowest category of radiological waste, so low in fact that much of it arises outside the nuclear industry, originating instead from activities in laboratories, universities, hospitals, and so on. Broadly speaking it can be disposed of in two ways, with its appropriate route dictated by a combination of waste volume and specific radionuclide constituents. For small amounts, activity levels must comply with limits measured in Becquerels per cubic meter, whilst for bulk quantities the measurement is based on concentrations per ton. Whatever the overall quantity, achieving VLLW status is determined by the presence of particular isotopes, such as carbon-14 and tritium.

Small volumes, typically from non-nuclear industries and with an appropriately low radiological fingerprint, may be disposed of at municipal waste sites, whilst large volumes, with a higher radiological fingerprint, may be directed to specially licensed landfill sites. Materials in this category range from gloves and protective clothing, to soil, rubble and some metals.

13.2.2   Low level waste (LLW)

Although harboring higher activity levels than VLLW, this type of waste can invariably be manually handled. It is normally bagged or drummed at source and ideally compacted before being routed to a surface, or near surface disposal facility. Prior to being stored in its final home, LLW is usually overpacked, say in drums or ISO containers, and placed in excavated pits or large concrete vaults which, when full, are buried and mounded over.

Materials in this category can be similar to their VLLW counterparts, ranging from concrete scabblings, soil, rubble, and some metals, to all manner of decommissioning wastes. Where possible, rather than disposing of LLW it may be recycled. Metal melting, for example, can reclaim lightly contaminated metals for other commercial uses, leaving a relatively small quantity of LLW in the form of a slag.

13.2.3   Intermediate level waste (ILW)

This waste, which typically needs at least half a meter of concrete to shield it, includes some sludges, fuel cladding and other materials that have occupied reactors or radioactive caves, cells and canyons. If feasible, it is segregated from low level wastes that often accompany it and grouped with other ILW materials that share similar characteristics.

Initially ILW may be stored in its original unconditioned state, but at some stage it will be loaded into a suitable disposal container and immobilized, probably within concrete, a polymer matrix, or bitumen, in readiness for indefinite storage in a geological disposal facility. Of course such storage involves a myriad of environmental considerations, not to mention significant cost, so quite rightly there is considerable pressure on designers, plant operators and decommissioning teams to minimize this type of waste, concentrating instead on maximizing LLW and VLLW.

13.2.4   High level waste (HLW)

It is often assumed that spent nuclear fuel sits within this category. However, as we discussed in Chapter 1, where spent fuel can be retrieved from its storage facility the option always exists to reprocess it and realize latent energy harbored within it. So spent fuel does not necessarily count as waste.

In the UK then, as with several other countries, the vast majority of HLW stems from fuel reprocessing operations. Where it exists in a liquid state HLW carries a risk of leaking, which is far from ideal for long-term storage. In these situations it is solidified by vitrifying it in stainless steel containers, which makes it both easier to handle and safer to store.

13.3   Evolution of a waste management strategy

So we can see that there are logical hierarchies for the management of both conventional and radiological wastes, which is a good place to start, but how do we transition those principles into a live project? And indeed how does the wider picture of a waste management strategy evolve as a facility's design progresses?

In the previous chapter, we saw how a project's program of activities divides into six primary phases so, by way of illustration, we will home in on each of them to see what occupies the minds of waste management specialists at each step. In this way we can follow how the whole process gradually evolves and ultimately transitions into the reality of an operating facility.

13.3.1   Waste management—project inception

Project inception	Optioneering	Option refinement	Detail design	Construction & installation	Commissioning

13.3.1.1   Waste management strategy

Once a new facility is mooted, a plan is needed for how both conventional and radiological wastes will be dealt with. It goes by names such as *integrated waste strategy*, *waste management strategy* and *waste management plan*, but they all boil down to the same thing. Whatever the title, the plan's development must begin very early indeed, in fact on practically the first day that a project is even suggested as a possibility. Not only that, it must be integrated with a similarly titled waste management plan that covers the entire host site.

As we have seen, at this early stage of a project's life there may be several alternative technologies under consideration, so straightaway waste management specialists can begin an evaluation of the types and volumes of radiological waste that may be generated by each. It is far too early to produce accurate numbers, but at least a comparable analysis is possible. Apart from environmental considerations, dealing with radiological waste is expensive, so knowing one technology could create two or three times more than another will be a significant selection factor.

13.3.1.2   On-site waste treatment options

At this pre-design stage, one of the first steps is to determine if the host site can treat radiological wastes, whether solid or liquid, which would arise if a proposed facility were to become operational.

Even if appropriate treatment facilities exist, it is important to establish if they have sufficient capacity to handle the additional wastes that may be directed to them. Furthermore, for liquid wastes in particular, some consideration must be given to the distance between a proposed new facility and its prospective waste treatment plants. It matters because the active infrastructure needed to connect two facilities could be a major project it its own right, conceivably with costs even higher than the actual project being considered. We shall return to this subject in Chapter 15.

If appropriate treatment facilities do not exist, or do but are coming near to the end of their operational life, or simply do not have sufficient spare capacity, then it opens up a whole debate on justifying why the proposed new facility should even be considered. It may be, for example, that a host site has ample provision for the treatment of solid ILW, but not the additional liquid ILW that is destined to arise if a new facility were sanctioned. Factoring in the cost of a new liquor treatment plant, particularly if it has potential to serve other facilities in the future, will further complicate the debate.

13.3.1.3   National waste management strategy

As if all that wasn't enough to contend with, it is also necessary to observe constraints of a national radiological waste management strategy which, since it originates from guidance issued by the International Atomic Energy Agency (IAEA), will be enshrined in some form of binding statute (we shall discuss the IAEA's role in the next chapter).

At its highest level there must be certainty that a permanent home, such as a deep disposal repository, exists for projected radioactive wastes, or will do within an appropriate timescale. At a more detailed level, there must be confidence that wastes produced by a proposed new facility will conform to conditions for acceptance (CFA) imposed by prospective national storage facilities, otherwise those wastes will have nowhere to go, certainly in the long term.

13.3.2   Waste management—optioneering

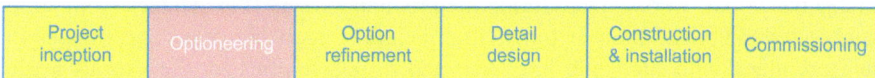

Project inception	Optioneering	Option refinement	Detail design	Construction & installation	Commissioning

Having given some consideration to strategic waste management issues during a project's inception phase, the process that will ultimately turn it into reality now begins. With no doubt several facility options to consider, it is important to build a thorough understanding of the long-term waste burden associated with each of them. Waste treatment and storage is a significant element of the option selection process, so we need to quantify differentiators and factor them in.

13.3.2.1 Cross-site integration

Importantly, before a design option can be offered as a potential solution, there must be at least agreement in principle that all downstream processing plants can indeed accept its wastes. This involves investigating CFA at proposed treatment facilities, confirming they can be satisfied and that sufficient spare capacity exists. Without all of these *building blocks* in place, it is far too risky to continue pouring effort into developing an option that could effectively be inoperable before it is even built.

13.3.2.2 Waste minimization

Unsurprisingly, the main emphasis from this point onwards is keeping waste arisings as low as possible, a task which is complicated by a need to balance comparative volumes of VLLW, LLW, ILW and HLW, not to mention their chemotoxic and non-hazardous companions. Unfortunately there are far too many variables to come up with a simple rule of thumb that can be applied during the option selection process. So design teams must work closely with specialists in developing proposals that satisfy the waste management hierarchy, both conventional and radiological.

13.3.3 Waste management—option refinement

Project inception	Optioneering	Option refinement	Detail design	Construction & installation	Commissioning

13.3.3.1 Hierarchy fine-tuning

In terms of waste generation, this is the time to revisit a waste management hierarchy to seek out ways of squeezing maximum benefit from themes that were set in motion during optioneering. There might, for example, be opportunities to replace solvent products with water based types that are less problematic to treat and dispose of, or recycle inactive cooling water so as to minimize demands on a site's water resources.

This is also the time to step up liaison with waste treatment plants, adding detail to arrangements that were agreed during the previous phase. This includes firming up batch sizes and waste transfer timetables, including probable peaks and troughs in demand, along with finalizing interfaces and responsibilities for the various waste handover operations.

13.3.4 Waste management—detail design

Project inception	Optioneering	Option refinement	Detail design	Construction & installation	Commissioning

13.3.4.1 Trials

Even at this stage there are still opportunities to hone plans that were set in train during previous phases. Let's say one of a facility's operations entails decontaminating stainless steel waste containers with high pressure water. No doubt blasting containers with plenty of water will do the trick but it might take several attempts and use an awful lot of water, which, let's not forget, has to be treated.

What we need here is to get the right balance between using as low a volume of water as possible, but with sufficiently high pressure to effectively remove any contamination that may be present. To do this well, engineers instigate decontamination trials, using varying pressure, different-shaped spray nozzles, and so on, then utilize the results of those tests to finalize their design proposals.

13.3.4.2 Embed waste management strategy

This phase also concentrates on ensuring facilities will not deviate from their prescribed waste minimization measures when they do become operational, so wherever possible automation is introduced to guarantee that decontamination and other processes are always conducted with monotonous waste minimizing regularity.

In addition to selective trials and automation, the project team also use this period to develop operating instructions which will further instill a facility's waste minimization practices into its day-to-day operations.

13.3.5 Waste management—construction and installation

Project inception	Optioneering	Option refinement	Detail design	Construction & installation	Commissioning

13.3.5.1 Waste management plan

Up to this stage, we have primarily been thinking about waste minimization in a radiological context, but that is not to say we are underestimating the contribution conventional waste can make to this same theme. This can be especially pertinent during the often lengthy construction phase which is required for many nuclear projects.

In the UK, there is a legal requirement that the build phase has its own *waste management plan*, which must be integrated with a site's overarching waste management plan and cover the full spectrum, from segregating wood, metal, plastics, and so on, to reusing excavated soil and rubble for landscaping, or as hardcore for roadways and other new build projects. As always the scale of many nuclear projects makes this an endeavor especially well worth pursuing.

13.3.6 Waste management—commissioning

Project inception	Optioneering	Option refinement	Detail design	Construction & installation	Commissioning

13.3.6.1 Refine waste management plans

This phase brings with it the opportunity to turn what has probably been years of waste management planning into a firm reality. Operating instructions are refined in order to embed good practice and ensure waste minimization is achieved with unfaltering regularity for years to come.

A key feature of this stage is the scaling up of earlier trials and experiments into a fully functioning plant. This task is not just about proving various processes will work, but is also time to investigate the possibility of making further improvements. So, for example, earlier trials may have demonstrated that a particular decontamination process can be satisfactorily conducted with 50 L of water, but then the commissioning folks will not be content at achieving this same result. They will run the process to establish if comparable results can be achieved with less water, say 40 L. It may not sound like much, but the cumulative effect of multiple minor improvements, spread over maybe decades of operations, can add up to significant reductions in the volume of waste generated by a facility.

Of course such waste management refinements take time to investigate, and where improvements are identified more time is needed to carry out the necessary modifications to plant, equipment and procedures.

If we hold that thought for a moment, and then think back to the previous chapter and our discussion on project planning, then contemplating just the single issue of *waste management* goes some way to explaining why project programs often run to tens of thousands of lines. It reinforces our recurring theme of how fantastically interwoven the whole design development process is.

Safety

14

Safety in the nuclear industry divides broadly into two categories: occupational or *conventional* safety and nuclear or *radiological* safety. If out of the blue we were asked to describe a safety issue, most of us would concentrate on the area of occupational safety. We would mention trip hazards, fire escape routes, adequate lighting, or the need to wear personal protective equipment (PPE) such as hard hats, steel toe-capped boots, and so on. These are all within the sphere of conventional safety and apply whether you are building a shopping complex, working in a warehouse, using power tools or indeed operating a nuclear plant.

On these pages we are mainly concerned with nuclear issues, but then from time to time we must branch off our main route to follow a subject which is closely related. This is one such occasion. Occupational safety is so inextricably linked to nuclear design processes that we need to examine the subject here, sufficient to gain an appreciation of the context within which nuclear safety resides.

14.1 Occupational safety

It is important to stress at the outset that safety is not just an issue for construction teams, where hard hats and the likes are on display; rather it must pervade a facility's entire lifecycle. This is particularly true for nuclear plants, where a focus on staying safe while dealing with tons of concrete and steel will one day be superseded by a regimen which is more appropriate to handling radioactive materials. In the United Kingdom, conventional safety is governed by the *Health and Safety at Work Act*. Similar statutes apply in other countries where, although details differ, broad principles are always the same.

Health and safety rules stipulate that the design of a facility should cater for safe installation, testing, commissioning, operation, maintenance, modification, and ultimately decommissioning. That covers most eventualities, so clearly these responsibilities are a wide-ranging and serious business. The good news is that designers are not alone in dealing with such matters; employers and employees also have responsibilities.

Employers are responsible for the health, safety, and welfare of their employees. So they are accountable for supplying safety information, providing training and supervision, and establishing emergency procedures. In addition they must ensure equipment is safe to use and that there is a safe working environment.

Employees must cooperate with their employer by adhering to health and safety rules which pertain to their workplace. They must follow safety procedures and must not interfere with safety equipment, or the safety features of any other plant and equipment. Interestingly, they are also responsible for letting their employer know if they spot something unsafe or downright dangerous in the workplace.

Nuclear Facilities. http://dx.doi.org/10.1016/B978-0-08-101938-2.00014-3

For example, an employee cannot ignore a loose cable or forget about a near miss; they must be proactive in delivering a safe working environment. So it works both ways and crucially is not even slightly optional for either party.

Clearly this is a theme that covers an awful lot of ground, everything from working at height to handling hazardous substances, to considerations of vibration, electricity, asbestos, machinery, construction, lifting equipment, transport, fire, explosion, and many more. And of course all of these conventional issues must be synchronized with the constraints of nuclear safety. To get a feel for the subject I have selected three issues for our examination: asphyxiation, confined spaces, and noise, but in truth others on the list would work equally well. For our purposes, these are enough to set the scene for the wider picture of how nuclear safety is addressed.

14.1.1   Asphyxiation

Take a hazard such as the use of solvent based paints or sealants. The build-up of fumes when they are being applied, particularly indoors, can induce a feeling of nausea or even risk asphyxiation for those involved. We can minimize risks by techniques such as segregating the area, installing temporary forced ventilation and ensuring operatives wear appropriate breathing gear, but dangers will always remain; it is far better to eliminate the problem's root cause.

Happily there is almost bound to be an alternative water-based paint or sealant which can meet the same performance requirements as solvent types, but without the fumes. By specifying products which are inherently safer, designers can make a danger go away, a feat which is always rather satisfying. Unfortunately, many safety concerns elude such ready solutions.

14.1.2   Confined spaces

Guidelines and regulations associated with escaping from a building are, in the main, based on risks to occupants from smoke and fire. So if you are inside an office block, shopping complex, an industrial facility, or any public building, you can be confident that should an emergency occur you could choose from at least two escape routes. Occasionally, the functions or processes within industrial facilities make it impossible to stick to this golden rule. A fairly classic example is a pump which must be located at a point lower than the processes it serves. In situations like this, gaining access to the pump may force designers to create a pocket of space below the surrounding floor. This is a confined space (Fig. 14.1).

If there is no practical alternative, then a confined space is allowable within the regulations, say for purposes of occasional maintenance and repair, but conversely it is definitely not permissible in an area regularly accessed for routine operations. Statistically, the main risk to personnel in a confined space is not fire but the atmosphere within it, so we need to be pretty careful how we go about arranging them and their surrounding areas.

Argon gas, which is sometimes used in the nuclear industry to mitigate a risk of fire, is a good example of a truly nasty hazard. It is odorless, invisible, and heavier than air. If it were to leak from a pipe that was within, or even close to, a confined space then it

Fig. 14.1 Confined space.
© Nuclear Decommissioning Authority.

would sink to the bottom of the space where an operator could be unaware that argon had replaced normal air, and suffocate in moments. To compound the danger a colleague rushing in to help, perhaps unaware of the menace, could very well suffer the same fate.

Clearly designers have some very serious responsibilities here. Wherever possible they must design out dangers to operators, but if hazards do remain then they must be minimized. Where argon is concerned, an obvious design consideration is how its pipework gets routed in the vicinity of a confined space. For sure, pipes carrying such gases must be kept well away from the area.

Ordinarily, if there is a possibility that asphyxiant gases may seep into a confined space, consideration may be given to installing a forced ventilation system which would stir up and dilute the gas. However, in the nuclear industry this can engender a new hazard, as blowing air around during maintenance operations could lift radio-active particles which would otherwise have stayed put, giving rise to an airborne radioactive contamination hazard.

There is a lot more that could be said about confined spaces, such as giving due consideration to how an injured person would be lifted out of the area, or dealing with fumes from say welding operations, but we get the idea. They really are best avoided, but when shown to be absolutely necessary then designers, safety specialists and plant operators must work in concert to develop the safest possible arrangement.

14.1.3 Noise

It may not be top of the agenda when starting out on the long road to developing and delivering a new building, but one fine day it will raise its uncompromising head. Questions such as: "What noise levels are we dealing with? What have you done to mitigate excessive noise? Why isn't that machine in a separate room?" will all be asked. The answers need to demonstrate compliance with legislation and where appropriate be backed up by evidence. If noise is not given the consideration it demands, we could well be faced with modifying equipment or even building fabric, probably during commissioning when time is bound to be at a premium. So it is best to deal with it in a timely manner and avoid the pain which invariably accompanies procrastination.

Noise is not just about volume, so when calculating its effects it is necessary to take into account three separate elements: volume, or *intensity*, which is measured in decibels (dB), frequency, measured in hertz (Hz), and duration, which is simply the time period of exposure to a sound. To simplify things for those of us who are not specialists in this arena, the combined effects of volume and frequency are wrapped up in what is known as a *dBA rating*. To make things even easier, regulations quote acceptable durations for exposure to sounds with a potentially harmful dBA rating. And unsurprisingly the higher the rating the less time we can spend exposed to it. Actually there are some parallels here to the way we deal with exposure to radiation.

A dBA rating reflects more closely the true human response to particular sounds; decibel (dB) relates to the volume factor, while "A" represents a weighting against which that volume is applied. This is where frequency enters the equation, because two sounds with the same volume but different frequencies will not be perceived by the human ear in the same way. The way it works is that although we can normally detect frequencies between 20 and 20,000 Hz, the human ear is more sensitive to sounds between 1000 and 3500 Hz. So if we listen to three sounds all with the same dB intensity, but one at 500 Hz, one at 2000 Hz, and the other at 7000 Hz, the sound at 2000 Hz will seem the loudest. Frequencies outside what we might term our core sensitivity range just do not register the same.

Thankfully, health and safety guidance makes all this good stuff easy to understand and legislation, in terms of protection, kicks in when noise levels in the workplace exceed 85 dBA. A facility operating below that level is deemed so safe that personnel are in no danger of suffering any impairment to their hearing during the working day. As noise gets further and further above 85 dBA, regulations demand increasing levels of protection for the workforce, ranging from ear protectors to suppressing noise at its source.

To give a feel for what 85 dBA sounds like, the very quietest sound is deemed to start at zero decibels which is known as the *threshold of hearing*, but most people can only perceive a sound when it gets to say 10 dB. Regular conversation is normally held at around 60 dB, while a jet engine screaming into the sky at take-off can reach 130 dB or more. We shall experience real pain when a sound reaches 130–140 dB and, even worse, our eardrums can be perforated instantly at a level of 160 dB.

An engineering community can tackle noise on two fronts, by controlling as far as possible any noise emanating from equipment, and by segregating noisy operations from those of a more tranquil nature. Let's take equipment first.

Manufacturers and suppliers of equipment should be able to provide a dBA rating for their standard range, which is fine if you need to order something "off the shelf." But then the nuclear industry uses an awful lot of custom built plant and equipment, so establishing a dBA rating during the design stage is not quite so straightforward. Fortunately the task can often be simplified somewhat, as similar equipment may already exist elsewhere. I know it sounds like a contradiction to say equipment can be unique and similar and the same time. The thing is, it may be unique in the sense that there is nothing absolutely identical, but oft-times new equipment is a variation on a previous theme. In situations where we can identify existing equipment with noise characteristics similar, if not identical, to the proposed equipment it helps considerably in early dBA assessments.

What about segregating those noisy activities? There is obviously a sliding scale with problem noises, from the 85 dBA borderline, right up to those that are literally deafening. Sources of really horrendous noise are easily identified; they will be recognized early in the design process and dealt with very comprehensively indeed. Noises that are less raucous but nevertheless still problematic may evade early detection. We can take two examples to illustrate the point.

Imagine a grinding machine or compactor, or maybe a large mechanical shredder, is specified for a facility. Everyone will know straight away that any of those will make a penetrating din when they are turned on, so when ideas for such machines are first incubated noise is bound to be on the agenda. It is so obvious it is guaranteed to get the full-blown treatment. All possible endeavors will be used to suppress noise at its source: a soundproof enclosure will be provided, whatever it takes. At the other extreme there are sounds which, although well within regulatory limits, are nevertheless quite unacceptable, so still need to be segregated in some way.

For example, I mentioned at the beginning of Chapter 4 that nuclear facilities often have an administration block appended to their main industrial plant; for the most part this is an arrangement that works very well. However, a situation to avoid here is locating noisy operations on the industrial side of a wall which divides these two buildings. If this does happen then even with sound deadening measures in place, office personnel may still be subjected to an unremitting drone in the background, or equally bothersome, a repetitive thump transmitted through the building fabric.

Of course such noises would be nowhere near the 85 dBA threshold, but would nevertheless be at least an unwelcome distraction and in extreme situations quite intolerable. Situations like these are not always too obvious during a project's early development phases, so designers must be alert to the possibilities and configure facilities to ensure office-based personnel are not tormented by nearby industrial operations.

We could continue discussing conventional safety ad infinitum, but for our purposes we have covered enough to recognize its wide-ranging nature and the importance of carefully weaving it into a nuclear facility's design, which is our primary area of interest. Now we can return to our main theme.

14.2   Nuclear safety

In the United Kingdom, nuclear or *radiological* safety groups are in the privileged position of existing because their role is effectively enshrined in law. It comes about because the International Atomic Energy Agency (IAEA) set a raft of principles by which its members must conduct nuclear related activities, with each cascading this guidance into some form of binding statute. Along with the United Kingdom more than 150 other nations are Member States of the IAEA, so they too adhere to similar legislative protocols.

14.2.1 Site license conditions

In the UK's case, the IAEA's top tier guidance is incorporated into an act of parliament, where it is translated into 36 site license conditions (SLCs) and as you might expect adherence to these conditions is mandatory for operators of all nuclear licensed sites. And indeed similar conditions apply to radiological facilities, such as research establishments and manufacturers of radiopharmaceutical products, which are not located on nuclear licensed sites. If we were asked to guess at what those 36 conditions might be, we could no doubt quickly list four or five, mainly concentrating on confining radioactive materials and ensuring operations on nuclear sites do not harm either people or the environment. Coming up with 36 could be a bit of a stretch, so clearly they must cover a lot of ground.

The full list of SLCs is shown in Fig. 14.2 and as many headings reveal little of their intent I have added a short summary against each. A study of their full text makes it clear that nuclear regulators are variously involved in issuing directives to site license holders, specifying exactly what their expectations are, approving proposals as the need arises, and ultimately, when they are entirely satisfied, issuing consents to proceed. In practice then, regulators are thoroughly consulted by potential site operators, and others, during the nuclear site license application process and, once granted, regulators continually monitor compliance with SLCs.

As their name implies, SLCs apply to the safe operation of a nuclear site. In that regard a new facility cannot simply appear in the midst of a site before license conditions kick in, so some of them apply to the design and construction of a new plant, as indeed they do to decommissioning later on. And crucially, at predetermined points along the way, nuclear regulators must grant formal approval to proceed. So not only are regulators always looking over our shoulders, but at every step of the way they have particular requirements that must be satisfied and very real power to withhold licenses, consents or approvals, or even withdraw those already granted.

Clearly, with so much at stake, it would be foolhardy to leave regulator assessments and liaison with site license holders to the whim of individuals, from any party. So as we might expect there is crystal clear direction on what it takes to satisfy regulatory requirements. This wide-ranging guidance reflects IAEA standards and, in the case of the United Kingdom, is embodied in a rather comprehensive suite of safety assessment principles (SAPs) which cover everything from safety management, to engineering principles, to radiation protection, fault analysis and many more. Furthermore, guidance on the interpretation and application of SAPs is contained in a set of technical assessment guides. All in all then SAPs, or their equivalent, play a crucial role in ensuring there is no room for misinterpretation of regulatory expectations, an attribute which is invaluable to all parties.

It is the *nuclear safety group* that has an intimate knowledge of all regulator requisites, including pertinent legislation and exactly what it takes to comply with it. Their expertise permeates every aspect of a nuclear facility's life, along with the operation of sites on which they reside. In many ways then, this group is the glue which binds all nuclear activities together and right at the heart of what they do is the nuclear safety case.

No	Site license condition	Summary
LC 1	Interpretation	Defines terms used in all other license conditions
LC 2	Marking the site boundary	Delineate licensed site boundary. Provide secure fencing and implement access control arrangements
LC 3	Restriction on dealing with the site	Ensure property transactions, possession, and occupancy are appropriately notified
LC 4	Restrictions on nuclear matter on the site	Make appropriate arrangements for bringing nuclear matter onto a site and for its safe storage
LC 5	Consignment of nuclear matter	If consigning nuclear matter to other sites, appropriately record the amount, its characteristics, packaging, and recipient
LC 6	Documents, records, authorities, and certificates	Maintain records to demonstrate compliance with site license conditions
LC 7	Incidents on the site	Make arrangements for the notification, recording, investigating, and reporting of incidents
LC 8	Warning notices	Provide notices on; meaning of warning signals, location of emergency exits, and measures to be taken in the event of a fire or other emergency
LC 9	Instructions to persons on the site	Ensure all persons on site receive instructions regarding risks, hazards, precautions, and action to be taken in event of an accident or emergency
LC 10	Training	Provide suitable training for personnel responsible for operations which may affect safety
LC 11	Emergency arrangements	Implement and rehearse accident and emergency arrangements, including integration with any off-site assistance. Ensure responsible persons are trained and appropriate equipment is available
LC 12	Duly authorized and other suitably qualified and experienced persons	Ensure that only suitably qualified and experienced personnel perform safety-related duties
LC 13	Nuclear safety committee	Establish a nuclear safety committee to oversee compliance with SLCs and arrangements for all safety-related matters, both on and off site
LC 14	Safety documentation	Produce, and ensure appropriate assessment of safety cases for design, construction, manufacture, commissioning, operation, and decommissioning
LC 15	Periodic review	Ensure safety cases are periodically reviewed and reassessed in a systematic manner
LC 16	Site plans, designs, and specifications	Maintain an up to date site plan showing the boundary and all safety-related buildings, along with a schedule that describes their operations
LC 17	Management systems	Implement management systems which ensure safety is given due priority
LC 18	Radiological protection	Ensure radiation exposure to nuclear workers is assessed and notify the regulator if dose limits are exceeded
LC 19	Construction or installation of new plant	Obtain consent before embarking on construction of new facilities or installing plant, and do not alter endorsed plans without approval
LC 20	Modification to design of plant under construction	Seek approval to modify any aspect of the design which may affect safety. Classify proposed modifications according to their safety significance and implement changes in agreed phases

(Continued)

LC 21	Commissioning	Seek approval for proposals to commission safety-related plant or processes and implement commissioning in agreed phases
LC 22	Modification or experiment on existing plant	Classify proposed modifications or experiments according to their safety significance and provide documentation to substantiate their safety. Seek approval before commencement
LC 23	Operating rules	Demonstrate the safety of proposed operations and develop operating rules to ensure safety is maintained
LC 24	Operating instructions	Ensure operations are conducted in the interests of safety, comply with operating rules and are carried out in accordance with approved written instructions
LC 25	Operational records	Maintain records of the operation, inspection, and maintenance of safety-related plant; plus the amount and location of all radioactive materials
LC 26	Control and supervision of operations	Ensure safety-related activities are supervised by formally appointed personnel, recognised as asuitably qualified and experienced person (SQEP)
LC 27	Safety mechanisms, devices, and circuits	Ensure safety-related plant is only operated, inspected, maintained, or tested when appropriate safety provisions are in place
LC 28	Examination, inspection, maintenance, and testing	Conduct regular examination, inspection, and maintenance of safety-related plant, by SQEP personnel and in accordance with the plant maintenance schedule
LC 29	Duty to carry out tests, inspections and examinations	In addition to LC 28, conduct any such tests, inspections, and examinations as deemed necessary by the regulator
LC 30	Periodic shutdown	Plant or processes shall be shut down, where it is necessary for the purposes of examination, inspection, maintenance, or testing
LC 31	Shutdown of specified operations	Shut down any plant, operation, or process as may be deemed necessary by the regulator
LC 32	Accumulation of radioactive waste	Minimise the rate of production and quantity of radioactive wastes and maintain records of accumulated wastes
LC 33	Disposal of radioactive waste	Dispose of accumulated or stored radioactive waste as directed by the regulator and in accordance with environmental permitting
LC 34	Leakage and escape of radioactive material and radioactive waste	Ensure radioactive material is adequately contained and cannot leak or escape. However, if a leak or escape does occur ensure it will be detected. Notify such incidents in accordance with LC 7
LC 35	Decommissioning	Make arrangements for the decommissioning of any safety-related plant or process. Divide decommissioning into stages and do not commence, or proceed from one stage to the next without regulator approval
LC 36	Organisational capability	Provide adequate financial and human resources to ensure the safe operation of a nuclear licensed site. And do not alter such arrangements, including organisational structure, without regulator approval

Fig. 14.2 Summary of site license conditions.
© Bill Collum. Based on data from the Office for Nuclear Regulation (ONR).

14.3 Safety case

As we make our way through these coming pages, we shall see what it takes to put a nuclear safety case together and discover it is a major project in its own right, one that can account for 10% of the entire design budget for a new facility. More than any other activity it is continually evolving through an almost unending cycle of iterations, each keeping pace with the current stage of a facility's life, while at the same time laying the safety foundations for what will follow.

In essence a nuclear safety case, normally abbreviated to "safety case," documents the radiological hazards associated with a facility, or a particular task and describes the measures which will be used to assuage them. It articulates how all relevant standards *will* be satisfied or, as it evolves, *have* been satisfied. And crucially a safety case demonstrates that risks to persons have been reduced to as low as reasonably practicable (ALARP), a subject we shall discuss later. The most notable characteristic of a safety case is the systematic way in which it identifies hazards and equally important how methodical it is in substantiating how a facility or a particular task will be safely managed.

For all of its attention to detail, the incubation and subsequent development of a safety case is quite a nebulous affair, with nuclear safety groups always working on several fronts and engaged in continual dialog with designers, constructors, plant operators, and a whole host of others, most notably the nuclear regulators themselves. But where does it all begin?

You will recall that in Chapter 6 we discussed *mass and energy balance*, which is developed primarily by the process engineering team, and established it is the basis on which a facility's design solution will be founded. Its development, this very early activity, is the point at which a nuclear safety group enters the fray. Straightaway they are assessing the primary hazards associated with materials which are destined to be handled, how radioactive they might be, whether explosive gases could be generated, might some materials be fissionable? and so on. And with that a safety case is born.

At this point I should be explaining that there are several categories of safety case, but part of that dialog touches on the role of safety committees, so we need to go there first. As usual sticking to a strict chronological order is pretty much impossible, but then an occasional diversion does no harm.

14.3.1 Safety committees

Arrangements for inculcating safety into all nuclear related activities varies a little from one nation to another and even among site licensees in the same country, but that said the same overriding principles are always observed. Namely, that in addition to complying with IAEA guidance, in the UK's case summarized in 36 SLCs, proposals with nuclear safety connotations must be scrutinized and approved by committees or personnel who were not involved in their preparation. The ways in which such oversight is structured varies, depending largely on whether the activities in question are conducted on a nuclear licensed site or a nonlicensed site. That being said, the principles are broadly similar so we shall follow the regimen adopted for licensed sites.

14.3.1.1 Nuclear safety committee

As if to underscore the all-pervading dominance of a safety case, the operation of this committee is governed by a specific SLC: in this case LC 13. It covers quite a lot of ground and stipulates, for example, that it must have a minimum of seven members, at least one of whom is independent of the site's operator. In fact, the words "independent" or "peer review" often find their way into the name of this powerful committee.

In addition, proposed members cannot join the committee until their qualifications, career history, and so on have been submitted to, and endorsed by the nuclear regulator. Its terms of reference must be approved by the regulator, as must any proposed amendments. Minutes of committee meetings must be submitted to the regulator within 14 days of taking place, and so it goes. All in all it makes for a comprehensive system, in that a widely experienced committee of what are sometimes termed "greybeards" keep a close eye on a site's safety related activities, and the nuclear regulator keeps a close eye on them.

14.3.1.2 Site safety committee

This is more of an internal governance board, again made up of highly experienced individuals such as facility managers, engineering managers, radiological protection advisors, those nominated as design authority's for the various engineering disciplines, and of course safety specialists. And whereas the *Nuclear Safety Committee* tends to concentrate on matters with clear safety implications, this group also scrutinizes proposals which, on the face of it at least, carry no radiological risk whatsoever. In other words, apart from doing what we expect safety committees to do, their remit also extends to "Let's have a look, just in case."

Now we can get back to the different categories of safety case.

14.3.2 Safety case categories

When we consider the vast range of tasks that may be conducted on a nuclear site, from simple equipment modifications, to operating a new reactor with all of its support facilities, it is clear that, in terms of a safety case, one size will not fit all. The main imperatives here are to ensure a safety case is subjected to an appropriate level of scrutiny and also suitable approval processes, which may include the safety committees we have just discussed. In recognition of the potential diversity of tasks some organizations establish safety case categories, typically four in number, with each attracting a level of rigor which is appropriate to a project's particular hazards, risks, and consequences.

Where safety cases are divided into various categories, different organizations have their own nomenclature to distinguish between them, but most opt for either 1–4 or A–D, and while some denote their most onerous level with a 1 (A) others use 4 (D). For our purposes, we shall take the numeric route (Fig. 14.3) and apply number one to the highest or most onerous category.

1	Potential to cause serious radiological harm far beyond a nuclear site's boundary
2	Potential to cause serious radiological harm in vicinity of a nuclear site
3	Potential to cause local radiological hazard
4	No inherent radiological hazard, but task conducted in a nuclear environment

Fig. 14.3 Safety case categories. © Bill Collum.

14.3.2.1 Category 1

This category is reserved for facilities with potential to cause serious and wide-scale radiological harm, therefore not limited to the confines of a nuclear site but way beyond its boundaries, possibly extending to a national or even international scale. Cleary these facilities must incorporate numerous layers of safety into their design and operating procedures. So if the *potential* for such far reaching consequences exists then this category will apply. An obvious example would be proposals for a new nuclear reactor, as unfortunately we know from historical precedents that they harbor potential to deliver catastrophe on a mighty scale.

All radiological safety documentation in this category is initially evaluated by a site's safety committee, and only when they are satisfied is it passed up to the nuclear safety committee for their weighty examination. As we might expect the minutia of detail developed and issued to them is considerable, all of which is thoroughly pored over, with no doubt further explanation, justification, and evidence called for until eventually every nuance has been satisfactorily addressed.

During its initial stages, the nuclear regulator is consulted on scope, methodology, and so on, then as each batch of safety case documentation passes its on-site scrutiny it is formally issued to them for a further round of meticulous analysis. Clearly the whole governance process is exhaustive to say the least, but then necessarily so.

14.3.2.2 Category 2

This category applies to facilities with a capacity to cause radiological harm, which although not quite at category one levels is not too far behind. As previously, we are considering *potential* consequences beyond a site's boundaries, but in this case of a more localized rather than far reaching scale. The majority of facilities that handle or store materials in the intermediate to high level radiation range are governed by this category, with those not included coming under the higher category one.

Rules of engagement, so to speak, for this category of safety case are essentially the same as those applied to category one. Any slight differences that do arise will result from inherent risks and their consequences, rather from the safety governance processes themselves. In practice then, there is little to differentiate the two.

14.3.2.3   Category 3

This category applies to facilities and operations with significantly lower radiological hazards than the two above it, so as a result we find a step change it governance processes. Ordinarily oversight is provided solely by a site safety committee, with a proviso that the nuclear safety committee and nuclear regulator both reserve the right to examine any or all aspects of documentation which is produced. Typically we could expect this category to be applied to facilities handling low level waste (LLW), such as a plant which sorts, compacts, and repackages LLW.

14.3.2.4   Category 4

This, the lowest of the categories, applies to tasks which of themselves do not carry any radiological risks. They are, however, being conducted within in a nuclear facility, or in some cases nearby, so for that reason may warrant some form of safety case. As with category three, oversight is provided by a site's safety committee with higher authorities able to step in at any time they wish, either for additional review or to conduct an audit. Tasks at this level include erecting new offices, relocating access platforms, creating a new door opening, and so on—as I say, all nonnuclear.

For all nuclear facilities, even these workaday tasks are tightly controlled by some form of facility modification procedure, with design authorities and other senior personnel being called on to review proposals and grant their approval. Invariably radiological safety considerations are incorporated into that process.

14.3.3   Evolution of a safety case

The evolution of a safety case is so multifaceted a process that describing it through a regular narrative would get very convoluted indeed, so to keep the subject manageable we shall examine it from three different perspectives. First up, and to set the scene, we shall take a high level view of its primary components: the main pieces of a safety case jigsaw, so to speak. Next, we shall look at each of those components in more detail, to see how a safety case steadily evolves throughout a project's entire lifecycle. Then later we shall take a wider view of how safety synchronizes with the bigger picture of design, construction, commissioning, and operation of a nuclear facility. Here we go.

Although somewhat of a précis, Fig. 14.4 serves a useful purpose in that it provides some context for what is about to follow. On its second row we have the familiar project phases, from inception onwards, with a safety case arching across the top and effectively binding them together. I have adopted the convention whereby a safety case changes its name to become an *operations* safety case when that stage is reached, although this is not always the case. The main message here is that the construction of a safety case begins on day one of a facility's life and is never ending.

I use the word *construction* because as we shall see, assembling a safety case is a very methodical process, with each component interlocked with those around it. It is an important image because it highlights the reality that if we do not get this right, if we do not build a rock-solid safety case, then the whole design process will ultimately crumble and our long-planned facility will literally never get out of the ground. It is a bit of a stark message, but for nuclear facilities, a "pretty good" safety case just won't cut it.

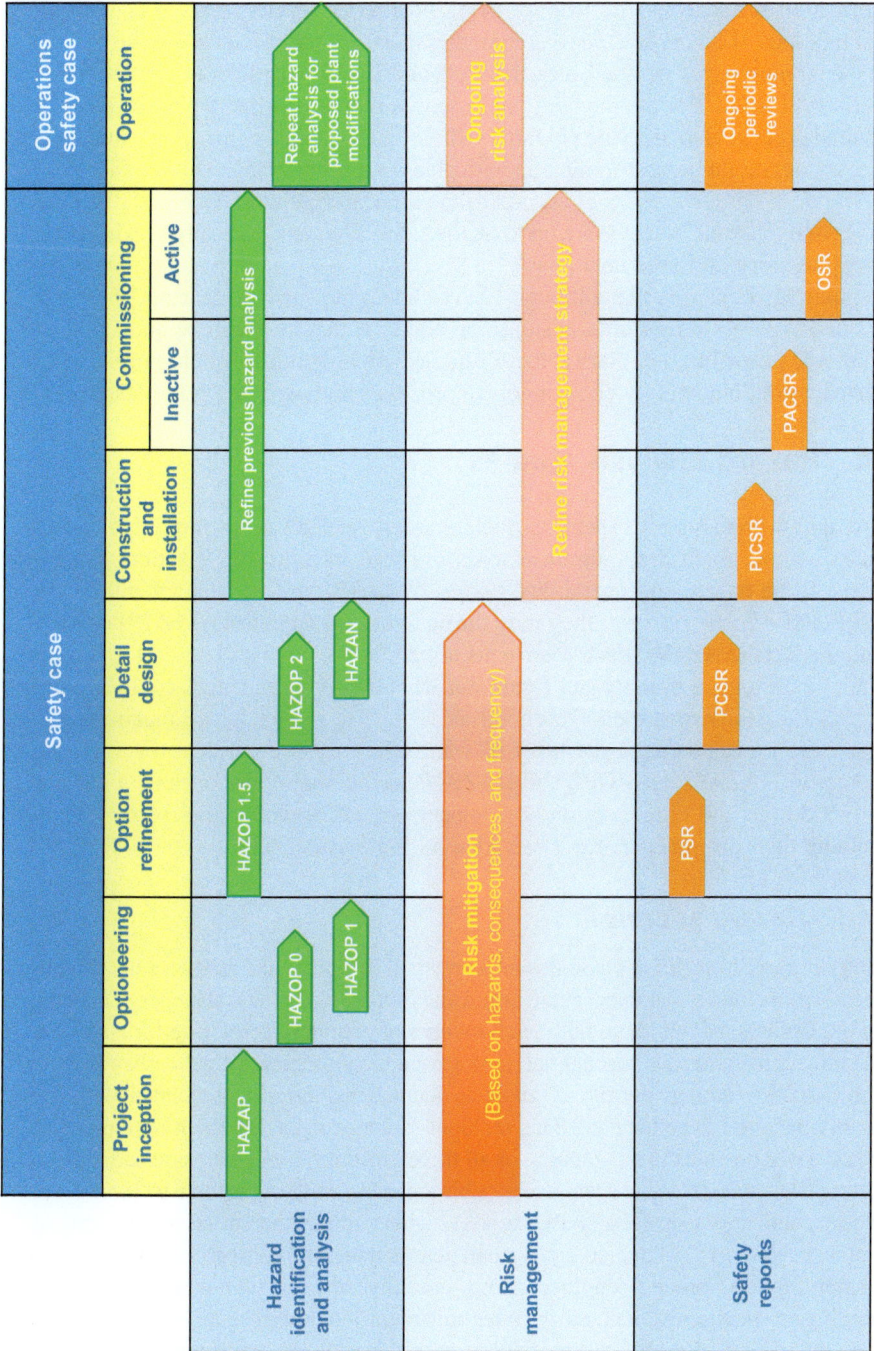

Fig. 14.4 Primary components of a safety case.
© Bill Collum.

Below the project phases in Fig. 14.4 are the big three of every safety case: hazard identification and analysis, risk management, and a series of safety reports. I am afraid that at this stage the story is a blur of acronyms, so beloved of the nuclear industry, but over the coming pages we shall pick them off one at a time and get to the bottom of what they all mean. For the moment, the essential message is that we must follow a structured process of identifying potential radiological hazards, figuring out what their consequences might be, and coming up with plans to minimize and manage risks. And most important, issue a series of reports to the nuclear regulator, each with the dual purpose of explaining what we are up to on the safety front and formally seeking their approval to carry on to the next phase.

I should also mention that although I have shown the various elements of hazard analysis (HAZAN) in neat orderly steps, the reality is that there can be quite a bit of overlap and repetition. For clarity, the figure depicts how things would unfold in an idealized world, but being such an involved process some tweaking is inevitable.

14.4   Hazard analysis studies

It is virtually impossible to speak to nuclear safety specialists for more than a few minutes without them dropping an acronym or two into the conversation. Top of the list will be hazard and operability studies (*HAZOP*), maybe supplemented by a number, and if really on a roll they may throw in hazard appraisal (*HAZAP*) or even hazard analysis (*HAZAN*). Each represents a methodology which is used to identify potential radiological hazards and figure out what their consequences could be, and in the case of HAZANs their likelihood as well. The reason for so much airtime is that these processes sit at the heart of what safety folks do. Taken together, the outcomes of HAZAPs, HAZOPs, and HAZANs have a major influence on a nuclear facility's design and ultimately on how it operates, all of which goes some way to explaining their prominence. Let's see what all the fuss is about.

14.4.1   Hazard appraisal

During a *project's inception* phase design teams will be poised and ready for the off, but have little more than a customer specification and first draft of a mass and energy balance to go on. Understandably there will be all manner of quandary clamoring for attention, from siting constraints and throughput, to potential technologies. To get early concepts off on a sound footing, priority is given to identifying hazards associated with the inventory destined to be handled. This includes factors such as levels of radiation and whether it will be alpha, beta, gamma, or all three? Might flammable gasses arise from any part of the process? What chemotoxic or biological hazards could there be? and so on.

The hazard appraisal (HAZAP) exercise, which makes an initial assessment of potential hazards, is conducted by a small team, typically comprising members of the safety group, process engineers and possibly one or two other specialists, depending on specifics of the facility under consideration. It will all be high level stuff, so we could think of a HAZAP as a first toe-in-the-water to point our eager design team in the right direction.

14.4.2 Hazard and operability studies

HAZOP studies, which analyze the safety implications of design and operation proposals, are not exclusive to nuclear projects but widely used in any industry where safety is a particular concern; in our case, we are concentrating primarily on radiological hazards. Unlike an initial HAZAP, they are a much more formal affair with precious little room for maneuver on improvisation. They are conducted in a workshop format and in essence are a structured brainstorm, led by an experienced chairperson from the safety group and assisted by a secretary of similar caliber.

Attendee competencies must satisfy criteria laid down by national standards, which are often supplemented by guidance from site-specific procedures. Apart from the safety team and process engineers, we can also expect to find customer representatives, a project's engineering manager, members of the wider engineering community, and in some cases various specialists. Most important, there will be at least one person with experience of operating comparable facilities on the host site, who may be the customer representative performing a dual role.

HAZOPs are conducted several times as the maturity of a facility's design evolves, starting during the *optioneering phase* with HAZOP 0 and moving on in later stages to the increasingly meticulous HAZOP 1 and HAZOP 2. At first they examine overall concepts but ultimately concentrate on discrete processes and even individual items of equipment. And each can run for days at a time, particularly in later stages when more detailed information is available for scrutiny.

External dose	Density
Internal dose	Pressure
Shielding	Viscosity
Temperature	Speed
Corrosion	Visibility

Fig. 14.5 Example key words.
© Bill Collum.

To ensure nothing is overlooked the chairperson uses a prescribed list of key words (Fig. 14.5) to stimulate discussion, with the whole tenor of debate based on attendees probing design and operating proposals with a series of "What ifs?" each followed by a seemingly endless stream of even more "What ifs?" To be honest even my safety pals would concede HAZOPs are a pretty tedious affair, so wise chairs keep spirits buoyed with a continual flow of hot drinks, cold water and excellent lunches. As with all HAZOPs, discussion is recorded in tabular format with any identified issues attracting actions, maybe scores of them, all of which must be resolved by a project team and subsequently examined and ratified by later HAZOPs.

14.4.2.1   HAZOP 0

A HAZOP 0 typically examines implications of the still-evolving mass and energy balance, along with the early process description which fleshes it out. In addition there may be a set of outline concept drawings, possibly at this stage for several options, which give a feel for how a facility's mechanical handling, chemical processes, and so on might be configured. The study will assess a facility's suite of early concept designs and be a key component of the option downselection process. Not only that, its deliberations also inform the design development which is about to follow.

The earlier HAZAP will have generated first thoughts on a facility's hazard management strategy, but it is here at HAZOP 0 that the strategy really starts to take shape. Most significant, from the point of view of a design team, the hazard management strategy is translated into a list of *safety functional requirements* (SFRs), a topic that we shall cover later when we discuss the *engineering schedule*.

14.4.2.2   HAZOP 1

Although design proposals are still at a concept, or optioneering stage, a HAZOP 1 team, including its appropriate specialists, will have much more information to explore. The mass and energy balance, and process description will have been further refined, and by now there will be suite of diagrams for process flow (PFD), mechanical handling (MHD), ventilation flow (VFD), and so on, along with all manner of supporting documentation. As with subsequent HAZOPs, before getting started on new material the team will ensure previous actions have either been properly closed out or superseded, say by association with a design option or technology which has since been discarded.

A HAZOP 1 methodology is largely the same as that previously employed, but for a few enhancements such as more searching key words. Once again there are clear rules of engagement, so a project team will know in advance what documentation is required and the level of definition it should have achieved. In terms of documentation, there is a fine balancing act here. With several options on the go it would be somewhat lavish to generate unnecessary detail for each of them, but then again there must be sufficient information for an informed safety analysis. Too much information and we have wasted time and money, too little and a HAZOP 1 will stall until gaps have been filled. This potential dilemma is one of many reasons why a prescriptive procedure is so important.

In addition to adding more detail to hazards previously identified, a HAZOP 1 report invariably highlights a host of new ones. This increasing level of definition and deeper comprehension of hazards is used to refine the evolving design and hazard management strategy, which apart from translating into new SFRs will no doubt result in subtle changes to many of those already recorded.

14.4.2.3   HAZOP 1.5

Conducted during the *option refinement* phase, this analysis is strictly speaking a continuation of the previous HAZOP 1 process. However, in recognition of increasing levels of definition it is often referred to, perhaps informally, as a HAZOP one and a half.

With a project reaching the stage of concentrating on a single design option, a HAZOP 1.5 study will have the luxury of examining considerably more detail than has previously been the case. By now the mass and energy balance, along with supporting diagrams, documentation, and facility layouts, will all be in robust shape and able to withstand greater scrutiny, which in turn translates into a more exhaustive hazard management strategy. Taken together then, this integrated approach to design development and hazard management allows us to confidently freeze a design in readiness for the next phase.

14.4.2.4   HAZOP 2

As hazard analysis progresses into the *detail design* phase it moves further and further from what we might term big picture issues and, with availability of increasing detail, begins to scrutinize discrete processes, particular items of equipment and ultimately individual valves, pipelines, and so on. So although membership of HAZOP 2 study teams is largely as before, we can also expect to see more specialist expertise which is appropriate to the topic being discussed. In addition, where earlier HAZOP teams will have included one or more personnel with experience of operating nuclear facilities, at this stage the team will be supplemented by operators with more specific, hands-on knowledge of the particular equipment and processes being analyzed.

The other change here is that whereas earlier HAZOPs will have concentrated on the "consequences" of hazards, with some judgement on how they might occur, a HAZOP 2 zooms into the detail of exactly what might "initiate" them. Previous HAZOPs, for example, may have considered risks associated with transferring radioactive liquors along a pipeline. At this stage, more key words will be added to the list, prompting assessment of the same pipeline in terms of risks that may arise from "no flow," "less flow," and "more flow," and not just local to the pipeline itself, but both upstream and downstream of it. It is just one example, but does give a feel for the minutia which is hunted down by a HAZOP 2.

This depth of analysis not only identifies individual initiators, but also delivers accurate data on how serious the consequences might be and the likelihood, or risk, of them being realized, which in turn enables engineering teams to focus their energies on specific initiators and find ways to stop them happening, or halt them if they do. This subject of risk, which you will have noticed is a recurring theme, is so multifaceted that it warrants a dedicated section, one we shall get to shortly.

14.4.3   Hazard analysis (HAZAN)

Once a HAZOP 2 program hits its stride, then for major projects they are conducted week in and week out for months on end, so one of the perennial difficulties facing project teams is how to deal systematically with the blizzard of data being generated, much of it repetitious.

For example, let's say a facility has three electric overhead traveling (EOT) cranes of the type discussed in Chapter 9: one with a safe working load (SWL) of 20 t, which is assigned to general purpose duties and covers most of a building's footprint; a second with a SWL of 5 t, which operates in a C3 (amber) workshop; and another with a 50 t SWL which is employed to move heavily shielded flasks in and out of a truck bay. In addition there are a dozen or so lifting beams dotted around the main building and its ancillary structures.

14.4.3.1   Fault schedule

The interesting thing here, from the information management perspective, is that all of these lifting devices will harbor common hazards, a factor which could make for inefficient, not to mention tedious, repetition in their evaluation. This is where a fault schedule (Fig. 14.6) comes in. Although it sounds like a project manager's secret dossier, it is actually a system used to rationalize information which feeds into the next stage of hazard analysis.

Using raw data generated by multiple HAZOP 2 studies, hazards or *faults* with common themes are identified and grouped into families, or *fault sequence groups*, which are transposed onto a fault schedule. This includes details of how each fault could be initiated and what the consequences might be. With that done, faults within each family can be considered together by the detailed hazard analysis (HAZANs) that follow.

In the case of cranes and lifting beams, generic hazards pertaining to, for example, dropped loads could be evaluated together, with the specific nuances of individual crane operations being considered separately. Another example might consider multiple cases of steam ejectors which are destined to drive liquors along pipelines. With these devices, a common fault might be the potential loss of steam supply, leaving liquors in transit to flow wherever gravity may take them—not something that can be left to chance. Again common themes are reviewed together by HAZANs and ultimately find their way into the safety reports which are issued to nuclear regulators.

Not only does grouping comparable faults speed along their analysis, it also guarantees commonality in how like-faults are assuaged. So a fault schedule, with its fault-family methodology feeding into consolidated HAZANs, has much to commend it.

14.4.4   Engineering schedule

From what we have seen so far it is clear that, particularly from the detail design phase onwards, there is a great deal of safety related stuff going on, from meetings and reports galore, to a continual stream of analysis, identification of risks and postulation of unwelcome consequences. So this is a good time to draw our breath, and remind ourselves that the whole point of the exercise is to figure out what we must do to ensure nuclear operations do not harm those within their facilities, or members of the public beyond a site's boundaries.

What we need therefore, is some way of marshaling all of this safety case information in a way that facilitates its transition from guidance and direction into hard risk mitigating reality, and for that we need some serious engineering. Not only that, we also need confidence that it will prevent identified risks being realized. Enter the *engineering schedule*. It is a crucial document which arches across this immense exercise and draws a whole host of safety related threads together.

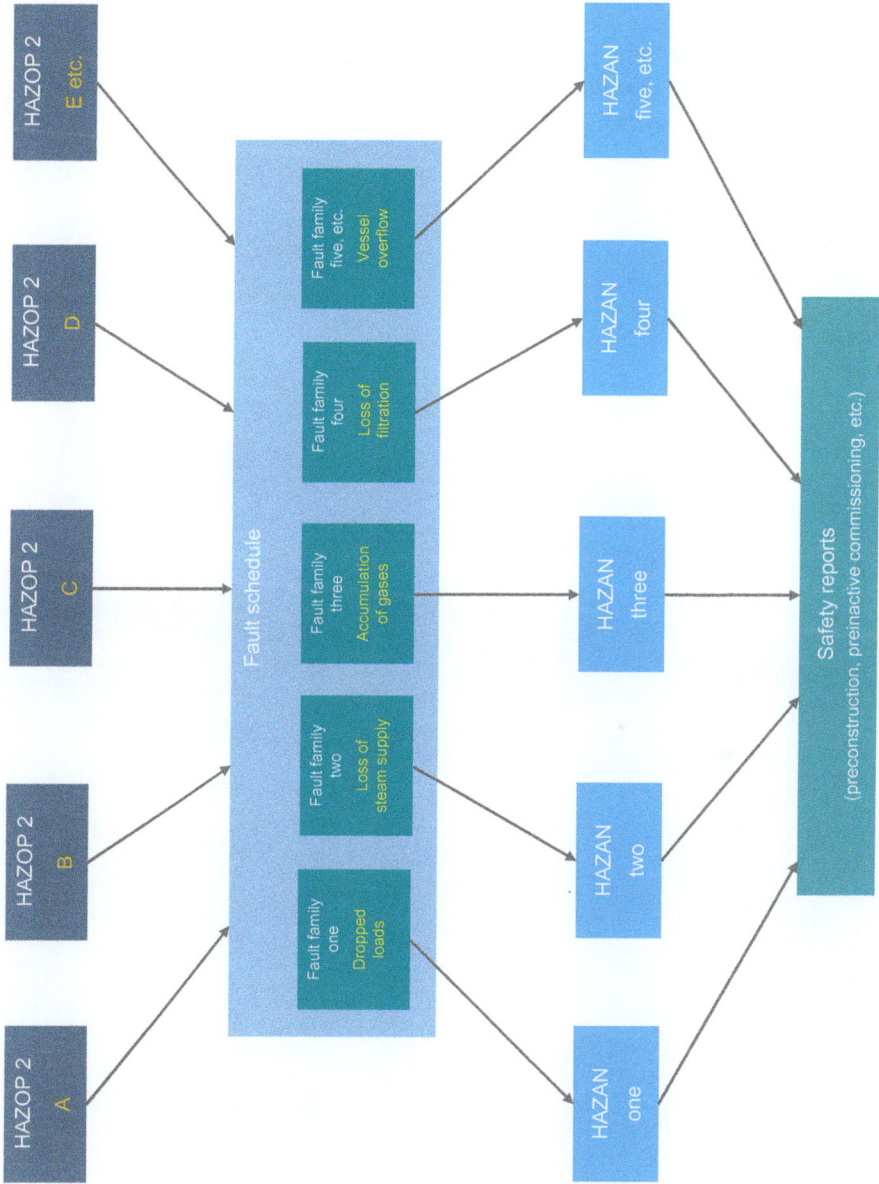

Fig. 14.6 Fault schedule.
© Bill Collum.

An engineering schedule lists all of the equipment which is classed as being "important to safety," known collectively as *structures, systems, and components* (SSCs), in other words anything that will be relied upon to prevent a fault, with its resulting hazard occurring, alternatively to halt it if it does occur, or mitigate its consequences. This database documents the lineage for each SSC, starting with the function it must provide, through to how claims made for its performance can be justified. Let's take the example of moving highly radioactive materials around (Fig. 14.7) to see how it works.

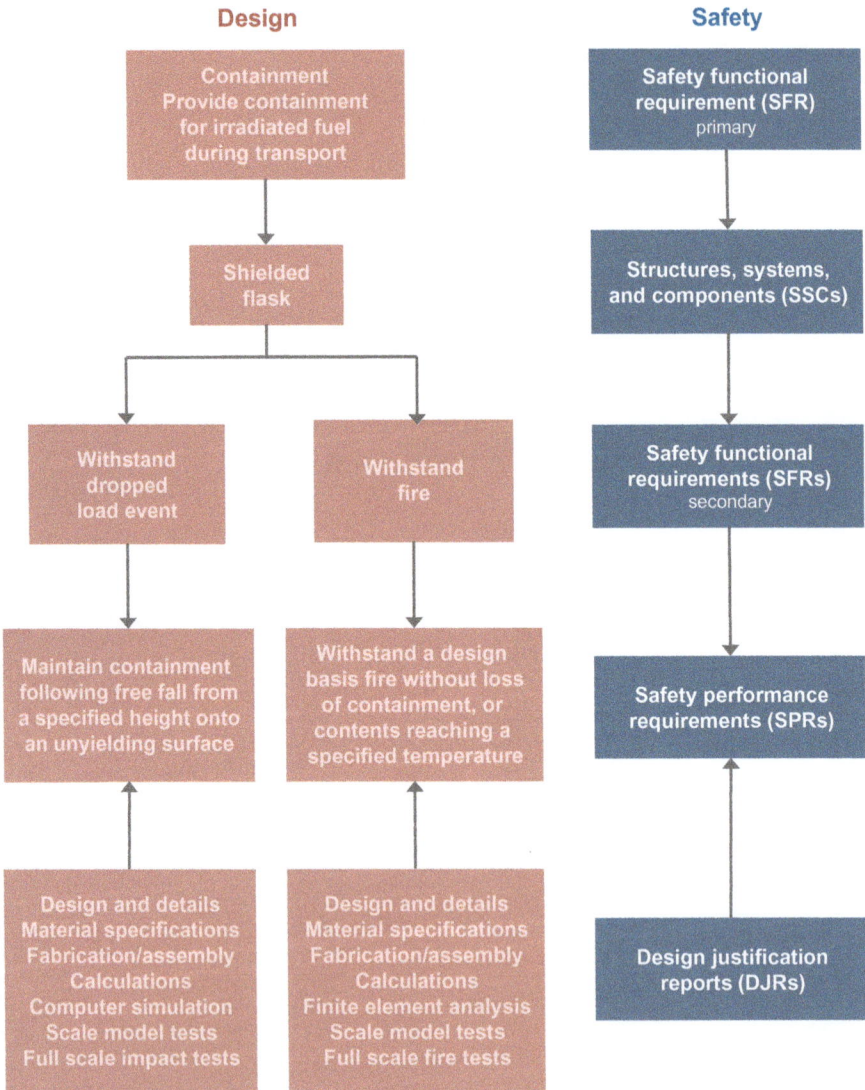

Fig. 14.7 Satisfying a safety functional requirement.
© Bill Collum.

14.4.4.1 Safety functional requirements

At its very highest level an SFR should ideally be a single word such as *containment* or *shielding*, so there is absolute clarity on the objective in question. From it cascades specific safety requirements and the means by which they will be achieved. In our example we see the not uncommon situation whereby a secondary SFR appears further down the chain, adding as it does more detail into the mix. In fact, as far as a safety case goes, this is the stage where engineering communities first receive something tangible to get their teeth into.

14.4.4.2 Structures, systems, and components

As mentioned just earlier, SSCs are responsible for fulfilling SFRs, in this case a heavily shielded steel flask which provides safe containment for irradiated fuel while in transit. Other, what are termed *passive* SSCs, include concrete shielding structures, reactor pressure vessels, and even buildings themselves, where they provide defense against extreme environmental conditions such as seismic events. The other family of SSCs is classed as *nonpassive* or *active* and pertain to alarms, sensors, and so on.

In some countries, such as the United States, SSCs may be subdivided into safety-class SSCs and safety-significant SSCs. When we see them categorized in this way, those designated safety-class attract somewhat higher evaluation criteria with regard to issues such as public protection. However, for most situations it is a point of detail, other than being aware that such a distinction can exist.

14.4.4.3 Safety performance requirements

This is where the specifics of SSCs really start to drive themselves into the design process. In effect, the SFR from which safety performance requirements (SPRs) derive is a mini design specification which spells out exactly what, in safety terms, is expected of a particular system or item of equipment. In the case of our example flask other requirements, in addition to those indicated in Fig. 14.7, may cascade out, such as an ability to withstand submersion in a specified depth of water for a given period, without loss of containment.

14.4.4.4 Design justification reports

You will notice that up to this stage the arrows on Fig. 14.7 have all been flowing downwards, with each step adding more detail to the safety requirements which must be fulfilled. Now, though, the arrows switch around, because this is where an engineering community steps up to the plate with their response. Design justification reports (DJRs) are developed to address each SPR. They encompass all of the information necessary to show what a proposed engineering solution is and, most importantly, evidence such as calculations, material specifications, analytical methods and results, test data, and so on, that demonstrate it is up to the task.

Of course it sounds quite straightforward to say it like that, in just a few words. The reality however, is that even a relatively modest DJR is a major undertaking in its own right, requiring considerable liaison between engineering teams, safety specialists, manufacturers, test facilities, not forgetting regulatory authorities and on and on. It is pertinent reminder of just how crucial the long-term planning is that we discussed in Chapter 12. It all pays off when, with the approval of each DJR, another satisfying block of the safety case slots into place.

14.4.5   Operating rules

In following the thread of activities which are used to build a safety case, from a structured HAZAN, through a fault schedule, to the engineering schedule, SFRs, SSCs, SPRs, and DJRs, we have concentrated on what might be termed its more tangible aspects. Important as they are, these elements can only succeed, can only contribute to meaningful and sustained safety, if they are enveloped within an appropriate operating regimen. Without such protocols it is analogous, potentially, to having a fabulous car equipped with an impressive array of safety devices, but then driving it badly.

We need rules, and more specifically we need a comprehensive suite of rules that govern how our painstakingly engineered safety measures will be managed within the context of an operating nuclear facility, or indeed during its subsequent decommissioning phases. In recognition of its importance this aspect of nuclear safety is covered by several site licence conditions (SLCs), most notably LC23, LC24, LC25, and LC 26.

In essence, an analysis of potential hazards associated with a facility or task, inform development of the *limits and conditions* which are necessary for the consistent application of nuclear safety. They are captured in operating rules, technical specifications, key safety management requirements, and other similarly named documents. And of course adherence to such processes and procedures is mandatory.

14.5   Risk

In normal parlance, the words "hazard" and "risk" are quite interchangeable, but since we are about to discuss them in quite a bit of detail it is worth being clear on how they are interpreted in a formal safety setting. A "hazard" applies where there is the potential to cause harm, such as oil on a tiled floor, whereas "risk" relates to the likelihood of occurrence and severity of the harm that the hazard presents.

In terms of building a safety case we have covered a lot of ground, but there is still a major component to consider, one which hinges on a combination of potential hazards, how frequently they might occur and what their consequences might be, in other words, *risk* (Fig. 14.8).

In our everyday lives, with the exception of sitting perfectly still in an ultra-secure environment, pretty much everything we do involves some form of risk, but for the

Consequences × Frequency = Risk

Fig. 14.8 Risk.
© Bill Collum.

most part we may not dwell on it overmuch. That being said, risk does occupy a sliding scale in our minds, starting in our subconsciousness and rising to center stage in our considerations, covering everything from making a cup tea, to taking a car journey, to skiing, and rock climbing.

On a personal level we can weigh up such situations, make our mind up on how to proceed and what, if any, precautions to take. In the nuclear industry, however, as with so many others, we need some form of system. It works on two parallel fronts, on the one hand doing all we can to make hazards go away, which is equivalent to telephoning a friend rather than driving around to see them. While on the other, and recognizing it is impossible to completely nullify all hazards, carefully analyzing how they might unfold and putting measures in place to stop them being realized or mitigate their effects.

14.5.1   Hazard control hierarchy

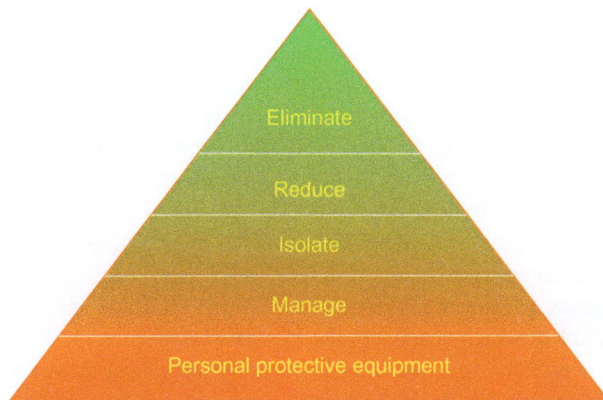

Fig. 14.9 Hazard control hierarchy.
© Bill Collum.

The first component is known as the hazard control hierarchy (Fig. 14.9). It comes in several variations, to the extent that headings can vary a little and even its trademark pyramid is sometimes shown inverted. However, its fundamental message is always the same, namely that the further we move from eliminating a hazard altogether, the more engineering and human involvement is needed to corral it, hence the greater its residual risks. When it comes to hazard control the key word to keep in mind is *passive*: the more passive, the better.

14.5.1.1   Eliminate

Clearly this is the ideal but as we might expect it is also the most difficult to achieve, particularly when we bear in mind that eliminating a hazard should not be done at the expense of creating another one, or more, somewhere else. A bit like having your groceries delivered rather than drive to a store yourself, where travel risks have not gone away but simply transferred to someone else.

In a nuclear context there are, for example, "dropped load" risks associated with picking up radioactive products and moving them around, which occasionally can be overcome by setting them down on a trolley or conveyor, and effectively skating the material from one workstation to another. In most cases, entirely eliminating a risk will result in inactivity, so we need to move to the next level.

14.5.1.2   Reduce

Also known as *substitute*, this category is all about finding ways to dilute risks so as to inhibit their ability to do harm. Oddly enough it is not uncommon to find two potential and equally valid solutions which are diametrically opposed to each other. On the one hand for instance, we may consider moving an increased payload of radioactive material from one point to another, reasoning that fewer transfers will reduce opportunities for a fault to occur, a strategy which is even more attractive if we can demonstrate that the potential consequences, even with increased inventory, would be broadly similar.

For instance, we saw in Chapter 1 that drums of intermediate level waste (ILW) often travel in four-drum stillages when being shipped to storage vaults. Not only does this improve their stacking abilities, but it also delivers a fourfold reduction in handling operations. We may even postulate that the robust construction of a stillage would help to protect drums in the event of a fall. On the other hand we could propose limiting the capacity of ILW drums to say 200 L rather than the usual 500 L, or more, arguing that although additional handling would be involved, the smaller quantities would minimize inherent risks.

Both approaches harbor the potential to reduce risk and so warrant careful evaluation to establish which is appropriate in a particular situation. Essentially then, this category concentrates on examining ways to perform hazardous operations with either reduced frequency or less radiological inventory.

14.5.1.3   Isolate

Also known as *engineering*, this strand of the hazard control hierarchy recognizes that if a fault were to occur, its consequences must be contained and not allowed to spread beyond the scene. Since we are primarily concerned here with radiological risks the most common, although not exclusive response, is shielding, usually concrete. Of course once we assign concrete an important role in a facility's safety case, then we must ensure it will not crack or crumble or otherwise let us down. So we can see a link here to the extreme environmental design criteria, particularly seismic, which we discussed in Chapter 5.

In addition, in terms of the safety case itself, isolation or *containment* of radioactive materials is classed as a primary safety functional requirement (SFR), with concrete becoming one "structure" within the group of structures, systems and components (SSCs) which are important to safety, and so on through the secondary SFRs, SPRs, and DJRs we discussed a little earlier.

Apart from the plethora of abbreviations, it is a good example of how interwoven the design development process is. It also highlights the importance, albeit counter intuitive, of not always concentrating exclusively on the particular task at hand. Rather it is a good idea to look up now and again, to ensure discrete tasks being undertaken by a project team fulfill a meaningful purpose in the grand scheme of things. In this instance, links between engineering and the safety case it must ultimately satisfy.

14.5.1.4   Manage

When dealing with significant hazards, with the exception of "eliminate," no single category from the hazard control hierarchy can be viewed as a solution in its own right. Manage, also known as *management* or *administration* is a good example. For minor hazards, such as those associated with handling very low level waste (VLLW) it can have an important role to play, but beyond that only features in support of the higher tier "reduce" or "eliminate." However, while management does not remove hazards it can make a significant contribution to risk reduction. This encompasses warning signs and training, to operating instructions and permit systems that control, for example, how maintenance tasks are carried out.

In essence this theme boils down to *behavior*, more specifically establishing processes that instill safe behavior into those responsible for the operation of nuclear plants. Occasionally you may see "behavior" listed as a separate heading in the hazard control hierarchy, but ordinarily it is incorporated within this category.

14.5.1.5   Personal protective equipment

As with conventional safety, PPE and indeed respiratory protective equipment (RPE) is worn in a radiological environment, either because a hazard most definitely exists or because the chances of one occurring are very real indeed. It covers quite a wide range, from respirators and dust masks to rubber boots, tough glasses, and gloves. Most of us would rather do without the constraints of wearing such kit, but of course when danger lurks we are eager to don the whole ensemble.

From the perspective of hazard control it comes as no surprise that this category is way down the list of preferences, so much so that proposed use of PPE or RPE for "routine" radiological operations is normally deemed unacceptable. In a nuclear context then, this kit tends to figure mostly in ad hoc decommissioning tasks, with just an occasional appearance in support of radiological safety within operating nuclear facilities.

14.5.2   Fault sequence

When it comes to combating risk we can see that the hazard control hierarchy has an important role to play, but on its own it can only achieve limited success. To get to the heart of this particular issue we need to uncover the anatomy of specific risks, so as to reveal how best to negate them. The ways in which a hazard may unfold are captured in a *fault sequence*, which is not to be confused with a fault schedule, and is a rather comprehensive way of examining hazards, determining their severity, and quantifying how to impede them. It is two pronged.

14.5.2.1   Unmitigated hazards

On their first pass examination, risks are assessed as though they were totally unrestrained and free to do their worst, with Fig. 14.10 illustrating how a typical sequence unfolds. I have shown five hazards progressing from a single initiator. However, actual numbers can vary considerably, depending on the initiator itself, characteristics of the inventory involved and particular circumstances of individual facilities.

As an example we could envision dropping a flask loaded with highly radioactive material, both liquid and solid. At this stage, analysts will assume the flask fails catastrophically, immediately releasing airborne contamination, allowing liquid effluent to seep out, radiation to shine unconstrained, and so on.

Having identified potential hazards the next step is to follow their malicious trail, beginning with radiological release paths. In other words how those hazards might spread. Typically, as in the case of dropping a flask, analysis will examine paths of aerial, liquid and radiation itself, with each hazard assessed separately against its on-site and off-site consequences, along with an estimate of initiating event frequency. Moreover the on-site category is further subdivided, firstly into consequences and initiating event frequency within facilities themselves, and then within the bounds of a site's perimeter. Remember all of this is unmitigated, so at this stage it is bound to paint a grim picture of the risks involved.

Of course everyone knows there will be an array of measures to halt such menaces, so on the face of it the whole exercise smacks of being a complete waste of time. The beauty of this exercise, and the reason it matters so much, is that it gives safety specialists and risk analysts a crystal clear picture of exactly what lurks within each hazard, not some notion which is blurred by vague assumptions about possible mitigations that might be in place. No, better to figure hazards can run amok, then we know for sure what we are up against.

14.5.2.2   Mitigated hazards

This is all very well, but the prospect of hazards leaping out from right, left, and center is unsatisfactory to say the least. So now we move to a structured process of doing something about them. The objective here is straightforward enough: put measures in place that will stop hazards occurring. But then be on the safe side, line up additional barriers just in case hazards get past the first, or subsequent ones. In other words we invoke the principle of *defense in depth*.

Fig. 14.11 shows how a risk profile is transformed by the introduction of appropriate mitigation measures. It illustrates how hazards resulting from initiating events are either stopped in their tracks or blocked by subsequent barriers. On-site hazards relate to individuals within the confines of a nuclear site, whereas off-site hazards are associated with members of the public and are assessed from two perspectives.

The first concentrates on individuals who live close to a facility so are at risk of being directly affected by a serious accident, while the second category is concerned with societal risk so takes a wider view of implications for the population as a whole. In addition to these two aspects of off-site assessment, we know from Chapter 2 that

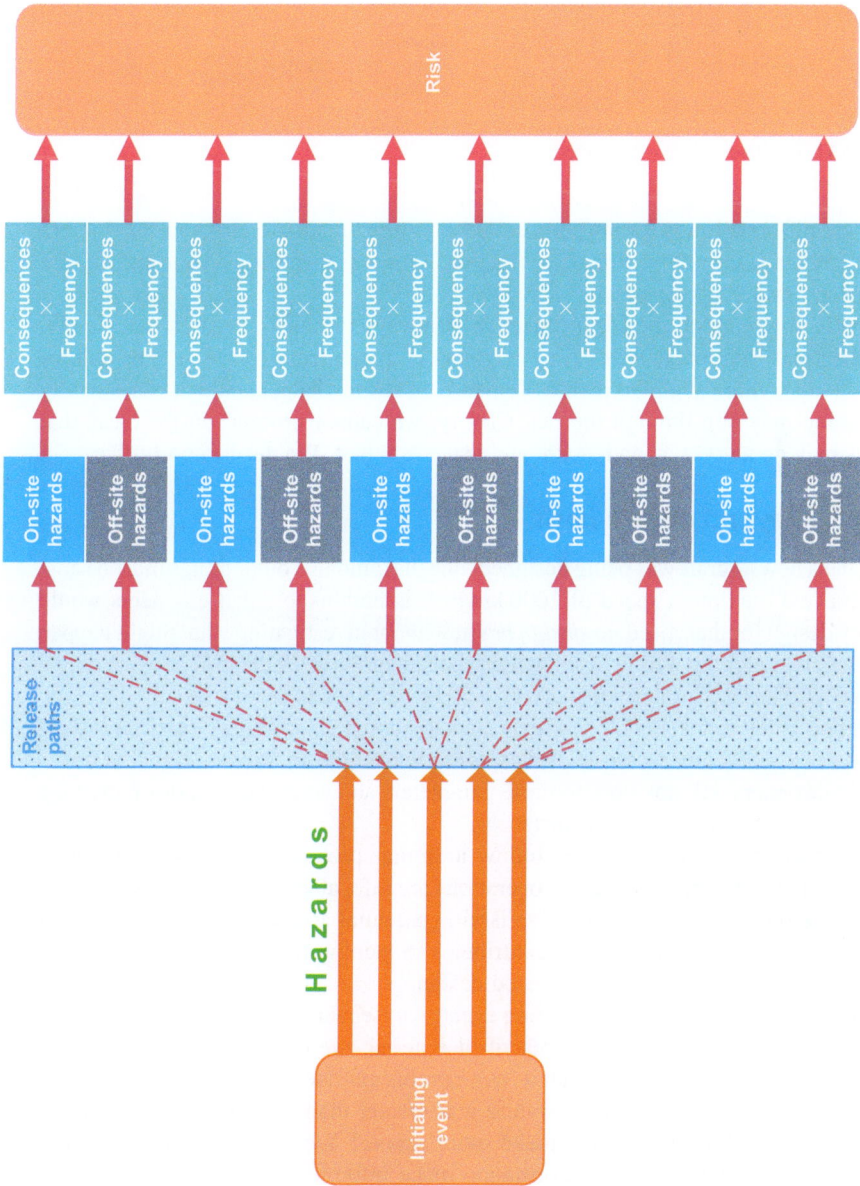

Fig. 14.10 Fault sequence—unmitigated.
© Bill Collum.

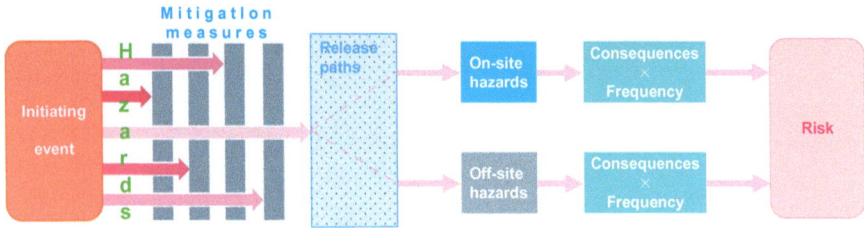

Fig. 14.11 Fault sequence—mitigated.
© Bill Collum.

there are different limits for the permissible exposure to radiation for nuclear workers and members of the public. For a combination of reasons therefore, on-site and off-site groups are evaluated separately.

Predictably enough the ideal objective is always to nullify hazards completely. But then as I mentioned earlier, the only way to guarantee zero risk in any walk of life is to do nothing at all. So realism demands an acknowledgement that occasionally something may slip through the net. Clearly, we cannot proceed on the basis that things may go awry now and again, and leave it at that. We need to go further.

14.5.3   Beyond design basis

Imagine a new aircraft was being designed, one that amongst other things must be able to achieve a maximum speed of 1000 km/h. It is unthinkable that engineers would simply design for that speed, its *design basis*, without investigating what might happen if it were to travel just a little faster. Might it disintegrate at 1050 km/h or would all be fine at 1200 km/h? Similarly the aircraft will be designed to carry a specified weight, fly in particular atmospheric conditions, and so on, but what if these conditions were exceeded? Clearly criteria are needed on which to base the aircraft's design, but just as clearly designers will not work to those parameters and wash their hands of anything even slightly outside those circumstances.

Designers of nuclear facilities follow a similar protocol, but rather than focus exclusively on ensuring a facility's operations are safe, albeit with some margin, they also delve further into the detail of satisfying potential accident sequences and their consequences. This level of rigor contributes to gaining the fullest possible understanding of the capability of a proposed design.

So far all of the postulated initiating events, consequences and frequency we have considered have been based on prescribed criterial, the *design basis,* all clinically assessed and documented. But what if future events unfolded in a way that was beyond the design basis? What if a temperature was greater than specified, a load heavier, radiation levels higher, an earthquake more severe? We need to investigate such matters, then armed with this information make informed decisions on whether or not to introduce additional mitigation measures.

In practice then a sensitivity analysis is conducted to establish, in particular, if a *beyond design basis* event could result in a *cliff-edge* effect. This occurs where just a small deviation from a plant's design basis conditions results in a sudden and large variation in its status. So, for example, loss of a cooling system combined with a higher than prescribed temperature, may result in spalling of concrete which could undermine its structural integrity and reduce essential shielding properties.

In a similar way, nuclear sites, as with any major industrial complex, must acknowledge the possibility that a severe accident may occur, one that by definition is unforeseen. So although the accident itself may be unexpected the response to it must be meticulously planned and well-rehearsed. In the United Kingdom, the response to such events is governed by site license conditions (SLCs), particularly LC 11 *Emergency Arrangements*. It, and its supporting guidance documents, stipulate that where necessary the local authority and off-site emergency services must be consulted in preparing plans, involved in rehearsals, and so on. In addition, a thorough analysis may determine that a nuclear site must provide its own emergency response capability, probably supplemented by off-site support. This may, for example, include a firefighting team and sophisticated medical treatment facilities.

14.5.4 Probabilistic risk assessment

Evaluating risks against both *design basis* and *beyond design basis events* provides essential information and informs development of appropriate risk mitigation measures. However, on their own they are not quite enough. To get the clearest possible picture of strengths and weaknesses inherent in the design and operation of a nuclear facility, we need a deeper understanding of what is more likely to happen.

For this a probabilistic risk assessment is conducted. Rather than examining risk at the limits of a design basis, or beyond design basis, it generates quantitative data on hazards, their frequency and consequences, but this time at a more "probable" level. Although of course probable here is not to be confused with "routine." The exercise of examining risk from multiple perspectives, on several tiers, is essential to ensuring risk mitigation measures are appropriate to the full spectrum of potential hazards, in all their guises.

The thing is, not all risks are equal, so we need some structure in how we evaluate and deal with them.

14.5.5 Risk classification

Imagine you are a coordinator for the fire and rescue services and that within quick succession you receive two emergence calls. First up is a garden fire that, as a result of too many damp leaves, is billowing smoke into the surrounding area. On questioning the caller you are satisfied there is no actual danger, just an unpleasant nuisance. The second call reports a serious road traffic accident involving multiple vehicles. Your response to each varies enormously. On a quiet day the garden fire may have warranted a visit, but not right now. For the road traffic accident, you launch a well-rehearsed plan that involves all of the emergency services, scores of personnel and an impressive array of highly specialized equipment.

What you are doing in this situation is responding to risk, with your actions being proportionate to the hazards involved. Exactly the same principles apply to all industrial operations, including those on nuclear sites. The prerequisite, however, when dealing with radiological hazards, is that the response to significant risks must be either permanently in place or ready to trigger at a moment's notice. And to do that, to implement a sensible pro rata response, we must first determine how serious, or otherwise, individual risks might be.

For our purposes here we can take it that there are two approaches to risk assessment. One is semiquantitative, so is used where numerical estimates of frequency are only partially available or not available at all. Likewise where there is insufficient data to measure potential consequences. In essence this approach assigns an order of magnitude to risk, albeit that, as we shall see in a moment, it is more sophisticated than a simple high, medium and low.

Quantitative risk assessments assign numerical values to frequency and consequences, so can be used to ensure specified risk-based targets are being satisfied. To achieve this they use specialized tools and techniques to identify hazards, estimate the likelihood that those hazards may be realized and the severity of their consequences.

For completeness I should mention that there is a third approach to risk assessment. Qualitative assessments are, as their name suggests, based on quality judgements so lack any kind of numerical underpinning. On the whole this type of assessment is quite a straightforward process, one based largely on informed judgement along with reference to appropriate guidance. A good example being the kind of risk assessment we might undertake before embarking on a car journey or prior to erecting scaffolding. Of course they have a place and an important one at that, but are not appropriate to the situation we are considering here.

Fig. 14.12 shows how a semiquantitative risk analysis is depicted, with each hazard rated in terms of its consequences and probability, which together determine a level of

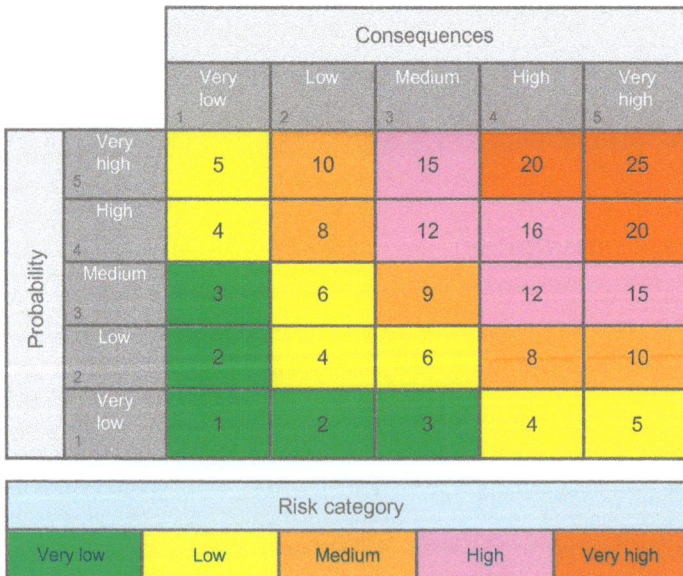

Probability \ Consequences	Very low (1)	Low (2)	Medium (3)	High (4)	Very high (5)
Very high (5)	5	10	15	20	25
High (4)	4	8	12	16	20
Medium (3)	3	6	9	12	15
Low (2)	2	4	6	8	10
Very low (1)	1	2	3	4	5

Risk category: Very low | Low | Medium | High | Very high

Fig. 14.12 Risk classification.
© Bill Collum.

risk. There are quite a few ways of configuring this exercise, with consequences often evaluated against terms such as moderate, serious and severe. And probability, or frequency, assessed against the likes of improbable, possible, anticipated, and so on. I have gone for a 5×5 matrix, from very low to very high, and incorporated the common practice of assigning a number to each of the escalating levels. Numbers on each axis can then be multiplied to give individual risks a numerical value between 1 and 25, which is a decent spread. So, for example, a hazard of medium probability (3) and high consequences (4) will attract a value of 12, which puts it in the high risk category.

If we now link back to the fault sequence (Fig. 14.10) this classification technique is first used to evaluate unmitigated risks and assign an all-important significance level to each of them—in our case between 1 and 25. Now we are really getting somewhere, because quantifying risks in this way allows us to accurately ascertain the array of obstacles needed to nullify them. It brings us to the point where we can turn to legislative guidance and site-specific procedures which specify exactly how we must mitigate particular types of risk. More specifically, we can determine how many measures will be needed to thwart each risk and how dependable, or reliable, those measures need to be.

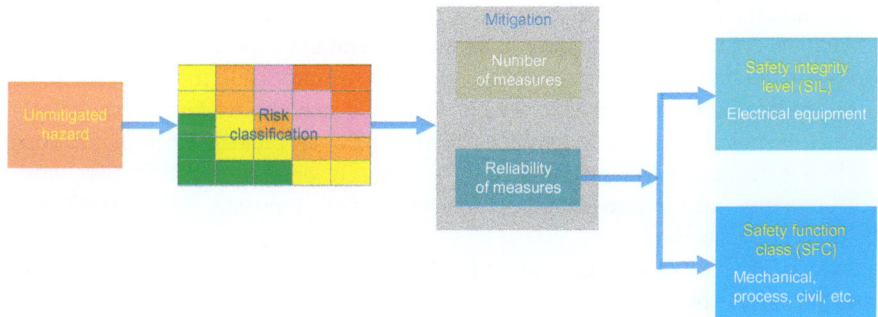

Fig. 14.13 Mitigation measures and reliability.
© Bill Collum.

This element of reliability, or "how good," divides into two distinct types (Fig. 14.13). Where reliance is placed on electrical equipment to fulfil a safety role, it is rated by its safety integrity level (SIL), discussed in Chapter 10. So, for example, hardwired equipment always scores higher than wireless systems. Everything else—mechanical equipment, process kit, civil structures, and so on—is rated via a protocol which is variously named, with *safety function class* (SFC) being a common enough example. And in both cases there are generally four levels of dependability. Whether it be a SIL, or SFC type methodology, claimed reliability levels must be corroborated by historical data and design justification reports (DJRs). And as we saw earlier (Fig. 14.7), DJRs sit within the hierarchy of safety functional requirements (SFRs), structures, systems and components (SSCs), and safety performance requirements (SPRs), so the whole risk analysis and mitigation exercise hangs together very well indeed.

To complete this stage, a mitigated fault sequence (Fig. 14.11) is run again, to ensure the proposed array of mitigation measures is up to the job. If not, the mitigation strategy is revisited, if necessarily several times, and the sequence rerun until eventually there is evidence that all hazards have been satisfactorily annulled.

As far as risk goes you would think that would be that; everyone involved would congratulate themselves on a job well done and figure it was time to go home early, but no, there is more to do.

14.6   As low as reasonably practicable

By now we have been through a comprehensive process of identifying hazards, postulating what their risks might be and putting measures in place to stop them being realized, but then up steps ALARP. It stands for *as low as reasonably practicable*, with similar processes going by names such as *so far as is reasonably practicable* (SFAIRP) and *as low as reasonably achievable* (ALARA). Whatever it is called, this technique is invoked to squeeze the last drop of diminution from residual risks that have not been entirely abated by earlier processes. In the nuclear industry, as with others, it is used to minimize conventional risks, but in our particular area of interest, namely a radiological context, is all about minimizing an operator's exposure to radiation.

To put this process into perspective we need to think back to Chapter 2 for a moment. You will recall that a nuclear worker's maximum annual exposure is limited, in most situations to 20 millisieverts (mSv) and that many organizations self-impose a limit of at least half that, so 10 mSv or less. If we convert those measurements into microsieverts (μSv) then they translate into 20,000 and 10,000 μSv, respectively.

As I say, ALARP in a nuclear sphere is all about minimizing exposure to radiation, primarily on the margins, so considers potential reductions measured in just a few microsieverts, often down to single digits. As you can imagine it is a process which is much used when considering ad hoc decommissioning tasks, but it also has an important role to play when developing proposals for new facilities, particularly (but not exclusively) in relation to maintenance activities.

That word "reasonable," which sits at the heart of an ALARP process, is a tricky one. The difficulty being it is so subjective it is bound to be perceived quite differently from one person to another. Thankfully, ALARP processes have been around for decades so by now the legal interpretation of "reasonable" is pretty well defined. In practice then, design proposals are subjected to ALARP reviews, and only endorsed when it would require "unreasonable" social and economic cost to further reduce exposure to radiation, or indeed to combat risks of a conventional nature.

You will notice that *cost* is part of the evaluation process, a factor which at first glance can appear quite inappropriate; after all we are talking about radiological risks here, so why give cost a voice? If we stand back and look at it objectively, design teams could totally disregard costs and develop proposals that cut down radiation exposure in a given situation, which sounds laudable enough. In reality however, they may have spent a fortune on reducing an operator's dose by less than would be received on a short flight. So although ALARP seeks to drive radiation exposure right down, it will draw a line when the sheer effort and cost involved would be grossly disproportionate. In other words, some common sense must be applied. Fig. 14.14 illustrates how it works.

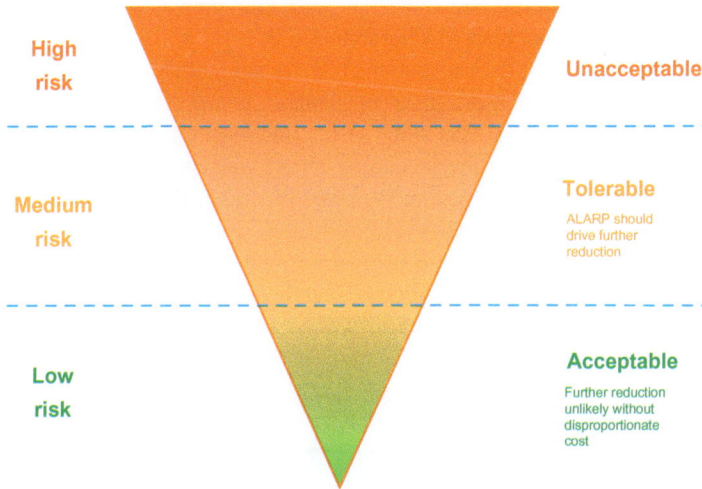

Fig. 14.14 Application of ALARP principles.
© Bill Collum.

When it comes to high risks economic factors do not enter the equation. They must be mitigated regardless of cost, so ALARP does not even make it onto the agenda. In fact if a situation arose where it was impossible to mitigate unacceptably high risks, then a proposed task, or even an entire project, would be declared unfeasible and halted straightaway.

At the other extreme, where risks are extremely low, it would be nonsensical to deploy an ALARP process that seeks to shave percentage points from radiation exposure which is barely discernible from background levels. That being said, regulators must be presented with evidence that the relevant risk estimates are valid.

In the middle, where things are not quite so clear-cut, ALARP finds much fertile ground. Here, several design options will be prepared for situations where operators may be exposed to radiation, with selection of the preferred option based on criteria such as potential to introduce additional hazards, dose to implement, reliability, along with results of a cost–benefit analysis and so on.

Looking back over recent pages, we can see there are robust processes in place for hazard identification and analysis, along with comprehensive techniques for the mitigation and management of risk. So far so good, but how do we keep nuclear regulators up to date on safety related developments? And crucially, how do we obtain their consent for what is proposed?

14.7   Safety reports

In the United Kingdom, it is a requirement of SLC 14 that safety case arrangements must be submitted to nuclear regulators for their scrutiny. They may not necessarily demand sight of every detail but, case by case, will stipulate exactly what they want to review and when they expect to see it. For significant projects regulators receive a suite of safety reports which are issued at particular points, beginning during the design phase and on throughout its operational life.

Safety reports are generally issued to the nuclear regulator's office towards the end of a project's prevailing phase: *option refinement*, *detail design*, and so on. They summarize what we might call the story so far, and highlight any pertinent changes that may have occurred since the preceding report was issued. Their main purpose is twofold. Firstly safety reports specify, stage on stage, the tests that will carried out and criteria against which performance will be measured. In other words they give regulators a clear picture of how confidence in safety measures will be robustly demonstrated. Secondly, and with a proviso that the first step has been achieved, they seek approval to embark on the next project phase. Many of their pages are therefore dedicated to presenting a safety case for what is about to follow.

Conventional	
Chemotoxic	Fire safety
Human factors	Environmental assessments
Construction management	
Radiological	
Radiological zoning	Radiological assessments
Dose uptake	Criticality assessments
Shielding	Hazard analysis
ALARP studies	Risk management
Aerial discharges	Effluent discharges
Waste management	Nuclear fire safety

Fig. 14.15 Safety reports—common themes.
© Bill Collum.

Clearly the titles and exact purpose of safety documents varies from one to another, but having said that they do embrace common themes, examples of which are shown in Fig. 14.15. And in concert with our recurring safety case tenet, these reports are not limited to the confines of a particular facility but will, where appropriate, range across a site and even way beyond its boundaries. Aerial and effluent discharges would be good examples.

14.7.1   Preliminary safety report

Issued towards the end of a project's option refinement phase, a preliminary safety report (PSR) represents the first foray into what we might term the safety case proper. Up to this point project teams have been agreeing the game plan with regulators, identifying high level hazards and developing first thoughts on a hazard management strategy. A PSR consolidates all of this information, along with provisional estimates of what the consequences of hazards might be and safety measures that may be applied to them. In addition, this is the stage at which a provisional category is assigned to the safety case itself, selected from the four levels we examined earlier. For our discussion here we are assuming the most onerous situations, so will take it that we are dealing with a category one or two.

14.7.2   Preconstruction safety report

In the United Kingdom, SLC 19 stipulates that before construction of a new plant can begin, consent must be granted by the nuclear regulator. The vehicle for obtaining that consent is a preconstruction safety report (PCSR), which is issued towards the end of a project's detail design phase. As with all such reports there is a lot riding on it being well received, so it is not the kind of document you want to issue at the last minute and hope for the best. With this in mind, before the report is issued there will have been months, maybe even years of liaison with the regulator, possibly including advance issue of some of the report's sections. Even so, it can still take 3 months or more to obtain the formal go-ahead. All in all then, there are plenty of reasons for putting a lot of effort into this report and getting it out on time.

Why all the fuss, though? After all digging large holes, pouring concrete, and so on, is all routine stuff for the construction folks. Yes there are lots of *conventional* safety issues to consider, but why are nuclear regulators so interested? Why, for example, do they even need to grant consent for the site preparation activities which precede actual construction? I know these are valid questions, because when I first heard of a PCSR, I went through similar thoughts myself. Regulators are looking for assurance that a project team has conducted a rigorous process of identifying potential *radiological* hazards, and evidence that there will be measures in place to circumvent them, or better still avoid them altogether.

In terms of radiological safety a regulator's interest divides into three areas, starting predictably enough within the confines a proposed new facility. Of particular relevance to our discussion here, their second area of interest focuses on the construction site, which will almost certainly be considerably larger than a facility's final footprint. A PCSR must present evidence of conducting comprehensive site surveys, which together identify all existing structures and services, both above and below ground. An important factor here is whether anything currently on a site, or passing through it, plays a role in the operation and safety case of facilities located elsewhere on the wider site. An obvious example would be active drainage systems, but water, steam and electricity can be equally important.

A regulator's third area of interest begins just outside the construction site and extends up to a nuclear site's boundary. Factors here will pertain firstly to whether construction activities may pose a danger to adjacent facilities or site infrastructure, such as the possibility of a crane toppling over, or a tall scaffold collapsing. Beyond this there will be wider site safety issues to consider, such as vehicle movements to and from the construction site. Interest here will center on a possibility that large, heavy payloads may collide with safety related plant and infrastructure, or indeed endanger buried utilities and active drainage systems.

A PCSR then is all about substantiating a project's grasp of the characteristics and purpose of everything on, or connected with a construction site and, for that which is *important to safety,* describing plans to ensure it will not be disturbed, or failing that, how it will be modified in a safety compliant manner. To make things doubly interesting, if there are plans to tinker in any way with something that contributes to the safety of other plants, then they in turn must demonstrate to regulators that their safety will not be compromised either during or after the modifications. And of course all of this must be accomplished within a timescale that suits a new plant's construction program.

Needless to say, once development of a PCSR is up and running there will be plenty of on-site liaising and coordinating to do, not least of which will be the vexing question of who pays when a new facility's project team is forced into upgrading infrastructure, which just happens to cross their site on its way to another plant. Apart from anything else, time taken to negotiate responsibilities and interfaces can delay the otherwise steady progress of a safety case.

To get to the stage of having sufficient confidence to submit a PCSR, a building's layout must be firmly locked down. That is not to say every detail must be frozen, but certainly all of its major structural elements must be. And to do that the rest of the design, from all disciplines, must be very well advanced. By implication, all HAZOP 2 studies must have been completed, which means a comprehensive risk management strategy must be in place. And so it goes.

Such an array of interlocking activities, all feeding into a PCSR, is one of the reasons design change procedures are so important; they play an essential role in ensuring that the safety intent of approved designs is not inadvertently undermined by subsequent modifications. Having said that, care is needed to ensure change control procedures are introduced at just the right time. Too early and they will hinder a design's healthy evolution, too late and uncoordinated changes can wreak all manner of havoc, and not just with a safety case. My personal view is that rigorous design change procedures should not be introduced until a project's optioneering phase is completed and a preferred design selected. Up to that stage things are far too fluid to make full-blown change control a practical proposition.

As I say, on the face of it, the notion of a PCSR does not sound too demanding; a few words about upcoming construction and we are good to go. In reality, it is a major undertaking, one that demands a mighty and carefully planned effort from all involved. The prize is regulatory approval, without which a project cannot progress beyond its detail design phase, which is not a great place to stop a job.

14.7.3   Preinactive commissioning safety report

As with the preceding PCSR, a preinactive commissioning safety report (PICSR) is submitted towards the end of a project's prevailing phase, in this case construction and installation. Again the nuclear regulator will need several months to assimilate and evaluate its content and, assuming they endorse it, grant a license to proceed with inactive commissioning of the, by now, largely constructed facility.

Before it launches into a discussion on proposed inactive commissioning, a PICSR will summarize how life has moved on since the PCSR was issued. This covers everything from design changes, to the status of a facility's ongoing construction and includes an update on safety significant equipment, such as construction cranes, which have now left the scene.

In the main a PICSR steers clear of regurgitating discussion on tasks which have been successfully concluded, as they will have featured in the earlier PCSR. Instead it concentrates on presenting a refreshed safety update for the present day. Where necessary this will include revisions to safety assessments, such as loss of containment, or dropped loads, which are yet to come to the fore. With the scene set, a PICSR moves onto a discourse on how upcoming inactive commissioning will be safely accomplished.

When you get right down to it, much of what happens during inactive commissioning takes place within the confines of the facility in question and is covered by conventional safety measures, but not everything. For example a temporary electric overhead travelling (EOT) crane, used solely for construction purposes, may need to be removed from a facility while inactive commissioning is underway. If this is the case then a PICSR must expound plans for how adjacent facilities and safety related infrastructure will be protected when that operation takes place.

Further afield, it is unlikely that the control systems for any new facility will be entirely isolated, or *standalone* from those round about. In fact it is common practice to mimic alarms and safe shutdown systems in distant facilities, or even a central emergency control room. In themselves such connections are inactive so can be performed during the forthcoming phase. It goes without saying that modifying satellite control systems, which are already fulfilling safety functions for other facilities, is a serious business. For this reason a PICSR must spell out how these tasks will be accomplished, and provide substantiated reassurance that existing safety cases will not be compromised.

As if inactive commissioning was not complicated enough, there are almost always a couple of additional and connected factors which add nicely to the challenge. As I mentioned in Chapter 12, once a project moves into its implementation phase it is unlikely that time will be in generous supply. As a result, one of the favored hurry-along techniques is to run with some form of concurrent construction, plant installation, and inactive commissioning. Then just to ratchet things up another notch (the second factor) it will also be necessary to phase activities so as to energize and test equipment in one area while still installing kit in another.

Such challenges are certainly not unique to the nuclear industry, but that said, concurrently constructing, installing and testing kit while in close proximity to operating nuclear plants, or infrastructure that supports their safety case, does impose some additional complications, all of which must be identified and comprehensively managed. As usual the regulator will demand reassurance, and a PICSR is where they must find it.

14.7.4   Preactive commissioning safety report

Following the now familiar pattern, a preactive commissioning safety report (PACSR) is issued to regulators during a facility's preceding inactive commissioning phase, in this case forming part of the submission which seeks a license to commence active commissioning. Once again, it will not revisit old ground which has already been thoroughly covered by the PICSR, or indeed the PCSR which preceded it. Safety cases can be hefty enough documents to begin with so regulators will not be too enamored by spurious repetition.

We can, however, be certain that inactive commissioning will have revealed a requirement to modify some elements of existing operational procedures and safety related equipment. So updates on these, along with any live safety assessments, such as say loss of the ventilation system, will legitimately be included in a PACSR. Regulators will have been kept in the loop on such developments, but it is this document which brings them all together and delivers a coordinated rationale on how recent developments feed into an evolving safety case. With the regulator bang up to date, a PACSR can move on to expounding plans for active commissioning.

Before we get into that, it is worth mentioning that development of PICSRs and PACSRs is not quite a one-size-fits-all activity. For major facilities with a significant radiological inventory, they will definitely be prepared separately and issued some months apart. However for small, relatively straightforward projects, with little by way of radiological hazards, there can be a case for streamlining the process and combining the two into a single document. That said, I must stress that such an approach cannot be presumed as a foregone conclusion, rather it must be formally agreed with regulators during the discussions which take place at a project's inception. Where regulators do agree to such a course they will almost certainly demand hold points between the two activities. For our purposes here, we shall continue with the more representative scenario of separate reports for each commissioning phase.

On day one of its active commissioning phase a nuclear facility is still radiologically benign, but by the end of the process, which can take 6 months or a good deal more, it will house radioactive materials and demand all of the paraphernalia necessary to keep them safely contained. Due to its nature the maxim throughout this entire phase is *incremental*. That is to say, we begin with limited sources of radiation in specific areas and gradually escalate their presence until finally a facility reaches its operational conditions. The methodology, including tests, by which these escalating steps will be negotiated is detailed in the PACSR. To close the loop, so to speak, the results of those tests are presented back to regulators, often in the form of a Post ACSR. Let's look at an example.

Let's say we have a shield door that during a plant's operational phase will segregate a highly active cave (C4/R4) from an adjacent equipment maintenance bay (C3/R3). One of the dangers here is that the door may inadvertently be opened while maintenance personnel are on the other side, potentially exposing them to life-threatening levels of radiation. As we know, in such circumstances it would be unacceptable to rely on a single safety measure to keep operators from harm's way, so we invoke the principle of multiple mitigation measures discussed earlier.

In this particular situation there will be mandatory operating procedures, including visual checks of the area, which must be satisfactorily concluded prior to opening the shield door. This is not a bad place to start, but the specter of human error demands more. In addition then, there will be interlocks that prevent entry via the maintenance area's personnel access door while its shield door is open. Better, but not quite good enough.

A thorough safety analysis will hypothesize that, for whatever reason, these measures have both failed and the shield door is instructed to open while personnel are in the area. To circumvent the possibility of harm, a gamma detector is installed within the adjoining cave and just behind the shield door. The detector is linked to systems which prevent the door opening, if it senses harmful levels of radiation while there is a possibility the maintenance area may be occupied. So now we have an acceptable spread of safety measures but to test this last one, the gamma detector, we need a source of radiation.

Fig. 14.16 Sealed radioactive source.
Science Photo Library.

During active commissioning a sealed radioactive source (Fig. 14.16) usually cobalt-60, will be introduced into our example cave and placed behind its shield door. Using in-cave equipment such as a manipulator, the container is opened and its radio-active source exposed. With preparations complete, the shield door is instructed to open then tests conducted to ensure the inquisitive gamma detector, and associated safety systems, stop the door opening if there is any possibility of harming personnel in the area.

In a similar way radiation sources are imported into caves and other shielded environments, starting at a weak level and gradually increasing their intensity. These sources are moved around to test systems, while at the same time monitoring radiation levels in areas immediately outside shielded enclosures. Quite apart from operator training, this incremental approach gradually builds confidence in a facility's containment, equipment, and operating procedures, until eventually it is deemed safe to transition into handling the actual inventory for which it was designed.

14.7.5 Operations safety report (OSR)

By now, several years will have passed since a new facility was first conceived, but finally the finishing line is in sight, so our weary project team must spruce themselves up for one last push.

Towards the end of active commissioning they will add another tier to the already burgeoning safety documentation, one which spells out how those operating a facility will adhere, day in and day out, to practices which are mandated by its carefully constructed safety case. This covers not just operating procedures, but also plant maintenance schedules, work permit systems, plant modification procedures, and in fact anything that contributes to ensuring a facility adheres to its safety regimen. And as we might expect, all of these processes are monitored by regulators during the two commissioning phases.

Once we reach this point an application can be made to the nuclear regulator for consent to operate the facility: a big moment to say the least. Understandably the process of granting consent can take around 3 months or so, during which our shiny new facility is raring to go. To take advantage of the lull there will be a full program of inductions, operator training, plant familiarization, and all of the other preparations necessary to get everyone ready up to speed. Finally then, when consent does come through, a facility's operations team will be on their marks and ready to go.

14.8 Periodic reviews

There is no doubt that gaining approval to operate a nuclear facility is a major milestone, but from a safety case perspective not too much changes. The emphasis simply switches from demonstrating a facility "will" operate safely when that day finally arrives, to proving it "is" doing so and will continue in that vein until operations cease and it transitions into the various phases of decommissioning, which of course will warrant further iterations of the safety case.

Once a plant is operating nuclear regulators continue to be ever vigilant and kept in the loop on developments. And in keeping with the preoperational philosophy of consolidating safety related matters into formal reports, an operations safety case is updated and reissued on both a short-term and long-term cycle. As we might expect this is not optional. In the United Kingdom, the requirement for periodic reviews of live safety cases is covered by SLC 15.

14.8.1 Short-term reviews

The main purpose of these reviews, which are normally held on a 3-year cycle, is to affirm that a facility's safety case, including updates, is still fit for purpose. To do this effectively it does not simply provide a snapshot of the prevailing moment in time, but looks back over the previous 3 years and forward for the same period. Crucially these reviews revisit the operational conditions and assumptions on which a facility's design was founded, essentially verifying that its safety case is not being undermined by subsequent modifications.

For example, are there any changes to the inventory being processed, particularly its radiological characteristics? Is throughput in line with original projections? Has it increased or equally valid decreased? Are cranes performing duties for which they were designed? Are assumptions on which the prevailing safety case is based still valid? and so on. If any of the original parameters have changed, or are about to, then an analysis of the potential effects must be initiated. As we might expect this includes consideration of safety related issues, which in turn may result in modifications to a facility's operations safety case.

Even though all such matters are addressed individually as they arise, a 3-year review lays out the whole picture in a single report and provides assurance that appropriate hazard analysis and risk management processes have been followed. Not only that, the review extends to a search for potential cumulative effects of multiple plant modifications, and again ensures the integrity of an operations safety case is not compromised.

14.8.2 Long-term reviews

While these reviews, which tend to be held at around 10-year intervals, visit ground covered by their short-term counterparts they also add a more strategic dimension. For example, everything associated with a now-operating facility, from its construction, to the technology deployed, to maintenance procedures and much more, will be have been based on design codes, standards and best practice which prevailed at the time it was conceived and assembled, but then such factors never stand still, so are continually being tweaked, upgraded, and variously improved. This major analysis considers the safety ramifications for a plant as it now exists, versus how it would differ if being developed at the present time, and makes recommendations on what, if any, action is required. There are four possible outcomes.

Where adopting current best practice, say slight enhancements in manipulator design, would have no effect on safety, then there is no imperative to make changes and all can stay as is. At the next level, say improvements in liquor filtration processes, it may well be appropriate to modify or replace existing equipment. However, where benefits are marginal it is not necessary to implement changes as a matter of urgency; rather a deadline is agreed with regulators, one that may, for example, coincide with a facility's annual maintenance shutdown. The reasoning goes that tackling nonurgent plant modifications as a one-off exercise could create more hazards than it assuages, so it is advisable to wait.

At the third level, which is thankfully a rare event, it may be that the potential consequences of not implementing newly identified best practice are so far reaching that a facility's operations must be halted until the necessary modifications have been made. This could, for example, result from lessons learned at other nuclear facilities. It may involve physical modifications to a plant, but could just as legitimately be a matter of implementing procedural changes. Whatever the root cause, everything stops until changes are implemented and regulators are satisfied it is safe to start up again.

Our fourth possibility does not quite fit the standard pattern of, (1) no action required, (2) action but nonurgent, or (3) stop until safe to proceed. This category covers situations where upgrading to current standards is physically impossible, a classic example being changes to design codes for structural steelwork or reinforced concrete. Clearly we cannot sensibly set about replacing steelwork members or modifying rebar, but what we can do is check whether the safety of an operating facility may be compromised by a structure which already exists.

The imperative here is to run calculations which take account of new design codes, while at the same time concentrating on specific requirements of an existing facility's operations safety case, such as maintaining its radiological containment, withstanding a specified seismic event, and so on. If, as is usually the case, it can be demonstrated that an existing structure is up to the job then there is no problem with a facility continuing its operations. One of the factors that tends to help in such assessments is the conservatism which will invariably have been built-in to original calculations. That said, if a facility were to fail its revalidation then it would need to either curtail activities in some way, maybe reducing its radiological inventory, or cease operations altogether.

Earlier in this chapter, I mentioned that the safety case can account for maybe 10% of a nuclear facility's entire design budget, a number which at first glance may appear somewhat disproportionate. Having peeled back a few layers and gained an appreciation of what it entails, I am sure we have to concede that it represents excellent value. This is particularly true when we consider that a safety case provides the framework around which a nuclear facility is built, then continues to support its operations and ultimately decommissioning. In the nuclear world, if we do not have an unshakable safety case then we do not have a facility. It's as simple as that.

14.9   Safety case integration

There we have it; we have covered the safety case in some detail and seen how it weaves its way into practically every aspect of a project's life, from early concepts through to a facility's operations and beyond. To complete the picture, I have prepared seven diagrams (Figs. 14.17–14.23) one for each of a project's primary phases. We do not need to discuss them in detail, since much of their content is covered elsewhere in this or other chapters and, with that context, the remainder is fairly self-explanatory.

Their purpose is to give some additional perspective to a safety case by showing it within the context of a project's overall hierarchy of activities. It reinforces the message that safety considerations dominate a project from the day it is first conceived. During the design period it drives a facility's engineering solution, then once operational governs a considerable spectrum of day-to-day activities. Yes, a safety case is all-pervading to say the least.

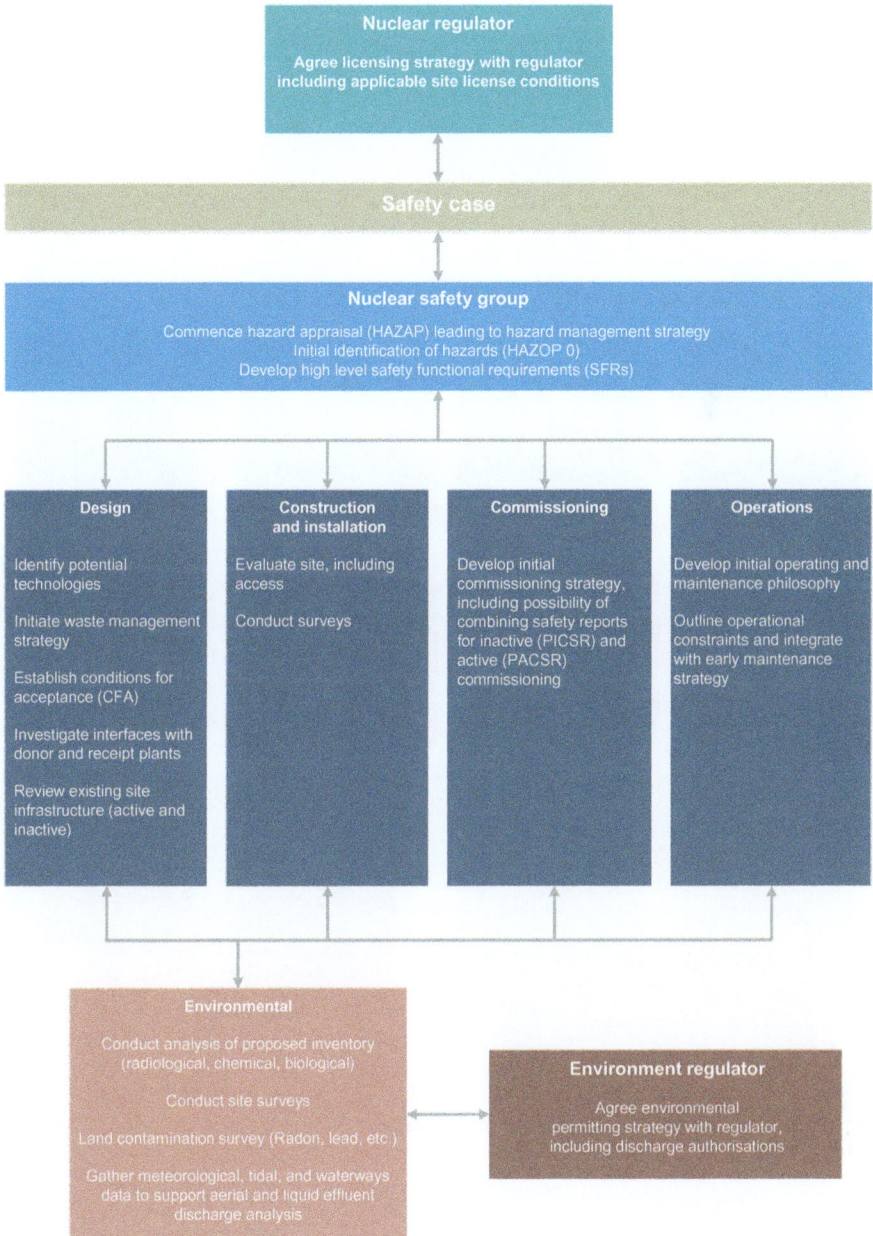

Fig. 14.17 Safety case development—project inception.
© Bill Collum.

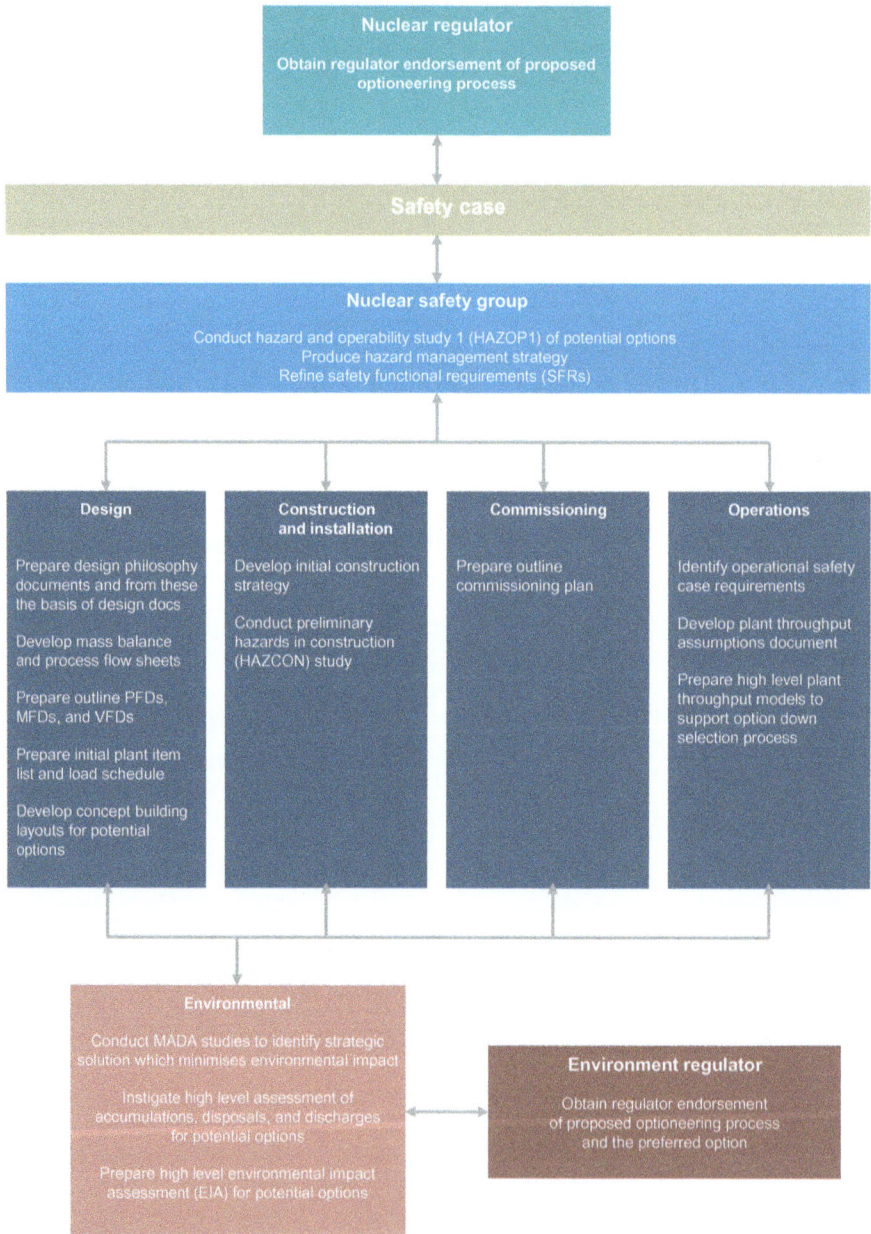

Fig. 14.18 Safety case development—optioneering.
© Bill Collum.

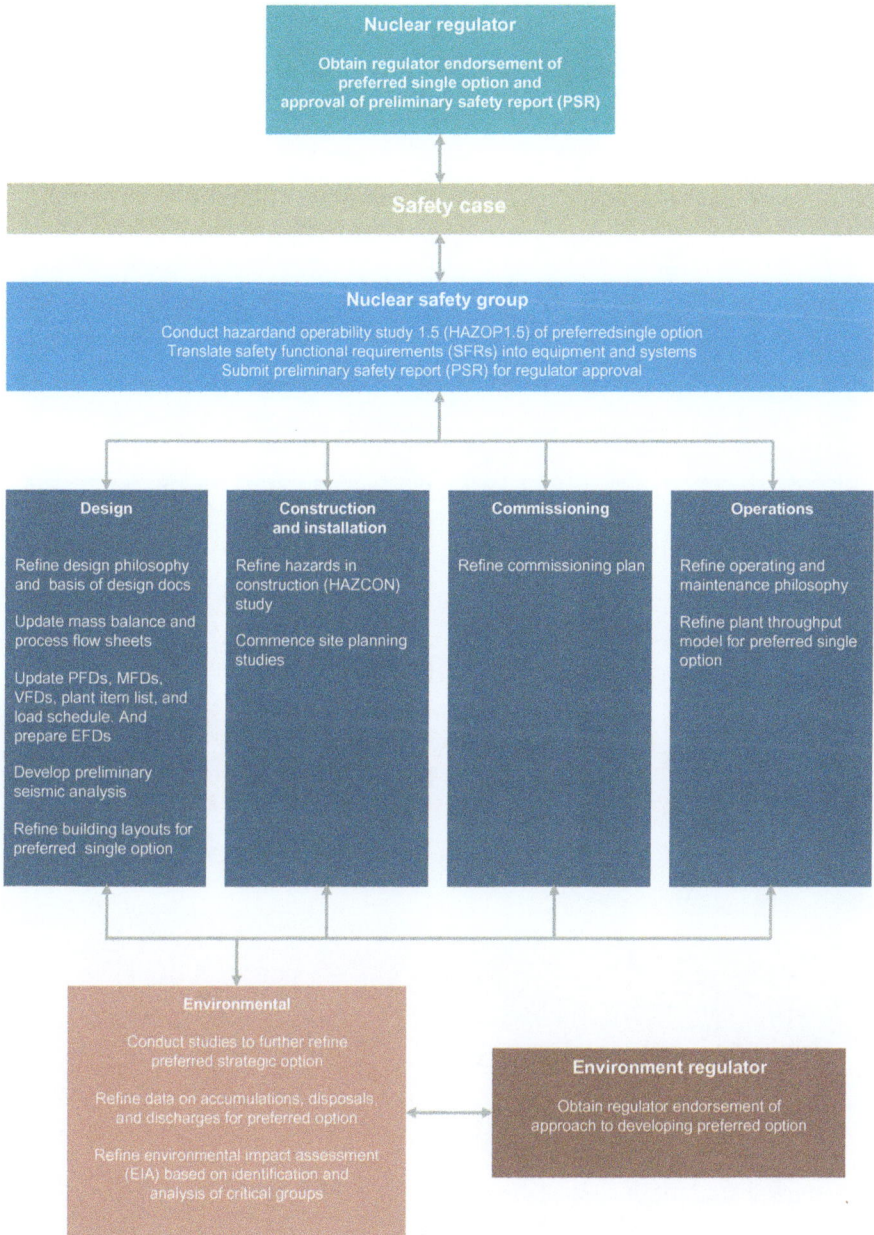

Fig. 14.19 Safety case development—option refinement.
© Bill Collum.

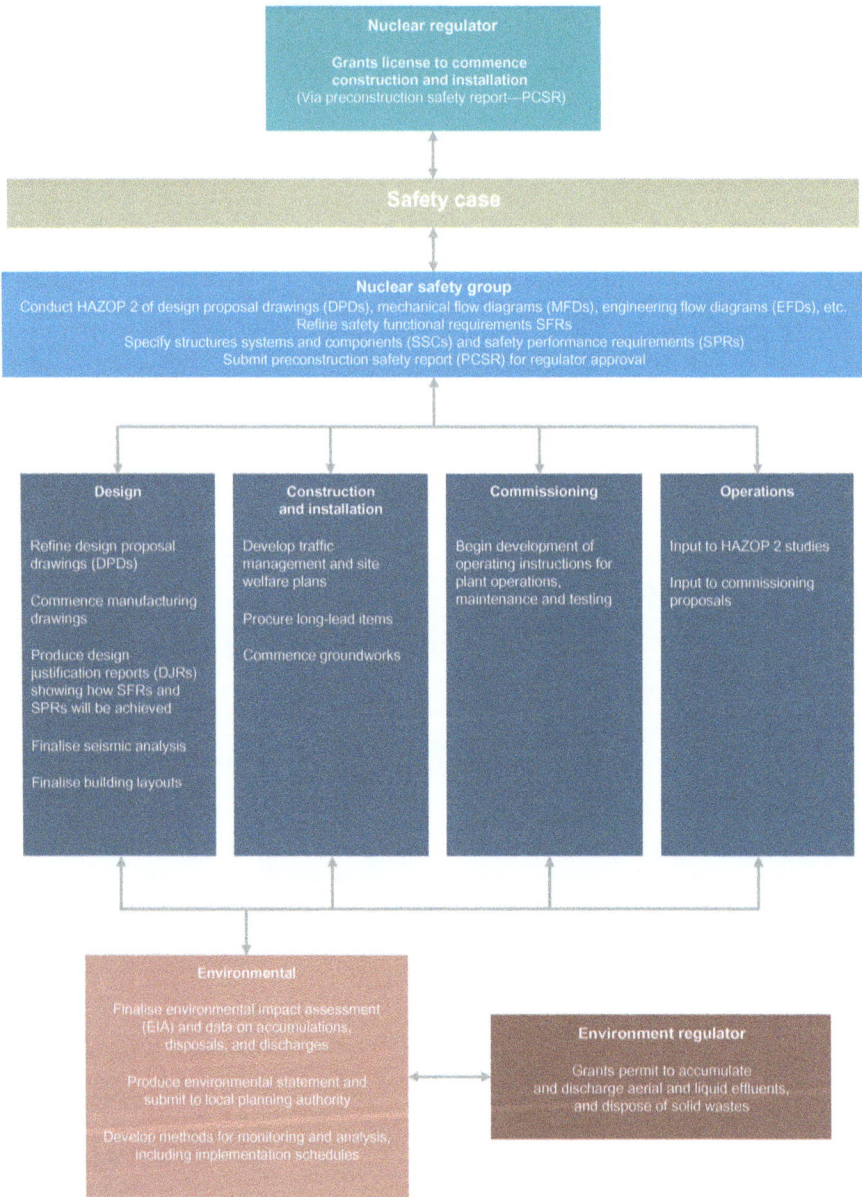

Fig. 14.20 Safety case development—detail design.
© Bill Collum.

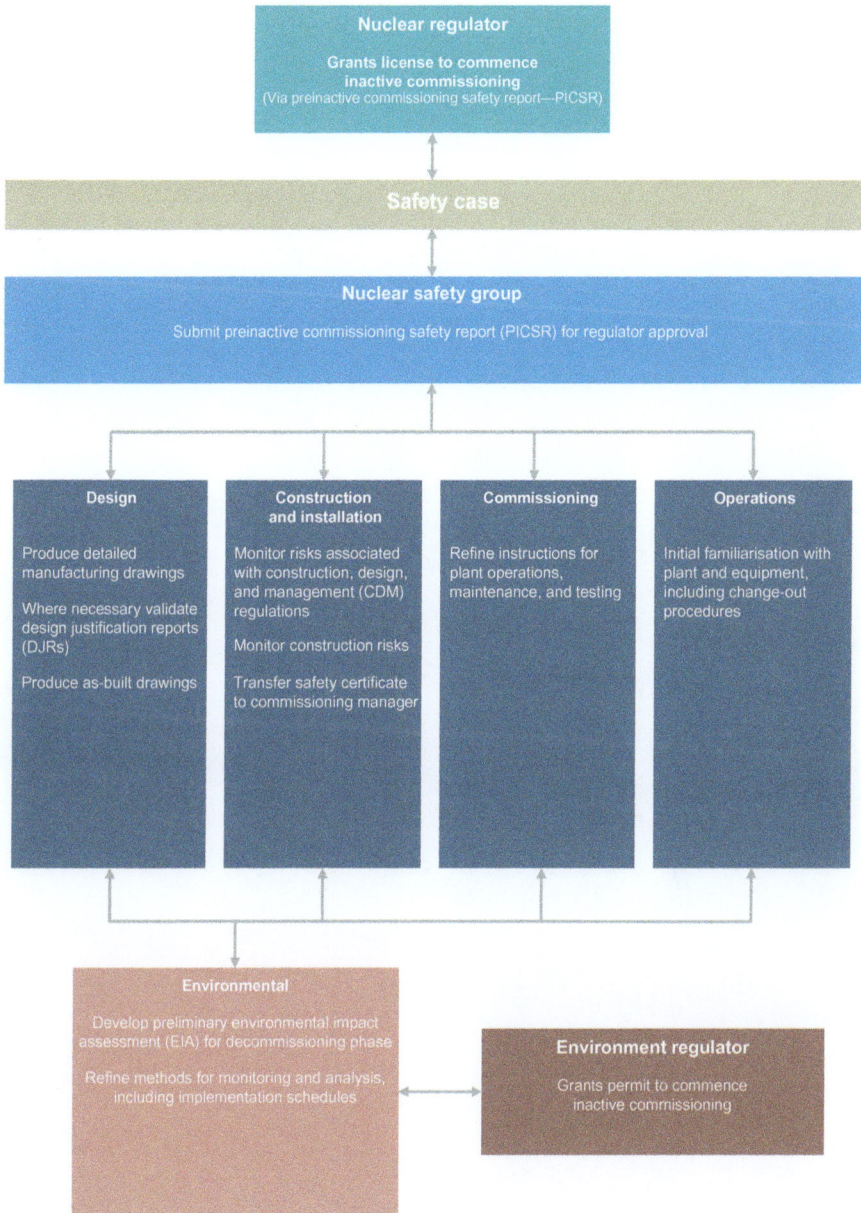

Fig. 14.21 Safety case development—construction and installation.
© Bill Collum.

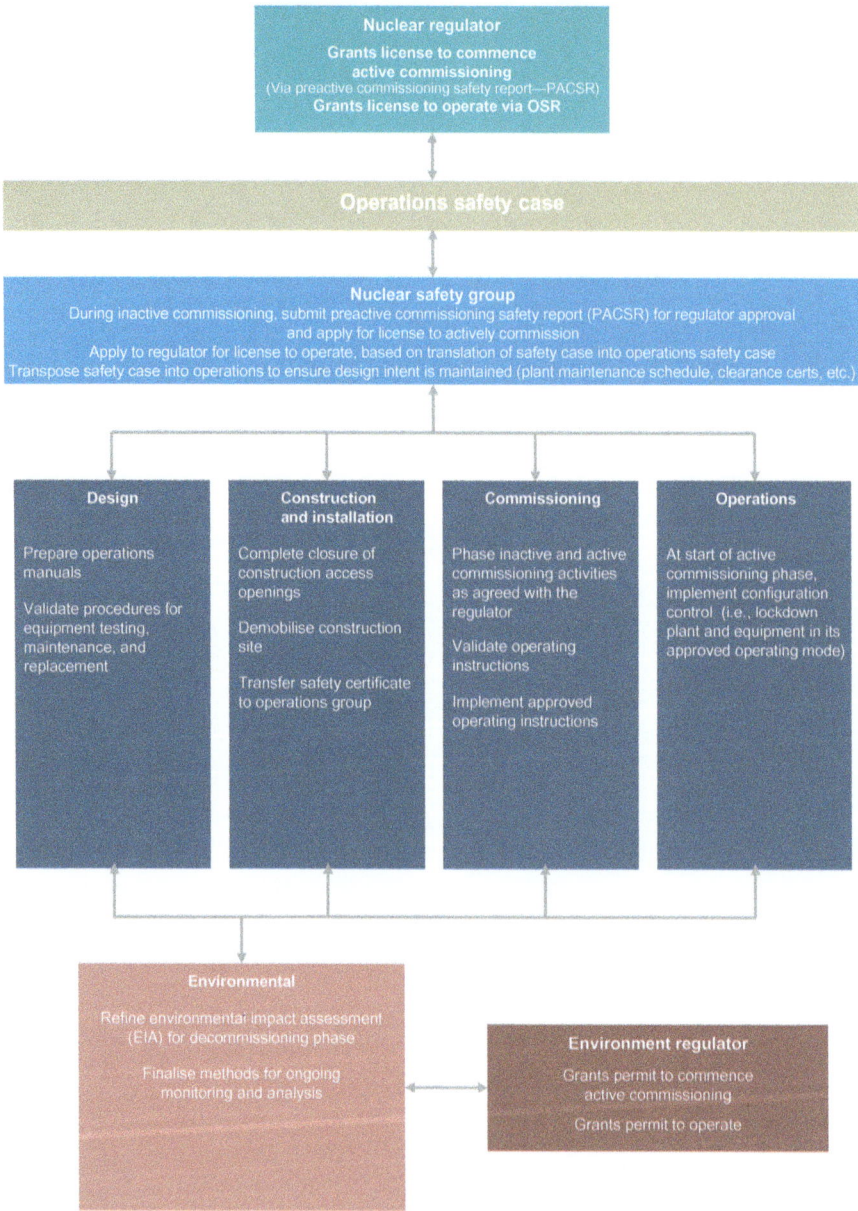

Fig. 14.22 Safety case development—inactive and active commissioning.
© Bill Collum.

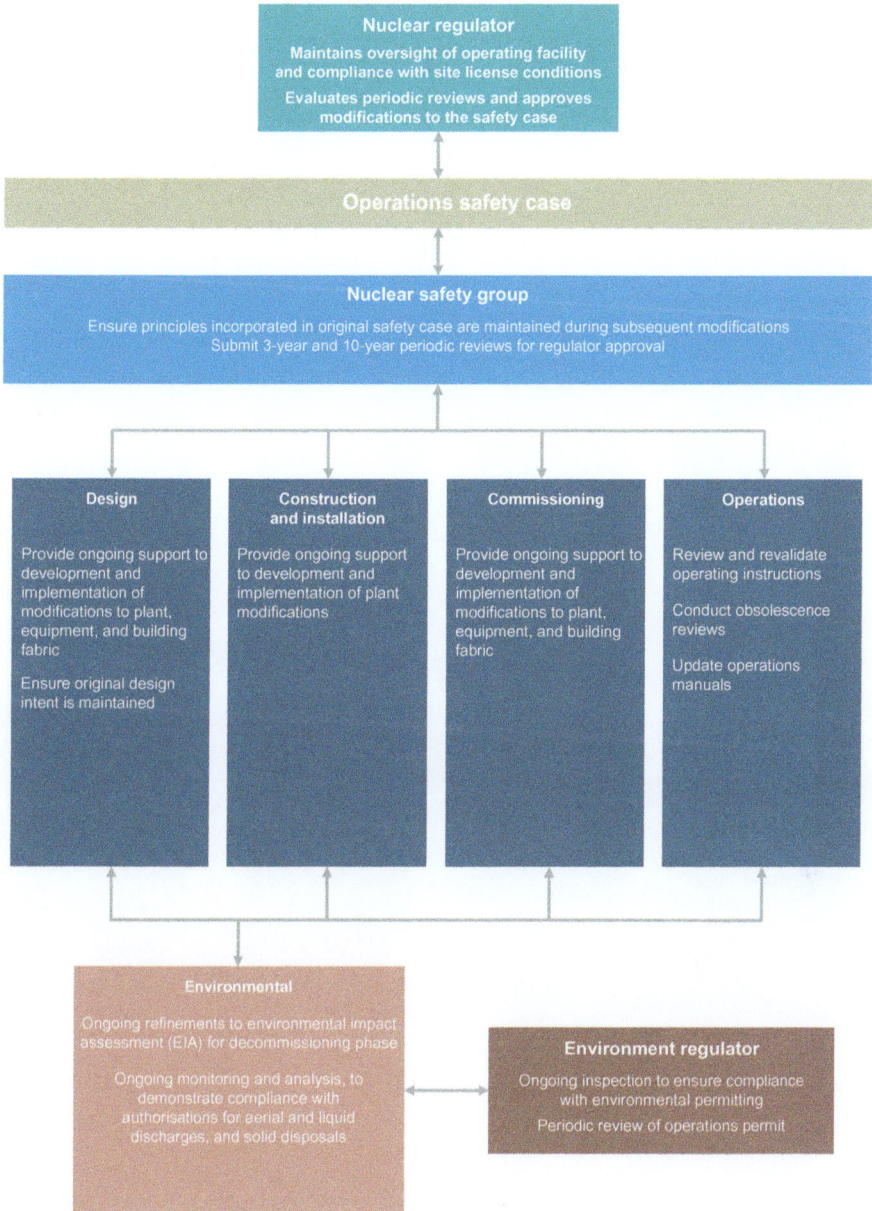

Fig. 14.23 Safety case development—operations.
© Bill Collum.

Decommissioning planning

15

Decommissioning planning is a particularly challenging task, for several reasons. Firstly, in its most basic form it covers periods of a decade or two and, even more testing, in its ultimate guise can take in a vista which gazes thousands of years into the future. In addition it is desperately multifaceted, grappling as it does with everything from constraints of dealing with radiation and contamination, to safety and environmental issues, availability of funding and waste disposal facilities, the need to retain valuable knowledge along with resources able to use it to best advantage, and on and on.

As if all that wasn't enough to contend with, each strand is intricately interwoven with the next, to the extent that the tangle appears quite impenetrable. Thankfully, if you stand well back, far enough to view the entire picture, there is a hierarchy, a logic to how the whole process evolves and is managed. To make sense of such a vast subject, we shall unravel it a thread at a time, beginning with the *planning factors* that lie behind all decommissioning programs, and from there work our way up to a strategic level where we shall see how the whole endeavor is drawn together, but first we need to touch on site license conditions (SLCs).

15.1 Site license conditions

When we consider that decommissioning nuclear facilities is at least as significant an undertaking as building them in the first place, and arguably even more hazardous, it comes as no surprise that regulatory scrutiny does not wane at this stage, in fact quite the opposite. To reinforce the point it is covered by legislation which, as we saw in the previous chapter, cascades from guidance issued by the International Atomic Energy Agency (IAEA). In the UK's case, decommissioning obligations are embodied in SLC 35, which is summarized in Fig. 15.1.

| Decommissioning | Make arrangements for the decommissioning of any safety related plant or process |
| | Divide decommissioning into stages and do not commence, or proceed from one stage to the next without regulatory approval |

Fig. 15.1 Site license condition 35.
© Bill Collum. Based on data from the Office for Nuclear Regulation (ONR).

Although SLC 35 is the primary condition it is not alone. So, for example, safety documentation (SLC 14), radiological protection (SLC 18), disposal of radioactive waste (SLC 33), and many more, which applied during a facility's operations phase, will still prevail.

Nuclear Facilities. http://dx.doi.org/10.1016/B978-0-08-101938-2.00015-5

As an added complication, by its nature decommissioning brings about a gradual reduction in regulatory controls until eventually they are released entirely. As a result compliance with SLCs does not remain static, but must instead anticipate and keep pace with a facility's evolving condition, until one by one they wither and are no longer relevant. So there is quite a bit of choreography involved in managing the dissolution of SLCs, most notably in terms of hazard management.

Decommissioning then is not just a good idea, something we can turn our attention to when we are good and ready, it is a legal requirement and regulators have clear expectations which must be satisfactorily fulfilled. So they, along with other interested parties must be consulted when plans are being devised and, most important, nothing of significance happens without their prior approval.

15.2 Planning factors

As I say, everything about decommissioning, from incubating initial thoughts to its actual implementation is highly interconnected, but that being said, the process has to begin somewhere, so this is the place. Analyzing the various planning factors (Fig. 15.2), which are essentially the building blocks that underpin all decommissioning programs, enables us to generate information which feeds into the implementation activities that come later.

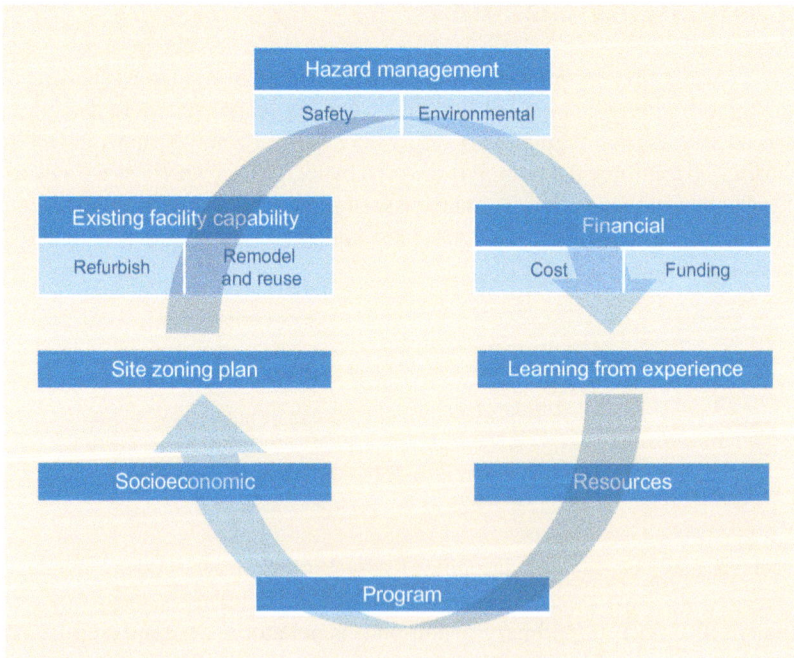

Fig. 15.2 Planning factors.
© Bill Collum.

Of course, decisions made here must be revisited as consequences of those activities come back down the line, but at least we can get started and make some meaningful progress.

Even here, at this constituent level, we see evidence of interdependencies between various themes. So, for example, there are obvious links between the need to learn from experience and issues related to resources. Similarly we can see that funding is closely associated with questions of whether it might be a good idea to refurbish an existing facility so that it can continue doing the same job for longer, or instead remodel a building for an entirely different purpose. In truth, and to varying degrees, each planning factor has tentacles which reach out and mingle with all of the others.

I could continue highlighting interdependencies at every twist and turn throughout this whole chapter, but I am afraid it would become very repetitive so I will spare you too much of it. You get the idea: when it comes to decommissioning nothing stands in splendid isolation. Let's get into those planning factors.

15.2.1   Hazard management

15.2.1.1   Safety

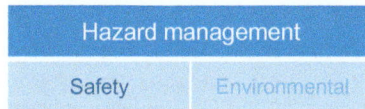

If we were asked to sum up succinctly what decommissioning is all about, then the term *hazard management* goes a long way to encapsulating it quite nicely. When you get right down to it, the whole point of decommissioning is to reduce hazards with an ultimate goal of eliminating them altogether. So management, active management is a key ingredient. Furthermore, on large nuclear sites it is neither possible, nor desirable, to apply a common decommissioning plan (DP) and identical timescale to multiple facilities. We need to prioritize the task and to a large extent it is the potential, or *risk*, of hazards being realized that dominates how those priorities are sequenced.

So where do these hazards come from? From a decommissioning perspective, any facility that houses a radiological inventory when it ceases operations automatically constitutes a hazard. It does not need to be imminent or have catastrophe written all over it, what matters here is that such hazards will not go away without some form of intervention. The objective therefore is to identify and quantify a site's hazards, including how they might escalate if left unattended, and register them on a sliding scale from high to low. With that done, we can draw up plans to apply a proportional and timely response to reducing or eliminating those selfsame hazards.

So far so good, but unfortunately a site-wide hazard analysis is not something that can be conducted once and then forever serve as a permanent reference; there are just too many variables and they do not stand still. So analysis is continual, with nuances of latest findings invariably resulting in revisions to a site's decommissioning plans.

It goes without saying that this is a serious business, so we need some formal structure to how safety related decisions are made and approved, and yes you have guessed it, this links straight into the all-pervading safety case which we discussed in the previous chapter.

As we know, a facility's safety case evolves throughout its design, construction and commissioning stages, after which it continues, in many cases being renamed as an operations safety case. What is of interest in this context is that the level of effort and rigor which went into developing an operations safety case continues unabated as it transitions again to become a decommissioning safety case. Moreover, the regimen of updates and periodic reviews continues, until all hazards have been entirely abated and a safety case is finally closed out. It is quite a marathon.

15.2.1.2 Environmental

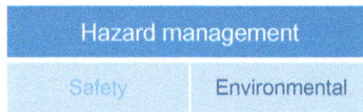

Hazard management	
Safety	Environmental

As with any other industry, nuclear sites must comply with environmental constraints of not harming people, wildlife, flora and fauna, and so on. That being said, decommissioning involves taking facilities apart, so there is no avoiding the fact that the potential for initiating environmental hazards is higher than during normal operations. On the other hand if we failed to decommission, then facilities could eventually degrade to the point that radiation would no longer be confined by adequate shielding and, potentially even more hazardous, contamination could escape into the environment.

Clearly then, we need to get on with decommissioning, but just as clearly we need to exercise caution in how we go about it. As a result, when nuclear teams contemplate decommissioning it is the specter of environmental consequences, along with their potential to cause harm, which are at the forefront of their minds. This is what they must guard against.

We have seen previously that nuclear organizations, across the world, must comply with environmental legislation which places a ceiling on aerial and effluent discharges from their sites. It is worth noting here that just as sites are not allowed to increase those discharges in order to accommodate a new facility, the same is true during decommissioning. In both situations nuclear sites must stay within the prescribed limits of their annual discharge authorizations, which incidentally tend to be driven downwards year on year.

In short, any proposed increase in discharges from a facility, including those associated with decommissioning, must be offset by reductions elsewhere on the site. Any departure from this protocol would be most unusual and only permitted if approved by a nation's environmental regulators.

15.2.2 Learning from experience

Learning from experience

Just about everything we do requires some level of skill, from making toast or operating a TV at one end, to flying a fighter jet or performing complex surgery at the other. As a rule we can expect that the more challenging a task is the longer it will take to master. And crucially, we can only become truly adept by advancing in incremental steps, with each requiring more knowledge, more practice than the last. Simply put, we progress by *learning from experience* (LFE). For relatively straightforward tasks we may not even notice, but it happens all the same.

In terms of decommissioning there may appear to be some bravado in straining at the leash to take on a site's most difficult challenge, some kudos in going for the big one straight away, but it does not work like that. Just as with any complex task, particularly one so multifaceted, decommissioning advances best in ascending steps. Happily a site-wide decommissioning strategy can instill an element of control into how the exercise escalates, so it is an opportunity not to be squandered.

On large nuclear sites in particular, there is the prospect of multiple facilities joining a queue to be decommissioned, all patiently awaiting their turn. None are going to be easy, but some will present less of a challenge than others. There is the prospect here of phasing the decommissioning challenge, of scheduling it with an increasing degree of difficulty. Facilities that pose similar challenges could even be grouped together and tackled by starting with the simplest first, then valuable LFE could be incorporated into decommissioning plans for those that follow.

At the level of decommissioning a single nuclear facility, the ideal approach to LFE centers on removing materials in step with their levels of contamination and radio-activity. This begins with free release, then moves through very low level waste (VLLW), low level waste (LLW), and intermediate level waste (ILW). (We can take it that high level waste (HLW) materials are removed before decommissioning gets underway.) This gradual raising of the radiological bar is necessarily mirrored by a similar transition from straightforward manual handling, to an increasing reliance on personal protective equipment (PPE) and respiratory protective equipment (RPE), and ultimately culminating in deployment of remotely operated equipment.

Of course this is quite simplistic, as there are other factors at play which are always poised to scupper such an idealized approach. For example it may not be possible to remove inactive ductwork too soon, because a plant's ventilation system needs to stay operational until contaminated materials have been dealt with. However, you get the idea: whether it be a single facility or a whole site, start with simple stuff and use accumulated learning to gradually up the ante.

It is worth noting here that of itself having decommissioning experience within an organization does not equate to prowess in LFE. This is particularly true of the timescales we are considering here, where building up knowledge, retaining it and handing it on requires a long-haul commitment. To capitalize on experience, decommissioning procedures, along with information storage and retrieval processes must be capable of gathering learning and disseminating it again at just the right moment. It is a highly specialized area and a major undertaking in its own right. The significant issue from a strategic point of view, is to recognize the importance of knowledge so that it can be harnessed like any other precious resource, otherwise it will be dissipated like gold dust in an ocean.

15.2.3 Financial

This planning factor divides into two components with one being actual costs, the price tag, and the other how those costs are funded. Although sides of the same coin, the constraints this pair impose on decommissioning planning are entirely different. That being said, we can be talking pretty big numbers here, sometimes very big indeed, so unsurprisingly both facets are subject to constant scrutiny and demands for robust justification. All of which is entirely as it should be.

15.2.3.1 Cost

Financial	
Cost	Funding

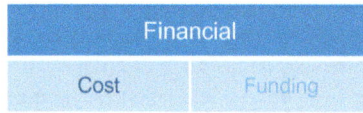

When we consider the price of an item, we normally concentrate on its upfront capital cost and tend not to dwell overmuch on the long-term, or lifetime costs which will ultimately be incurred. So with a new washing machine, for example, most of us budget for the sticker price and only investigate maintenance and disposal costs when the time comes. With nuclear facilities we need a totally different mind-set. Of course the capital cost is hugely important, not least because it will probably be sizable and the "hit," so to speak, is more or less immediate.

If we step back and take a wider view, it becomes clear that the headline grabbing number only gets us so far. It covers design, procurement, construction, installation of plant and equipment (P&E), commissioning, and maybe some research and development. Beyond that, there are operating costs to consider, with all that entails in terms of overheads, upkeep and ongoing maintenance. Even further down the line there will be decommissioning, which as we are seeing is a complicated and costly business.

All of this money has to come from somewhere, so unlike a washing machine or refrigerator we cannot simply turn our attention to the issue of lifetime costs at the last moment. Its magnitude, as a percentage of the original capital cost, is just too high. Instead, full life-cycle costs must be determined and factored into a facility's optioneering, evaluation and selection processes from the outset. Unsurprisingly, customer organizations, regulators, financial institutions, and indeed governments can take a keen interest in the numbers, and indeed the methodology by which they have been determined.

Decommissioning costs can vary considerably from one facility to another, so to be truly representative they need to cover a very wide spectrum. We must include expenditure on decommissioning design proposals, more specifically refinement of high-level plans developed during the initial design process, all decommissioning activities, such as dismantling, sorting, segregating and processing radioactive waste, storing those wastes, no doubt for a long time, and finally demolition of the original building and possibly remediation of the site.

In the next chapter, we shall discuss designing nuclear facilities with their eventual decommissioning in mind, so for the moment just need to keep in mind that decommissioning must be considered practically from day one of the design process. We must plan for it with a vengeance, have firm plans for how it will be accomplished and a credible grasp of the financial implications.

15.2.3.2 Funding

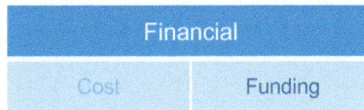

So cost is one thing, the all-up bottom line, but on its own does not get us too far with the exercise of decommissioning planning. Funds for decommissioning individual facilities, or indeed entire nuclear sites, are not delivered to suit the demands of site operators, but rather there is choreography to how the money arrives. The source of funds varies from one nation to another but ordinarily derives either directly from governments, maybe via a tariff on energy companies, directly from energy companies themselves (self-funded), or possibly from governmental financial institutions who allocate funds on behalf of a group of countries.

In some cases time may be a needed to accumulate funds, so there could be a period of several years where little or no cash is available. In others, the level of funding may fluctuate quite markedly over time, while some nuclear site operators are in the enviable position of receiving funds at a steady rate, year on year. Whatever the funding model, it is crucial that operators know years in advance how much they can expect and what the delivery pattern is destined to be. That way they can plan around anticipated funding, and the more notice they have the better those plans will be.

The upshot is that a site operator's preliminary decommissioning plans can provide an indication of how much money is needed and when, but actual working plans must be programmed around reality of those twin financial constraints of cost and phased funding. It would be too much to suggest that money, in its various guises, totally dominates decommissioning planning, but it is certainly hugely influential.

15.2.4 Existing facility capability

Every nuclear plant is designed with a particular purpose in mind, so on the face of it, it would seem that once its operations phase is complete, a facility's life is over and decommissioning awaits, but not necessarily. Equally, a facility may have operated for so long that it will soon be too tired to carry on. Maybe it is time to call it a day and move operations to a splendid replacement, but then again maybe not.

Whatever the circumstance, where the useful life of a facility seems to be within sight, we must be careful not to assume decommissioning is a foregone conclusion. After all, these buildings take a fair sized investment and are invariably tremendously robust, so maybe we could hold off with the *liability* name tag and instead help them preserve their *valuable asset* status. To achieve this there are broadly speaking two alternatives: refurbish, or remodel and reuse.

Before we examine them, it has to be said that both activities can appear somewhat detached from anything connected with decommissioning planning. However, as we shall see, they do manage to weave their way firmly into the process, not least because refurbishing or remodeling a facility requires some level of decommissioning, maybe quite extensive at that. And of course this *partial* decommissioning must be incorporated in future plans.

15.2.4.1   Refurbish

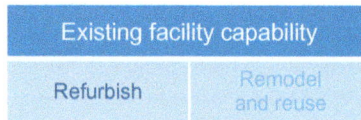

This option is all about enabling a plant to continue doing the same operational activities but for longer, recognizing that if this proves to be a nonstarter, going for a new build replacement instead. It is a tricky quandary and one that may come about for several reasons.

Plant and equipment (P&E) may be reaching the stage where its burden of maintenance and downtime, with associated costs, will make it too unproductive to be viable. P&E could be approaching the point where regulatory authorities will no longer certify it. Alternatively, a building's fabric or even its structure could be causing concerns. It could be a combination of all three, or some other form of looming aggravation. Whatever the reason we have a facility that can only perform for a few more years, after which its functions will still be essential to continued operations of its host site. So what do we do, give it an overhaul or build a replacement?

The kneejerk reaction will almost certainly point to new build. We all love a blank sheet and the prospect of creating something fresh. Of course there is too much at stake to rely on little more than a hunch, so in reality these decisions must be based on cold hard facts, careful analysis and a firm rationale. But what goes into the evaluation?

There are no prizes for guessing that cost is a major factor, so we need comparable estimates of the sums involved for refurbishing an existing facility versus building a new one. Initially a rough order of magnitude, or ROM estimate will do nicely, since there is little point in spending time and money on detailed estimates if we can quickly establish there is a financial chasm between the two alternatives. That being said, even a ROM estimate needs something on which to base the numbers, so there is still a fair bit of work to do.

Fig. 15.3 Refurbishment.
© Nuclear Decommissioning Authority.

For an existing facility we need a feasibility study to establish whether it is even possible to carry out a refurbishment (Fig. 15.3): can equipment be upgraded, the safety case revalidated, environmental regulations satisfied, and so on? If the signs are reasonably positive, then the study moves on to outline the scope of work involved, and from that we can generate a ROM estimate and an indicative schedule. For the new build option we need to establish if a suitable site is available, or maybe multiple options at this stage, and prepare a concept design; armed with that we can get to work on a comparable estimate and schedule.

In addition we also need to investigate more peripheral issues such as: if an existing facility was refurbished could its host site cope with the disruption to operations, which may last for a long time? If a facility's life was extended, would its location still align with long-term plans for the host site? For example would its relationship with donor and receipt plants, which may also be due for replacement in the years ahead, still be viable? Even though refurbishment is being evaluated as a potential primary objective, might it also be possible to enhance a facility at the same time: maybe improve throughput, expand operational abilities, and so on?

As usual, there is plenty to think about but with sufficient forward planning, no preconceptions and a thorough analysis, an unambiguous answer will eventually reveal itself.

15.2.4.2   Remodel and reuse

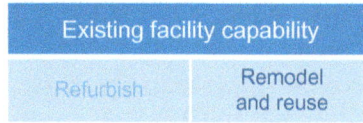

Existing facility capability	
Refurbish	Remodel and reuse

Unlike refurbishing, this option is all about converting what would otherwise be an obsolete facility into a new use. This may be similar to its original purpose but then again could be entirely different. There are no rules.

In a domestic situation we are quite prepared to turn our garages into kitchens, attic spaces into bedrooms, even a home into a shop or vice versa. In the commercial world, office blocks are continually being revamped and adapted for new purposes, likewise for factories and other industrial premises, wholesale changes are a familiar occurrence. Even huge ships, which look quite obdurate to me, are routinely refitted, sometimes even chopped through like a Swiss roll and extended.

It has to be said however, that if you were to wander around practically any operating nuclear plant and contemplate thoughts of remodeling it, the whole notion is bound to feel quite daunting. But if we weigh this against visions of stripping out the internals and demolishing a perfectly sound building, then it spurs us on to get more creative, more determined and to take remodeling in our stride. Yes the nuclear industry faces unique challenges, doesn't everyone, but not to the extent that remodeling is impossible, far from it. Happily there are lots of opportunities to give outmoded facilities a new and different lease of life.

If following the UK type approach to design then, in terms of remote operations, mechanical handling caves hold out the best prospect of being converted to a new purpose. With sufficient forethought and planning, including configuration of their import and export routes, caves have an inherent flexibility that we can capitalize on. It may entail stripping out much of their original kit, but existing manipulators, windows, cranes, shield doors and bogies, not forgetting concrete shielding and ventilation systems, all have potential to be put to good use.

Closed cells on the other hand tend to be more bespoke and tailored towards treating the particular liquors and sludges for which they were originally designed. That is not to say we should give up on reusing them. If processing equipment cannot be adapted to take other feedstock, there may still be opportunities to strip out their kit and refit the empty shell for other purposes. For example, as part of the original build we could form penetrations in cell walls for future shielded windows, manipulators, and so on, then seal the openings until they were needed, maybe decades in the future. As long as we plan for such augmentation during the early design process, then much can be done.

When considering reuse, the US canyon approach affords more flexibility than cells since existing vessels and processing equipment can be removed and replaced. However, some considerable forethought and careful planning is needed, to ensure connection points for detachable jumpers are located to cope with a new configuration of process pipework and utilities. This may mean installing unused connection points when a canyon is first constructed, but it is a small cost in comparison to the benefits that can be realized later.

As with refurbishing, the prospect of remodeling and reusing facilities has a significant influence on how decommissioning plans are drawn up. Apart from the partial decommissioning that both require (which must be planned) it also means that rather than making provision to fully decommission a building, we can shelve those plans until later years and instead concentrate our efforts on something else.

So there is potential to reuse nuclear plants for new purposes, maybe even partly reuse them, but there is something else. Examining potential reuse opportunities is not terribly effective if viewed in isolation, say from the perspective of a single facility. In other words, studying the evolving design of a new building and dreaming up possible alternative uses will not get us too far. Reuse is a site-wide issue, one where future requirements of a site must be identified, as far as possible, years in advance. Armed with this information, new build design teams can rise to the challenge of satisfying those needs within precincts of their upcoming building, thus eliminating the prospect of yet another new facility.

What this tells us, is that remodeling and reuse is not a matter that simply crops up each time the scope or functionality of a new building is being contemplated, or indeed as one nears the end of it primary operations, rather in links directly into a site's long-term development plans; it has to, for this is where we get the *shopping list* of future needs. Furthermore, for major sites in particular, identifying future requirements is a fluid activity that demands constant attention and careful coordination between the here and now and the far horizon, truly interface management on a grand scale, so there is a cost. However, by approaching decommissioning planning on the basis that decommissioning may not necessarily be the right answer, at least not just yet, we can take huge strides towards reducing future liabilities, saving money, minimizing radioactive waste arisings, curtailing land usage and on and on. Enough said.

This kind of forward planning, where facilities are designed with the future in mind, is a huge subject in its own right, one we shall examine thoroughly in the next chapter.

15.2.5 Site zoning plan

Site zoning plan

If you were to take an aerial view of any large industrial building or suite of buildings on the day they first open for business, you would no doubt see a well-ordered road layout, car parks, neat fencing, and maybe immaculate lawns and landscaping. Presented with such an orderly and seemingly obvious scene, it would be easy to assume it was simply a spruced up version of the arrangement put in place when construction first began. However, if you could rewind to those early days, when mighty earthmoving machines were at work, temporary utilities being rolled out, portable office accommodation being set down, and so on, you would witness a very different scene.

In fact if you hung around long enough you would see the whole site transition through several iterations. For example, a concrete batching plant might appear for a few months, marshaling areas would be established to store construction materials, tower cranes would swing into action for a while. It may all appear quite disorganized but in truth the whole sequence will be carefully choreographed, with each setup serving its purpose and bringing a site closer to its final steady-state configuration.

Exactly the same thing happens in the nuclear world. For major sites with multiple facilities the exercise must be repeated many times, with "local" arrangements for each facility being carefully integrated with other buildings nearby, and beyond that with the wider site. Just as the area around individual facilities must transition through several phases, so must an entire site; in fact, things seldom stand still.

Of course we need to exercise control so all large industrial sites have a grand plan. In practice then, either an individual or a committee of some sort oversees developments and arbitrates on priorities. Apart from authorizing land usage they take an interest in proposals for transport links, utility arrangements and, in the case of nuclear, the networks which carry radioactive liquids between buildings. Individual facilities must "plug in" to this ever-evolving big picture, so it is quite a coordination exercise.

All of this is complicated enough for any industry, but for nuclear sites in particular is only half the story. In parallel with plans for how sites will expand and evolve over time, we also need a clear vision of how they will retreat in the future, which may be decades or even well over 100 years away. As a minimum we must establish the broad principles by which, area by area, facilities and infrastructure will gradually disappear, until eventually a site is returned to greenfield, or at least brownfield conditions, or possibly a combination of the two.

New build design teams need this information because, apart from the imperative of putting buildings in the right long-term place, they must also take into account future plans for a site's infrastructure and other buildings with which their new facility must interact. Without this long-term vision we face the prospect of making decisions today, of developing sites in ways which will hinder their orderly disentanglement in the years ahead. Simply put, buildings must harmonize with their surroundings throughout their entire existence, not just on the day they begin to operate. Let's take a couple of examples to illustrate the point.

Many nuclear facilities are connected to treatment plants which process their active liquid effluents. As a result the location of effluent treatment plants (ETPs), along with the network of arteries connected to them is more crucial than most. If an ETP is due to be replaced at some stage; moving to a new location, it is imperative that facilities dependent on it can maintain their active connections. Conversely, from a site zoning point of view, it makes sense to erect new ETP-dependent facilities close to the location of a future ETP. There are all manner of scenarios, what matters is to recognize dependencies of one plant on another and ensure interactions will be viable for as long as they are needed.

It is worth noting that if a plant finds itself stranded some distance from an unforeseen ETP replacement, it is not just a matter of running a pipeline between the two buildings. Active connections between facilities may take the form of an above ground duct, a trench or a pipebridge, depending on the terrain they must cross. Whatever the solution, they are well shielded and invariably seismically qualified. Pipebridges in particular (Fig. 15.4) are impressive feats of engineering in their own right, so much so that a lengthy one could easily cost more than the facility it serves. Definitely not the sort of news you want to sneak up on you.

Fig. 15.4 Shielded pipebridge.
© Nuclear Decommissioning Authority.

When it comes to long-term zoning of a site, storage facilities can be equally problematic. Many nuclear sites store radioactive wastes within the confines of their site boundary. Arrangements vary from one country to another, but can include the full spectrum from VLLW and LLW, to the more challenging ILW and HLW. In addition to wastes, reactor sites may also store their spent fuel, which of course is highly radioactive.

Very often storage arrangements are categorized as an interim measure, where materials are held while awaiting shipment to a national repository which will be their permanent home. In the nuclear world, temporary can be a long time, maybe well over 100 years. As a result we can confidently predict that storage facilities, of one type or another, will be the last buildings standing on many nuclear sites.

It would be a pity to spend decades diligently decommissioning a nuclear site, only to find that towards the end of the exercise we found multiple stores dotted around it, each within their own enclave. Far better to congregate all stores together so that we are unconstrained in remediating the rest of a site. If other constraints rule against this ideal, it should at least be recognized as an important site zoning objective and efforts made to collocate as many stores as possible.

So whether it be connections between plants, locations of storage facilities, arrangements for utilities and infrastructure or any other site-wide issue; just as a site's long-term zoning plans have a tremendous influence on proposals for new build facilities, they are equally important when it comes to drawing up decommissioning plans. Neither aspect can sensibly be viewed in isolation.

15.2.6 Socioeconomic

Socioeconomic

Nuclear sites employ a lot of people; even small ones may have a permanent workforce of a couple of hundred, and for major sites this can easily rise to several thousand. Even so, most sites are located in areas that are some distance from the main centers of population. I am generalizing here, but the pattern tends to be one of a site in a rural, often costal setting (Fig. 15.5) with one or two small towns nearby. So far so good, but when a decommissioning team, or come to that a construction squad, rolls into town those steady-state numbers can multiply considerably, and very rapidly indeed.

Fig. 15.5 Nuclear site in a costal location.
© Nuclear Decommissioning Authority.

Let's say that over a period of 2 years, 5000 people move into the area surrounding an isolated nuclear site. That might sound like a make-believe number, but we are not just talking here of site-based personnel. It takes large teams of designers, planners, cost estimators, and so on to sustain major decommissioning projects, so very often in the interests of efficiency and communications many of them are temporarily located close to the site. In addition, small industries and businesses that support decommissioning activities will also be drawn into its catchment area. When we include dependents of all those either directly or indirectly engaged by a site, then such a number is by no means out of the question. In some situations, the number could be considerably higher, but let's stick with 5000.

As decommissioning work starts to tail off, we could find that a couple of years after its peak the number drops back to where it started. Then maybe a year or so later a similar cycle begins again, this time reaching a peak of say 3000. And so it may go, year after year. Imagine the strain those comings and goings put on local communities. Pick any element you like, schools, housing, hotels, restaurants, medical facilities, road networks, utilities, and many more. You can picture the scene.

These small communities will have been down this "influx" road before, certainly when the site was first established and maybe several times since. So they can be quite adept at absorbing the waxing and waning of large workforces descending in a rush and leaving again in a hurry. However, this does not mean it is easy to deal with, that it just happens, far from it, so site operators need to recognize the difficulties and give their local communities a helping hand. Wouldn't it be far better to phase decommissioning activities across major sites so as to smooth out the peaks and troughs? In this way we can expand the local population by say two or three thousand and then hover steadily around that level, possibly for decades.

If you are thinking ahead, you will have already concluded that there is another reason for maintaining a steady workforce: their expertise. With continual comings and goings it would be virtually impossible to nurture a pool of decommissioning expertise. If specialists are starved of decent work every few years, they cannot survive. The whole industry, both within a site and around its boundaries will be in a state of perpetual learning, forgetting and re-learning, which undermines the imperative of LFE which we discussed earlier. In such a situation the nuclear site, local community and businesses will find it difficult to engender collaborations.

By recognizing socioeconomic issues, decommissioning planning can make a major contribution to establishing a stable environment in which everyone can thrive. So it turns out that what is best for the local community is also ideal from a decommissioning perspective. Perfect.

What I have just described relates to large nuclear sites with lots of facilities and the prospect of decommissioning activities stretching many years into the future. There is, however, an altogether different scenario, where a site has a small number of nuclear plants, say one or two reactors along with their support facilities. Once again we are probably talking of a rural setting, one where the site may well be the area's dominant employer. This time, when decommissioning is complete we do not have a queue of other facilities waiting for the workforce to move onto. Instead the site will be closed down and its employment cease. Clearly this is a worrying situation for a site's employees and understandably the prospect of decommissioning their livelihood away is not going to be well received. We need a different approach.

In these situations it would insensitive to flood the area with decommissioning hotshots and crack on urgently with ridding a site of its facilities. Providing there are no overriding safety or environmental factors, we can afford to slow things down and tackle the task at a different pace.

For a start the site's workforce, who will be skilled in its operation and maintenance, should, where feasible, be given the opportunity to retrain as decommissioners. And not just its implementation, but also organizational activities such as management, programming, satisfying regulatory constraints, and so on. True, a number of specialists will be needed to guide them, but many of the skills needed to operate a nuclear site are, to varying degrees, transferable to the decommissioning arena. In addition to retraining existing personnel, rather than invest in sophisticated equipment we should wherever possible adopt a more labor-intensive approach, obviously with an eye to personnel safety, particularly the as low as reasonably practicable (ALARP) principles discussed in Chapter 14.

Beyond the actual decommissioning, site operators also have responsibilities in relation to longer term prosperity of the local community and those who have served them well. Of course this type of planning involves a great deal that is literally outside a site's jurisdiction, so many other stakeholders need to take an active role in developing plans that will endure long after a site has closed. This includes the local community, employee representatives, politicians, government agencies, financial institutions and many more; all need to pull together to secure the future.

Fortunately, there will be an abundance of highly skilled technical and scientific expertise to draw on, so if well managed the closure of a site and a transition to new enterprises can be smoothed considerably. Crucially, planning must begin many years before a site's closure, otherwise the outlook will be unnecessarily bleak for all concerned. So we need to get busy.

15.2.7   Resources

Resources

It would be naive to postulate that nuclear sites have an infinite pool of resources, skilled in every capability, and even more improbable to envision surplus specialists sitting around on the off chance they may be needed 1 day. Just as with every other industry, numbers match demand. Furthermore, with the proviso that it does not compromise safety, a site's personnel are always deployed to reflect their skills, availability and business demands, along with other considerations such as career development, training needs, succession planning, and so on. Clearly managing resources is both challenging and hugely important, but we could say much the same of any industry.

What we can observe is that the more specialized an occupation or an industry is, the more critical its resources become. And just to ratchet up the pressure, the longer the timescales an industry is dealing with the more emphasis there must be on maintaining a capable workforce, one that can stay at the top of its game year after year, decade after decade, and even into the millennia ahead. The nuclear industry certainly fits that profile, and within it the decommissioning arena ticks every resource-critical box.

Now we could argue that when it comes to decommissioning planning, the resources factor is already covered within LFE and the socioeconomic issues we have just discussed. However, it is far too important to be tacked onto other factors and risk the diminution that could bring. No, the issue of resources needs to grab the headlines and the phrase most often employed to spell it out, particularly in nuclear circles, is *suitably qualified and experienced person*, widely referred to as SQEP. To compound resourcing matters the industry's learning curve is both steep and very long, which is not a great combination when you need SQEP resources in a hurry.

What all of this tells us is that, more than most, the nuclear industry needs to put considerable effort into long-term development of its personnel. Furthermore, with the timescales we are considering here succession planning is always going to be crucial. In fact, in many ways the determination that goes into planning nuclear facilities and the farsightedness which is applied to decommissioning planning, must be mirrored with equal vigor on the SQEP resources front. As potential showstoppers go, it's right up there.

15.2.8 Program

Program

We examined programming, or *scheduling* in Chapter 12 so know this is the means by which much of a project's evolution is coordinated. Their pages list all of the activities that must be performed, sometimes numbering tens of thousands, provide a timeline against each of them and crucially keep track of dependencies between one activity and another—all good stuff, and as we have seen fundamental to the successful delivery of an operating nuclear plant. In our previous discussions we saw how a program helps to harmonize activities from a project's inception and through the various design phases, then onto construction, installation of P&E, and up to the point where a facility is commissioned and handed over to its operations team.

However, when we take our knowledge of programming and switch to a decommissioning context, it is immediately clear that its essential role must continue unabated as a facility transitions into this phase, and then stay one step ahead of the action until demolition is complete and all follow-up activities have been concluded.

Of course, early on we do not need a detailed program of goings on that may be decades away, but we certainly need an outline. As a consequence of peering so far ahead, a facility's decommissioning program is bound to go through lots of revisions, not least because it must interact with a site's overarching decommissioning program, which itself will be subject to continual change.

A site's suite of programs is influenced by, and has an influence on all of the planning factors we have just discussed; furthermore they are all live issues which are constantly shifting. Site zoning is an obvious one, as is hazard management, a factor which by its nature is perpetually adjusting priorities. But then all of the others must be thrown into the mix, with everything from decisions on whether to refurbish an existing facility or build a new one, the possibility of remodeling a facility for a new purpose, the influence of cost and funding constraints and on and on, all competing and pulling in opposing directions and all having repercussions on a site's ever-evolving programs. It is a tall order to keep on top of it, so I hardly dare mention that there is even more to come.

We shall see shortly that decommissioning is rarely an enterprise that starts one day and continues unremittingly until it is completed. Rather it goes through several distinct phases, normally with lulls in between, all of which must be carefully coordinated and programmed. Beyond that there are choices in the strategic approach to how decommissioning is effected, each having a huge influence on how programs evolve. As I say, there is more of this to come but clearly it is a very long-haul and, if we were to dwell on the totality, quite overwhelming, so you will be pleased to hear there is some good news.

For a start, time is on our side; this is not to be confused with "decommissioning won't happen for decades so we don't need to do anything." No, we need to make the time buffer work for us. The big concession is that initial decommissioning programs, developed while a facility is being designed and built, do not need to be too detailed. In fact overdoing the detail too soon would stifle the whole process to a standstill. The imperative is that we must do enough to be confident we are on the right track, that a site's decommissioning programs are integrated and most important that we are not leaving future generations to deal with the aftermath of a programming abyss.

In addition to using time to our advantage, every nuclear facility must have a DP, the details of which we shall get to later. For the moment it is worth noting that its first draft is prepared during a facility's design stages and thereafter the plan is continually updated as the years roll by. The idea is that as we move towards preparations for actual decommissioning, the plan is bang up to date and ready to inform pretty much everything that happens next, including refinement of the decommissioning program.

So although the challenge is immense, it is good to know that we can tackle programming in incremental steps and base much of the exercise on an evolving plan which is continually pointing the way. No matter what aspect of decommissioning we consider, one of the overriding principles is to exercise control, not to let it run away from us; in this regard, programming can make a major contribution.

We have already covered quite a lot of ground, so it is worth drawing our breath for a moment before we move on. What we can observe, is that if viewed from a distance decommissioning planning could be perceived as an exercise in coordinating technical and engineering matters, but by getting closer we discover those elements cannot be addressed in isolation.

We can see that to successfully execute decommissioning it takes a much wider view; it demands incorporation of constraints related to cross-site programming, cost and funding, priorities of hazard management, a site's long-term zoning plans, availability of SQEP resources, and so on. Without this added dimension decommissioning plans would quickly be undermined, to the extent that those charged with their implementation will flounder when the time comes. It turns out that rather than having a supporting role, planning factors are central to keeping the decommissioning show on the road.

15.3   Operations phase

We shall shortly examine decommissioning's various phases and how the whole sequence fits together. However, to put that discussion in context we need to step back for a moment and concentrate on a facility's operations, more specifically two essential precursors to decommissioning which are performed as this phase draws to a close.

Clearly a facility's primary purpose is all about conducting the operations for which it was originally designed and built, but that's not all. We cannot run a nuclear plant at full tilt until the day it stops operating, then turn out the lights and walk away—no surprise there. We need to prepare for an orderly shutdown, during which we must ensure safety is not compromised, and that neither people nor the environment will come to any harm as we make our retreat. This is where those precursors come in.

First up, for those facilities that house say plutonium (Pu) products or a highly active inventory, we can take it as read that such materials will be removed. So a nuclear reactor, for example, will be defueled at the end of its life and fuel moved to a cooling pond (Fig. 15.6). As we know from Chapter 1 irradiated fuel needs to cool for quite a while before it can be transferred to a longer term home. With this

Fig. 15.6 Fuel cooling pond.
© Nuclear Decommissioning Authority.

constraint in mind it is not unusual to find that decommissioning sometimes includes retrieving fuel from a reactor's cooling pond. So the operations phase includes removal of what we might call a facility's primary inventory, but what next?

15.3.1 Post operational clean out

The final operation (Fig. 15.7) of all nuclear plants is *post operational clean out* known throughout the industry as POCO. It may not feel like it, but strictly speaking POCO has nothing at all to do with decommissioning; in fact, if you want to be picky, you could even argue it should not be included in this discussion. POCO however, is so inextricably linked to decommissioning and such an important precursor to it, that it we must give it an airing.

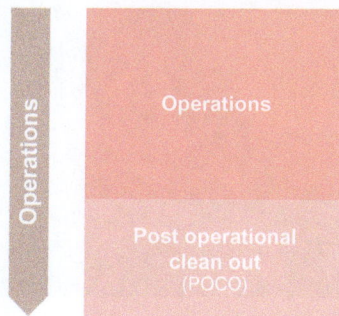

Fig. 15.7 Operations—primary phases.
© Bill Collum.

So what is it? We get a fair impression from the title. When you get right down to it POCO could be described as a major spring-cleaning exercise, conducted at the end of a nuclear facility's life. Over time, for example, residues from active liquors may build up within pipework and process vessels. From an operational point of view such deposits should not present much of a problem, since either the original process design will have anticipated their accumulation, or modifications will have been made during a plant's operational life to abate potential problems they may cause. It is a different matter during decommissioning, because unlike a clean pipe or vessel those deposits will emit radiation. During a plant's working life they are well shielded, but by definition decommissioning involves dismantling and removing those selfsame radioactive items.

Obviously everyone wants to minimize levels of radiation and contamination before decommissioning gets underway. It makes the job inherently very much safer, maximizes opportunities for hands-on dismantling rather than slower, more costly remote methods, and reduces volumes of active waste generated along the way. In cases where process residues remain, one of the POCO activities will be to flush systems through with a solution such as nitric acid, which will dissolve residues and carry them off for treatment elsewhere. Apart from process residues POCO also seeks to remove accumulated solid wastes, including powders, from a facility and generally leave the place spruced up and ready for what follows.

It is no surprise that POCO can be simplified considerably by adopting a comprehensive housekeeping regimen while a facility is operating. A bit like regularly wiping down the oven at home, rather than allowing deposits to accumulate into an industrial scale problem, one that can take a fair chunk of your weekend. But unlike a domestic situation, dusters and rubber gloves will not make too much of an impression.

As a result, the wherewithal to keep a building in tip-top condition is considered during the design stage, with input from operators, and incorporated into a facility's operating and maintenance (O&M) procedures.

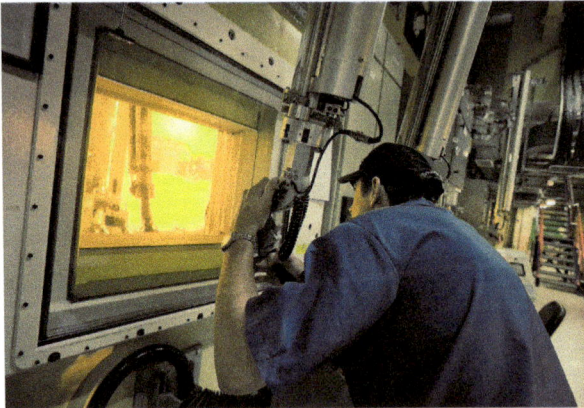

Fig. 15.8 POCO—preparing a cave for shutdown.
© Nuclear Decommissioning Authority.

The easiest way to draw a line under POCO's remit is to recognize the one thing which above all differentiates it from decommissioning. True POCO is carried out using operational P&E, so that which already exists (Fig. 15.8). That being said, a case may be made for minor enhancements to a plant so as to simplify and improve safety for the decontamination and dismantling (D&D) phase which follows at a later date. In addition POCO may also make use of facilities, such as an active liquor treatment plant, used by the facility when it was operating. And whereas decommissioning normally requires the introduction of more kit as it progresses, POCO needs none of that: you just finish normal operations and get on with it. All right, some preparations are needed, but you get the idea.

Apart from using existing kit, the other most recognizable trait of POCO is how it ends. Unlike an operating facility it should be passively safe. In other words, although there will no doubt be areas harboring radiation and contamination, there should be no prospect of having to rush to a facility to deal with a safety related incident. Yes, its ventilation system may still be running, monitoring equipment continuing to analyze the environment, and so on, but POCO's end state should ensure any form of urgent intervention will be unnecessary.

The beauty of POCO—and this is crucial—is that in the main it is performed by a facility's operations personnel, those who know their building intimately and are familiar with every inch of every mile of pipework under its roof. They know every vessel, every valve and motor, every item of equipment. They are as familiar with the building as we are with our own homes. If POCO is delayed for some reason, then those eventually charged with carrying out the task, maybe years later, will spend a long time familiarizing themselves with their inheritance, so there are lots of imperatives for getting on with it in a timely manner.

It is worth mentioning that to underscore the fact that POCO is an operational activity, it should normally be included in a facility's operations safety case and, strictly speaking, not afforded prominence in subsequent iterations which cover the various phases of decommissioning.

15.3.2   Operations—Secondary phases

In many cases, we can expect nuclear plants to follow the well-worn path of fulfilling the purpose for which they were designed, transitioning through POCO and then moving onto decommissioning, but not always.

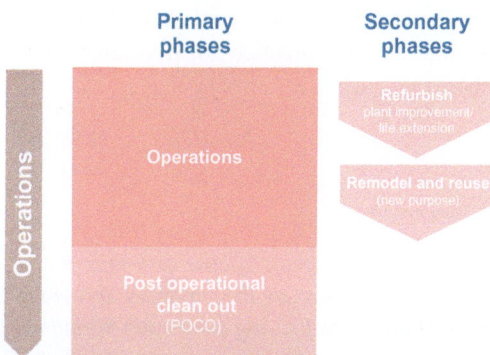

Fig. 15.9 Operations—primary and secondary phases. © Bill Collum.

We know from our discussion on decommissioning planning factors that as a facility nears the end of its design life there may be a possibility of refurbishing it so as to extend its useful operating period. Likewise a decision may be made to remodel an aging plant and dedicate it to a new purpose. Fig. 15.9 highlights the fact that these alternative paths, or *secondary phases* are firmly rooted in a facility's operations period, yet, as we know, if invoked will have major implications for long-term decommissioning planning. And not just for the facility in question but potentially across its host site. Once again it demonstrates the all-pervading nature of decommissioning's tentacles, and reminds us that no matter what aspect of decommissioning we are considering the sooner we get to work on it the better.

15.4 Decommissioning phases

As I mentioned earlier, when it comes to decommissioning it is not necessarily a task that is performed as one continuous activity; actually such an approach would be quite unusual. The normal practice is to take things one step at a time, biting the challenge off in predefined phases. In fact the requirement to divide decommissioning into stages is embodied within SLC 35, so regulators would need some convincing if an alternative strategy were proposed.

The various phases are spread over anywhere from a few years to several decades, maybe even beyond 100 years. However, we are not talking here of a process which runs continually for all of that time. Some periods will indeed be marked by frenetic activity, but others can make an equally valid contribution with apparently little discernible progress, what we might call a managed wait. As we shall see, there are many reasons for moving so methodically; suffice to say for the moment that taking a nuclear facility apart, one which initially at least houses radioactive materials (albeit not its primary inventory) is at least as challenging as building and kitting out a shiny new building in the first place.

As if to prove the point, there are several variations in the number of decommissioning phases and the titles given to them, along with a plethora of descriptions on the detail of what each entails. Happily, there are common themes which everyone adheres to and can be rationalized into four primary phases (Fig. 15.10) in other words the ones which pretty much always happen.

We shall take these top-tier steps first and then look at two secondary phases which, although prevalent enough, are more dependent on specifics of individual facilities so may not necessarily be invoked. To complete the picture, we shall also take in two tertiary phases which are somewhat outside standard practice so, if conducted at all, must be justified on a case by case basis.

15.4.1 Primary phases

15.4.1.1 Surveillance and maintenance

You know what it is like if you leave your home for just a couple of weeks while you go away on holiday; there are jobs to be done to secure the premises and leave everything in good order. This covers everything from locking doors and windows,

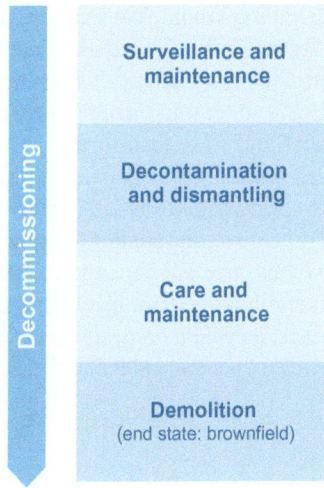

Fig. 15.10 Decommissioning—primary phases.
© Bill Collum.

to turning off the water supply, canceling newspapers, maybe arranging for someone to cut the grass, and so on. Likewise, you cannot walk away from a nuclear plant, or come to that any other industrial building, for a prolonged period and trust everything will be fine when you return.

The surveillance and maintenance (S&M) phase is invoked where POCO has been satisfactorily concluded and a facility has ceased its operations, but is not yet ready to be taken apart. So although the primary radioactive inventory has been removed, there will still be areas of radiation and contamination which must be contained, and in many cases shielded, until they are dealt with by the phase which follows.

As usual, exact details vary from case to case but typically include continuing operation of a facility's nuclear ventilation system, or at least some elements of it, along with running radiological monitoring systems, retaining the wherewithal to conduct sampling campaigns, and so on. The list goes on, but essentially the S&M phase ensures that a facility's remaining radionuclide inventory remains in a safe, secure and stable condition. In addition there will be routine maintenance of a building's fabric which, although a significant enough task in its own right, is doubly important because it contributes to that primary objective.

15.4.1.2 Decontamination and dismantling

This phase is often referred to simply as *dismantling*, but this can conjure up an image of taking stuff apart as freely as we would in a nonnuclear environment. So it is preferable to use the fuller title of *decontamination and dismantling*, as it gives more of a clue to what is really involved.

The easiest way to understand what this phase encompasses is to visualize the state of a facility when it has been completed. If, after D&D, you were to wander around what was once an operating nuclear plant, you would find that virtually all trace of its nuclear heritage had been erased. Only the building shell and its main structural components would remain, posing for visitors the conundrum of what purpose its past life had served.

At a more detailed level D&D, at least in its purest sense, entails stripping out and disposing of pretty much anything in a facility that is not totally immovable. It operates on four levels, which result in four corresponding classes of material. The first is classed as exempt, also known as *free release*, which is no different to materials found in a nonnuclear environment so can re-enter the supply chain or be disposed of at a regular landfill site. The other three are defined in radiological terms as very low level (VLL), low level (LL), and intermediate level (IL). Depending on its potential future use, each type is either recycled or designated as waste and disposed of elsewhere. If categorized as waste they are badged with the now-familiar descriptions of VLLW, LLW, and ILW.

VLL materials

As you can imagine the range of dismantled materials is wide and varied. Typically we find electrical cable and the trays that support it, ventilation ductwork that once carried fresh air, doors and their frames, all manner of fixtures and fittings, and lots of steelwork, concrete, and rubble. Radiologically speaking VLL materials are just a touch above the exempt category, so can be disposed of at specially licensed landfill sites. Alternatively some VLL materials may undergo a period of radioactive decay, after which they satisfy exempt criteria and go that same route.

LL materials

Moving up to the next category we have LL materials. In the main this covers items which although contaminated are within a range that allows them to be handed manually. In practice this embraces the wide gamut of consumables that pass through any nuclear plant, such as the PPE worn by operators in C3 (amber) and C4 (red) environments, materials used to construct temporary enclosures, some HEPA (ventilation) filters, and so on.

These materials arise routinely during a plant's operational years, are normally categorized as waste (LLW) get bagged or drummed (Fig. 15.11) and dispatched down familiar routes to an appropriate store (Fig. 15.12) or disposal site.

Once we enter the D&D phase and start taking things apart, the volume and variety of LL material increases considerably. That being said, the approach to its handling, packaging and disposal, although a little more ad hoc than during normal operations, is largely unchanged.

A good example of a LL material that arises during D&D occurs where we find contamination embedded in some of the concrete structures that once enclosed active environments. The bulk of such concrete, and certainly its reinforcing bars, will satisfy criteria for off-site disposal, but before that we may need to remove its contaminated surface.

The standard practice is to scabble concrete (Fig. 15.13) using a machine that pounds its surface with an array of tough steel pins, or alternatively abrades it with a rotary cutter. Either way the top layer is pulverized and falls away, with the process

Fig. 15.11 Drummed LLW.
© Nuclear Decommissioning Authority.

Fig. 15.12 Low level waste store.
Science Photo Library.

repeated until ingrained contamination has been removed. The contaminated debris that results is gathered up and packaged, with VLLW scabblings shipped to a licensed landfill site and LLW to an appropriate disposal site. Once concrete has been cleaned up in this way it can be removed, when the time is right, by employing conventional nonnuclear techniques.

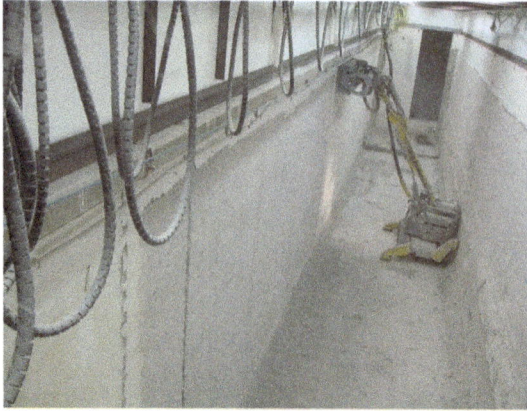

Fig. 15.13 Scabbling concrete.
© Nuclear Decommissioning Authority.

IL materials

Finally we have IL materials which demand more stringent controls and, with the exception of those at the lower end of this category's radiological spectrum, are too active to approach. As a result, where personnel do get involved in dismantling operations their exposure is governed by the ALARP principles we discussed in Chapter 14, essentially ensuring it is *as low as reasonably practicable* In addition they will almost certainly need some form of PPE or RPE (Fig. 15.14) in this case a fire-retardant suit.

Fig. 15.14 Manual D&D operations.
© Nuclear Decommissioning Authority.

Once again materials in this group cover quite a spectrum, with most originating from the heavily shielded C4/R4 (red) radiological environments we have discussed previously. As a result practically everything is fabricated either mostly or entirely from stainless steel. This includes mechanical equipment, vessels and pipework that in their day were relied on to perform impeccably year after year. So no doubt this kit will be terribly robust, which is not great when it comes to taking it apart.

Earlier maintenance constraints will have seen to it that modularization has been maximized. This helps somewhat in taking things apart, but there will still be much that is necessarily rooted to the spot and unwilling to yield its position. Added to this, it is likely that the entire floor and at least part of the walls in these areas will be lined with stainless steel, all expertly welded in place. When we put the whole scene together it is clear that disassembly and size reduction would be difficult even if we could stroll in at our leisure, but then when we factor in radiation and contamination the challenge escalates exponentially.

As usual, with sufficient forethought and planning the dismantling task can be streamlined considerably, but it will still be a slow and meticulous process, one that demands sophisticated cutting techniques and careful choreography of handling equipment, possibly with much of it conducted in-situ by decommissioning personnel.

As we know there is always an imperative to minimize waste volumes, particularly at the radiological levels we are considering here, so wherever possible retrieved IL components are decontaminated to get them down to the LL category. With this in mind decommissioning operations are always arranged to minimize possibilities of cross-contamination, and also configured to segregate the various waste streams at the earliest opportunity. Unsurprisingly, most IL material is unsuitable for recycling so is categorized as waste (ILW) and, after suitable conditioning and packaging, shipped to an appropriate waste storage facility.

Fig. 15.15 Remote dismantling. © Nuclear Decommissioning Authority.

With its difficulties of getting close to materials and associated widespread use of remotely operated equipment (Fig. 15.15) it is no surprise that dismantling and handling IL materials is the exercise most commonly associated with the term *decommissioning*. It is certainly the most painstaking of all decommissioning tasks, the one with what we might call the highest degree of difficulty, so understandably grabs most of the attention.

Ideally D&D should be conducted as an extension of the POCO exercise, in that a facility should draw on its own P&E and resources to the maximum extent possible. This is an essential element of our drive to minimize costs and curb creation of future liabilities, and is a theme we shall return to in the next chapter when we discuss *self-decommissioning*.

15.4.1.3   Care and maintenance

The fundamental difference between this and the earlier S&M phase is that buildings no longer pose a radiological hazard, so there is nothing of note to keep an eye on, certainly nothing that needs radiological monitoring. In this regard care and maintenance (C&M) is more akin to tending to a regular industrial facility while it awaits demolition. Speaking of which, you would think D&D would be followed more or less immediately by a facility's demolition and indeed sometimes it is, but it might be better to wait.

It would be an understatement to say that nuclear sites must take safety more seriously than most. The thought of demolishing a sizeable building close to an operating nuclear plant will not be greeted too enthusiastically, likewise if close to a facility awaiting D&D, so still housing nuclear materials. It can be done, but not in the vigorous way we might have witnessed at a regular industrial complex. Everything revolves around ensuring the continued safe operation of neighboring facilities. For example, it is notoriously difficult to make a safety case for proposals to use tower cranes close to nuclear plants; they might fall over. The chances are slim but when it comes to drawing up demolition plans, that slight possibility can be enough to rule them out of contention. In case you are wondering, use of explosives in the vicinity of nuclear plants is so rare that we can pretty much assume it is out of the question.

By the way, a nuclear facility in this context covers more than just buildings with a radiological inventory. Utilities, or services, and the infrastructure that binds nuclear plants together must also be considered and protected. After all, it would be a pity to spend months figuring out how to protect a safety-critical facility while nearby demolition is carried out, only to have one of its essential service lines severed on the first day. The utilities receiving most scrutiny will of course be those directly related to a plant's safe operation. These vary from case to case but typically include electrical power, water, active liquor connections, and so on.

So we can see that C&M fills the void while waiting for demolition to begin. As for the actual C&M task itself, in the grand scheme of things there is not that much to it. The main objective is to ensure a building's fabric does not pose a hazard either to people in the vicinity or to other buildings nearby, especially those which are either operational or housing nuclear materials. As we might expect, a facility's maintenance regimen during the C&M stage will be in marked contrast to the cosseting received during its operational years. In those heady days and with a radioactive inventory under the roof, maintenance will have been a serious contender when annual funding was divvied up. It is a different matter during its twilight years, when such expenditure would ordinarily be totally unjustified.

There is however one factor that may weigh in favor of maintaining a building's external appearance: public perception. Most industries pay attention to how they are viewed by the public. For the nuclear industry, which is subjected to more scrutiny than most, appearances matter a great deal. It may be that a perfectly serviceable building shell is spruced up because none of us are impressed by peeling paint or a stained façade. Strictly speaking it may not be essential from the safety or maintenance perspective, but other imperatives may deem it worthwhile.

For completeness I should mention that where we are considering uncomplicated facilities, probably dealing with just VLL and LL materials, it may be possible to bring a facility to a passively safe state through POCO alone. In other words to skip the S&M and D&D phases and go straight to C&M or even demolition. Everything hinges on the exact inventory, potential hazards, and so on, but under the right circumstances it is occasionally possible for POCO to achieve radiological cleanliness.

15.4.1.4 Demolition

With the earlier phases concluded, demolition of a redundant nuclear plant (Fig. 15.16) should really be no different to that practiced in the nonnuclear world, albeit with the possibility of provisos I have just mentioned, so maybe some restriction on use of tower cranes or even mobile cranes—everything depends on potential hazards and local circumstances.

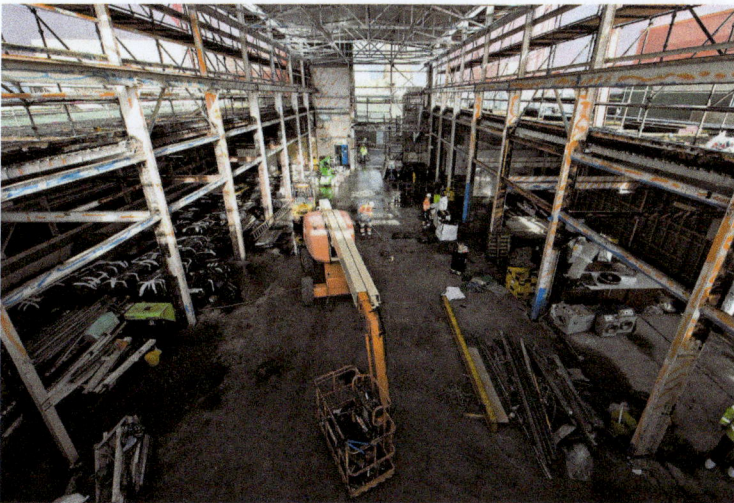

Fig. 15.16 Demolition of redundant nuclear plant.
© Nuclear Decommissioning Authority.

There is however a difference in the way nuclear and nonnuclear demolition is defined, more specifically what the end state is. On a regular industrial site we can take it that demolition infers taking a building down, uprooting its base slab and foundations, and carting the whole lot away. In the nuclear world, when we speak of demolition in a decommissioning context we need to be clear on exactly what we mean. Ordinarily the interpretation is that a building's structure will be taken down and no doubt removed, but its concrete raft and foundations remain in place.

Fig. 15.17 In-ground sampling.
© Nuclear Decommissioning Authority.

In decommissioning terms the action of breaking up and removing in-ground concrete and other materials, including soil, is known as *remediation*. We shall discuss this in a moment when we look at secondary phases, but essentially remediation kicks in where sampling (Fig. 15.17) shows there is either known or possible contamination within a facility's concrete base, or the ground beneath it.

Where a facility is demolished but its raft, or elements of it are left in place, then the area is defined as being in a *brownfield* condition. To achieve *greenfield* status, where a site is returned to its original pristine condition, we need to conduct a thorough remediation campaign.

Independent of whether or not remediation is planned, there is a necessary precursor to demolition, namely a thorough radiological survey must be conducted beforehand. This provides final confirmation that remaining building materials comply with conventional disposal criteria, so can be recycled or dispatched to a landfill site. With the survey done and approvals in place, heavy equipment is free to move in and raze the scene.

15.4.2   Secondary phases

In the real world it is not always possible or indeed sensible to stick rigidly to the decommissioning sequence I have just described, so in practice we must be prepared to exercise some flexibility. As I mentioned earlier the four primary phases are pretty much always conducted, but they may not necessarily be alone. Outside of those top flight activities there are all manner of potential diversions, but two of them are common enough to merit *secondary phase* status (Fig. 15.18).

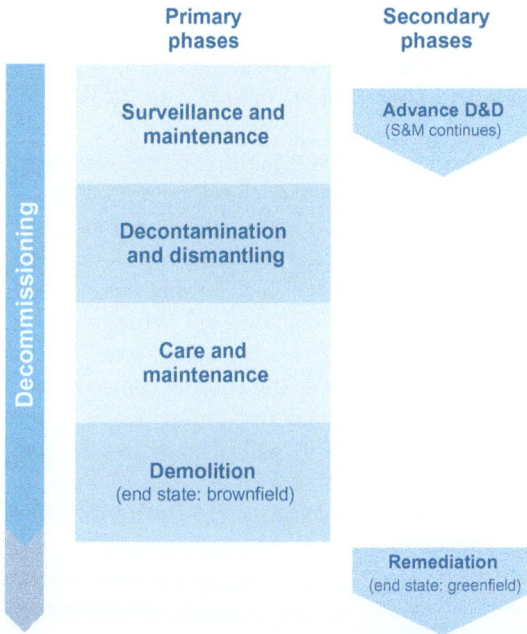

Fig. 15.18 Decommissioning—primary and secondary phases. © Bill Collum.

15.4.2.1   Advance D&D

In the normal run of things we can expect the S&M phase to be a quiet time, with little going on apart from overseeing a facility's radiological wellbeing and generally keeping the place ticking over. However, from time to time, it may make sense to make a foray into the D&D phase which is due to follow.

The reasons for an early start can vary considerably. It may be that members of the original operations team are awaiting redeployment, so before moving onto new roles are free to put their expertise to use on D&D tasks, an opportunity too good to miss. Perhaps a hazardous or potentially hazardous situation needs to be dealt with and prudently should not be made to wait for all-out D&D, which may be some years away. Or it could be that upcoming D&D entails some particularly difficult or even unique challenge, such that a targeted start would generate valuable LFE which can be used to inform detailed planning for full D&D. Whatever the reason, we should be alert to the possibility of launching some preemptive D&D and reaping benefits when the time comes to begin in earnest.

15.4.2.2   Remediation

We have seen that in a decommissioning context, unless explicitly stated to the contrary *demolition* is taken to mean removing a building down to the ground. Furthermore, such demolition is conducted without need of special radiological controls. Where we are certain that a facility's concrete raft and foundations do not harbor radiological hazards, then demolition's remit can be extended to include in-ground materials. However, if a facility's base or the ground around it is known to harbor contamination, or the possibility exists, then we must make a decision on how to proceed.

For some facilities agreement may be reached with regulators and other interested parties, that removing a contaminated concrete base, or surrounding soil, may not make a lot of sense. Of course on the face of it that sounds a bit lax to say the least, so we need to understand what is going on. As usual everything comes down to specifics. What levels of contamination are present? Will it do any harm if left undisturbed? If removed, where would materials be disposed of? What might be gained and how much would it cost? You can imagine how the debate goes.

Let's take two extreme cases. If in-ground concrete or soil is badly contaminated, sits cheek by jowl with an occupied area and poses a risk of harming people or the environment, there is little to ponder and such materials will be uprooted and sent on their way. On the other hand, on large nuclear sites such as many of those in the United States, facilities may be dotted around the landscape like Australian sheep farms. In such situations, assuming contamination poses no immediate hazard and will not do if left alone, it would be difficult to find merit in digging it up and moving it, particularly when we factor in the mandatory ALARP principle of minimizing radiological exposure to site operators. For other cases, between those at opposite ends of the spectrum, there are most definitely no ready answers. So exhaustive analysis and widespread consultation will be needed to reach a consensus on how to proceed.

When we put the whole picture together it becomes clear that decommissioning does not necessarily result in a site being returned to greenfield conditions. In some situations in may make sense to bring a site to brownfield status and reuse it for purposes which are compatible with those conditions. Obvious contenders are a new nuclear plant or waste storage facility but, depending on exact conditions which can vary a great deal, other industrial or commercial premises may also be viable.

15.4.3   Tertiary phases

There is no doubt that in the normal run of things, decommissioning's four primary phases are routinely conducted; in addition, the two secondary phases are fairly commonplace. Beyond that, if we go a little off-piste, we may encounter two tertiary phases (Fig. 15.19) which occupy territory visited on more of a *needs must* basis.

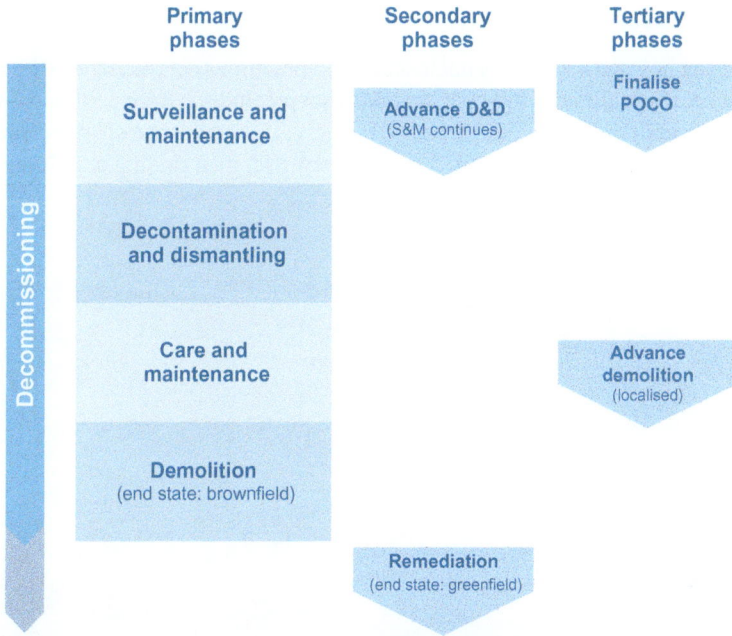

Fig. 15.19 Decommissioning—primary, secondary, and tertiary phases.
© Bill Collum.

15.4.3.1 Finalize POCO

This activity recognizes that POCO may not have been fully implemented at the end of a plant's operational life, so deals with elements that missed their operations-phase timeline. Having said that, in recognition of fact that POCO is most definitely an operational activity and one best executed seamlessly at the end of that phase, delayed POCO is not afforded the prominence of qualifying as a secondary phase. So here it is under the *tertiary* banner.

Occasionally then we may make a place for some elements of POCO during what should really be pure S&M but there are two provisos. Firstly there must be as little delay as possible between concluding operations and returning to POCO, otherwise we risk undermining the established rationale for conducting the exercise in a timely manner. And secondly, any delay is bound to complicate POCO: loss of knowledge, reenergizing equipment, and so on. This may have safety implications and will certainly drive up costs, so the case for delay needs to be a compelling one.

We know by now that when it comes to decommissioning and the preparations for it, there are multiple factors all pulling in opposing directions so POCO, or elements of it, may on occasion be legitimately postponed. It is, however, far from ideal, so we do need to think long and hard before going down that route.

15.4.3.2 Advance demolition

We have seen how demolition on nuclear sites is sometimes delayed due to the risk of damaging nearby facilities. However, this does not preclude us from getting on with at least some demolition, as long as we can do it safely. If we think back to our discussion on remodeling facilities and putting them to a new use, then ridding a building of its internal structures can have quite a lot going for it.

Fig. 15.20 Advance demolition.
© Nuclear Decommissioning Authority.

One of the best opportunities arises where a redundant building is covered by a portal frame, so has no internal steelwork and therefore plenty of elbowroom. Of course we could do this type of localized demolition within any kind of building, but portal framed structures are particularly attractive as, apart from being relatively unconstrained often have an electric overhead travelling (EOT) crane tucked under the roof, a piece of kit which can be very useful to the demolition and remodeling folks. As the main steel structure and outer cladding, or *siding* remains in place, this can be a good way of making headway on demolition (Fig. 15.20) while not raising radiological safety concerns.

There are plenty of other permutations where advance demolition may contribute to reducing a site's liabilities, minimizing environmental impact, keeping costs down and generally being viewed as a good idea. To maximize opportunities, we just need to keep in mind that demolition need not necessarily be an all-or-nothing activity.

It would seem that once we are familiar with decommissioning's various phases (Fig. 15.21) and how they relate to closing out a facility's operations, we are up to speed and so ready to draw up decommissioning plans and get on with it. However there is another level, a strategic dimension, that sits above individual phases and without which

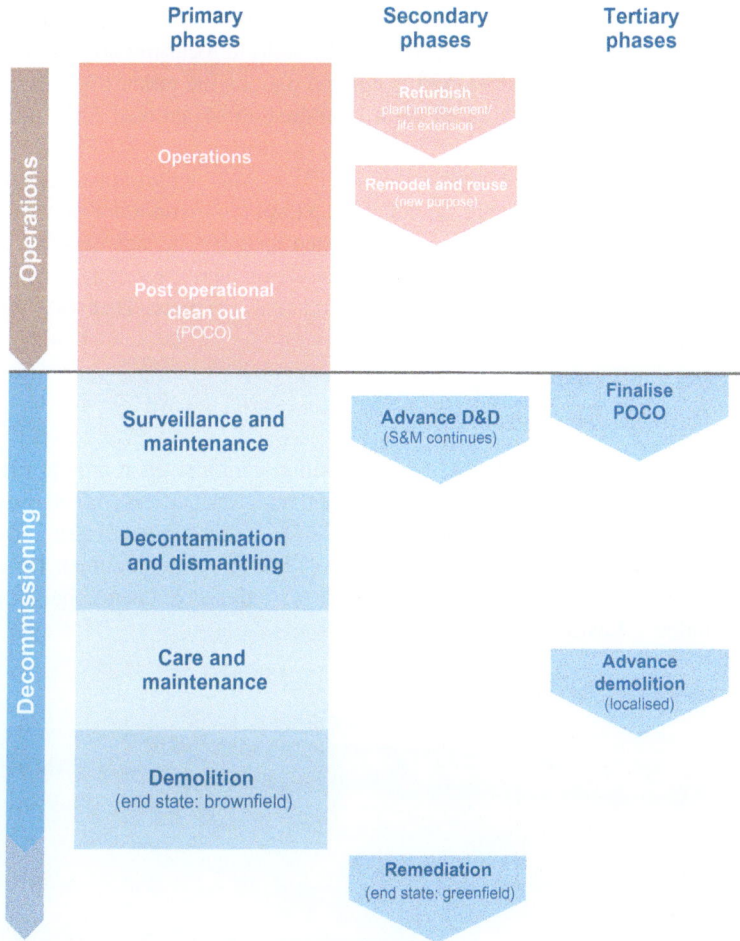

Fig. 15.21 Operations and decommissioning phases.
© Bill Collum.

we cannot proceed. In other words, although we have pieces of the puzzle to hand they can be assembled in different ways. It takes a strategic view to figure out which arrangement works best in a given situation, so this is where we shall go next.

15.5 Decommissioning strategy

In essence there are three strategic approaches to decommissioning, which between them dictate how long the process will take and what the final result will be. As an indication of how important this element is, for most nations it is unlikely that individual nuclear sites will be allowed to select a decommissioning strategy for their own facilities. They will have an input certainly, but ultimately it is governments, through their regulators and expert advisors, who formulate strategy and impose it on sites under their jurisdiction. This type of strategy tends to focus mainly, although not exclusively on redundant reactors, and is applied nationwide rather than piecemeal. And as we might expect it is based on a combination of national influences and international best practice.

Broadly speaking the three alternatives are *immediate dismantling*, get on with decommissioning straightaway and do not ease up until it is complete. *Safe enclosure*, execute some elements of decommissioning then stop for several decades before completing the task. *Entombment*, execute some elements of decommissioning then secure what remains of a facility and leave it permanently in place. That was the quick tour, let's look a little closer.

15.5.1 Immediate dismantling

With this approach (Fig. 15.22) a decommissioning team arrives at around the time POCO is being concluded, bypass surveillance and maintenance (S&M) and promptly get busy on decontamination and dismantling (D&D). If this strategy is being invoked in its purest sense then care and maintenance (C&M) is also sidelined. However, recognizing that demolition may be delayed for safety or operational reasons, I have included C&M in the diagram.

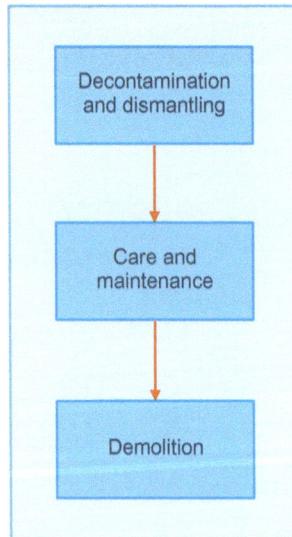

Fig. 15.22 Immediate dismantling.
© Bill Collum.

One of the main factors influencing whether or not to press on with this fast-track strategy hinges on a facility's radiological inventory. In fact there is a clear divide between beta-gamma plants and alpha plants, those dealing with plutonium (Pu) products. We have seen previously that fresh Pu-241 emits precious little radiation, so with appropriate protection from inhalation, usually a glovebox, is quite safe to handle. However as Pu-241 decays, through a process known as nuclear transmutation, it transforms itself into americium-241, an isotope which is radioactive. Transmutation of Pu-241 is gradual, so takes around 50 years to reach its peak and then remains there for maybe a couple of hundred years. However, levels of radiation will be sufficient after say 5 years, 10 at the most to rule out manual handling.

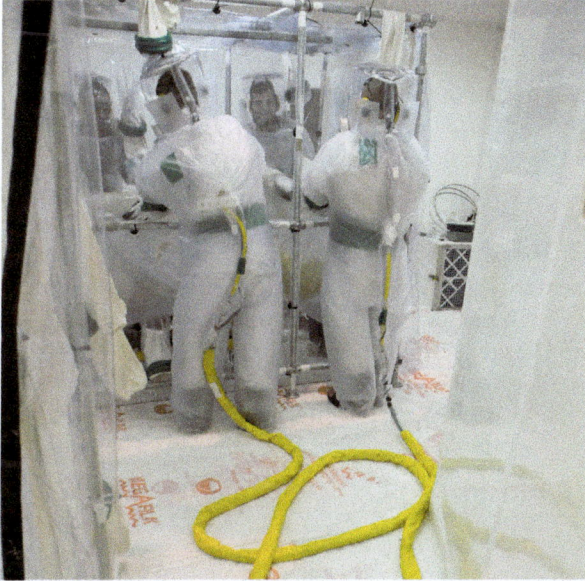

Fig. 15.23 Glovebox dismantling.
© Nuclear Decommissioning Authority.

What this tells us is that when decommissioning alpha plants (Fig. 15.23) with their potential for a residue of Pu-241 powders and impending Am-241, time is of the essence. If we delay there is the prospect of having to introduce shielding and deploy remotely operated equipment. All of which is more complex, costly, time consuming and hazardous than would be the case if we tackled the job early on. So alpha plants may be the front runner for immediate dismantling, but beta-gamma plants are by no means precluded.

Whatever the facility, immediate dismantling holds out the promise of several distinct advantages over other strategies, not least many of the operators who have just completed POCO may be able to stay on, joining decommissioning specialists and feeding their detailed knowledge of a plant into its demise.

Just as helpful, the existing operations phase P&E including the ventilation system, along with utilities and infrastructure will all be in good order. So rather than being resurrected some years later, all this can readily be put to use on D&D. This may even extend to conditioning, packaging and exporting the various waste streams. In addition, by getting on with decommissioning so soon, the regulatory framework which governed the operations phase will almost certainly still prevail. This may not sound like much, but the effort and cost involved in catching up on new rules at some time in the future can be considerable.

So lots of advantages, but they do not necessarily outshine the drawbacks. If we get on with decommissioning beta-gamma plants without allowing for a decay period, radiation will be at maximum levels, with all that entails in terms of shielding and remote operations. Importantly, a decommissioning team would face the prospect of incurring a higher collective dose than would be the case if D&D was delayed for some years. Remember, site operators must adhere to the ALARP principles of ensuring dose is *as low as reasonably practicable*, which is a serious responsibility.

In essence, then, and from a purely radiological perspective, when dealing with alpha plants it makes sense to get on with D&D straightaway, while for beta-gamma plants there are clear advantages in delaying, but this is only part of the story. If we look beyond the pros and cons of implementing D&D and think back to site-wide planning factors of hazard management, site zoning, funding, resources, and so on, it becomes clear there is much more at play here than relative merits of whether or not to decommission a particular facility urgently, or hold off for several decades. Not least, decommissioning is only feasible where stores are available for the various waste streams, so there we have something else to factor in. As ever there is a great deal to consider and a significant effort required in terms of forward planning.

15.5.2   Safe enclosure

Also known as *deferred dismantling* and *safestore*, this strategy (Fig. 15.24) comprises the familiar decommissioning phases but mixes them up a little. As always, we assume that materials in the radiological high-level category were removed towards the end of normal operations and that POCO has been satisfactorily concluded.

So we are ready to get on with decommissioning, but unlike immediate dismantling conclude it would be prudent to hold off on going too far. It could be due to our old friends the decommissioning planning factors elbowing their way in with higher priorities, but the main reason, the one that makes safe enclosure particularly attractive, centers on radioactive decay.

As we know, with the notable exception of Pu-241 (which decays to become americium-241) most radioactive materials handled by the nuclear industry become less active with the passage of time. Let's take the example of activated metal, a condition which is particularly relevant to application of this strategy. We saw in Chapter 1 that when stainless steel occupies the inside of a nuclear reactor it becomes radioactive, more specifically, *activated*. It happens due to a process known as *neutron capture*, which hinges on the way cobalt-59 isotopes within stainless steel "capture" an additional neutron. Once this happens the previously benign Co-59 isotopes transform into Co-60 which is unstable and therefore radioactive.

What is of interest in this context is that Co-60 has a half-life of 5.3 years, which means that within 30 years its initial activity will have decreased by a factor of 64. So you can see what is coming here. If instead of taking a reactor apart soon after it ceases

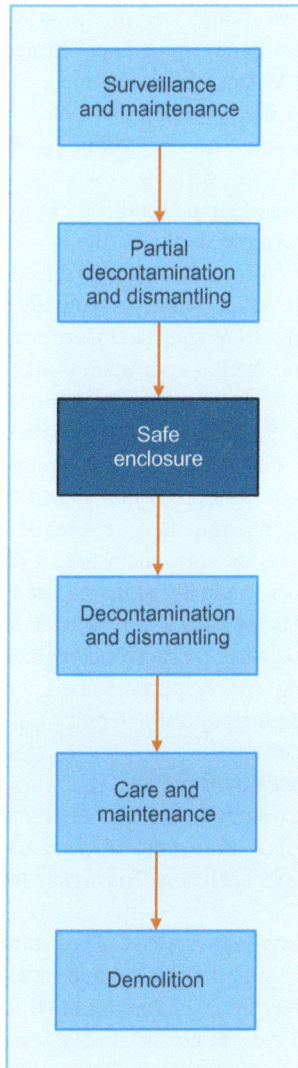

Fig. 15.24 Safe enclosure.
© Bill Collum.

operations we wait for a generation, activity from activation will have reduced considerably and the whole task will be simpler, less hazardous and an altogether more appealing prospect. Of course activation is just one strand of this particular debate, so in practice a safe enclosure period may last up to 100 years.

So delaying can be a good idea, a useful tactic, particularly from an ALARP perspective, but that does not mean we can kick our heels for a few decades before deciding to make a start, quite the opposite. In fact a great deal of preparation is required before a safe enclosure period can begin. As with immediate dismantling it is possible to bypass the surveillance and maintenance (S&M) phase. However, since we have dispensed here with the "immediate" nametag we can assume there will be at least some period of S&M, but it does not have to be.

With a title of "safe enclosure," we can anticipate that D&D, initially at least, is not fully implemented, so sure enough, following S&M, we move into partial D&D. However, there are too many variables for a one-size-fits-all approach to this phase. As a minimum, partial D&D must be sufficient to achieve a prolonged period of safe enclosure before returning to complete the task.

The goal here should be to attain, as far as is possible, a *passive* safe enclosure state, rather than an *active* one—in other words, minimize reliance on ongoing S&M. Broadly speaking facilities in *active* safe enclosure are accessed regularly, maybe daily, for purposes of keeping P&E in an operational condition and for ongoing surveillance. At the other end of the spectrum, with *passive* safe enclosure, facilities become dormant or close to it. So access is only required a handful of times a year to check all is well, essentially to inspect surveillance equipment and the building fabric. One measure of achieving passive status is that a facility becomes "standalone," so severs all of its operational connections to other plants.

Whatever the starting point for safe enclosure, with the passage of time and radioactive decay, the emphasis gradually shifts to become more and more passive. However, with some elements of D&D still outstanding, at least some form of surveillance will still be required, albeit that the exact regimen may well change over time. As with immediate dismantling, for our purposes here we can take it that completion of D&D is not followed straightaway by a facility's demolition, so can assume a period of C&M will probably follow.

Apart from reaping the benefits of reduced radiation levels for D&D operations themselves, adoption of this strategy also shifts the emphasis on radioactive wastes downwards, generating less ILW and more in the LLW, VLLW and free release categories. So a period of safe enclosure has much to commend it, but then the converse is also true.

With several decades elapsing between POCO and returning to complete D&D, it is unlikely that decommissioning specialists will have access to personnel who operated the facility. And those who may be available might struggle to recall nuances of the precious details that occupied their minds in the days they walked its floors. Further-more, the partial D&D that is carried out must be conducted when a plant's radiation levels are at their highest, or close to it. So as we have just discussed, some careful thought must be given to how much upfront work we are prepared to undertake in order to make the safe enclosure period as passive as possible.

In addition, any P&E that may be called on to complete D&D must be kept operable, or at least in a condition that enables it to be reenergized when needed, or alternatively new P&E must be installed. Just as significantly, any P&E that is important to a plant's safety during the safe enclosure period must be kept in tip-top condition.

We can see that although costs of completing D&D have been postponed (therefore benefiting from effects of *discounted cash flow*), by stretching the overall timescale, some additional costs are incurred. But then D&D is simplified considerably, so will cost less when it is performed. Aside from finances there are ALARP principles and environmental considerations, so we are not short of things to think about. Exact details vary from case to case and much depends on whether the safe enclosure is passive, active or somewhere in between. Whatever the detail, there is much to factor in when deciding whether or not to opt for a strategy of safe enclosure.

15.5.3   Entombment

I need to say straightway that strictly speaking entombment sits on the fringes of being recognized as a "strategy." It is, though, sometimes adopted, so for completeness I have included it here. The reason for its lowly status is that entombment, also known as in situ *disposal* and *on-site disposal* is really only acceptable within a narrow set of circumstances, so is most definitely not mainstream.

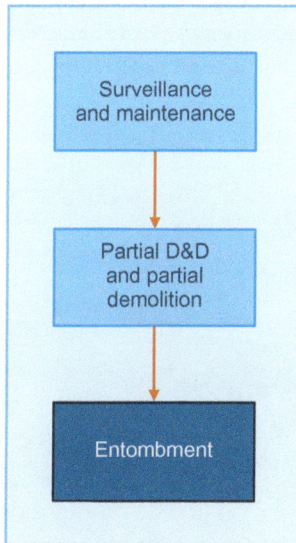

Fig. 15.25 Entombment.
© Bill Collum.

To get entombment (Fig. 15.25) in perspective we need to think first about countries with a significant nuclear program. They will typically have several nuclear licensed sites and scores of nuclear facilities. In these circumstances it makes perfect sense from an environmental, financial and security point of view, to develop a nationwide waste management strategy and establish national repositories for long-term storage of radioactive wastes. In this way all wastes, both operational and from decommissioning, can be congregated in centralized locations which are much easier and less costly to manage than the dispersed alternative.

If we contrast this with countries that have quite a small nuclear program, those with just a single research reactor or maybe a couple of nuclear plants, then thoughts of establishing centralized national waste storage facilities are open to question. In these situations there may be little merit in embarking on the challenge of gathering up wastes from one or two locations, conditioning and packaging it, then shipping it somewhere else. Maybe, after careful scrutiny, the consensus will point in another direction.

For those countries with one, or just a small number of nuclear facilities, or perhaps in situations where nuclear plants are particularly isolated, there may be case for taking a more minimalistic approach to decommissioning. When we weigh up environmental consequences, potential hazards, ALARP implications and financial factors, then in some cases the conclusion may be to leave wastes in situ and make them safe for permanent on-site entombment. So how do we go about it?

If selected as a strategy, then the whole emphasis must be on ensuring an entombed structure remains radiologically secure for as long as its particular isotopes may pose a risk. With this in mind, the starting point assumes all irradiated fuel is removed and shipped to a facility that can either process or store it safely. If fuel must remain on site, then it should be stored in a way which is acceptable to the relevant regulatory authorities. Depending on its quantity and exact characteristics this may be in concrete casks (Fig. 15.26) or within a bespoke storage facility.

Fig. 15.26 Dry fuel storage cask.
© EDF Energy.

With fuel dispatched POCO is conducted, or at least some elements of it, following which we shall assume the facility enters a period of surveillance and maintenance (S&M). However, as with other strategies there is always the option to bypass this phase. Once preparatory works are complete the main task is to ready a facility for its permanent entombment. Achieving this requires a handful of steps, but that being said, there is some flexibility in their exact running order and indeed some activities may legitimately overlap, so we just need to keep that in mind.

As far as is practical, all remaining radioactive materials are gathered together in a common location, which will almost certainly be one of the building's existing shielded enclosures. If necessary, more than one area may be used, but ideally they should be adjacent to each other. If the building has a basement area then ordinarily this is preferred to storing materials at a higher elevation. To get to this stage, or in parallel with it, will require at least some element of D&D but this can be tailored to correspond with a building's upcoming entombment configuration.

Next up, the nub of this whole process entails pumping grout into areas selected to store radioactive materials; once solidified, this locks a facility's radiological legacy permanently in place. With this accomplished it is time to minimize structural volume of the building that remains. This includes reducing its height, which may be preceded by extracting an EOT crane, and also removing any parts of the building that can safely be demolished.

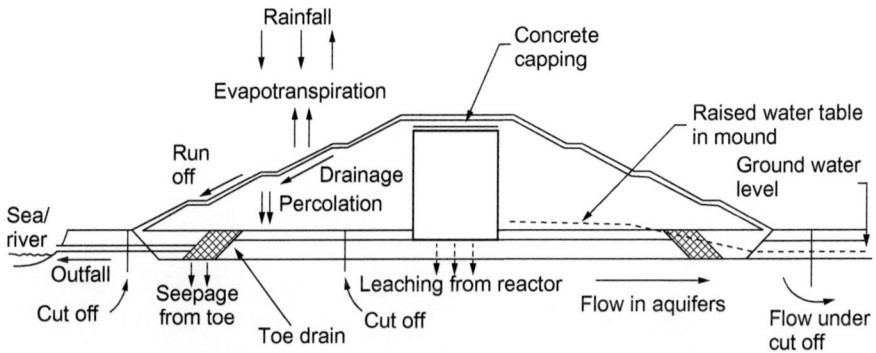

Fig. 15.27 Entombment concept.
© Woodhead Publishing.

With wastes grouted in place and a building's volume reduced, the remaining mass can be further encapsulated within more concrete, which in turn may be mounded over and landscaped with a mixture of stone, gravel, and soil (Fig. 15.27). Whatever the agreed end state, adequate drainage is required to minimize any possible ingress of water.

The obvious advantages of entombment are that it can be accomplished relatively quickly and is less expensive than alternative strategies. However, its other attributes are not quite as attractive. Any D&D that is performed must be done while radioactivity levels are high, albeit that D&D is kept to a minimum. In addition, land occupied by an entombed facility cannot be used again, a factor which in some cases will represent a significant loss, and not to be underestimated, there will be a reduction in the volume of material available for recycling.

Most significant, and the factor that weighs heavily against widespread adoption of this approach, is that entombment creates a permanent surface, or near surface nuclear waste disposal site. As a consequence the site must be subject to an environmental monitoring regimen for generations to come, which is not an inconsequential task.

And understandably this approach will generate considerable interest from regulators, the public and other interested parties. Specifics are dependent on geographical location and environmental factors, but whatever the detail entombment can only be implemented after an exhaustive program of stakeholder consultation.

15.5.4   Strategy selection

In many nonnuclear situations when it comes to making a decision or selecting the strategic direction for a particular endeavor, there may be little to choose between alternatives on offer, with each being variations on pretty much the theme. As a result, making a clear-cut choice can be quite difficult, dare I say a little subjective.

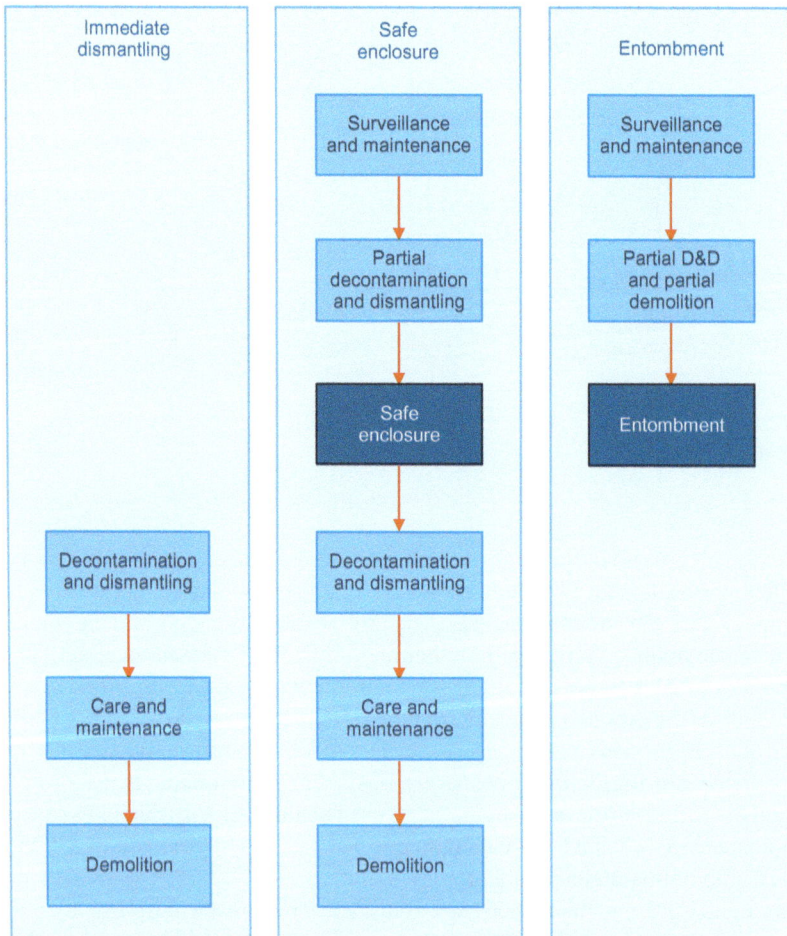

Fig. 15.28 Decommissioning strategies.
© Bill Collum.

By comparison determining a decommissioning strategy appears quite straightforward. After all the choices (Fig. 15.28) could hardly be more diverse so how hard can it be? Well, by now we know how deceptive appearances can be; we realize that behind those banner headlines are a myriad of subtleties, a subnarrative bound up within each planning factor, within each potential phase, with every nuance vying for recognition and demanding to be heard.

Clearly this is tricky stuff, not something for a hastily arranged half-day workshop. So long, long before it comes time to make such momentous strategic decisions, we begin our preparations, we start to plan.

15.6 Decommissioning plan

On the day a new nuclear facility is conceived, when tentative lines first breathe form onto a blank sheet of paper, thoughts of its spent hulk being picked clean and ignominiously demolished would be too much to bear. The reality of course, is that this is the fate which awaits every nuclear facility in the world and crucially planning for that day must weave its way through the entire design process. Indeed it must also be sustained throughout construction, commissioning and the years of operations that follow. But how do we make this happen, what mind-set do we need to adopt?

Think of any moderately lengthy activity, one that requires a fair bit of coordination—let's say planning a year-long round-the-world trip for yourself and a few friends, no doubt a well-deserved treat. Timing is important because several generous employers have granted you all the time off, but none of them will be amused if you take liberties and are late back, which is fair enough. Long before setting off you all get together to determine the route. So which countries you will pass through, cities to be visited, how long each stopover might last, something of sights to be taken in, things to do along the way, and so on, all calculated to match your budget. With the itinerary agreed and everyone signed up, it is time to get on with the actual arrangements.

At this stage you would concentrate on the first few weeks, so book flights, trains, ferries and other transport, along with accommodation for that initial period. You would also secure tickets for any tourist attraction or events that tend to sell out well in advance. Beyond these preparations you would have an outline plan for what followed, but be prepared to firm up on details closer to the time. In essence you would have a grand plan with lots of clarity and arrangements in place for the near-term, but less and less detail for the months that stretched ahead. That being said, if someone asked where you would be and what you might be doing during week 48 you would have a fair idea.

Imagine this situation, though. Let's say you do as I have described, but only for the first 3 months. What you have discovered is that there is quite a lot of effort involved in drawing up such plans, so 3 months is plenty and after that you will improvise. Friends question the wisdom of this laissez-faire approach, suggesting you risk missing out on a great deal and predicting that any form of budget management is destined to go out the window. When employers hear the news, eyebrows are raised and concerns expressed about whether you might be stranded in the Amazon rainforest, or marooned on some far-flung tundra when you are supposed to be back with them. All in all onlookers are impressed by what you are taking on, but underwhelmed by your preparations. There is consensus is that a round-the-world trip needs an element of adventure, but setting off with no clue whatsoever about three quarters of the journey is going a bit too far.

Back in the nuclear world the analogy is obvious enough, although we can skip the adventure bit. When we set out to design a brand-new nuclear facility it is unthinkable that we would plan up to the point of becoming operable and disregard the monumental challenge of decommissioning that lay ahead, particularly as there are endless opportunities during a facility's design and operation to streamline decommissioning when the time comes. This is a theme we shall return to in the next chapter when we discuss designing facilities with their eventual decommissioning in mind.

For new build design teams, the best way to grasp what is required of this particular challenge is to picture the day, as yet over the far horizon, when a decommissioning team will turn up and ask for your plan, the one they are to follow when taking your facility apart. Of course they will not be looking for fine detail, but to execute decommissioning efficiently they need a helping hand from the forethought and planning that only you can provide. As we know from historical precedents and many of us from our own experience, such plans are imperative to say the least.

So we need a plan, we can all see the value of it, but we are talking here of decades, possibly over a 100 years before it takes center stage. In the meantime, generations will come and go, key players change, umpteen times; the whole undercurrent to any long-lived nuclear plant is constantly shifting. By any measure this is a tall order so a great deal hinges on how we go about its planning. The vehicle for coordinating this whole endeavor is widely known as a *decommissioning plan (DP)*, an unambiguous title if ever there was one, but when does it start to take shape?

The one thing about a DP that we do not need to ponder is timing, since its development begins on practically day one of a project's life, certainly once the design starts to roll. It may feel counterintuitive, unnecessary even, to get cracking on plans to erase a facility before we even understand fully what it is. The way it works however, is that just as a facility's evolving design influences plans to take it apart, the opposite is also true, such that decommissioning proposals inform the design process. Granted it is barely noticeable at first, but as both elements mature, as design and decommissioning plans both pick up speed, they will reveal ample opportunity to influence each other.

A DP is the primary document, the spine to which all others attach. Regulators in different countries specify exactly what they expect it to cover, what supporting information is required, but in one form or another everyone produces a DP.

Contents typically comprise: details of a facility's various areas including major equipment and systems, interactions with other facilities, the original radiological zoning and containment philosophy along with any changes made during operations; details of construction materials; ground conditions; waste management strategy; operating history; regulatory requirements; identification of all relevant documents, drawings and electronic data and where it can be found; a description of a facility's hazards and links to its safety case; details of features that have been designed to aid the decommissioning process, an explanation of the chosen decommissioning strategy along with details of how it will be implemented; a definition of the required decommissioning end state, and much more.

All in all a comprehensive depository, providing a description of the original facility, changes made during its operational years and finally how, step by step, decommissioning will be realized. That sounds like a pretty comprehensive mandate, but for a DP it is not enough.

You will recall our earlier discussions on decommissioning *planning factors*, where we saw that while planning the decommissioning of individual facilities is important, crucial even; it can only be truly effective when integrated with much wider influences. An obvious candidate is a site's zoning plan which lays down principles for its orderly expansion and contraction (decommissioning). So the DP for an individual facility must harmonize with factors such as the imperatives of site-wide hazard management, financial constraints, influences of socioeconomic issues, maximizing opportunities to learn from experience, and so on.

As a result, the DP for an individual facility must reach beyond the confines of its walls and shake hands with long-term plans for other facilities with which it interacts. In addition it must take heed of long-term proposals for its host site and indeed affairs beyond a site's boundaries. Clearly, strategic integration of such matters is way beyond the jurisdiction of the DP for just a single facility. This only serves to highlight how important it is to coordinate what we might call local DPs with a site's long-term strategic direction, with the top-tier DP which embraces an entire site.

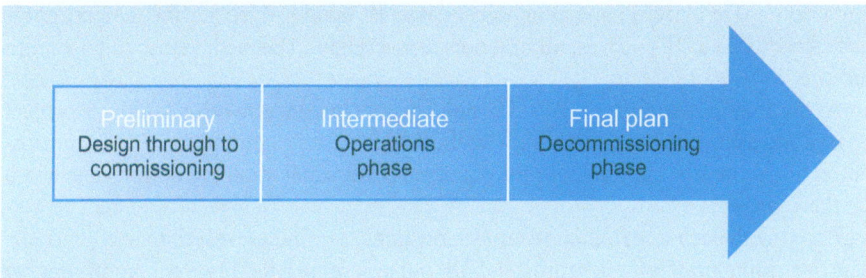

Fig. 15.29 Decommissioning plan.
© Bill Collum.

Having such a long incubation period, a facility's DP goes through lots of iterations, but happily they all fall neatly under three distinct headings (Fig. 15.29). The first, a *preliminary* plan is prepared almost from day one, so evolves through design, construction and commissioning. This plan establishes the decommissioning intent. Next up is the *intermediate* version which is continually updated throughout a facility's operational life, so apart from recording pertinent developments may legitimately influence the original decommissioning intent. The *final* edition firms-up on details of how decommissioning will be realized and begins its preparation some time, maybe several years, before a facility ceases operations. Remember, regulatory approval is needed before a DP can be formally adopted, so it needs to be demonstrably robust, which takes time.

To keep up the impetus and maintain an appropriately high profile, DPs must have a named owner throughout their life, a person with clearly defined responsibilities for their upkeep. Ownership mirrors the three phases of a plan's development, beginning with a project manager for the preliminary plan, a facility's incumbent owner during the operations phase and finally the decommissioning manager, the person responsible for implementing the DP. As with other of their duties, named individuals may delegate their responsibilities but ultimately they are accountable.

With so much to cover and over such a long period, we may envision DPs stretching to thousands of impenetrable pages, so some careful thought needs to go into how they are assembled, how much effort we should sensibly put in. If we make a half-hearted attempt (lip service) the result will be useless so we may as well not bother. But then if we throw untold resources at a DP and fiddle with it every day, the plan would cost an unjustifiable fortune and generate a blizzard of information which would ultimately be impossible to fathom. Again, what's the point? The upshot is that some pragmatism needs to prevail; we need to get the balance right.

To prove the point, for uncomplicated facilities, probably small scale and with a radiological inventory which is relatively benign, regulators may well conclude that a single document will suffice. But for major facilities and certainly when considering site-wide DPs, we need to go much further. We still need a clear story, the actual *decommissioning plan*, captured in a single document, but the bigger the task the more reference material there will be. Crucially, gathering this supporting documentation is not a bow wave activity, one which is always just up ahead; rather the data accumulates steadily as a DP evolves, all of it poised and ready to be retrieved so as to smooth the way in years to come.

As summarized in Fig. 15.30, a DP must evolve in concert with a facility's project phases and as it does keep a tight rein on the intricacies of planning factors, implementation of decommissioning's various phases and the wider strategic options discussed on these pages.

Faced with such a marathon we must constantly remind ourselves to stay on track. The overriding principles here are that a DP must: capture a facility's decommissioning intent and ensure it is always clearly visible; describe how decommissioning will be executed in a safe, environmentally responsible, and cost-effective manner; be practical and readily understood, up to date, and ensure that all necessary reference material is readily to hand. If it matters to decommissioning, then this is where we find it.

To be honest, thoughts of grappling with the *entirety* of a task so multifaceted, so desperately complicated, and over such a protracted period are enough to make most of us feel quite faint. Fortunately, as I have said before, time is on our side, so thankfully we do not need to bite off the whole enchilada in one sitting. We do, however, need to roll up our sleeves and tackle decommissioning planning with some enthusiasm, structured enthusiasm at that, otherwise things will only become even more convoluted, more hazardous, more costly, more time-consuming, and ultimately

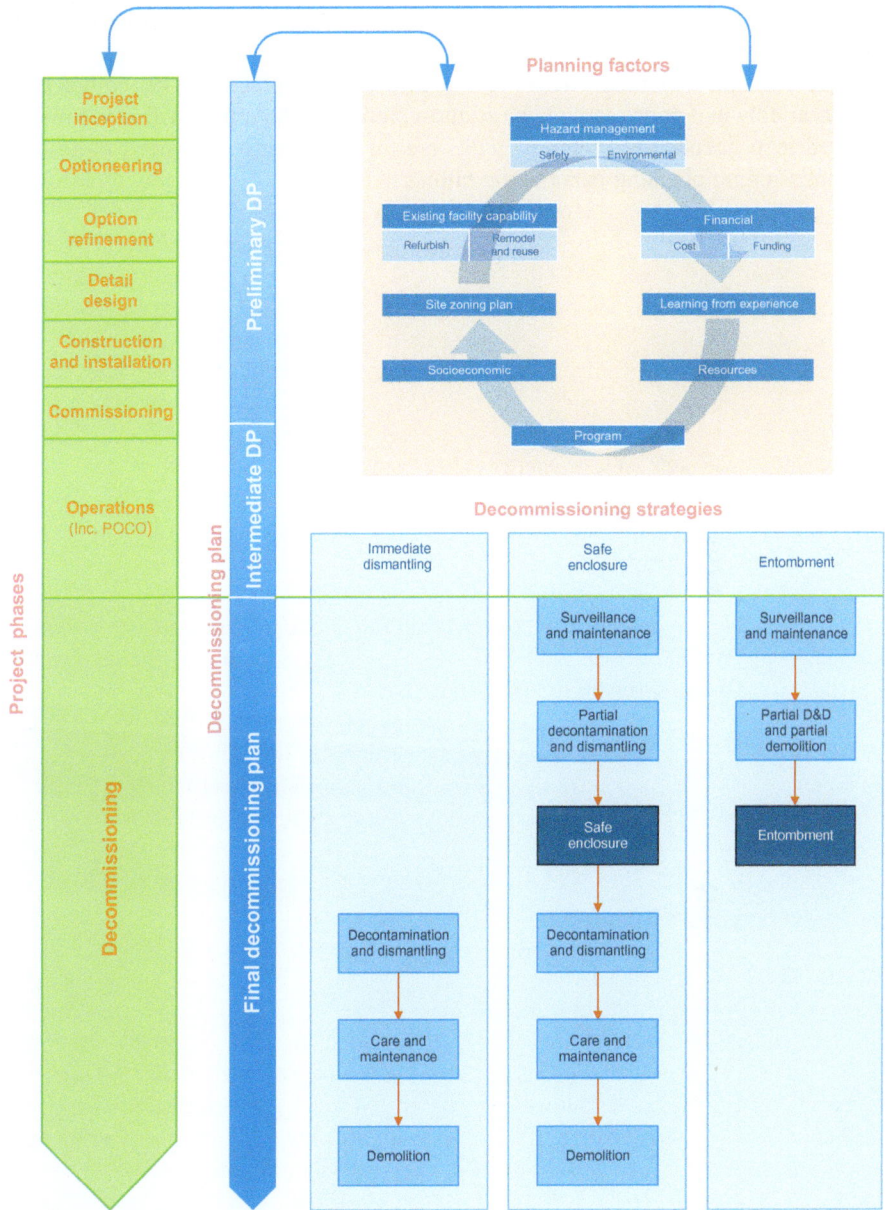

Fig. 15.30 Decommissioning planning.
© Bill Collum.

make it extremely difficult for coming generations to maintain tight control of the decommissioning process.

Future decommissioning teams will seize on a well-prepared DP, embrace it wholeheartedly and pour gratitude of all those, most of whom they will never meet, involved it its preparation. Such thoughts are all the incentive we need to ensure decommissioning planning gets the top billing it deserves.

Future-proofing

16

The term *future-proofing* has wide-ranging meanings, variously referring to anything from electrical components to entire cities. In our context it is all about extracting maximum value from buildings throughout their entire life. To achieve this, we must integrate the wherewithal to efficiently decommission them when the time comes, and wherever possible aim to remodel facilities for new purposes when their primary operations are concluded.

By definition nuclear facilities are radioactive and tremendously robust: altogether an excellent combination whilst they are operating but far from ideal when it comes time to take them apart. As a result, and as we have seen in the previous chapter, it is mandatory that we draw up comprehensive decommissioning plans and obtain regulatory approval before commencing the task.

Furthermore, to execute decommissioning as efficiently and safely as possible, radioactive plant and equipment (P&E), along with the building itself, should be configured to facilitate the process. In particular, radioactive materials must be ready to cooperate in being decontaminated, dismantled, packaged and shipped off to either an interim storage facility or permanent disposal repository. To accomplish all of this, to ready a facility for its inevitable destiny, we must embrace a process known as *design for decommissioning* (DfD).

16.1   Design for decommissioning (DfD)

As with just about all things *decommissioning* this is not something we can turn our attention to at the last minute, so DfD must weave its way through a facility's design development from practically the very first day. It features prominently during the design optioneering phase, when various schemes are jockeying for position, and has a significant voice at the point of freezing a design and selecting the preferred option. Once we pass that stage, apart from details, a design is effectively locked down so opportunities to incorporate meaningful DfD become quite limited.

To highlight the importance of DfD, it is worth considering what can happen if it is neglected or only invoked at a superficial level. For our purposes, we shall assume we are dealing with a beta–gamma plant, but do not need to dwell on its details since the outcome would be broadly similar for all such facilities. As always, we can take it that before decommissioning gets underway the primary radiological inventory has been removed and post operational clean out (POCO) satisfactorily concluded.

Nuclear Facilities. http://dx.doi.org/10.1016/B978-0-08-101938-2.00016-7

16.1.1 Self-perpetuating cycle

In the absence of DfD, the arrival of a decommissioning team is to all intents and purposes unexpected, so they would be forced to start from scratch in developing their own plans. In essence the whole decommissioning exercise would be one of improvisation, which is not great.

Fig. 16.1 In situ dismantling—manual.
© Nuclear Decommissioning Authority.

From this point onwards the *degree of difficulty* hinges on exact characteristics of a facility's radiological inventory, how extensive it is, how shielded areas are configured, how complex the building is, and so on. If the chips fall nicely, then with help of a little enabling works, such as installing tented enclosures, it will be possible to manually conduct a great deal of in situ decontamination and dismantling (D&D) (Fig. 16.1), size reduce waste, condition it, load it into disposal packages and send it off to a storage facility.

Moving up to the next level, where things are more radiologically challenging, a redundant beta–gamma plant will almost certainly require substantial internal works to prepare it for decommissioning, including the design and deployment of bespoke remotely operated equipment (Fig. 16.2). That being said, decommissioning could still be accomplished within confines of the facility being taken apart.

In the most extreme situation, where radiological challenges are particularly acute, significant internal building works and remotely operated equipment are again required, but this time retrieved intermediate level waste (ILW) materials, which may be solid or liquid, must be routed to a new specially built facility for processing and packing into disposal containers. Simply put, to decommission a redundant nuclear plant we find ourselves having to build another one, which itself must one day be decommissioned.

Fig. 16.3 shows what could happen if plant after plant was built without due regard to the principles of DfD, with a succession of new builds needed to rid a site of their

Fig. 16.2 In situ dismantling—remote.
© Nuclear Decommissioning Authority.

predecessor's radioactive legacy, the *self-perpetuating cycle*. Granted I have depicted a particularly extreme scenario; what it highlights, however, is the importance of avoiding any path which even remotely resembles that shown. In some situations we may legitimately make a case that one iteration of the cycle is unavoidable, a good idea even, particularly where a new clean-up plant will serve multiple decommissioning projects. Beyond that however, we cannot countenance being drawn further and further into the spiral.

Unsurprisingly, funders take a keen interest in this theme, and are unlikely to be impressed by any suggestion that their investment will need another sizable cash injection to decommission it when the time comes—some funding, yes, but within limits. With such a prospect, one way or another they may need to set aside considerable funds to cover the inevitable liability: the kind of news that is guaranteed a frosty reception. Even more disconcerting, the absence of robust future-proofing proposals would make it virtually impossible to bound what those long-term liabilities might turn out to be: which would be altogether far too open-ended.

What funders expect, quite rightly, is evidence that a proposed new plant will be able to deal with its own demise, or at least make a major contribution towards achieving it. They will also be looking for confirmation that anything beyond that primary objective has been identified, that there are comprehensive plans in place to deal with it, and that the whole exercise has been accurately costed. Wherever possible, they will also welcome news that their new facility will be able to contribute

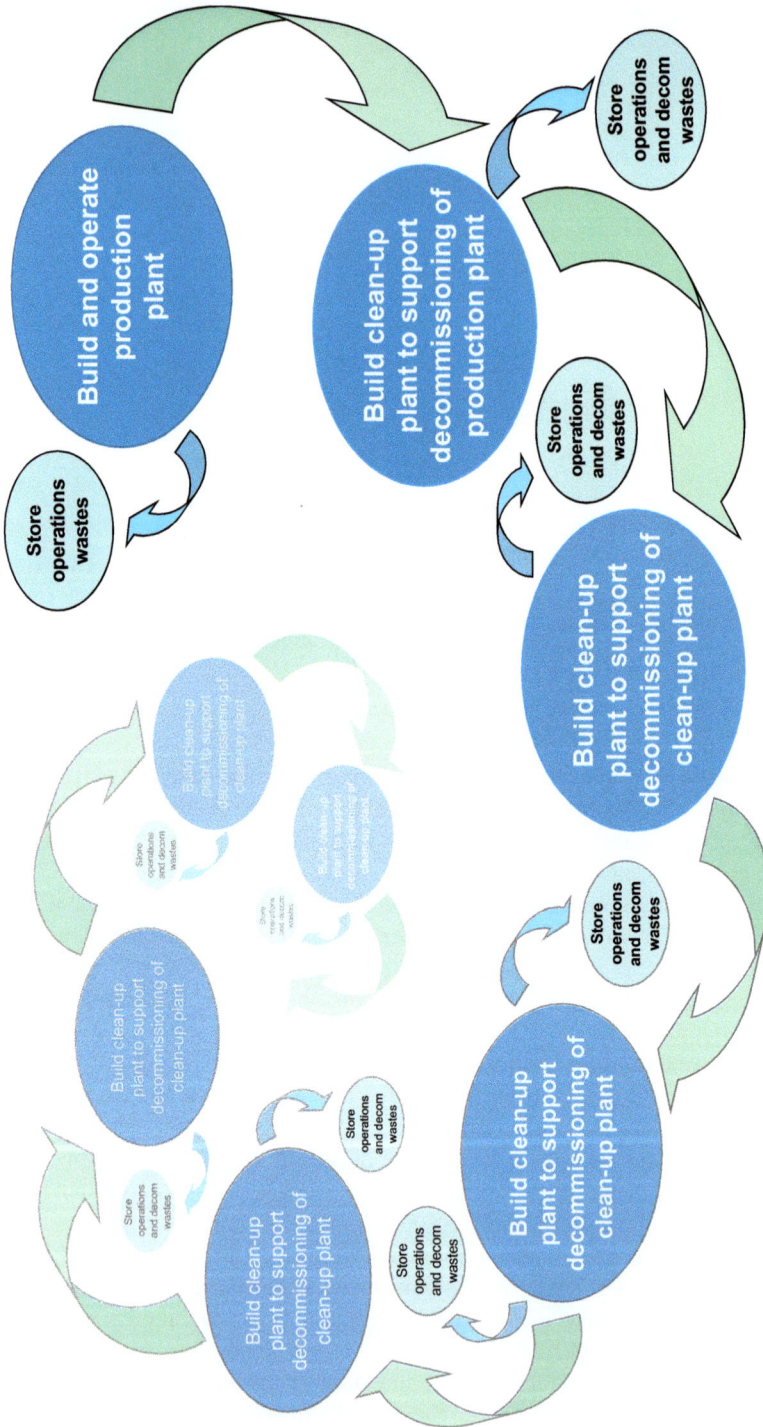

Fig. 16.3 Self-perpetuating cycle.
© Bill Collum.

to the demise of others on its host site, those less able to decommission themselves. And of course funders will demand a coordinated, site-wide strategy, one that ideally negates the need for new build plants which are dedicated exclusively to cleaning up legacy wastes.

Clearly then, designing nuclear facilities with their eventual decommissioning in mind is a serious business, arguably on a par with designing them to fulfill their primary operational objectives. Apart from the financial and environmental consequences associated with each new clean-up plant, any prospect of an open-ended build sequence would undermine our ability to confidently predict future financial liabilities. Ideally then, and this really is the *holy grail* in terms of future-proofing, nuclear plants should be designed to, as far as possible, self-dismantle, or *eat* themselves when the time comes. This is a theme we shall get to later when we discuss decommissioning enablers. So the case for DfD is a compelling one, but there is something else.

Thinking back to the preceding chapter, we already know that when it comes to designing with the future in mind DfD is not alone. In fact, our previous theme of remodeling facilities and using them for new purposes is so closely allied to DfD that we must group the two together.

16.2   Design to remodel and reuse (DRR)

Considering the cost of nuclear facilities and time taken to bring them into existence, it comes as no surprise that the industry is reluctant to participate in any kind of throwaway ethos. So wherever possible the emphasis is on retaining facilities which have seen out their original purpose and putting them to work on a new endeavor. In other words, wherever possible we *design to remodel and reuse* (DRR).

Happily DfD processes coincide, in large part, with the approach taken when designing facilities with reuse in mind. So although the task is not exactly a breeze, its implementation is propelled along quite nicely. The coincidence occurs because before a facility can be put to work on a new purpose it must be modified in some way, and much of that remodeling exercise is in effect advance, or partial decommissioning, particularly in the areas of D&D and partial demolition. Essentially then, reuse always adopts at least part of the path which is determined by DfD.

As a consequence, since DfD should always be incorporated in evolving design proposals, we have an opportunity to harmonize it with DRR and insert a new purpose between a facility's primary operations and its decommissioning (Fig. 16.4). That is not to say all manner of desirable reuse opportunities will automatically reveal themselves; we must work to make it happen, certainly in any meaningful way. What matters is to recognize the synergies between DfD and DRR so that we can set about developing complementary proposals.

Where a specific purpose can be identified for a facility's reuse, then a project team can focus on it and embed the wherewithal to smooth the way for its transition.

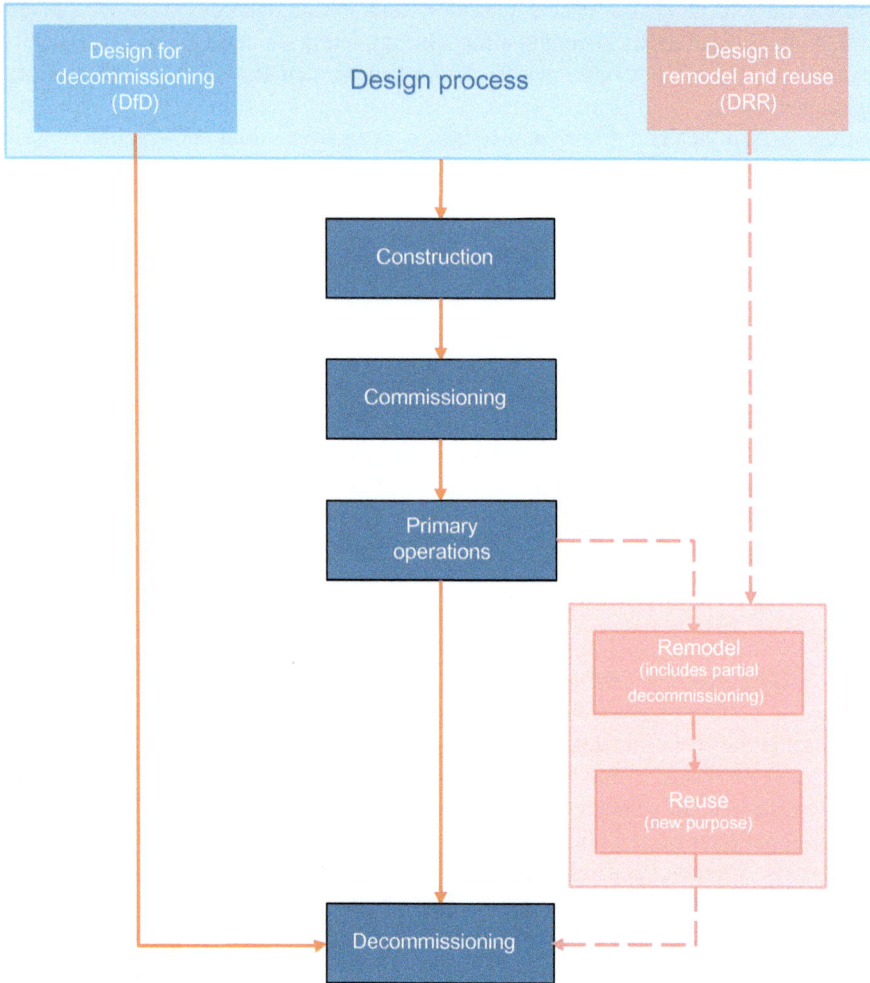

Fig. 16.4 Incorporation of DfD and DRR in design process.
© Bill Collum.

However, where it is not possible to establish a specific reuse, we must not give up on the notion. Instead we should investigate opportunities for non-specific, or generic, purposes. By leaving the door open to such possibilities we automatically place facilities in a healthy future-proofed state, whereby they are poised to contribute to a site's long-term sustainability. And crucially, since DfD eases the way to reuse, this generic approach need not be overly expensive.

Before moving on, it is worth mentioning that reuse of a facility is often visualized as an exercise in retaining a building's external envelope, its weather protection, along with some portion of its internal structures, then concentrating on what we might fit within the available space. Crucially, in what we might call this entry-level scenario, the candidates for what may be newly introduced are limited to assignments which are standalone or close to it. So contenders might include a storage area of some sort, a maintenance workshop, or maybe a handy place to locate a small test rig. All of this is absolutely fine and totally legitimate, but will only get us so far in our quest to cut through the self-perpetuating cycle and endow facilities with an ability to deal with their own decommissioning. We need to go further.

The key here is to recognize that operating nuclear plants are vibrant places. They are most certainly not standalone but rather act in concert with interfacing facilities, along with the on-site infrastructure that comes up to their doors and a plethora of rather diverse utilities. In fact, if we were to take the plan of practically any nuclear facility and overlay all of its cross-site interactions, both permanent and transient, it would be quite a spider's web, one that we can use to our DRR advantage.

The list of useful, reusable assets varies from one plant to another and is potentially very long indeed. It may include shielded pipework connections to facilities which can treat radioactive liquors: utilities such as compressed air and various types of gas and water, along with steam, reagents, and so on. In addition we can expect electricity to be delivered at several voltages, possibly with "A" and "B" (backup) supplies. We can also anticipate that road links will be excellent, maybe supplemented by rail connections and that a facility will be well served by on-site drainage systems.

Then let's not forget that within facilities themselves there will be a sophisticated nuclear ventilation system and radiological changerooms, In some cases we also find well equipped mechanical handling caves, active liquor processing cells, decontamination and monitoring equipment, electric overhead travelling (EOT) cranes and lifting beams, amber workshops, apparatus for sampling and analysis. The list goes on.

So when we think of reuse, we are not necessarily thinking of a glorified shed where "space" is the only commodity on offer. In our context here, reuse is all about recognizing the full gamut of what a facility has to offer—yes, including space, but also its existing P&E and the arterial network which enables it to function. For exemplary reuse we seek to retain these, if necessary instigate plans to reinvigorate them, and capitalize on their latent value long after a facility has concluded its primary operations.

16.2.1 Reuse—self-decommissioning

The one reuse about which there is never any question is self-decommissioning, although before getting to that point facilities will ideally make a detour to be remodeled and fulfill another useful purpose. Ultimately, all nuclear plants must draw on their own resources to the maximum extent possible, in order to accomplish their own decommissioning. In effect this is an extension of the POCO principle examined in the previous chapter, where a plant's existing equipment is used to conduct its *post operational clean out*, essentially the exercise which draws its primary operations to a close.

Once we move beyond POCO, when we begin decommissioning proper, our overriding objective is to maximize use of a facility's existing P&E and other resources to achieve its own demise. This is particularly relevant during D&D

(Fig. 16.5). If the original capability needs to be supplemented, this will have been identified by the DfD exercise and either installed as part of the initial build, or be readily incorporated when the time comes.

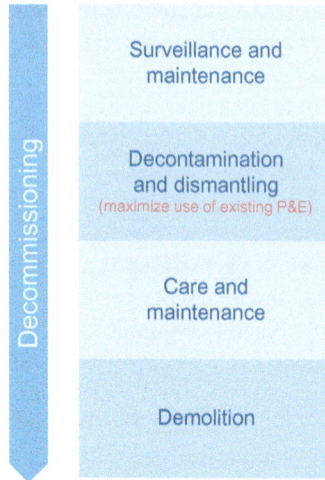

Fig. 16.5 Maximize ability to self-decommission.
© Bill Collum.

Self-decommissioning then, particularly D&D, should always be seen as a plant's primary reuse objective. This is the gold standard against which all proposed new facilities are evaluated. The last thing any nuclear site needs, is to find that when a plant concludes operations it stands helpless and unable to make a meaningful contribution to its own demise, instead sucking in valuable external resources and maybe even calling on a new building to send it on its way. Therein lies the path to a self-perpetuating cycle, which as we have seen is wholly unsustainable.

16.2.2   Reuse—specific purpose

Beyond the prerequisite of self-decommissioning, imposing an arbitrary future use on a facility is unlikely to hit the spot when time comes for it to transition, so ideally we need to match reuse to real demand. Happily then we can turn to a site's long-term zoning plan for guidance. If particular functional requirements exist then they will be captured here; this is where we find the *shopping list* of future facilities discussed in the previous chapter. Armed with this information, new build project teams can rise to the challenge of capitalizing on DRR to the maximum extent possible.

In terms of reuse objectives, everything comes down to specifics of a site's long-term needs and the ability of upcoming facilities to accommodate them, so there is no prescriptive formula. That being said, there are two broad reuse themes which are perennially to the fore.

When nuclear facilities are being decommissioned, particularly beta–gamma plants, two capabilities are always in demand. First is an ability to handle radioactive waste and get it into disposal packages, such as stainless steel containers, an activity which may require some form of size reduction. And second, is somewhere to store waste packages until they can be shipped off to a permanent disposal facility. Consequently, when considering opportunities to remodel and reuse facilities, this twosome are almost always bound to get an airing.

We shall see how such reuse might be achieved when we discuss *enablers* later in this chapter. So for the moment we can just observe that any facility which needs a mechanical handling cave to fulfill its primary operations is well placed to contribute to its own decommissioning and possibly that of other facilities. Furthermore, facilities with a reasonably large and uncluttered floor area, say a crane hall, hold out the prospect of providing interim storage for radioactive waste packages.

As I say, these two themes may figure more than most, but not to the extent that they are the only show in town—far from it. The objective here is to continually scour a site's register of future requirements, and always be alert to potential synergies between upcoming facilities and their ability to satisfy those demands. The overriding aim is to drive down a need for new build facilities to the lowest extent possible, particularly those concerned exclusively with decommissioning activities. And remember, every success in this regard has the potential to minimize radiological wastes, reduce environmental impacts, and save considerable sums of money. As exercises go, this one certainly deserves our unrelenting attention.

16.2.3 Reuse—non-specific purpose

It goes without saying that achieving perfect symmetry between what a proposed new facility can sensibly be remodeled to do and a pre-defined supplementary role is not always achievable. So rather than impose an arbitrary future use we need to embed a more flexible approach. In the absence of explicit functional requirements, we should at least configure new plants to satisfy principles of non-specific, or generic reuse. In other words, embed the means to transition into an as yet undefined new role. However, we must be wary of adopting this approach too freely, since it would rather undermine our strategic objective of instilling predictability into a site's demise.

We know that in many ways mandatory DfD sets the scene for a facility's remodeling and reuse; it brings us part-way there. So as opportunities go it is simply too good to miss, but how do we go about it?

In the normal run of things, particularly where industrial buildings are designed with a single purpose in mind, their external envelope pretty much wraps around whatever is going on inside. As a result the profile of many buildings becomes somewhat irregular; it is not always the case but it happens often enough. To be honest, strict adherence to a mantra of minimizing building volume in this way is not always such a good idea, since adding complexity to a building's façade only serves to drive up costs. And certainly from a long-term perspective, such geometry can literally eat into the objective of maximizing our options.

In our context here, of non-specific reuse, we need to lay on as much flexibility as possible, and one way of achieving this is, wherever possible, to utilize a portal frame structure, a system which as we know from Chapter 5 has to be rectangular. The beauty of this approach is that a building's structural steelwork is arrayed solely around its perimeter, opening the opportunity to clear away some or all of its internal structures and replace them with something entirely different, maybe housing a pilot plant, assembling large fabrications, maintaining shielded flasks, storing waste packages whilst they await onward shipment, and so on. Equally valid, it may be that creating a large uncluttered area is part of our DfD strategy, where the space is earmarked for activities that will contribute to a facility's objective of self-decommissioning. As always, options are wide and varied.

In assessing viability of the portal frame tactic, one of the main considerations is seismic performance demanded of a structure during both its primary phase and that which follows. Where a combination of the span across a portal and seismic accelerations point to adopting a more traditional *stick build* approach, we should not necessarily abandon our plans. It may, for example, be possible to configure a building in such a way as to facilitate a combination of stick build and a smaller portal framed area. Alternatively there may be merit in adopting a structural concept which utilizes steelwork in some areas and concrete in others.

In some situations then, embedding the ability to create a large open area within a building can be an attractive proposition. In others, however, it may be impossible to achieve, or simply not required for a prospective reuse. This may occur, say, where we wish to bestow a new lease of life on facilities which house large concrete structures, such as caves, cells and canyons. This is a subject we shall get to later, when we discuss making provision for remote handling of decommissioning wastes.

There are many ways in which we might configure a building's structure, but by giving equal prominence to immediate and long-term uses there is potential to develop an arrangement which satisfies both. What we are looking for here, is to avoid any tendency towards designing nuclear plants which could be described as a *one-trick pony*. Rather, from a generic reuse perspective, our goal is to equip buildings to become a workhorse for whatever might reasonably be expected to come their way. Of course this is not always possible, but it should be recognized as an important objective, particularly as DfD always cracks open the door to potential reuse opportunities.

16.3   Integration with decommissioning strategy

As with so much of what goes on when designing nuclear facilities, DfD and DRR adhere to the recurring theme of needing to be addressed early in the process. To highlight how crucial this is, we need only think back to the previous chapter and our discussion on decommissioning strategy.

You will recall that broadly speaking there are three alternatives: immediate dismantling, safe enclosure also known as safestore, and entombment. The significant factor here is that a facility's post operation activities and sequence in which they are conducted will vary enormously depending on the strategy with which they must coincide. And this in turn has major implications for how DfD and DRR are developed and implemented. To realize maximum benefit from our investment in future-proofing, it must be tailored from the outset to a plant's prevailing decommissioning strategy (Fig. 16.6). In this way we can harmonize DfD, DRR and strategy, and smooth the way for appreciative future generations.

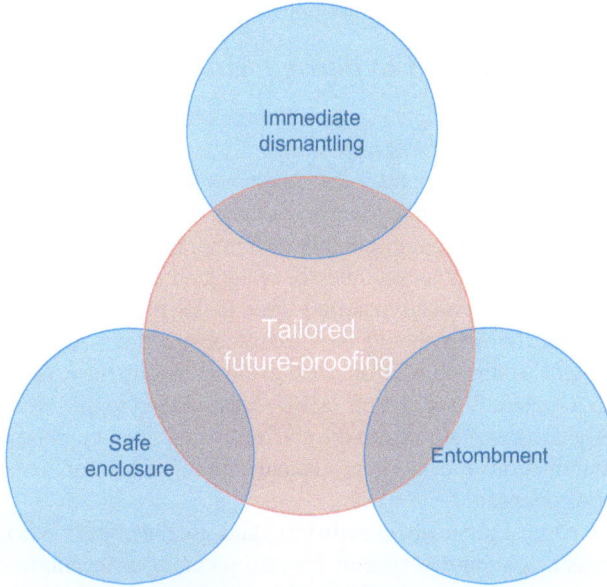

Fig. 16.6 Integration with decommissioning strategy.
© Bill Collum.

16.4   Principles of DfD and DRR

Designing nuclear facilities to comply with all relevant regulations, be delivered on time, at the right cost, satisfy operational requirements, be readily maintainable, and so on is challenging enough. So we need to be careful not to smother the process by imposing layers of complex processes for what is, initially at least, distant reuse or decommissioning.

So a key factor here is getting the balance right on how much is done upfront and what can follow closer to the time. What we need is a thorough approach to DfD and DRR and certain knowledge that when the time comes a facility is ready to embrace both. As I say however, we can do without losing our way in a blizzard of procedures, so can distil the fundamentals into just three essential principles.

16.4.1   Do not compromise primary operations

A facility's operators must be able to go about their day-to-day duties, without being hampered by DfD or DRR features which will not be required for many years. So, for example, if future-proofing components are added to an EOT crane and as a result complicate its maintenance regimen, if as yet unused ductwork restricts access to P&E, if workshops are *temporarily* far from where they need to be, then those elements would not be deemed a success.

We shall discuss future-proofing *enablers* shortly, so for the moment just need to note that whilst their tangible features are inevitable, and indeed to be welcomed, they must literally not get in the way.

16.4.2   Simplify transition to decommissioning or reuse

Decommissioning teams or those charged with remodeling a facility, will no doubt be delighted to hear that DfD and DRR have been fully incorporated within the building they are about to remodel or take apart. So designers must not disappoint, they must ensure the shift to post operation activities is as seamless as possible.

For example, if during decommissioning, a temporary C2/C3 (green/amber) sub changeroom will be needed to access a newly designated C3 area, say one earmarked for size reduction of contaminated materials, then its location should be identified in advance. In addition the area should be clear of obstructions and, because of the change in radiological zoning, provision made to either modify the existing ventilation system or install a mobile filtration unit (MFU) later. Added to this, routes should be assigned to safely transfer materials to the size reduction area, appropriate provision made for decontamination processes and adequate access laid on for import and export of waste disposal packages.

So although the transition is not exactly plug-and-go, the objective is to streamline it as much as possible. With this type of forward planning a decommissioning team will arrive, be handed a predetermined game plan, and find the scene at least partially set for them to implement it. This allows them to focus their specialist expertise on finalizing details and executing the plan, which is infinitely preferable to the alternative of starting from scratch and improvising what is bound to be a suboptimal solution.

16.4.3   Maximize use of original P&E

We only need to wander around nuclear plants for a short time to realize their P&E does not come cheap. So when it comes time for remodeling or decommissioning, the last thing we want is to rip out existing kit and replace it with something else, especially when the newly arriving shiny P&E is not too dissimilar to kit going out the door. Such a path consumes an awful lot of time, costs a great deal and generates unnecessary waste, so all in all has nothing whatsoever to recommend it. We adopt a different tack.

If an operating facility needs kit such as EOT cranes, bogies, manipulators, shielded windows, shield doors, pumps, valves, decontamination equipment, assay equipment and a whole lot more, we must look for opportunities to keep it gainfully employed after its primary operations have been concluded. As I say, decontamination and dismantling (D&D) demands exactly the same type of kit, so designing P&E solely for its initial purpose is only a partial solution. Instead project teams must look to the future by ensuring, as far as possible, that P&E locations, configurations, and functionality are either suitable for a facility's entire lifecycle, or easily adapted to it.

From a DfD and DRR perspective the objective is to squeeze maximum value from all P&E, along with the building itself and infrastructure which serves it. Nothing should be sidelined until we are convinced it is unable to contribute further to a facility's remodeling, its reuse for new purposes, or ultimately to its own demise.

16.5   Funding future-proofing

There is no doubt that whilst the twin objectives of DfD and DRR are entirely laudable, to the extent that it would be difficult to argue against them, the question of demonstrating value for money and paying for their implementation is bound to arise.

As we might expect, to secure funding we need more than broad statements about how excellent the return will be, so sure enough DfD and DRR proposals are evaluated during a facility's design optioneering phase and only incorporated when shown to be worthwhile. Actually, analysis of future-proofing is not limited to financial aspects, but also considers factors such as programming implications, waste minimization, environmental impact and the all-important *as low as reasonably practicable* (ALARP) considerations. One way or another, the availability of upfront funding weaves its way through most debates, which just serves to highlight how very important it is.

Deciding how much to pay for future-proofing is a particularly knotty problem, for two reasons. The first revolves around effects of *discounted cash flow* (DCF), essentially a debate on tying up cash immediately rather than handing it over at more or less the last minute. Thankfully there are formulas for that one. The big one, though, is how much we are prepared to spend today, in order to minimize financial liabilities which will not manifest themselves for maybe several generations. Granted this one has elements of DCF, but the main factor here is one of *appetite*, in other words a client's willingness, and indeed ability to invest in the future. Unfortunately this one is formula-free, so instead we turn to pragmatism.

As a starting point, we have to accept that no matter how attractive the return on upfront investment is forecast to be, no funder is going to sanction eye-watering sums to future-proof a facility. Just imagine asking for a 50% increase in the cost of bringing a plant to the point of becoming operable. The response, quite rightly, will be that either the concept design or future-proofing proposals, or both, are flawed and must be revisited.

The good news is that when it comes to DfD and DRR a great deal can be achieved for relatively little cost, say routing services with an eye to the future, or reusing P&E as we have just discussed. Such tactics will only add fractions of a percent of a facility's entire budget, so we can all sign up to that. However, this will only get us so far. As we are about to see, delivering truly meaningful DfD takes more than loose change. So having drawn up their proposals, future-proofing practitioners must be ready to back them up with a convincing cost benefit analysis.

When determining how much to spend upfront, one thing that focuses the mind is historical precedents that demonstrate what can happen when plants are designed with little or no regard for DfD. In some situations this has resulted in decommissioning costs being equivalent, in real terms, to the original price tag of a facility or even several multiples of it. Not to mention the years of considerable effort that unplanned decommissioning demands—clearly an unsustainable proposition and one to be strenuously avoided.

As I say, faced with deciding how much to invest in DfD and DRR it comes down to appetite, with different governments and other funders having their own particular threshold. My personal view is that where proposals can be clearly shown to deliver significant lifecycle savings, allocating 5% of a facility's upfront capital cost to future-proofing should be deemed acceptable. And by that I mean, not unexpected and viewed as entirely appropriate. Beyond that, if a case is sufficiently compelling then additional funds should not be withheld.

Of course such funding must not be viewed as an *additional* cost. Rather it is akin to preparing a safety case, environmental impact assessment, an operating and maintenance philosophy, seismic analysis and a very long list of other essential exercises. Future-proofing is just another of the crucial elements which contribute to delivering a viable project. Although in relative terms 5% is quite a small number, in reality it can be quite a considerable sum of money, particularly where major nuclear projects are concerned. So we need to give it a little context.

In previous chapters, we saw that by the time a nuclear facility reaches the point of becoming operable, its design activities may have consumed around 30% of the entire budget. Looking at where that 30% goes, we know from Chapter 14 that 10% can be consumed by a safety case, so 3% of the entire budget, a figure we shall return to in a moment. The remaining 27%, or thereabouts, being spread across everything from design philosophy and basis of design documents to programs, cost estimates, functional specifications, project management and scores of other tasks, all resulting in hundreds of documents, thousands of drawings and a plethora of sophisticated 3D computer models. So far so good, but what about planning for the future?

To deliver robust future-proofing proposals we need to develop clear, coordinated and implementable plans for how decommissioning, and potentially remodeling will be accomplished. To achieve this, DfD and DRR must weave their way through all of a project's top-tier documents. They will, for example, have dedicated sections within design philosophy and basis of design documents and feature prominently in scores of others.

Programs and cost estimates will not stop at the point of a facility becoming operable but must continue, albeit not to the same level of detail, to show how things evolve over the many years that follow. In addition, we shall have a 3D computer model and suite of facility layout drawings that illustrate how a facility will evolve once its primary operations are concluded. In effect showing how a building will transition through various configurations as it is decommissioned, particularly through the D&D phase. Where appropriate, there will also be options for how a facility may be remodeled and used for other purposes.

By taking this approach, then by the time a facility becomes operable a site's owner will be in possession of a very comprehensive suite of documents, including programs, cost estimates, and drawings, which together providing clear direction on what happens once a facility has fulfilled its primary purpose. In addition, the necessary enabling works (which we are about to discuss) will have been implemented. So along with the necessary documentation, a facility will also be physically prepared for its post operations future.

16.5.1   Decommissioning cost estimate

To give a feel for what we are striving for, it is worth comparing the cost estimate for building a new facility with that of its decommissioning. As we saw in Chapter 12, at the conclusion of a facility's detail design phase, the cost of bringing it to the point of commencing operations must typically be shown to have an accuracy of plus or minus 5%. And as we also saw, the effort required to generate an estimate of that quality is truly monumental.

At that same point in time, at the end of detail design, we cannot expect to quantify a facility's decommissioning costs to the same level of accuracy, but neither can we proceed with a vague notion of what those costs might be. What we are looking for, what customers need before releasing the cash to build a new facility, is a decommissioning cost estimate which is accurate to plus or minus 10%. Of course this is a tall order, but without that level of assurance, without a clear appreciation of future financial liabilities, it is impossible to develop an investment justification.

Clearly this level of preparation takes time, and in the case of on-plant features that will aid decommissioning, needs materials. So there has to be a cost. If we stick with my estimate of 5% of a facility's upfront cost, then bearing in mind that the exercise is two-pronged—namely forward planning, backed up by appropriate enabling works— we can expect around half to go on design documents, drawings, programming activities, computer modeling, preparing the decommissioning cost estimate, and so on, leaving the remaining half for tangible enablers, either within a proposed new building or on the site itself.

In other words if we set aside on-plant preparatory works, then the level of *office-based* effort required to future-proof a facility, to ready it for decommissioning or reuse, is comparable to that demanded of the safety case I mentioned a moment ago, say around 2.5–3% for each, which puts the whole exercise into perspective quite nicely. And just as with a safety case, an upfront investment in DfD and DRR delivers a fabulous return; both, in their different ways, instill confidence in the future.

16.6   Future-proofing enablers

We can plan for decommissioning or remodeling a facility, we can analyze, strategize, and optimize, and indeed we must, but ultimately what matters is how effective our upfront planning will be when it comes time for implementation out on a plant.

If future-proofing is to realize its full potential we need more than fine words and diagrams. We must smooth the way and for that we need to make tangible preparations. Granted the features provided may lay dormant for decades, but when called upon will make the kind of difference that can scarcely be overstated. To ease transition through a plant's post operation phases we need dependable enablers waiting in the wings. Although there is no such thing as a *standard* nuclear facility there are common DfD and DRR themes, so we can take a tour of those.

16.6.1   Radiological zoning and ventilation

We know from Chapter 3 how radiological zoning permeates nuclear plants and plays a major role in how they are configured and operated, and we recognize that *rad* zoning and nuclear ventilation are so inextricably linked that neither can sensibly be discussed without the other. Being so ubiquitous, it would seem that once rad zoning has been established it is pretty much locked down for a facility's entire life. Happily however, it is more malleable than appearances might suggest.

To recap, rad zoning divides into two categories of contamination (C) and radiation (R), with guiding principles being to segregate contamination zones so as to stop cross-migration of airborne particles, and where radiation is present to confine it behind adequate shielding. Furthermore, nuclear ventilation is primarily concerned with maintaining containment at the *interfaces* between contamination zones. A feat which is achieved by establishing a relative depression between zones and ensuring air moves across zone interfaces, such as sub changerooms and shield doors, with appropriate velocity.

16.6.1.1   Contamination

Armed with our existing understanding, it becomes clear that whilst facilitating changes to contamination zones is not exactly a click of the fingers, it is eminently achievable. However, it can only be done efficiently if dual zones are investigated whilst facility concepts are being developed, and then synchronized within a whole-life rad zoning philosophy. The challenge is to identify areas which will ultimately be rezoned, due to say a proposed reuse, which includes the D&D operations that take place during decommissioning.

With that done we can rationalize those areas and identify future-proofing requirements for containment enclosures, doors, lobbies, and so on. This includes provision of ductwork, dampers and if necessary additional HEPA filter cabinets. Once we know what is needed, we can decide which elements should be installed as part of the original build and what can wait until some years down the line. For the latter, enabling works should be carried out so as to smooth the way for their subsequent implementation.

So in terms of contamination we can plan to modify boundaries between future, newly created "C" zones, and equip ventilation systems to maintain their containment. So that just leaves radiation to deal with.

16.6.1.2 Radiation

Once again we need a grand plan, a vision for how decommissioning or indeed remodeling will be accomplished. From this we can identify areas of a facility which will require enhanced shielding when its primary operations are concluded. This may be due to reuse for new operations, or could just as likely be an enabler for decommissioning's D&D phase.

For areas which are already shielded but destined to need more, say a cave that requires 750 mm of concrete for its primary operations, but rising to 1 m in later years, then the solution is usually to provide additional concrete from day one. Undoubtedly this will result in extra concrete and rebar and, depending on specifics, maybe longer through-wall liners, deeper shielded windows, and so on. But if the rationale is sound, we can create a shielded environment which will still be in use when its original operations are a distant memory. In these situations additional costs are funded from the future-proofing budget we discussed a moment ago.

Fig. 16.7 Rebar.
Shutterstock.com.

For areas that do not need to be shielded initially, but will do at some point in the future, we must decide, case by case, whether to provide it on day one or construct it at a later date. If it is to come later, then ideally we should prepare the area by pre-installing rebar (Fig. 16.7) or starter bars which can be used to couple rebar when the time comes. In truth, installing rebar and pouring concrete in a radiologically controlled environment is a time consuming and expensive process, so if at all possible we should strive to include all such construction in the original build phase.

Where a steelwork structure is perfectly adequate for a facility's primary operations, say to enclose a large marshalling area and support an EOT crane, but where that same area could usefully be put to work later, if only it were shielded, there is the option to consider constructing the enclosure in concrete from the outset. It may sound a touch lavish, but very often the costs of comparable steelwork and concrete structures are broadly similar. As a result, and if we get creative enough, it is possible to construct a large shielded area without necessarily eating into a facility's future-proofing budget, or certainly not using too much of it.

Fig. 16.8 ILW stored in concrete overpacks.
© Nuclear Decommissioning Authority.

Incidentally, if a future use will introduce significantly higher loadings onto a concrete slab, say to store overpacked containers of ILW (Fig. 16.8) then we must construct an appropriately strengthened slab during the original build. The alternative of backfitting a heavy-duty slab is pretty much guaranteed to be a non-starter.

16.6.2   Utilities and active pipework

In many ways the lifeblood of nuclear plants flows from their utilities, for it is they that keep it energized and support many of its day-to-day operations. In Chapter 6 we saw that not only is the list of utilities quite extensive, but each comes in several guises. So, for example, there are potentially more than four types of gas, three variations for both water and electricity, two for reagents and steam. And compressed air can be directed to at least three different purposes.

In addition, many nuclear plants have an infrastructure of active pipework, which emanates from activities such as processing radioactive liquors and operating wet decontamination systems. This pipework may be routed directly to an on-site effluent treatment plant or, where small volumes are involved, to a browser connection point located on a building's external perimeter. And just as utilities and active pipework sustain the primary operations of a nuclear plant, they are equally important during its decommissioning.

With this in mind, when a facility's suite of utilities and system of active pipework is first identified, sized and installed, we need to coordinate the exercise with long-term plans for decommissioning, and in some cases for a plant's remodeling and reuse. A crucial element of this task is to establish the *point of need* for each utility and active pipework connection throughout a facility's entire lifecycle. And with that to ensure an appropriate distribution network is either permanently in place, or can easily be accommodated when the need arises.

In the case of non-active utilities, additional elements can often be incorporated quite economically. However, the case for installing active pipework, which may not be needed for several decades, will come down to specifics of the distances involved, levels of radioactivity, relative elevations of various liquor holding tanks, and so on. Where it is not economically viable to install a full lifecycle network as part of the original build, then it is imperative that active connection points are provided for installation of the necessary pipework at a later date.

16.6.3   Remote waste handling

The starting point for this enabler is to recognize that any area which must be shielded during a facility's primary operations has potential to house remote decommissioning tasks in the years that follow. For beta–gamma plants in particular, this sits at the heart of our overriding objective of performing self-decommissioning to the fullest extent possible. The solution does not even need to be particularly elegant; piecemeal will do and ad hoc is absolutely fine. Procrastination is the enemy of a site's clean-up campaign, so wherever possible it is preferable to get on with self-decommissioning at a sedate pace rather than spend years, and no doubt considerable sums, preparing to perform the task with a more grandiose solution.

The ideal here is to make do with P&E we have. If not, then to either supplement it or provide enablers which will make its enhancement possible at a later date. In addition it may also be appropriate to import modules, pre-planned of course, to aid the decommissioning process, say to immobilize containerized wastes. Beyond that we begin to stray away from our objective of self-decommissioning, maybe even to the extent of requiring a new building to accomplish the task.

The most complex challenge in this context is dealing with ILW. So extracting it from where it is and, in the case of solid materials, possibly conducting some form of size reduction. Next we must load raw waste, either solid or liquid, into disposal containers, immobilize it, say within concrete and lid the containers. With that done, waste packages will almost certainly need to be monitored to ensure they comply with the conditions for acceptance (CFA) of their receiving plant. And since we are monitoring, we need an ability to decontaminate the external surface of packages which do not satisfy CFA criteria.

Finally, after navigating all of those steps we must provide the means to store waste, either temporarily within its originating plant, or alternatively have the capability to load ILW packages into a shielded flask, or overpack and ship them off to a storage facility elsewhere. It is quite a marathon.

Every nuclear plant harbors its own particular nuances, to the extent that the short summary above gives just a glimpse of what it can take to deal with ILWs. Whatever the detail, we can be sure this is not an enabler that can be satisfied with a few wall brackets and an extra coat of paint. This one takes real investment, so for beta–gamma plants in particular, will have no problem devouring a future-proofing budget, both in terms of design time and on-site implementation.

Clearly we need a shielded area to house decommissioning's remote operations, so nuclear facilities which already have a cave, or US type canyon, for primary operations are automatically well placed to conduct their own waste handling. For others, project teams need to work a little harder. Wherever possible then, we should configure at least one of a facility's pre-existing shielded areas to perform remote waste handling duties when its primary operations are concluded. This is done as part of the DfD exercise, when facility layouts are configured with an eye to the full lifecycle rather than concentrating exclusively on near-term operations.

If they do not exist as part of the original build, or do but are in the wrong locations, then it may be necessary to install cave-face workstations to aid D&D, essentially the shielded windows and manipulators we discussed in Chapter 7. We know that such kit is hugely expensive, so it is unlikely a sensible case could be made for purchasing and installing it during the original build. Rather, the pragmatic and most cost-effective enabling strategy is usually to install through-wall liners which can accommodate the necessary equipment, but then block those liners and only purchase windows, manipulators, and so on when they are needed.

Apart from workstations and other paraphernalia necessary to handle solid and liquid decommissioning wastes, there is another consideration, one which is quite multifaceted. We need routes to get ILW materials into the remote handling area, routes to get empty disposal packages to the point where they are filled, and routes to export filled ILW packages to either a conjoined storage facility, or into shielded transport flasks which can carry wastes to a more distant location.

As if all of that wasn't enough, we must also develop a pre-planned choreography for the order in which decommissioning wastes will be extracted from their existing location and dispatched to a remote handling area, essentially the self-decommissioning activity by which a facility gradually eats itself until all of its wastes disappear. Once again all of this is determined as part of the DfD exercise, when a proposed new facility is configured at the outset to accomplish demands of its entire lifecycle.

Where it proves impossible to accommodate all of decommissioning's remote handling activities within structures which were essential to a plant's original operations, then the facility must *take the hit* so to speak and provide the wherewithal to make it possible. Simply put, the decommissioning problem cannot be passed onto someone else.

In these situations a facility may be enhanced by incorporating a small shielded "shell" which is pre-designed for decommissioning tasks and can be kitted out when needed. Alternatively a location may be set aside, supported by the necessary utilities and infrastructure, to install a pre-designed remote handling module when the time comes. Whatever the detail, provision must be made, to the maximum extent possible, for a facility to handle and process its own decommissioning wastes. This may be permanently installed or designed and ready to implement at a later date, or a combination of the two. As I say, conducting primary operations without robust plans for what happens next is not an option.

It has to be said that as enabling works go, developing a remote handling philosophy years in advance and providing the means to achieve it is quite a tall order. Particularly as we must cajole a plant originally designed for other purposes into doing it. However, historically the lack of such inbuilt flexibility has resulted in extraordinarily complex and costly decommissioning challenges, and been a major culprit in forcing site owners to construct facilities solely for decommissioning purposes, the *self-perpetuating cycle*.

So the case for this branch of future-proofing is overwhelming and its measure of success quite straightforward. Long before a nuclear plant is built we must know *exactly* how it will be picked apart and put preparations in place to do it.

16.6.4 Fire compartmentation

One of the many elements invisible to us as we walk around any large commercial or industrial complex is ways in which fire safety has been incorporated into them. We may spot emergency exit signs and fire extinguishers, but what is not immediately obvious is how all such buildings are divided into zones, or *fire compartments* to be precise.

The rationale goes that if a fire were to break out and there was nothing in place to impede its progress, then it would be free to spread rapidly throughout an entire building. This would be a dangerous situation for anyone inside and, from an investment protection perspective, make it difficult for firefighters to save parts of the building, even areas some distance from the seat of a fire's initiator. Fire compartments are there to at least slow the progress of a blaze and ideally, maybe aided by firefighters, stop it spreading from one zone to the next.

Details of fire safety practices and relevant legislation vary from one country to another, but underlying principles are pretty much the same. Everyone adopts a philosophy of fire compartmentation, with construction between zones being rated to withstand a fire for half an hour, one hour, and so on, rising in 30 minute increments, typically to a maximum of three hours. And of course if a wall is rated to say one hour, then doors within it must deliver the same performance. In fact if you look closely at doors in many buildings, such as schools, hospitals, and airports you will notice small plates signifying they are fire doors and variously labeled FD30, FD60, and so on.

On the face of it, all of this is far removed from anything at all to do with future-proofing a nuclear plant. However, if we plan to remodel a facility for new purposes or adjust its configuration during decommissioning, then we must take account of fire compartmentation. We must ensure it remains steadfastly in place and is not compromised in any way. As with other enablers we must take a whole-life approach and either establish future-proofed fire zones at the outset, or embed an ability to modify them at a later date.

Fig. 16.9 Fire sealing. © Nuclear Decommissioning Authority.

One area to pay particular attention to is fire sealing and fire dampers. The way it works is that where pipework, cables, ductwork, and so on pass through a firewall (Fig. 16.9) say one designated at 60 min, penetrations must be designed to withstand a fire for the same period of time. With this constraint in mind, we must synchronize proposed modifications to radiological zones with arrangements for fire compartments, particularly as the two often coincide with each other. So, for example, we should consider installing fire sealing in locations which accord with a facility's entire lifecycle, as it can be difficult to backfit at a later date.

16.6.5 Segregation

If you had time on your hands, say a few days, and were given free rein to familiarize yourself with a large industrial complex, or an airport, hospital, and so on, you would discover how fantastically intertwined all of its various constituents were. Choose any element, ductwork, cables, pipework, access routes, even the structure itself. Behind their snapshot appearance each would form a network that stretched across the entire facility. Even more impressive, all of the components would be designed to harmonize with each other and together form a labyrinth that sustained the whole complex.

For most industrial and commercial premises, including *operating* nuclear plants, this type of integration is exactly what we need, couldn't be better. But when we get around to decommissioning, to picking a plant apart, it can work against us. Even more problematic, if we intend to remodel and reuse a building for other purposes, then such entwinement is the last thing we need. For nuclear facilities then, we need a different approach.

With regard specifically to decommissioning, everything hinges on the predetermined sequence in which a building will be decontaminated and dismantled. So, for example, if an operating facility houses decontamination systems which could be employed during D&D, then we should ensure the necessary utilities stay in place for as long as they will be needed. Likewise, where those decontamination processes employ wet techniques, their active drainage systems should remain operable even as other parts of a facility are being steadily disassembled.

The same constraints apply to ventilation systems which, as we know, spread their tentacles into every corner of a nuclear plant. The objective is to ensure these systems, including their finely balanced cascade arrangements, remain operable until the radiological hazards they constrain have been satisfactorily abated.

Of course decommissioning teams *could* develop ad hoc arrangements, such as installing their own decontamination systems and bringing in mobile filtration units (MFUs), but far better to configure a building so that its various networks can provide useful service for as long as they are needed.

If rather than decommissioning a facility straightaway we plan to remodel and reuse it, then the issue of segregation has two dimensions. The most straightforward, which is challenging enough, is to develop service distribution networks, access routes, and so on which are compatible with both a facility's primary operations and a new purpose which is planned to follow. And whereas this may be described as primarily a micromanagement issue, the second dimension is certainly one of macro proportions.

The quandary here is that when we draw up plans to reuse a building, there is a fair chance we shall not wish to retain all of it; maybe we just want to employ a half or one third. However, if a facility has been assembled as a single integrated entity, we shall find ourselves unable to unpick unneeded elements and instead be forced to preserve the whole thing, and look after it. Ideally the portion we wish to retain should be isolated and equipped to standalone; that way we can get on with decommissioning and maybe even demolishing the remaining building, ridding a site of an unnecessary liability.

As usual, this is an exercise which can only be successfully accomplished by integrating it within a facility's design from the outset, not least because its structural concept must satisfy two entirely different building configurations. In essence the proposed long-lived portion must be housed within a segregated structure, one which also includes all of the necessary utilities and supporting infrastructure, so potentially the ventilation system, electrical supplies, radiological changerooms, workshops, and so on.

Let's take the relatively simple example of a crane hall which, during primary operations, encompasses a large marshalling area, along with an adjacent workshop and a small concrete structure which is used to export items from the adjacent "main" building. In later years the crane hall is destined to store concrete overpacks holding containers of ILW, the concrete structure will provide its import and export route and the workshop will act in support. The remainder of the facility is not required There are three objectives.

First is to configure the crane hall, concrete structure and workshop so that they can readily switch to their new role in the future. Second is to arrange the necessary utilities in a way that enables them to "stand alone" when the time comes. Thirdly, and most important, is to ensure the area is structurally independent of its adjoining nuclear plant. Otherwise as I say, a site's owner will be forced to retain the whole building, maybe for decades after its work is done, which is an expensive business.

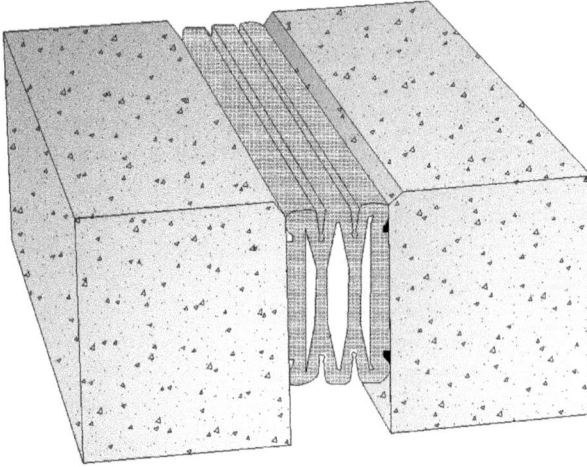

Fig. 16.10 Movement joint.
© Nuclear Decommissioning Authority.

We may elect to position the crane hall, concrete structure and workshop away from the main building, say by at least 5 m, and during primary operations provide a ground level link to accommodate transfers of personnel and equipment between the two separate areas. Alternatively we may unite the two constructions but use movement joints (Fig. 16.10) to keep them structurally independent. Whatever way it is achieved, setting up a building so part of it can be demolished at a later date is more complex than constructing a single fully integrated facility. But then it does hold out the possibility of creating an excellent storage facility for a fraction of the cost that would be incurred if designing and building one from scratch.

In truth, in this particular example, the structural folks may decide for their own reasons that the crane hall area and main building must be segregated in some way, say for independent seismic movement. With such a happy coincidence, the future-proofing dimension will contribute to a decision on how that separation is best achieved. Then with segregation almost for free, so to speak, the remaining task is to ensure utilities and so on are appropriately arranged for the long-haul.

16.6.6 Space

When decommissioning is underway in earnest or when a facility is being remodeled for new purposes, we can be confident the demand for exploitable space will be very high indeed. As a minimum, somewhere will be needed to laydown materials as they are stripped out, such as ventilation ductwork, cable trays, pipework, redundant P&E, and so on. And if the task is to remodel a facility, materials will need somewhere to sit whilst waiting to be installed. In addition, space will almost certainly be required to buffer waste disposal containers, either empties arriving at a building or filled containers awaiting export. And in many cases a clear area will be needed to locate modules which are destined to aid the decommissioning process.

The list of potential space-seeking candidates is long and varied, but that said they are bound by a common and obvious constraint. All must be located close to whatever it is they serve. For example, there is little point in providing a well-proportioned laydown area where there is precious little to lay down, or setting aside space in a location which is difficult to access. Supply must precisely meet demand.

The key here is to identify candidate areas in advance, during the DfD exercise, and ensure adequate space is allocated. Where feasible, this may be achieved by removing lightweight walls or reopening construction *blockouts*, but in other cases the necessary long-term space may be a permanent feature. Admittedly it may look a touch lavish during the years of a facility's primary operations, maybe even cause a little puzzlement. But one day, when remodeling or decommissioning kicks in, the waiting space will repay a hansom dividend. Before that, any suggestion of providing what may be perceived as *extra* space will need to be justified by an unshakable rationale.

In common with many other industries, the subject of *space* in nuclear circles can be a particularly vexing one, often accompanied by an almost obsessive desire to minimize it come what may. It stems from a notion that increasing space, or indeed reducing it will have a more or less pro rata effect on costs. So, cutting a building's footprint and hence its volume by say 10%, is assumed to see costs tumble by around the same amount. However, as we saw in Chapter 12, the spend profile for nuclear plants in particular undermines this theory. Typically they expend more than most on P&E, design activities and various types of management. As a result the *building*, or civil and structural element of a nuclear plant, may account for just one quarter of its all-up cost.

If we run the numbers, then increasing a building's footprint by say 10% would see its bottom line expand by 2.5%. Of course those figures are not precise, as there are far too many variables, but the underlying principle of a significant disjoint in the percentages holds true. What this tells us is that when it comes to future-proofing, to designing with decommissioning in mind, we need not shy away entirely from increasing the size of a building. Obviously I am not suggesting a blanket 10%, it is just an illustration. What we can observe is that increasing the width of a corridor, adding a little to the length of a crane hall, expanding the size of a workshop, and so on, will not break the bank, or to be more precise, will not consume too much of our future-proofing budget.

Ideally we would like to squeeze maximum benefit from every square meter of a building throughout its entire lifecycle, but in practice the value of a given space will shift over time. What matters, is to ensure we do not skimp on creating space for activities which may be decades away. One day it will be needed, and if a facility is found wanting those charged with its decommissioning will literally have nowhere to go.

16.7   Responsibilities

Potentially there could easily be 50 years, perhaps more than 100, between the time future-proofing proposals are first developed and their actual implementation. The challenge therefore is to ensure that as the decades pass, plans for DfD and DRR maintain an appropriately high profile and stand ready to be rolled-out in their preordained sequence. If not we face a prospect of the whole endeavor unraveling, quite possibly to the extent that little benefit will ever be realized from our well-laid plans. What we might call, a lost opportunity.

It could come about because a plant is unwittingly modified in ways which undermine its inbuilt ability to self-decommission, or a site's infrastructure is reconfigured in ways that inadvertently pull the rug from under a facility's capacity to be reused. In fact, over the years all manner of well-intentioned actions may scupper future-proofing measures that lay dormant whilst awaiting their time. Whatever the potential spoiler, we must guard against it.

16.7.1   *Client role*

Customers, whether they be government agencies or heads of energy companies, do not design nuclear facilities themselves. They specify what it is they want, with input from nuclear practitioners, and then rely on either in-house specialists or the supply chain to design and build it for them. Beyond that, clients may use their own team, or alternatively appoint other organizations to operate their facilities and indeed sites on which they reside.

In the context we are discussing here, of future-proofing nuclear plants, the role of a client is threefold. Firstly, to be crystal clear about what their expectations are. Simply put they must demand future-proofing in both its guises, DfD and DRR, and not be satisfied until presented with irrefutable evidence that both have been incorporated to the maximum extent possible. After all it is they, or their successors, who will ultimately pick up the tab for any shortcomings. Unsurprisingly then, a customer's interest in future-proofing is microscopic to say the least.

Secondly, having satisfied themselves that proposed future-proofing measures represent a sound long-term investment, clients must be prepared to sanction the necessary funding. Of course money is always tight, but as we have seen readying nuclear plants to embrace their demise is arguably on a par with designing them to fulfill their primary operations. So the case for funding both ends of the spectrum is equally compelling. Neither can sensibly be viewed, or funded in isolation.

Thirdly, over a nuclear plant's lifetime it can expect to witness many changes in personnel. Facility managers and operators will come and go, the organization charged with operating a site may change, and it is even possible that ownership of a facility's host site may transfer to another organization. The one constant here, is that no matter where a facility is in its lifecycle, its owner, whoever that may be, will always have a vested interest in the continued wellbeing of its future-proofing plans.

It is the client therefore, a site's owner, who takes the lead in ensuring plans for a facility's DfD and DRR maintain their prominence and are not undermined by uninformed actions. Owners may delegate this responsibility to their incumbent site operator, but must maintain momentum on future-proofing by embedding it, as a priority, within their operations procedures. In this way, even as generations pass, a future-proofing ethos will remain at the core of individual nuclear facilities and indeed entire sites until finally their decommissioning is complete.

16.7.2   Design team role

Clearly a design team sits at the heart of developing future-proofing proposals. It is they, along with their site-based advisors, decommissioning specialists, and so on who must maximize a facility's ability to decommission itself and wherever possible be reused for other purposes.

Before that however, before a nuclear plant can begin to operate, there is a long period of commissioning and training, during which designers gradually pass on essential knowledge to personnel who will own and control it. Part of this process sees a huge suite of indispensable documents transferred from the care of a design team's project manager to a facility's new owner: everything from operating procedures and plant maintenance schedules, to a hefty set of safety case documents.

In among this documentation will be a comprehensive description of how the facility has been designed and equipped to face its long-term future, all of which is integrated with the decommissioning plan and strategy we discussed in the previous chapter. This covers any proposed reuse, how the facility will contribute to its own decommissioning, the enablers that have been provided to smooth the way, and so on. Altogether a mine of information, all carefully compiled and ready to guide future custodians, step by step, through a facility's post operations future.

The onus here, the responsibility from day one, is on design teams to take the long view, to shun short- or even medium-termism and to root their proposals as firmly in the future as they do in the present day. So in the same way that safety, operability, throughput, and a long list of other factors weave their way into the design process, so must future-proofing, both DfD and DRR. For a sustainable nuclear industry there is no other way, it is as simple as that.

Design development

17

Let's say a project has reached the stage where the challenge has been defined, maybe not every full stop and comma but sufficient to get the building design underway, enough to start developing schemes for the *optioneering phase* we discussed in Chapter 12. Maybe the task is to design a new reactor complex, or a processing facility of some sort; maybe a plant is needed to encapsulate waste in concrete, or vitrify a radioactive liquid. Or it could be that a new facility is required to store radioactive waste. Whatever the challenge, funding has been released and a design team assembled (at least the first wave), all raring to go. Finally we are ready for the off.

But then imagine the first day, all gathered around a table and examining a blank sheet of paper. Where on earth do we begin? How do we develop a solution that satisfies a customer's specification, keeps the various engineering specialists and prospective plant operators happy, one that regulators will approve of, incorporates a sound approach to future-proofing, can be built and operable on time? And of course how do we ensure that the whole endeavor can be accomplished safely and within budget? It is quite a list, but we are undaunted by the task. Let's see how we pull the whole thing together.

Clearly there is no shortage of things to ponder, but happily for our discussion here, there is one activity that draws many of a project's diverse threads together: the facility layout. Whether it be construction, mechanical handling, chemical processes, radiological zoning, throughput modeling, operations and maintenance, the structure itself, decommissioning planning, future-proofing, or a whole host of others; one way or another, all are depicted and coordinated via a facility's layout. So this is where we shall concentrate our scrutiny.

As with our discussion on programming in Chapter 12, we shall assume the assignment is to create a new facility, but keep in mind that similar processes apply when we set about decommissioning those that exist already.

17.1   Layout preparations

It might seem that since here we are embarking on the creation of a new nuclear facility, maybe a monster one at that, then the first step must surely involve production of a few building layout sketches, some doodles to get our ideas flowing. But then tackling a facility en masse may look like a bit of a stretch, so thoughts may occur that it makes more sense to take it a bite at a time. Maybe the first step should be to concentrate on arranging a facility's nuclear core, say focusing on a reactor and its immediate environs, or the configuration of shielded caves and cells and the likes. Then with that sorted we could consider positions of supporting functions such as maintenance workshops, decontamination areas, electrical switchrooms, and so on.

Nuclear Facilities. http://dx.doi.org/10.1016/B978-0-08-101938-2.00017-9

Yes, such an approach can be made to work, just about, but it is a tortuous route and adds layers of complexity, and precious time, to a process which is complicated enough to begin with. And to be honest, if layouts are developed in such a piecemeal way they will always fall well short of the ideal. The main problem being that developing individual areas, no matter how key, in advance of others, imposes a fait accompli on arrangements for areas which will follow. In other words, first up best dressed, which is not a great technique. What we need is a plan, a structured approach, some way of getting to an ideal building configuration as quickly and efficiently as possible.

Building flow diagram	Areas of facility and how they relate
Constraints	The *givens* by which it will be bounded
Assumptions	Basis for initial design development

Fig. 17.1 Precursors to layout development.
© Bill Collum.

As far as building layouts are concerned, the reality is that at this stage we are not yet ready to draw a line, certainly nothing too meaningful. If we are to launch layout development on a sound footing we have other priorities, crucial preparations which must be made before we can even think about arranging the facility itself.

Whilst we are busy with the preliminaries, we may secretly be thinking about how our new building may be configured, but we must be patient and hold off until we are properly prepared. In fact rattling off designs at this stage will, I'm afraid, follow a predictively expensive circle that ends up back where it started. The precursor to layout development, I believe, comprises three elements as shown in Fig. 17.1, all of which are best tackled in parallel. Let's see what each involves.

17.1.1   Building flow diagram (BFD)

You will no doubt have seen diagrams of transport systems such as national bus routes, or rail networks, and noticed they bear no resemblance whatsoever to real geography. They are not drawn to scale and convey no hint of the twists and turns between one location and another. Their purpose is to distil complex, sometimes multi-leveled networks into an easily understandable image, so that we can find our way around. We can see at a glance how one location, one step along a journey follows the next, where we can switch between different legs of the same network and where there are links to other transport systems, such as airports and ferry terminals. In short, we can identify primary points within the system in question and also those with which it connects.

Before we can start the exercise of putting a nuclear facility together, we need to develop a similar image. It is called a BFD and is the key to establishing exactly what needs to be achieved when our pens finally touch down and we begin developing the building layouts themselves.

BFDs vary enormously, with some being quite high level and portraying interactions between a suite of related facilities. Whilst others concentrate on a specific operation, say loading a transport flask, and delve into considerable detail. Most though, the type we are particularly interested in here, explore a whole facility and are a level of detail sufficient to inform its early layout development.

Fig. 17.2 shows an example, where we can see boxes for the various activities along with arrowed lines which show how they interact. It also depicts the usual practice of color coding radiological zones. My personal preference is to stick with the standard rad zoning colors we discussed in Chapter 3: namely C2, green; C3, amber; and C4, red. And use yellow for C1 areas, simply to differentiate them from the white paper background. In this particular case I have used pale gray for zones designated as C3 (S). For nuclear sites that elect to subdivide the standard radiological zones, a BFD palette simply becomes more particolored.

To keep this exercise manageable it is best to develop BFD's through a few workshops, each attended by just a handful of key players from the project team. Attendees vary depending on the nature of a facility, but a representative from the host site's operations team should always be present so as to share their intimate knowledge of local practices with other members of the team.

The methodology adopted for BFD workshops first identifies the primary materials and equipment which are destined to enter a facility, such as new fuel for a reactor, radioactive liquor, transport flasks, and so on. Then picks them off one at a time and follows each stage of their journey through the building, until eventually they exit again. And whilst some will re-emerge within a matter of days or even a few hours, often in a changed condition, others may remain in a facility for many years, particularly those dedicated to storing radiological waste.

A BFD workshop team examines in detail what must happen at each step along the way: how materials will be lifted, moved, handled, pumped, buffered, stored, and so on. Whether cranes, bogies, manipulators, pumps, valves, shielding windows, cameras, and a myriad of other equipment may be involved and what role each will play. And crucially a first assessment is made of the radiological zoning which will pertain to each activity, normally color coded, as I mentioned a moment ago.

Developing a BFD is quite an exercise, but when done thoroughly provides a comprehensive overview of the primary activities which will be conducted within a proposed new facility. And most important, ensures there are no loose ends that could cause an unwelcome surprise when work starts on layouts themselves. Of course with pressure mounting to see a building's actual design materialize, developing a BFD may appear to be an unwarranted diversion. So I will give you an example to illustrate why it matters so much.

17.1.1.1 BFD process

Let's say a highly radioactive liquor of some sort is going to be treated in our new facility, and that it will enter from the end of a shielded pipeline, one which originates from another building nearby. Once inside, the liquor will pass through several processes

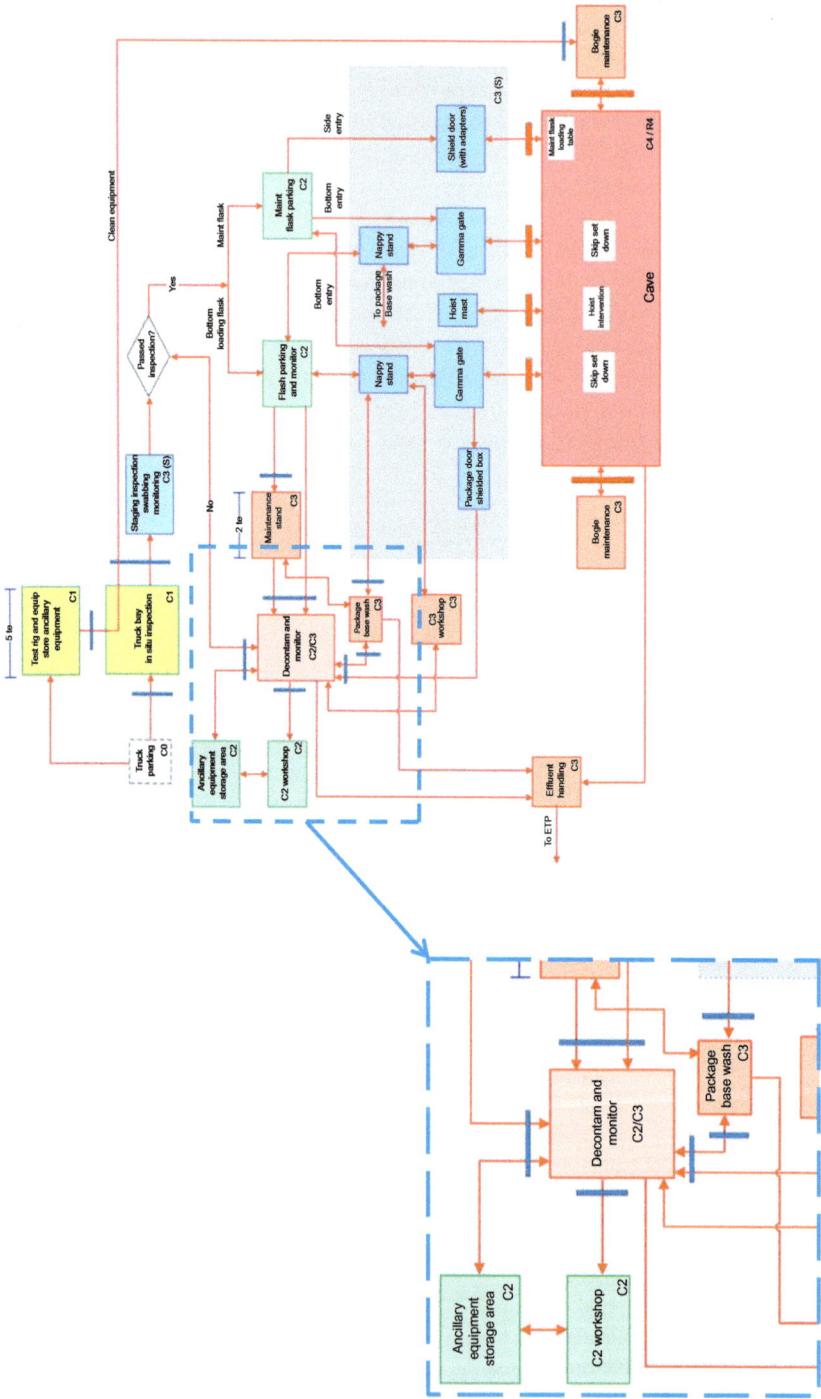

Fig. 17.2 Building flow diagram.
© Bill Collum.

until eventually its most active constituents are separated out and blended with molten glass, poured into stainless steel containers and dispatched to a storage vault. The remaining constituents will be routed to other facilities for further processing.

Our task here, in this example, is to figure out what needs to happen around the area where liquor enters a proposed new building, so that we can develop a layout. The BFD will of course cover an entire facility, but this one area will be enough to illustrate thought processes involved in the exercise. Incidentally, a BFD must steer clear of replicating the process flow diagrams and mechanical handling diagrams we discussed in previous chapters. They are useful contributors, but BFDs are more concerned with spatial requirements and interfaces, rather than capturing every nuance of a facility's processing or mechanical equipment.

In essence liquor will flow along a pipeline, be routed into a waiting receipt vessel, and from there to the first of several processes which are queuing up to deal with it. How difficult can it be? But look a little closer, with the relentless probing of a BFD exercise, and a few questions come to mind, the answers to which will determine how this particular corner of our shiny new building must be configured. For example:

- Would it be preferable for the radioactive feed liquor to enter our building from a pipeline below ground level, or would it be better to start its journey from a height of several meters, so that liquor can flow by gravity, rather than be pumped to the top of its receipt vessel?
- Does the donor plant have any option as to what elevation it dispatches radioactive liquors, or is the pipeline exit point already prepared and therefore a *given*, one that should to be added to our list of constraints?
- If we were to opt for a below ground pipeline, are there any obstructions such as existing utilities, other active pipelines, road or rail infrastructure, and so on that may require modification, or even diversion. And are such works technically and economically feasible?
- Could we accommodate a receipt vessel below ground or is it preferable, due to say ground conditions, to construct our new facility entirely above ground level, or *grade* as they say in the United States?
- If we were to opt for a shielded pipebridge, what height would it need to be to suit our receipt process? Would it be high enough for vehicles to pass below it? (Emergency vehicles normally need access to the entire perimeter of a nuclear plant.) And are there suitable locations for pipebridge supports?
- Will liquors be moved by fluidic devices, mechanical pumps, gravity or a combination of all three? For each of these systems we need to consider the constraints associated with a barometric head, maintenance requirements, and so on, described in Chapter 6.
- Is there a minimum incline at which the transfer pipeline must fall in order to, for example, accommodate liquor flow and drainage? And if so what are the implications for relative elevations of dispatch and receipt vessels in both facilities? The issue here is that a fall of say 1 in 20 between two buildings which are 50 m apart, amounts to a difference in elevation of 2.5 m, something that may be difficult to accommodate.
- Will liquor transfers be in batches or via a continual feed? If batches, what size will they be, will they always be the same and what is their frequency?
- If liquor is received in batches, do we need sufficient buffer storage capacity so as to smooth delivery into a continual feed for plant in the new facility?
- Does the liquor transfer pipeline need to be flushed, either from time to time or after every batch? Which facility will be responsible for the flushing process? Will the flushing medium, say nitric acid, be used just once or several times? Where will it be stored?

- Will the newly arrived liquor need to be kept agitated until it is moved on, or must it be allowed time to settle and stratify so that the supernate can be drawn off before processing begins? If so, how much supernate will there be, will it be returned to the donor plant, could it be used to flush the transfer pipeline and must it be held at a particular elevation to aid the transfer process?
- If stratification of the liquor is required, is the aspect ratio of its settling vessel important? How long must it be left, and is the duration such that several settling vessels may be required adjacent to the receipt tank? Alternatively, might it be preferable to have several receipt tanks where settling can take place?
- Is sampling required and if so will it be carried out pre, or post settling? What is the turn-around time for sampling results, and is it such that buffer vessels will be needed to store liquor whilst awaiting analysis? Is there a pass/fail as a result of sampling and are the results required to determine subsequent processing? If there is a "fail" whereby liquor does not satisfy the new facility's conditions for acceptance (CFA), where does the out-of-specification liquor go, returned to its donor plant, or other destination?

Although we have touched on several BFD type issues it is by no means a fully comprehensive list, in fact most of the questions posed would have follow-ups, supplementaries that would shed even more light on what must be accommodated around the liquor receipt area in question. We have, however, covered enough to give a flavor for how a BFD analysis goes. One question after another, picking up threads and following leads, until eventually we are satisfied that this particular corner of our proposed new facility has yielded up all we need to know of it, for the time being at least.

This is a good start, but clearly on a live project we would need to continue the BFD exercise until the entire facility had been covered. Armed with this starting point, we know exactly what our layouts must achieve when we begin their development—well, almost. You will recall there are two other activities which must be tackled in parallel with a BFD.

17.1.2   Constraints

This particular exercise may seem somewhat removed from developing a workable design as, rather than examining a building's operations, it focuses instead on the environs it will eventually occupy. However, without this analysis there is a strong possibility that our initial layouts will need significant revision when these non-negotiable influences are finally recognized.

For this reason it is important to identify, as far as possible in these early stages, any constrains, or *givens*, by which our new facility's design will be bound. Examples are innumerable but may include:

- Pipework connections already exist at the donor plant for a proposed new liquor transfer line. They are at an elevation of 8 m and on the facility's south side.
- Liquor will be delivered to the new facility in batches, not a continual feed.
- Liquor must be kept agitated at all times, as settling would result in a heel forming which would be difficult to resuspend.
- There is no provision on the host site to maintain contaminated mechanical pumps. If mechanical pumps are proposed, then the new plant will need to provide in-house maintenance facilities, or alternatively be prepared to dispose of and replace any pumps that fail.

17.1.3   Assumptions

To complete the triptych of elements necessary to set layout development on the right path, we need to determine assumptions on which initial optioneering will be based. I guess the notion of needing to *assume* anything at all could be taken to infer we are not yet ready to begin the design process, but we are. And the reason for needing to work, initially at least, with assumptions, is twofold.

For a start nuclear projects can be extraordinarily complex, with layer upon layer of competing considerations all vying for recognition, added to which the sheer scale of some facilities only serves to complicate matters further. As a result, if we were to hold off until every scrap of pertinent information was available before getting our toe in the water, we would face the prospect of monumental delays, to the extent it is doubtful any project would ever get started. Necessarily then, we need to employ a little pragmatism to get the ball rolling.

More crucially, developing designs for new nuclear facilities is the type of iterative process that demands a judicious start. Radiological zoning is a good example, where radiation and contamination levels will, to a large extent, be dependent on equipment and processes which will ultimately be deployed within a facility. The two facets evolve together, until finally we are in a position to meld our design and radiological zoning philosophy into a solution which complements both. It is impossible to determine either one in isolation.

As I say, there comes a point when we need to take a deep breath and jump in. For the example we are considering here, figuring out arrangements around the area where radioactive liquor enters a building, assumptions may include:

- Batch size will be 500 L, delivered twice a day, 5 days a week.
- Existing on-site ion exchange facilities will have the capability and sufficient spare capacity to treat supernate which arises from the new plant.
- There is a viable in-ground pipeline route between the new plant and existing on-site ion exchange facilities.
- On-site laboratories will not have sufficient capacity to analyze liquor samples, so the new plant will need in-house analytical facilities.
- Ground conditions are such that subterranean construction is feasible.

Of course we cannot forge ahead for too long with unproven assumptions, so some members of the team will get busy on the urgent task of verifying them. If they turn out to be incorrect, it is still early days so there is no harm done and we can modify our initial concepts. The main priority being that we are up and running, we have a sound three-pronged basis for what we are doing, the whole design team is moving forward in a coordinated manner and some meaningful progress is being made.

17.1.4   BFD validation

Eventually, then, we shall have our first draft of a BFD along with its documented constraints and assumptions. Up to now the package will have been developed by quite a small team, so must be validated before being declared an appropriate basis on which to launch layout optioneering. Ideally then it should be presented to the wider project team, along with other key players who have not been involved in discussions so far.

The advantage of gathering such an extended group is that everyone scrutinizes a BFD and its supporting documents from their own perspective, ensuring the interests of their particular area of responsibility have been suitably addressed. Of course there will be lots of debate and conflicting interests to resolve, but this is the time to do it, to agree what we aim to achieve before committing to drawing lines on paper. With this forum's endorsement and no doubt a few tweaks to our BFD, we can pin it on a wall, admire it for a while and then crack-on with developing the layouts themselves.

17.1.5   BFD iterations

I am often asked if it is necessary to keep updating a BDF as information firms up and design concepts progress, sometimes in ways contrary to that originally anticipated. The short answer is, no. The fuller answer is along the lines that, developing a BFD package is a structured process that gets layout development and hence much of a facility's design off on a sound footing. In addition it is a good way of demonstrating to interested parties that we have a clear audit trail behind designs we are proposing, that they did not just pop out of thin air so to speak.

The thing is, once a design starts to roll and evolve, a process that can take anything from several months to a few years, the founding BFD gradually becomes superseded. It has served its noble purpose and gets left behind. In effect the mantle of design coordination shifts to building layouts, so updating the original BFD would serve no purpose.

However, there is one exception. Occasionally the design process will be running at full tilt: layout options flowing nicely, design philosophy documents well advanced, equipment looking more tangible by the day, when out of the blue a fundamental change of direction is announced. Maybe the facility needs to take on additional processes that had been destined for another plant, or the building must relocate to a different area of its host site. Whatever the reason, a monumental upheaval has occurred and our steady progress has been upended by the shockwaves.

When this happens, we cannot tweak our way out of it—trust me, it is far too complicated to even think of going there. My recommendation, I'm afraid, is that faced with such tumult a project has no alternative but to go back to the BFD process and start again. The good news is that the learning and design development already completed stands a project in good stead, so recovering back to the same position should be a much quicker affair than the first time around. Apart then from a project suffering an unwelcome curveball experience, we do not need to revisit its original BFD.

Before moving on, I must mention that the methodology I have just described, developing a BFD along with identifying constraints and assumptions, is by no means a mandatory element of the nuclear facility design process. It is, though, my personal preference. I like its structured approach and the way it gives key contributors a timely voice in proceedings. What matters is that whatever system is used to get from a client's specification to a frozen layout, we must be able to demonstrate that the end result is not simply "a" solution but, with all factors considered, is the "best" that can be achieved. Customers, not to mention regulators, take a keen interest in such things, so one way or another there must be substantive justification for where all the money is going and ways in which regulatory constraints will be satisfied.

17.2   Layout development

So far we have been discussing preparations for the design proper, getting ready to tackle the wide expanse of a blank sheet, but finally the day dawns when our pens touch down for the first time. What are we thinking as we start on the layouts themselves? Quite a lot, really. At this stage we have the BFD and its supporting documents to refer to, but none of this represents the reality of a building's geography. On top of that everyone on the team, every contributor, has their own particular objectives and constraints, all of which must be satisfied and incorporated into evolving schemes. So there is plenty to keep us busy.

To get a feel for the thought processes involved in developing layouts for a nuclear facility, we shall take a look at a couple of examples and analyze what needs to be considered as we mold our building into existence. To underscore just how convoluted this particular task is we shall start with an area which is barely nuclear, then later on examine an item that combines the additional layers of complexity which are introduced when radiological considerations enter the fray.

17.3   Vehicle bay

Practically every major nuclear facility has at least one road or rail bay attached, a place where large consumables or payloads, such as shielded flasks, enter and leave the building. At first glance, there is absolutely nothing to these areas: one or two large industrial doors to keep the elements out, an area big enough to park a large vehicle and the wherewithal to load or off load whatever is being shipped. Peel back a few layers, though, and there is much more going on than meets the eye, all of which the layout folks need to take into account and configure. Some of this will no doubt have been captured during the BFD exercise. So the task here is to raise it to the next level, to add three dimensions and move the facility a step closer to becoming a reality.

Let's take a look at what needs to be considered as the layout of a vehicle bay begins to take shape. We shall start with issues which apply specifically to dedicated rail or road bays, then move onto a wider view of factors that can be equally relevant to both.

17.3.1   Layout considerations—rail only

We can see at a glance that rail track is inherently inflexible to say the least, but the full extent of its geometrical limitations can be even more onerous than appearances suggest. Regular rail vehicles depend on adhesion to give their smooth wheels traction. As a result track must be laid at quite a gentle incline, otherwise drive wheels will spin when attempting to travel up hill. Worse than that, they can skid if brakes are applied when negotiating a downward slope.

Maximum track incline varies, depending largely on engine power and the weight of loads being hauled. That said, if we assume a relatively steep rail track incline of 1 in 50, then over a distance of 100 m the change achieved in level would be just 2 m, which is not great if you are dealing with a site location where ground rises at quite a significant incline. And of course as we get to distances of 200 m, 300 m and more the constraint becomes even more problematic.

For many sites, depending on their topography, such grading would demand a major and very expensive groundwork exercise, or simply be impossible to achieve. So before we begin planning a rail-served building we must establish whether it is even feasible to lay track to its door.

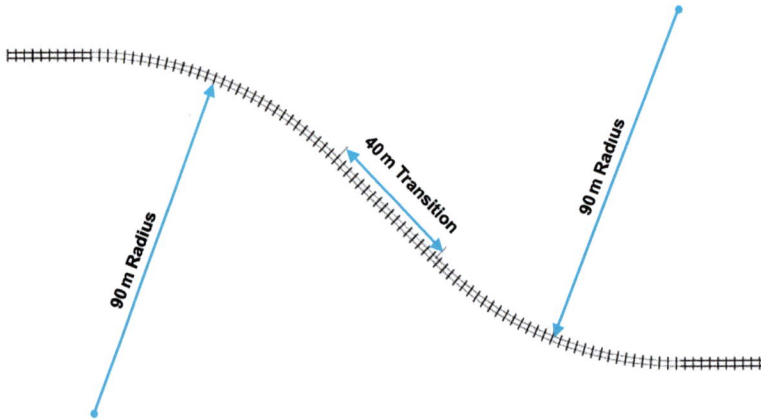

Fig. 17.3 On-site rail track geometry.
© Bill Collum.

The other big constraint, one we cannot excavate our way out of, applies to limitations on how rail track can be made to twist and turn (Fig. 17.3). As with most of the world, UK track has a width or *gauge* of 1435 mm. For fast-moving trains, such as those used on public transport systems, this results in a minimum bend radius upwards of 120 m. On nuclear sites, where speeds are typically limited to around 15 km/h (10 mph) this curvature can be reduced, but ideally should not be any less than 90 m, which is still quite a swathe. Not only that, rail track must transition out of a curve before it enters another one that bends in the opposite direction, and for this we may need a distance of 40 m or more.

It is possible to undercut these limitations a little, but unyielding steel wheels and track will both protest by screeching loudly as they are forced to bite into each other—not a pleasant sound. In addition, there is a predictable penalty in terms of increased maintenance and repair of wheels and worn track. All in all then it is best to stick to the rules, but it does take a lot of space.

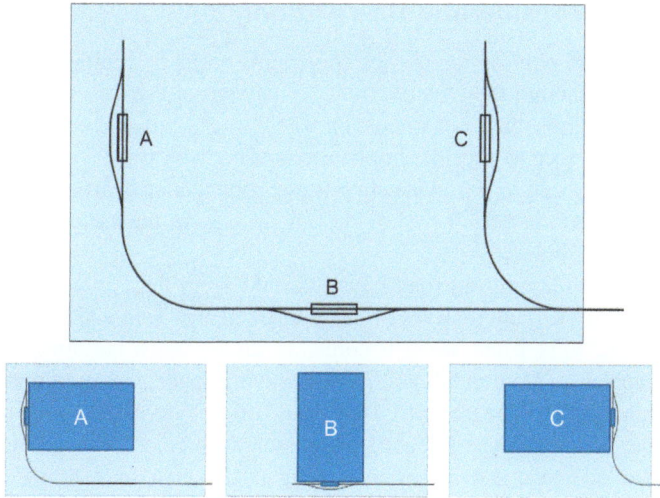

Fig. 17.4 Rail track influence on facility location.
© Bill Collum.

The composite image at the top of Fig. 17.4 shows how, when we transpose these restrictions onto an example site, there are really only three viable locations for a new facility's rail bay. Although it would be *possible* to locate a rail bay towards this particular site's upper boundary, the double bend needed to get there would sterilize so much land as to make it an unlikely contender. Facility "B" on the other hand could be moved to the right, freeing up land for other purposes, but then there may be higher priority considerations which rule this out. So there is plenty to think about.

It is also worth mentioning that this example assumes the facility's site is served by a dedicated rail spur, approaching from the right. If instead of a spur we needed to branch off, and re-join a site's trunk line, then options for track arrangements would be even more constrained.

The example above, in all three variations, assumes we have opted for flexibility delivered by a drive-through rail bay with parallel sidings. However, if we opted for a dead-end bay the alternative locations would still be pretty much the same. The one addition that becomes necessary where track ends within a rail bay is a buffer stop, the kind of thing we see in many railway stations, and which may need around 5 m to accommodate it.

Out of interest, I have come across situations where we have been dealing with sizeable building sites, apparently offering considerable flexibility in how they are used, but after investigation discovered there was only one single sensible location on the entire site for a rail bay, with all the implications that has for its attached facility. As I say, we only know for sure when rail's geometrical constraints and other influences are overlaid on our prospective site.

Clearly the location of a rail bay has a major influence on how a facility is configured, particularly when we factor it in with other constrains such as road networks, active drainage systems, topography and a very long list of others. It is most definitely not a topic we can home in on when a facility's layout is well advanced.

17.3.2 Layout considerations—road only

As with rail, the road only option carries constraints borne by limitations on vehicle maneuverability; although they are not quite as uncompromising as those applicable to rail, they nevertheless place constraints on where trucks can gain entry to a facility. In addition, we may need to consider implications of one way traffic routes on the host site and whether we need to introduce similar directional control around a new plant. Once again, these are matters that should be addressed in the early days of layout development.

When it comes to laying out road bays it would seem that, as with rail, we can choose between the twin alternatives of drive-through and dead-end; indeed, strictly speaking, we can. Very often when it comes to laying out a building, a dead-end road bay can appear to have much going for it: saving space, reducing road network infrastructure, and so on. The trouble is, to access these bays vehicles need to reverse, either in or out. Yes, we can put all sorts of safeguards in place to ensure reversing is carefully controlled and as safe as possible, using a banksman, for example, but it still carries a risk.

In common with other industries many nuclear sites take a view that reversing, particularly in a confined area, is such an inherently dangerous maneuver that it should only ever countenanced it in the most extreme of circumstances. Limiting ourselves to a drive-through option may add to the layout challenge, but is definitely well worth the effort.

17.3.3 Layout considerations—rail and road

We have touched on some of the issues that apply specifically to dedicated rail or road bays, so are ready now to see how the list widens somewhat when we examine issues that can apply to either situation, or even a combination of the two.

17.3.3.1 Buffering

First up, it is far too risky to assume vehicle movements will somehow coincide with the ideal timetable demanded of a new plant. A site's road and rail vehicles will almost certainly be busy serving several facilities, so have competing priorities to contend with. And certainly any notion of a driver dropping off a package, then hanging around for an hour or two whilst waiting for the next one to be loaded up, is likely to get short shrift from a site's transport manager.

Even when transport vehicles are available, it could be that a new facility's donor or receipt plants are not ready at their end. What we need is a certain amount of independence, some flexibility that stops our plant being at the beck and call of transport systems and other facilities. And for that we need a buffer, a place where inbound or outbound packages can be parked until external influences are aligned and we are good and ready to deal with them. Broadly speaking there are two alternatives, one external to a facility and one within it.

For road going trailers we can opt to provide a parking area close to a facility, or for flatrols provide rail sidings, both of which allow us to break direct links with donor and receipt plants. However, unless dedicated vehicles are purchased and a facility employs its own drivers, something which could be difficult to justify, it will still depend on a site's trucks and trains to shuffle payloads between their temporary parking areas and the facility itself. So it is a partial answer but may not deliver the flexibility which is required.

The other option is to provide buffer storage inside a facility, say parking positions for two or three transport flasks. All it takes then is for a building's electric overhead travelling (EOT) crane to transfer flasks to or from their transport vehicles, and either park them in their designated area or send them on their way. On the downside, if flasks or other packages are buffered they will need to be double-handled, which adds to a building crane's duties. In addition providing internal buffer areas adds to a facility's footprint which, assuming space is available, incurs a cost.

Ultimately then, a combination of operational considerations, throughput modeling and cost analysis will determine if buffer storage is required, and if so how much there should be and whether it is preferable to provide it inside or outside a building. Once again, as with all of the factors we are discussing here, timely consideration is crucial.

17.3.3.2   Vehicle characteristics

It is important to establish details of the actual vehicles which will be used to ferry packages to and from a facility, as this will have a direct bearing on dimensions of the vehicle bay itself. It may sound like a straightforward enough matter but in some cases, such as uncertainty over exactly which transport flask will be used, it can be surprisingly difficult, at least in the early stages, to establish which trailers or flatrols may need to be accommodated. And it is not just a matter of knowing what a vehicle's overall length will be, as its other attributes must also be factored into the design process.

17.3.3.3   Platforms

Very often, plant operators need to gain access to arriving and departing transport flasks and the likes, in order to conduct inspections, connect lifting beams, and so on, and for this they need platforming installed in a vehicle bay. Where just one type of trailer or flatrol is destined to be used, then it is simple enough to design platforms which match its dimensional characteristics. However, things become somewhat complicated when platforms need to marry up to several transporter types, as we are unlikely to find that one-size-fits-all.

If the switch between different trailer types is an occasional occurrence, then we may be prepared to contemplate installing adjustable platforms. If however it is due to be a frequent, even daily routine, then the time consuming prospect of continually adjusting platforms will be unpalatable to plant operators and may also risk compromising a facility's throughput. Alternative solutions can include stretching a vehicle bay to accommodate two or more separate platform configurations. Or the conundrum may even contribute to a decision to opt for more than one vehicle bay.

17.3.3.4   Means of escape

The role performed by platforms dictates that they are normally fairly lengthy, and because of space limitations are often tight-up against walls that run behind them. The thing to ensure here is that platforms do not impede emergency escape routes, either within the bays themselves or for those who may be hurrying through from an adjoining facility. This is particularly important at times when vehicles are occupying the bays, as pedestrians will be unable to get past obstacles caused by a combination of stationary trailers or flatrols and the platforms which run alongside them.

17.3.3.5   Consumables inspection

In addition to the comings and goings of nuclear payloads there will also be incoming consumables, the kind of thing needed by any large industrial facility. Potentially more problematic is that some packages, such as empty stainless steel waste drums, may need stringent quality checks before they can be accepted into the facility itself.

The reason for conducting checks prior to import is that if materials are rejected *after* entering the main plant, they may well be categorized as radiological waste and incur significant disposal costs. So it is much more cost effective to identify and reject noncompliant goods whilst they are still in the radiologically clean environment of a vehicle bay. It may therefore be necessary to provide a segregated area where these inspections can be carried out. Not only that, but we may also need space to temporarily store transport packaging, pallets, and so on, whilst awaiting removal from the area.

17.3.3.6   Local office

One of the repercussions of handling nuclear payloads is the need for a material tracking regimen, so waste containers, for example, always have a unique identifier. As a result vehicle bays normally need a small office where the necessary computer and so on can be accommodated. It may not sound like much, but within the tight confines of these bays it can be surprisingly difficult to free up an appropriate spot. It is one of those low-order issues that should never find itself in the spotlight, so we need to nail it early.

17.3.3.7   Engine fumes

Whether it be a locomotive or a truck, the chances are it will be powered by a diesel engine. This has a number of consequences which, although primarily safety related, can have a subtle influence on how vehicle bays are configured. The most obvious is dealing with fumes, more specifically the carbon monoxide and harmful particles which emanate from these mighty engines.

It can be addressed in one of two ways: either a flexible hose that connects to a vehicle's exhaust outlet, or more of a fixed system which may be installed under a vehicle bay's soffit or accommodated under an access platform. Whatever the system, we must ensure fume concentrations do not reach dangerous levels.

17.3.3.8   Fire compartmentation

The risk of a vehicle fire dictates that these bays are often designated as a segregated fire compartment. This tends to be primarily an issue of construction detailing and selection of materials used in the area, but it can easily extend to other spheres such as specification of an overhead hoistwell door. This in turn may determine how the door's drive mechanisms are configured, how much space it occupies and what its maintenance access requirements will be. So we need to keep an eye on these more subtle layout influences to ensure they are appropriately accommodated.

17.3.3.9   Wind turbulence

By their nature, the type of bay we are considering here always demands large industrial door openings, maybe six meters square, where vehicles gain entry. Bearing in mind that many nuclear facilities are located in exposed locations, this does make such doors susceptible to damage from strong winds. This may appear to be purely an architectural detailing matter but it is also an issue that should be considered during the layout optioneering phase.

Where possible then, these doors should not face into the prevailing wind, should avoid areas that will be subject to particularly violent wind swirl or turbulence, and steer clear of nearby facilities where strong suction forces may be created by a "wind tunnel" effect. Of course such considerations do not dominate the positioning of a vehicle bay, but where possible should be factored into the debate.

17.3.3.10   Floor level

Construction detailing dictates that the floor of a vehicle bay is normally around 250 mm lower than that of its adjoining facility, similar to the way domestic properties are always stepped at the front door. It is worth keeping in mind that we can capitalize on this essential difference in level by increasing it to say a meter, or even a good deal more. This can be particularly useful in minimizing groundwork on sloping sites and, apart from introducing a few additional steps, can generally be achieved without compromising access to the facility itself.

17.3.3.11   Changeroom access

As we have seen, there is a fair bit for operators to do in a facility's vehicle bay: everything from conducting inspections to connecting lifting beams, in fact anything to do with the receipt and dispatch of packages, particularly nuclear payloads. Having said that, it is unlikely there will be enough activity to justify having personnel dedicated exclusively to that area. Normally they are based next door, in the main facility, and make their way through to its adjoining vehicle bay as the need arises. The difficulty here is deciding how operators will move back and forth between the radiologically controlled environment of a nuclear plant, and a vehicle bay which has direct access to the world outside.

We know from Chapter 4 that everyone entering or leaving a nuclear facility must do so via a radiological changeroom, so recognize that exactly the same constraint applies to personnel moving between a nuclear plant and a vehicle bay. That being the case, this would appear to be a routine matter of simply providing an additional changeroom.

However, the issue is that changerooms take up space and equipment within them, including its upkeep, incurs a significant cost. For a facility's main access point there is no debate, we need one and we shall get our money's worth out of it. It can be hard to contemplate though, if a changeroom is destined to be located in some far flung corner of a facility and used just occasionally by a small number of personnel.

To make a decision on whether or not a satellite changeroom is justified we need to consider two factors: what distance a vehicle bay is from a facility's main changeroom, and if a satellite changeroom was provided, how much demand it would experience. Remember some nuclear facilities are vast, maybe bigger than a soccer pitch, so it could easily take 30 min or more for an operator to get to the main changeroom, cross it, get changed into their outdoor clothing and walk around the building to a distant vehicle bay. Not to mention retracing those steps to get back to their regular duties.

If this is even a moderately regular occurrence then we definitely need to provide an additional changeroom. On the other hand, if a facility is more compact and access to its vehicle bay is a relatively infrequent affair, then a second changeroom is bound to be perceived as an indulgence. It can be a tricky debate, so is a matter of carefully weighing up the pros and cons of individual situations before deciding whether or not to splash out on duplication.

17.3.3.12 Crane hook approach

You may one day find yourself standing in a vehicle bay, taking in the scene and speculating, privately of course, that it seems a good deal wider than it needs to be, even to accommodate large trucks or trains. The kind of apparent *faux pas* that project managers like to home in on in their energetic quest to shave expenditure. Happily those responsible can rest easy, because the seemingly generous expanse is not only deliberate, it is essential.

If we take the typical arrangement for a vehicle bay, then it will be served by an EOT crane located high above in the main plant, with package transfers being effected via a hoistwell in the bay's soffit. If you think back to Chapter 9 you will recall that all cranes have a side and end approach beyond which their sheer size makes it impossible for them, more specifically their hook, to reach. And the bigger the crane the greater these distances will be.

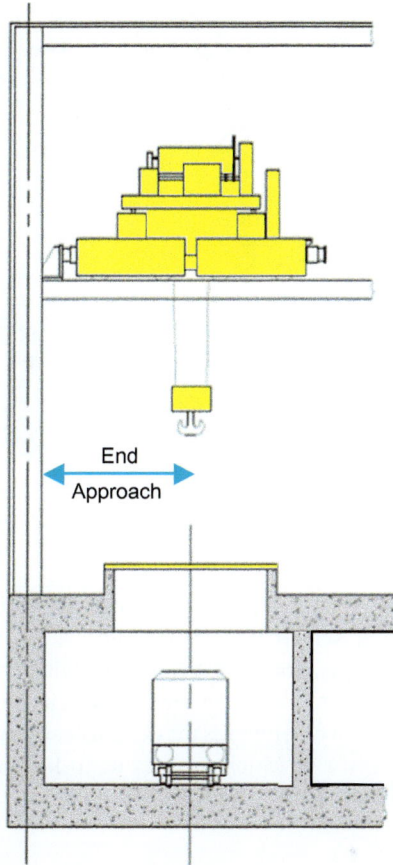

Fig. 17.5 Influence of crane geometry on a vehicle bay.
© Bill Collum.

For the kind of crane we are considering here, often with lifting capacities in excess of 100 tonnes, this can equate to a perimeter of between 3 and 5 m which is effectively out of reach. Fig. 17.5 shows how it can frequently be a crane, unseen from a vehicle bay, which determines its width and there is precious little we can do to claw back the space. We can, though, make use of it by earmarking the area for the inspection activities, temporary storage, and so on that I mentioned earlier.

17.3.3.13 Flask lifting

We know from Chapter 7 that all flasks are assigned a safe lifting height, above which they may fracture if an accident occurred and they plummeted onto an unyielding surface. If that height happens to be 3 m but other considerations dictate that a flask must be hoisted to say 8 m or more, then we need some way of ensuring its integrity will not be breached if the flask should fall. Broadly speaking there are two options: we can install the type of crush pad discussed in Chapter 7, which will cushion a fall but can add to the width of a vehicle bay, or we can install a rather splendid hoistwell door.

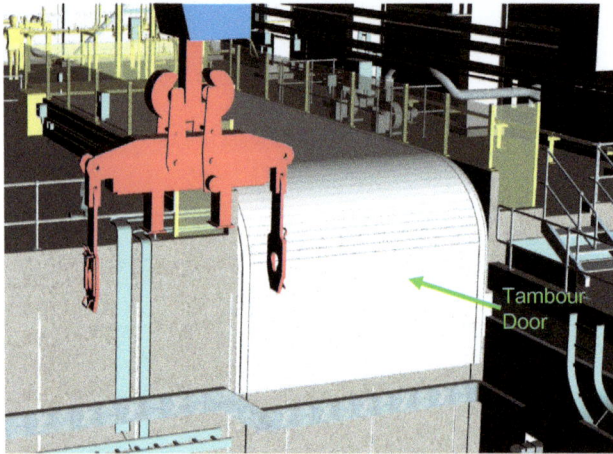

Fig. 17.6 Tambour door.
© Nuclear Decommissioning Authority.

A traditional hoistwell tends to be square or rectangular in profile and covered by either a horizontal roller shutter door or a sliding steel plate, which may come in one or two sections. There is, however, a more elaborate roller door, an industrial version (Fig. 17.6) of the tambour doors we see on Victorian writing desks, or storage cupboards in the workplace. In other words they can roll around bends. With this type of door we can form a hoistwell which enables us to keep a flask within its safe handling parameters and negate the need for a crush pad.

17.3.3.14   Future-proofing

We have seen in the previous chapter that designing nuclear facilities to conduct their primary operations is only half the challenge. To properly execute the task we must consider the whole lifecycle, so including the decommissioning phase and before that potential reuse for other purposes.

Beyond primary operations, one of the focal points for what happens next is a facility's vehicle bays. They are almost certainly the locations at which empty disposal packages enter a building, and through which decommissioning wastes will exit on route to their next destination. In addition, modules containing equipment designed to aid decommissioning processes will probably enter via these same bays. Of course we can anticipate that the majority of post operations payloads will have different handling characteristics to those processed during a facility's initial phase. In addition there is a possibility they will weigh more and have greater dimensions. All in all then, the brief for these areas can evolve considerably over time.

So here we have something else to factor into the design and configuration of vehicle bays. We must take the long view, by ensuring they contribute towards smooth running of a facility from their first days of operation until the last tranche of material exits their doors. Effectively then, and more than any other area, vehicle bays have an important role to play right up until the moment a facility stands ready for its demolition.

As I said at the beginning of this discussion, a vehicle bay is barely on the fringe of what constitutes a nuclear facility. Added to which their simple boxlike appearance seems to suggest they are an area of minimal challenge and almost no consequence whatsoever. The reality is that even here we find layer upon layer of factors, often conflicting, all clamoring for recognition and their rightful place in a final, integrated design solution. The apparent simplicity of a vehicle bay, serves us well in illustrating just how multifaceted the entire design coordination challenge can be. Let's take another example, one that crosses into a radiologically controlled environment.

17.4   Ventilation stack

On the face of it a stack is even more detached from a building than the vehicle bays we have just discussed, but then their tentacles do originate deep inside the facilities they serve, some of them from its most radiologically hostile areas. So the innocuous looking stack will do nicely as our second example of what we need to consider when developing layouts for nuclear facilities: but first we need to get a bit of terminology straight.

In normal parlance we tend to choose from just a couple of words when describing structures which emit smoke, exhaust gases, and so on. *Chimney* is normally applied to short structures such as those on domestic properties, whilst *stack* is reserved for the taller versions which attach to many industrial facilities. Now, I cannot say there is absolute consensus on the correct naming convention, but many will argue that the tall variety should be referred to as *windshields*, or even *chimneys*, and that tubes within them (which we shall get to in a moment) should be described as *stacks* or *ducts*. For our purposes, we shall adopt *stack* for the tall outer structure, and *flue* for their internal tubes (Fig. 17.7). As I say there are no hard-and-fast rules on this, but from my own observations these titles are the most commonly employed.

Before we get into our subject, it is worth setting the scene by examining why a stack is needed in the first place, which will go some way to explaining why they are appended to so many nuclear facilities. In earlier chapters, we saw how nuclear plants are divided into radiological zones, with each area categorized according to the highest levels of radiation and contamination that it could conceivably experience during normal operations. In addition, we have examined how ventilation systems move air around and how it is filtered before being discharged back to the environment.

Typically then, air from C2 (green) areas of *beta–gamma* plants is expelled without the need for any filtration, whereas C2 air originating within *alpha* plants is filtered prior to discharge. In both cases C2 air exits via either a short through-roof *stub* stack, or louvers on an external wall. Air from areas categorized as C3 (amber) and C4 (red) is cleaned up by the various stages of HEPA filtration we discussed in Chapter 8.

However, because it still has potential to carry minute radiological traces, this air must be discharged via a tall stack from where it is dispersed to the troposphere and beyond.

So we can see why stacks are needed, but what are we thinking as we set about determining where they will be located, their height and how a stack will be integrated with the building it serves?

17.4.1   Layout considerations

17.4.1.1   Flue arrangements

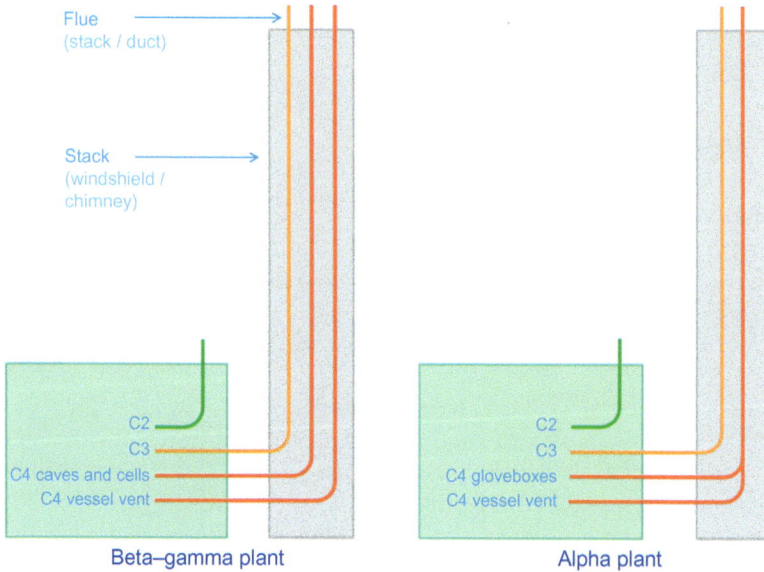

Fig. 17.7 Air discharge flues.
© Bill Collum.

Tall stacks, of the type we are considering here (Fig. 17.7) ordinarily contain at least two flues. In the case of beta-gamma plants one from C3 areas, a shared flue for C4 caves and cells, and a separate flue for ventilation of C4 vessels. For alpha plants there is again a flue for C3 areas, but this time just one more flue takes air from C4 areas, normally gloveboxes, along with any vessels that may be located within them.

If you look closely at the top of many stacks you will notice their flues protruding, a sight that is not exclusive to nuclear facilities, as many other industries segregate their aerial discharges in similar ways.

We shall see shortly that the speed at which air leaves a stack is very important, so using separate flues for each radiological airstream makes it much easier to manage their escape, or *efflux velocity*. In addition if airflows through one flue need to be shut down, say for maintenance, the velocity in those still operating is unaffected. In essence then, the role of a stack is simply to support tall flues and protect them from the elements, particularly buffeting by fierce winds, which is the reason they are sometimes referred to as a *windshield*.

17.4.1.2   Aerial dispersal

When it comes to expelling air from nuclear facilities, we know from Chapter 8 that HEPA filters do an excellent job of capturing particulate contamination, even down to submicron sizes. In addition, we have seen in Chapter 6 that off gas treatments are used to scrub species, such as iodine and tritium, from noxious gases before they are routed to a ventilation stack. Even so, minute traces remain. A stack's monitoring systems (discussed later) ensure such traces are within regulatory limits, but to comply with environmental discharge authorizations they must be appropriately dispersed.

We might assume this could be achieved by simply building a tall enough stack, but then it turns out the science in this area is extraordinarily complicated, even to the extent that a stack's height is not necessarily all it seems. Simply put, appropriate dispersal is achieved by limiting the concentration of radiological species expelled from a stack, its height and the velocity at which exhausts are ejected.

Aerial dispersal from nuclear plants is controlled by legislation which governs the dilution of radiological species in the atmosphere. The most important factor in determining permissible dilutions is derived from what is termed the *critical group*, sometimes referred to as the *representative person*, namely individuals whose location and lifestyle gives rise to the highest potential annual dose via ingestion and inhalation. And in this a stack's height has an important role to play. However, as I alluded to a moment ago, *height* in this context cannot be determined by simply getting out a tape measure, not even a lengthy one.

To get a full picture on how radionuclides are dispersed and therefore diluted in the atmosphere we must also factor in the speed, or *efflux velocity* at which exhausts exit a stack. For the situation we are considering here the minimum velocity is generally around 10 m/s (36 km/h/22 mph) but depending on variables such as the surrounding topography, can be as much as 40 m/s (144 km/h/89 mph).

Once efflux velocity gets to around 17 m/s, the rush on air creates sufficient noise at a stack's opening to be audible at ground level, and of course as velocity increases further so does the accompanying din. For facilities with a proposed location close to a site's perimeter, or one that may have office-based workers nearby, the specter of noise is a serious consideration, one which must be factored into selecting a stack's location, and potentially even the location of a facility itself.

A simple yet effective method of increasing efflux velocity is to narrow the neck of a flue in its final meter or two—the same principle we see applied to tapering nozzles which attach to some hairdryers. Funneling the neck down in this way can force air traveling at 10 m/s to emerge from a stack at say 15 m/s, which is a pretty good increase and contributes nicely to achieving both the required velocity and effective height (coming next). Having said that, HVAC engineers often develop their designs with an initial assumption that flues will either not taper at all, or only do so slightly, which is quite a prudent move. It effectively leaves some reserve capacity which, if necessary, can be called on later in the design process, when flue cones can be narrowed further and fans slowed down in order to regulate performance and reduce running costs.

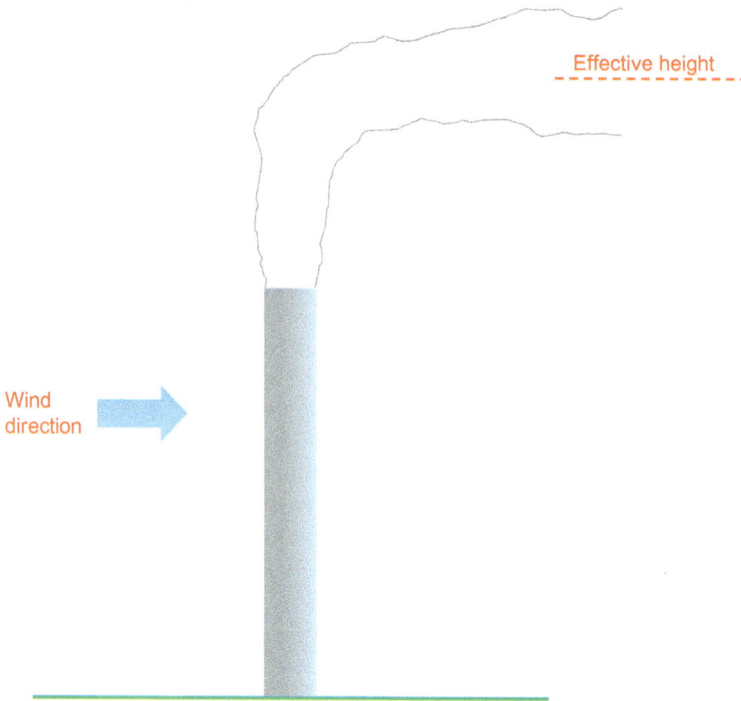

Fig. 17.8 Effective stack height.
© Bill Collum.

As for the "height" of a stack, there is of course its actual elevation, but the one that really matters is known as its *effective* height (Fig. 17.8), essentially the horizontal centerline of its plume. It is calculated through a combination of physical height, stack diameter, efflux velocity, average wind speed, temperature and a whole lot more. All of this contributes to determining characteristics of a stack's local plume, which is just one element of its dispersal modeling. Incidentally, if air is not expelled with sufficient efflux velocity it can *dribble* downwards, hugging the side of a stack and resulting in an effective height which is lower than its physical height.

It is no surprise to find dispersal modeling is a task that demands powerful computers capable of handling calculations required of the highly specialized computational fluid dynamics (CFD), in many cases supported by wind tunnel tests. Of course, until this complex work is complete, any *notion* of where a stack might be located or indeed its physical height is just that. We can make informed assumptions, but we must wait for real answers.

17.4.1.3 Prevailing winds

In parallel with developing aerial dispersal models we must also consider historical data on prevailing winds which, in its simplest form, is captured by a site's bespoke *wind rose*. It provides a useful indicator of the predominant directions and distances that plumes travel, so makes a valuable contribution in assessing potential exposure of the all-important critical group.

17.4.1.4 Stack location

Where a building needs a tall stack, there are broadly speaking two alternatives when deciding where to place it. Build a freestanding stack alongside the building, or place it within the building and have a portion, say 20 or 30 m poking through the roof. For stacks that are not too high, emerging from inside a building can have a lot going for it, whereas those that top out at say 100 m or more are almost certain to find themselves occupying a lonely spot outside. In between, where relative building and stack heights are reasonably close to each other, there is a grey area where design teams must work long and hard to come up with the right answer to a stack's location, and crucially be able to justify it.

Freestanding stack

There are several ways of constructing freestanding stacks. Normally they are formed from concrete (Fig. 17.9) or steel and completely enclosed: the *windshield* versions. Occasionally, they are constructed as an open steel framework through which their flues are visible.

Fig. 17.9 Freestanding concrete stack. © Nuclear Decommissioning Authority.

No matter how robust a nuclear facility may be, the possibility of a tall stack falling across it is simply unthinkable. So in addition to their considerable height, freestanding stacks also need a substantial foundation, not only to support their weight, but also to keep them from succumbing to effects of earthquakes and fierce winds. Not only does this place constraints on how far a stack must stand away from its parent building and others in the vicinity, but also brings characteristics of local ground conditions into the equation.

Soil structure interaction is always on the agenda when deciding how much space to leave between two seismically qualified facilities, but applies equally to a building and a nearby freestanding stack. During an earthquake the interaction can result in soil becoming mobile, incurring a potential loss of ground stability which needs to be borne in mind when locating a stack and designing its foundations.

Through-roof stack

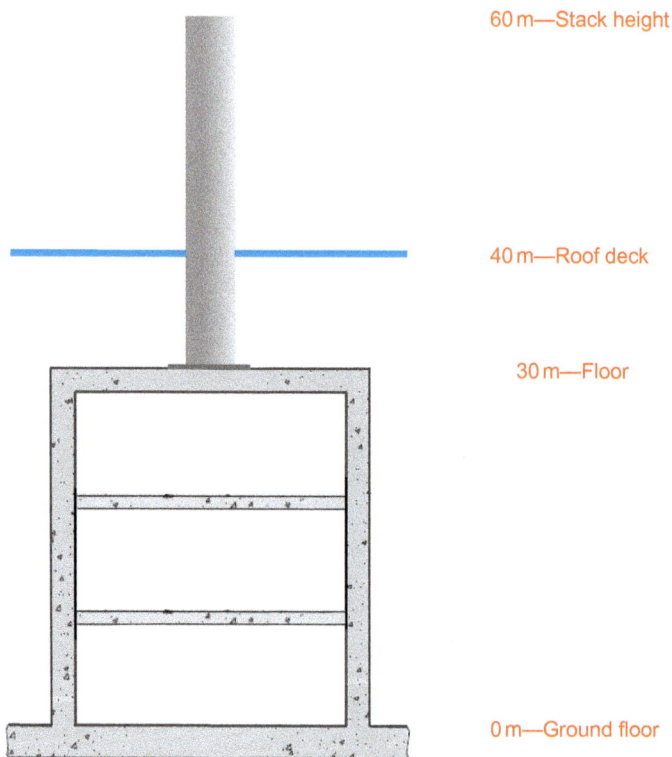

60 m—Stack height

40 m—Roof deck

30 m—Floor

0 m—Ground floor

Fig. 17.10 Through-roof stack.
© Bill Collum.

Rather than a freestanding stack we can go instead for their through-roof counterpart. Fig. 17.10 shows an example where a 30 m length of stack achieves an overall physical height of 60 m, which sounds like a good moneysaving deal. However, there are other less agreeable consequences to consider.

With a stack located inside a building all sorts of issues come into play, including how it will affect seismic performance of the structure supporting it. In addition there will be movement at the junction between a stack's circumference and the roof decking it passes through. So careful detailing is needed to accommodate differential movement, but still keep the structure weather tight. Not least there are complexities of installing a lengthy stack, such as that shown in Fig. 17.11, which is fabricated in one piece. Bear in mind that nuclear sites are notoriously reluctant to deploy cranes, especially the mobile variety, anywhere near facilities housing a radiological inventory. As a result, if there are any operational or even mothballed facilities nearby, then demonstrating it will be safe to swing a stack around is bound to be a challenge.

17.4.1.5 Strakes

Sticking with Fig. 17.11, the eagle eyed among you may have noticed spiral shaped fins attached to the top of many steel stacks. They are called strakes or helical spoilers and it is often assumed their function is to accelerate stack exhausts upwards, but they have quite a different purpose.

Fig. 17.11 Strakes.
© Nuclear Decommissioning Authority.

From a distance strakes may appear to be an indeterminate helix of some sort, but in reality there are rules governing their geometry. Exact configurations can differ a little depending on specifics of individual stacks, but generally it goes something like this. There are always three of them; they are equally spaced, every 120 degrees, around a stack's circumference, and their fins protrude one tenth of a stack's diameter. Their pitch, in other words the distance needed to cover a complete circumference, is generally five times a stack's diameter, and so it goes.

The purpose of strakes is to counter a phenomenon known as *vortex shedding*. Wind hitting a smooth sided stack applies substantial pressure to the surface facing it. After pushing around a stack, wind swirls on its lee side, creating negative pressures which pull a stack even further in the same direction as the wind. These pressures can cause a stack to vibrate, a bit like a giant tuning fork, which if unchecked can ultimately be quite destructive. Strakes interrupt that potentially harmful airflow and abate the effects of vortex shedding.

As with most things in life it is pretty unusual to get something for free, so strakes are no different. Winds buffeting against the fins, which increase a stack's diameter by say 20%, create additional loads, so a stack's supporting structure must be strengthened to withstand them. When considering the implications of a stack's location it is worth bearing in mind that vortex shedding is most prevalent in exposed areas, which is where we find most nuclear installations.

17.4.1.6   C2 stub stack

It is no accident that chimneys on domestic properties are usually located along a roof's ridge. Winds passing over the apex of a pitched roof can swirl around and create a downwash on their lee side. As a result, if chimneys are located away from a roof's ridge they must be built to stand taller than it, otherwise spiraling eddies can force smoke or steam down a flue and back to where it started. Similar problems are visited on the stub stacks which expel air from C2 (green) radiological zones. Although since we are normally dealing with flat, or near flat roofs, the downwash is mainly due to eddies from nearby parapets and higher buildings.

The challenge here is to identify a stub stack location which is subject to minimal turbulence, but also correlates to preferred airflow routes inside a building. With that done stub stacks generally rise to around 3 m above the roof deck they penetrate.

17.4.1.7   Link bridge

Ductwork emerging from the side of a building and heading for a freestanding stack needs to be supported in some way. Quite often it makes sense to design the supporting structure as a link bridge, so enabling it to double up its usefulness by providing an access route for maintenance personnel.

Short bridges can be *simply supported* at both ends, whilst longer ones may need intermediate supports; either way care is needed to deal with differential movement between a stack and its parent building. Where additional supports are required their ideal location may well clash with in-ground services that would be expensive to reroute, so potential obstructions must be investigated before finalizing a stack's location. And if a road or rail route passes under a link bridge, then sufficient clearance is needed to ensure vehicles can pass beneath it, a factor that can have a bearing on how ductwork is routed around the adjoining facility, particularly its elevation.

17.4.1.8   Stack monitoring

When it comes to locating a ventilation stack, whether it be freestanding or penetrating the roof of a building, arguably the most significant constraint originates from deep within its parent facility. As we know, air that has visited contaminated areas such as a C4 (red) cave is HEPA filtered. However, before it can be expelled from a stack, filtered air must be monitored to ensure its cleanliness conforms to aerial discharge authorizations.

Ductwork, from which air samples are taken, normally passes through a room that is dedicated exclusively to stack monitoring. That said, it is possible to perform the same activities without enclosing them in a room, but either way the principles are exactly the same. The most notable feature of a stack monitoring room is that exceptionally long ducts must pass through it in a perfectly straight line. In fact if you happened on one of these rooms without realizing its purpose, it would appear to be a spectacular waste of space.

With stack monitoring taking so much space, there is limited flexibility in deciding where it will be conducted. Furthermore, its arrangement has a major influence on options for a stack's location. Ultimately then, the position of both monitoring equipment and a stack itself must be considered together, and the two carefully synchronized.

The monitoring process requires several isokinetic probes, essentially small bore tubes, to be inserted through the wall of ducts which are heading for a ventilation stack. In terms of aerial discharges they form part of the last line of defense, so are linked to instruments which can halt airflows if noncompliant particles or elevated concentrations are detected. The problem is, if air is swirling around as it passes these sampling probes it cannot be claimed as truly representative of exhausts which are about to be expelled. As a result, airflows must be calmed down and streamlined and the easiest way of achieving this is to direct air into a long, straight length of duct.

Before we go any further I must explain the meaning of the term *hydraulic diameter*. We are all familiar with the regular use of diameter, meaning a straight line across a circle or sphere, one that passes through its center point. When describing ductwork, which can be circular, square, or rectangular in cross section, the convention is to quote its hydraulic diameter (D_h).

If a duct is circular in section, then diameter and hydraulic diameter have the same meaning. If square, then hydraulic diameter is equal to the length of any side and if rectangular is equal to the length of its shortest side. Having said all of that, in this particular monitoring situation ductwork invariably has a circular cross section, simply because it is less prone to accumulating contamination than the alternative angular profiles. Still, it is good to know these things.

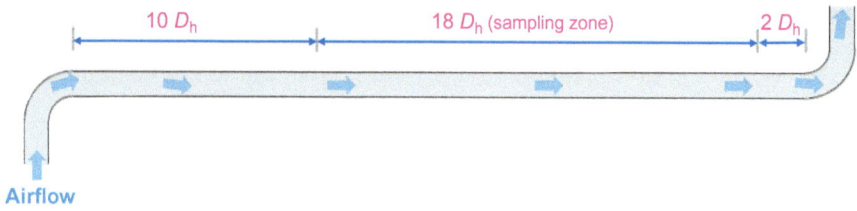

Fig. 17.12 Stack monitoring ductwork.
© Bill Collum.

The straight length (Fig. 17.12) needed to calm airflows is 10 hydraulic diameters, following which the sampling probes are spread over a length equivalent to 18 hydraulic diameters. It is not permissible to introduce a bend immediately after the last sampling point, as air hitting the bend will create turbulence back in the sampling zone. To abate this particular problem, ductwork needs to continue in a straight line for at least two hydraulic diameters beyond the sampling zone.

If we make a reasonable assumption that at least one of the ducts being sampled has a hydraulic diameter of 1 m, then add together the distance required to regulate airflows (10 m), take samples (18 m), and avoid turbulence near the last sample point (2 m), then we can reckon on a stack monitoring room being over 30 m long, or about half the width of soccer pitch.

If accommodating such a straight length is impractical, it is possible to insert an airflow straightener to calm airflows within ductwork more quickly. If viewed end on, straighteners are nothing more than a simple matrix of thin metal channels arranged in a square or honeycomb grid, and with an overall length of around 1 m. They work on the simple principle that turbulent air, forced into them, will be considerably better behaved when it emerges again at the other end.

Where space is particularly constrained it is possible to compress the stack monitoring process even further, although the specialized equipment required does cost more to design and manufacture. Having said that, the kit can be delivered in a module which is pre-tested and certified ready for use, which speeds along the commissioning process quite nicely. Whatever approach is used to accomplish stack monitoring, incorporating it into the design of a facility is most definitely not something that can wait for another day.

Identifying the ideal location for a ventilation stack is always going to be well down the list of a project team's competing priorities, simply because for all its complexity it is a workaday matter that should be dealt with almost imperceptibly. That said it so multifaceted a task that recognized best practice is to appoint a *stack coordinator* early in a project's life, someone to keep one step ahead of looming complications and smooth the whole process.

For all its outward banality, it turns out that if a stack is not given due consideration and in a timely manner, then the knock-on consequences could potentially derail a project's otherwise steady progress. It just goes to show that when it comes to designing nuclear facilities nothing stands in splendid isolation, not even a ventilation stack.

17.4.1.9   Future-proofing

In the previous chapter, we discussed some of the enablers which can contribute towards future-proofing a nuclear facility. In the context of a stack this has relevance to the sequence in which a building will be decontaminated and dismantled, whilst at the same time keeping its ventilation system running for as long as possible. Potentially more problematic are those facilities which are destined to switch to a new purpose when their primary operations are concluded. As we have seen this may involve partial demolition of a facility, so as to retain just those portions which are needed for its new lease of life.

The issue here is that if a ventilation stack is required to support a facility's new purpose—and there is a strong possibility it will be—we must ensure its location coincides with plans to partially demolish the original building. And it is not just a matter of ensuring a stack is not stranded some distance from a reconfigured facility; we must also see to it that its primary interfaces, such as stack monitoring, are also located with the long haul in mind. As always there is a long list of conundrums to keep us occupied.

The two examples we have just considered, of planning a vehicle bay or figuring out where best to locate a ventilation stack, give an insight to what goes into developing designs for new nuclear facilities. Of course for each example we could add much more, say on their interactions with adjacent areas, safety case influences, programming constraints and on and on. Essentially, we could continue weaving in themes covered by all of the preceding chapters.

We get the idea however, enough to recognize that similar issues and comparable thought processes pervade every corner of a proposed new facility, all looking for recognition, jockeying for position, and demanding our undivided attention. All of which goes some way to explaining why the whole design development process can take quite a few years.

If we do our job well, though (and of course we must), then when a facility is finally ready and poised to begin operating, we should be able to walk to any location or item of equipment within it and explain *in considerable detail* what is going on, why it ended up in that precise location and how everything around is humming along in perfect harmony. Wonderful.

And with that, we are off to our next project.

Index

Note: Page numbers followed by *f* indicate figures.

A

Activation. *See* Neutron activation
Advanced gas-cooled reactor (AGR), 11, 12*f*
Agitation systems
   fluidic, 172–173, 173–174*f*
   mechanical, 170–171
Air conditioning, 256–257
Air handling unit (AHU), 264–266, 264*f*
Air in-bleed filters, 248–250, 249*f*
Air quality, 266–267, 267*f*
Air sparge, 173–174, 174*f*
Alpha plants, 269
Alpha radiation, 48–49
As low as reasonably achievable
   (ALARA), 436
As low as reasonably practicable (ALARP),
   436–437, 437*f*
Asphyxiation, 406
Automation, 330–332, 331*f*

B

Backup electricity
   'A' and 'B' electrical supplies, 318–320
   firm power supply, 318–320, 319*f*
   guaranteed interruptible power supply,
     320–324, 321–323*f*
   guaranteed UPS, 324–326, 324–325*f*
   non-firm power supply, 318, 319*f*
Becquerel, 66
Beta radiation, 49–50, 49*f*
Beyond design basis, 432–433
BFD. *See* Building flow diagram
Bogies, 209–210, 210*f*
Boiling water reactor (BWR), 13, 14*f*
Bourdon gauge, 344, 344*f*
Bracing
   arrangements, 112, 113*f*
   braced cube, 112, 112*f*
   types, 112, 113*f*
   unbraced cube, 111–112, 111*f*
Breakpot, 161–164, 162–163*f*

Bubbler system, 336–337, 336*f*
Building flow diagram (BFD)
   iterations, 540–541
   process, 535–538, 536*f*
   validation, 539–540
   workshops, 535
Burnable poison, 19

C

Cable color coding, 317–318, 317*f*
Cable handling, installation, 326, 326*f*
Canyon concept, 141
Capacitance sensors, 340, 340*f*
Carbon dioxide, decontamination, 213
Cascade philosophy
   air pressure, 235–236, 236*f*
   reliability, 234–235, 235*f*
   velocity, 236–238, 237–238*f*
Casks. *See* Flasks
Cell and cave definition, 141–142
Centrifugal pump
Centrifuge, 178
Changerooms. *See* Radiological changeroom
Client role (future-proofing), 530–531
Client specification, 369–370
Closed cell processing facility
   computer model, 141, 141–142*f*
   dark cell/passive cell, 140
   maintenance-free processing vessel,
     140, 140*f*
Combined concrete and steel structures,
   121–122, 122*f*
Commissioning
   feedback, 394
   inactive and active, 392–393
   phased, 393
Concrete banding, 120–121, 120*f*
Concrete structures
   concrete banding, 120–121, 120*f*
   in situ concrete, 117–119, 118*f*
   structural information, phased release, 121

Concrete structures *(Continued)*
  precast concrete, 115–117, 116*f*
  wall penetrations, 119–121
Conditions for acceptance (CFA), 146–147,
    211, 349, 401–402, 523, 538
Conductive sensors, 339, 339*f*
Confined spaces, 406–407, 407*f*
Construction programming, 391
Contamination zones
  C1 (white), 66
  C2 (green), 67–68
  C3 (amber), 68
  C4 (red), 68–69
Control systems
  automation, 330–332, 331*f*
  basic control, 328, 328*f*
  control hierarchy, 327
  remote operation, 332, 332*f*
  SCADA, 332–334, 333–334*f*
  sequence control, 328–329, 329*f*
Conventional cranes
  annual examination, 290
  components, 275–278, 276–277*f*
  design standards, 278
  hook approach, 287–289, 288*f*
  lifting accessories, 280–284, 281*f*
  load limiting devices, 290–291
  operation, 278–280, 279–280*f*
  performance, 284–287, 286*f*
  polar crane, 296–297, 296*f*
  protection against dropped loads,
    291–293
  safe working load (SWL), 289–290
  seismic performance, 294–296
  zoning, 293–294
Conventional waste management hierarchy
  disposal, 398
  energy recovery, 397–398
  minimization, 396
  prevention, 396
  recycle, 397
  replace, 397
  reuse, 396
Cost breakdown structure (CBS), 374
Crane hook approach, 548–549, 549*f*
Cranes
  conventional cranes, 275–297
  high integrity nuclear cranes, 275–297
  in-cave, 297–312

Critical excursion. *See* Criticality
Criticality, 56–57
Cross-flow filtration, 179
Cross-site integration, optioneering, 402
Cuboidal and cylindrical flasks, 219–221,
    219–221*f*

D

D&D phase. *See* Decontamination and
    dismantling (D&D) phase
Decay chain, 22
Decommissioning planning
  definition, 500
  entombment, 495–498, 495–497*f*
  immediate dismantling, 490–492,
    490–491*f*
  operations phase, 472–476, 473*f*
  planning factors, 456–472, 456*f*
  primary phases, 476–484
  safe enclosure, 492–495, 493*f*
  secondary phases, 484–486
  site license conditions, 455–456, 455*f*
  site's zoning plan, 501
  strategy selection, 498–499, 498*f*
  tertiary phases, 486–489
  timing, 499
Decontamination
  carbon dioxide, 213
  dry pellets, 213, 213*f*
  submersion, 212–213
  swabbing, 211–212
  water, 212
Decontamination and dismantling (D&D)
    phase
  IL materials, 480–482, 480–481*f*
  LL materials, 478–479, 479–480*f*
  VLL materials, 478
Deep geological disposal, 41, 41*f*
Deferred dismantling and safestore,
    492–495, 493*f*
Demolition, 483–484, 483–484*f*
Design basis earthquake (DBE), 126–127
Design development
  assumptions, 539
  building flow diagram (BFD),
    534–538, 536*f*
  constraints, 538
  iterations, BFD, 540–541

layout preparations, 533–541
precursors, layout development,
 534, 534*f*
validation, BFD, 539–540
vehicle bay, 541–551
ventilation stack, 551–561
Design for decommissioning (DfD)
 client role, 530–531
 importance of, 505
 integration with decommissioning strategy,
  514–515, 515*f*
 maximize use, original P&E, 516–517
 primary operations, 515–516
 self-perpetuating cycle, 506–509,
  506–508*f*
 simplify transition, 516
Design justification reports (DJRs), 425–426
Design to remodel and reuse (DRR)
 client role, 530–531
 incorporation, design process, 509, 510*f*
 integration with decommissioning strategy,
  514–515, 515*f*
 maximize use, original P&E, 516–517
 non-specific purpose, reuse, 513–514
 operating nuclear plant, 511
 primary operations, 515–516
 self-decommissioning, reuse, 511–512,
  512*f*
 simplify transition, 516
 specific purpose, reuse, 512–513
DfD. *See* Design for decommissioning
Differential movement, 122, 123*f*
Direct on line (DOL), 329–330
Discounted cash flow (DCF), 517
Distribution board, 316, 317*f*
Double block and bleed, 152–153, 153*f*
DRR. *See* Design to remodel and reuse
Dry pellets, decontamination, 213, 213*f*
Ductwork distribution network,
 262–264, 263*f*

E

Electrical and hydraulic manipulator
 vs. MSM, 199
 rail mounted, 197, 198*f*
 underwater, 198, 198*f*
Electricity supply
 backup electricity, 318–326

cable color coding, 317–318, 317*f*
cable handling, 326
distribution board, 316, 317*f*
site ring main, 313–314, 314*f*
transformer, 314–316, 315*f*
Electromagnetic spectrum, 46–47, 47*f*
Electronic personal dosimeter (EPD),
 59–60, 59*f*
Embed waste management strategy, 403
Enablers, future-proofing
 fire compartmentation, 525–526
 radiological zoning and ventilation,
  520–522
 remote waste handling, 523–525
 segregation, 526–528, 528*f*
 space, 528–529
 utilities and active pipework, 522–523
End effectors, 202, 202*f*
Engineered gaps, 238–240, 239–240*f*
Engineering schedule, 422–426
Engineered store, 39
Engineering schedule
 DJRs, 425–426
 safety functional requirement, 424*f*, 425
 SPRs, 425
 SSCs, 424–425
Entombment, 495–498, 495–497*f*
Equipment transfer lobby, 103
Evaporation, 175–176, 176*f*
Extreme environmental events
 rain, 136–138
 seismic, 126–129
 snow, 134–135
 temperature, 135–136
 wind, 129–134

F

Fast breeder reactor (FBR), 25–26, 25*f*
Fault schedule, 422, 423*f*
Fault sequence
 mitigated hazards, 430–432, 432*f*
 unmitigated hazards, 430, 431*f*
Feedstock analysis
 conditions for acceptance (CFA),
  146–147
 organics, 148
 orphan streams, 149
 stratification, 147–148, 147*f*

Filter disposal, 254–255
Filtration
    alpha plants, 246–247, 247*f*
    beta-gamma plants, 245–246, 246*f*
    HEPA filters, 244–245, 244–245*f*
    mobile filtration unit (MFU),
        247–248, 248*f*
Financial planning
    cost, 460–461, 460*f*
    funding, 461, 461*f*
Fire compartmentation, 525–526, 547
Fire resistant cable, 326–327, 327*f*
Firm power supply, 318–320, 319*f*
Flasks
    design durations, 229
    operational requirement, 219
Flasks-bottom loading
    mobile gamma gates, 228, 228*f*
    nappy, 228–229, 229*f*
    permanently installed gamma gates,
        227, 227*f*
Flasks, nappy, 228–229, 229*f*
Flasks-top loading
    cuboidal and cylindrical, 219–221,
        219–221*f*
    fire testing, 223
    impact testing, 221–223, 222*f*
    lid removal, 223–226, 224–225*f*
    lifting above assigned drop height, 223
Flow measurement
    orifice meter, 346, 346*f*
    pitot tube, 347–348, 347*f*
Fluidic agitation, 172–173
Fluidic pumping
    barometric head, 158, 158*f*
    external RFD, 159, 160*f*
    internal RFD, 157, 157*f*
    propellant, 160
    pumping sequence, 159
    RFD, 156, 156*f*
Freestanding stack, 555–556, 555*f*
Fuel cooling pond, 472–473, 473*f*
Funding future-proofing
    decommissioning cost estimate, 519
    DfD and DRR, 517–518
    discounted cash flow (DCF), 517
Future-proofing
    client role, 530–531
    design team role, 531

    design for decommissioning (DfD),
        505–509
    design to remodel and reuse (DRR),
        509–514
    enablers, 519–529
    funding, 517–519
    integration with decommissioning strategy,
        514–515, 515*f*
    maximize use, original P&E,
        516–517
    primary operations, 515–516
    simplify transition, 516

G

Gamma radiation, 50
Gas-filled detectors
    components, 350, 351*f*
    detection process, 351–352
    gas flow proportional counter, 352
    gas types, 353
    G–M tube, 352–353
    proportional counter, 352
Gas flow proportional counter, 352
Geiger–Müller (G–M) tube, 352–353
Germanium detectors, 359–360
Glovebox
    maintenance, 273
    ports, 271–272, 272*f*
    purpose, 269
    radioactive materials, 268–269, 269*f*
    shielding, 272–273
    ventilation, 269–271, 270*f*
Gravity flow systems, 161, 162–163*f*
Guaranteed interruptible power supply,
    320–324, 321–323*f*
Guaranteed uninterruptible power supply
    (UPS), 324–326, 324–325*f*

H

Hardwiring, 334–335
Hazard analysis (HAZAN)
    fault schedule, 422, 423*f*
Hazard and operability studies (HAZOP)
    example key words, 419, 419*f*
    HAZOP 0, 420
    HAZOP 1, 420
    HAZOP 1.5, 420–421
    HAZOP 2, 421

Hazard appraisal (HAZAP), 418
Hazard control hierarchy
  eliminate, 427–428
  isolate, 428–429
  manage, 429
  personal protective equipment (PPE), 429
  reduce, 428
Hazard management
  environment, 458, 458*f*
  safety, 457–458, 457*f*
Heating ventilation and air conditioning
    (HVAC). *See* Nuclear ventilation system
Heat recovery, 257
HEPA filtration, 270. *See also* Manual *vs.*
  remote filter change
High integrity, crane, 275
High level waste (HLW)
  conditioning and packaging, 35, 35*f*
  interim storage, 35, 36*f*
  permanent disposal, 41–43, 41*f*
  radiological waste management
    hierarchy, 400
HLW. *See* High level waste (HLW)

I

ILW. *See* Intermediate level waste (ILW)
Immediate dismantling, 490–492, 490–491*f*
In-cave cranes
  Cartesian, 297–298
  guidance systems, 311–312, 311*f*
  modularization, 299
  polar jib, 298–299, 298*f*
  power supply, 306–311
  recovery, 300–306, 302*f*
  written scheme of examination, 299–300
In situ concrete, 117–119, 118*f*
In situ dismantling
  manual, 506, 506*f*
  remote, 506, 507*f*
Installed personnel monitor (IPM), 89–91,
    90–92*f*
Instrumentation
  flow measurement, 346–348
  intelligent instruments, 335
  level measurement, 335–341
  pressure measurement, 342–345
  temperature measurement, 341–342
Integrated breakdown structures, 375, 375*f*

Intermediate-level (IL) materials, 480–482,
    480–481*f*
Intermediate level waste (ILW)
  conditioning and packaging, 36–38, 37–38*f*
  interim storage, 39–41, 39–40*f*
  permanent disposal, 41–43, 41*f*
  radiological waste management
    hierarchy, 399
International Atomic Energy Agency
    (IAEA), 409
Ion exchange (IX), 179–181, 180*f*
Ionizing radiation
  alpha, 48–49
  beta, 49–50, 49*f*
  gamma, 50
  shielding, 48, 48*f*
IPM. *See* Installed personnel monitor
Island site, 74
Isolated diaphragm gauge, 344–345, 345*f*

J

Joggle boxes, 169–170, 169*f*

L

Learning from experience (LFE),
    459–460, 459*f*
Level measurement
  bubbler system, 336–337, 336*f*
  capacitance sensors, 340, 340*f*
  conductive sensors, 339, 339*f*
  radar systems, 338–339, 338*f*
  ultrasonic, 340–341, 341*f*
  vibrating forks, 337, 337*f*
LLW. *See* Low level waste
Low-level (LL) materials, 478–479, 479–480*f*
Low level waste (LLW)
  permanent disposal, 42–43, 42–43*f*
  radiological waste management hierarchy,
    399

M

Manipulators
  decontamination, 210–213
  electrical and hydraulic, 197–199
  MSM, 193–197
  power, 199–202
Manual *vs.* remote filter change, 250–253,
    251–252*f*

Mass balance
 definition, 143
 fruit cake, 143–145, 144f
 off gas, 145
 process flow diagram (PFD), 145
Master slave manipulator (MSM)
 change-out, 196–197, 197f
 components, 193, 193–194f
 gearing, 194–195, 195f
 human factors (HF), 195
 versatility, 195–196, 196f
Mechanical agitation
 guiding principle, 170
 pump-round arrangement, 171, 172f
Mechanical engineering
 cave arrangements, 214–218
 drum handling grab, 185–186, 185f
 flasks, 219–229
 handling cave, 186–187, 187f
 manipulators, 192–202
 remotely operated mechanical equipment,
  186, 186f
Mechanical handling cave, 186–187
Mechanical pumping
 centrifugal pump, 150–154, 151f
 progressive cavity pump, 150
Mitigated hazards, 430–432, 432f
Mobile gamma gates, 228, 228f
Motor control center (MCC)
 DOL, 329–330
 soft start, 330
 soft stop, 330
 VFD, 330
MOX fuel
 gloveboxes, 24, 24f
 plutonium, 23
MSM. See Master slave manipulator

N

National waste management strategy,
 401
Neutron activation
 chemical element variations, 20, 20f
 cobalt-59, 21, 21f
 cobalt-60, 20–21, 21, 21, 21
Non-firm power supply, 318, 319f
Nonionizing radiation, 47
Nuclear factors, programming
 division of costs, 378–379, 378f
 regulatory constraints, 379

Nuclear fuel cycle
 atom, 5–6, 5–6f
 burnable poison, 19
 control rods, 18–19
 decay chain, 22
 enrichment, 7–9, 7–9f
 fast breeder fuel, 25–26
 fuel fabrication, 9–11, 10–11f
 half-life, 18
 high level waste (HLW), 35
 intermediate level waste (ILW), 36–41
 low level waste (LLW), 399
 MOX fuel, 23–25
 neutron activation, 19–22
 neutron capture, 16–17
 nuclear fission, 17, 17f
 nuclear reactor types, 11–16, 12f
 permanent disposal, 41–43
 plutonium creation, 22–23, 22–23f
 reprocessing, 32–34
 spent fuel removal, from reactor, 26–29
 spent fuel routing, 29–32
 uranium mining and purification, 3–5, 3–4f
Nuclear ventilation system
 AHU, 264–266, 264f
 air conditioning, 256–257
 air in-bleed filters, 248–250
 air quality, 266–267, 267f
 cascade philosophy, 234–238
 containment, building perimeter, 241–243,
  242–243f
 containment, truck bays, 240–241, 241f
 ductwork distribution network,
  262–264, 263f
 engineered gaps, 238–240
 filter, change, 253–254
 filter disposal, 254–255
 filter life, 253
 filtration, 243–256
 gloveboxes, 268–273, 269f
 heat recovery, 257
 manual vs. remote filter change, 250–253,
  251–252f
 push-through filters, 255–256, 255f
 radiological zoning, integration with, 233
 solar heat gain, 258
 ventilation plant rooms, generic
  arrangement, 259–262, 260f
 ventilation sequence, 259–264
 vessel ventilation, 267–268

O

Occupational safety
  asphyxiation, 406
  confined spaces, 406–407, 407f
  health and safety, 405
  noise, 407–409
Off gas treatment, 182–183, 182f
  packed column, 183
Open sub changeroom, C2/C3 (S), 102–103
Operating basis earthquake (OBE), 127
Operating rules, 426
Operations phase, decommissioning planning
  fuel cooling pond, 472–473, 473f
  POCO, 473–475, 473–474f
  secondary phases, 475–476, 475f
Operations safety report (OSR), 444
Optioneering
  cross-site integration, 402
  design philosophy documents, 387
  layout development, 387
  mass and energy balance, 387
  preferred option selection, 385–386, 386f
  waste minimization, 402
Option refinement
  refine preferred option, 388
  review criteria, 388
  waste management, 402, 402f
Organic feedstock analysis, 148
Organization breakdown structure (OBS), 374
Orifice meter, 346, 346f
Orphan streams, 149
Overflow protection, 174–175
Overpacks. See Flasks

P

Packed column, 183
Periodic reviews
  long-term reviews, 445–446
  short-term reviews, 445
Permanently installed gamma gates, 227, 227f
Personal dose measurement
  EPD, 59–60, 59f
  TLD, 57–58, 58–59f
Personal protective equipment (PPE), 429
Personnel access (PA) doors, 204, 204f
PFD. See Process flow diagram
Phased commissioning, 393
Photomultiplier tube (PMT), 354–355, 355f
Pintle, 284, 285f

Pitot tube, 347–348, 347f
Plutonium creation, 22–23, 22–23f
POCO. See Post operational clean out (POCO)
Polar crane, 296–297, 296f
Post operational clean out (POCO)
  operations phase, 473–475, 473–474f
  tertiary decommissioning planning
    phase, 487
Power manipulator
  decontamination and maintenance,
    201–202, 201f
  deployment, 199–200, 200f
  end effectors, 202, 202f
Preactive commissioning safety report
  (PACSR), 442–444, 443f
Precast concrete, 115–117, 116f
Preconstruction safety report (PCSR),
  439–440
Preinactive commissioning safety report
  (PICSR), 441–442
Preliminary safety report (PSR), 439
Pressure measurement
  Bourdon gauge, 344, 344f
  isolated diaphragm gauge, 344–345, 345f
Pressurized water reactor (PWR), 11, 12f
Primary decommissioning planning phase
  care and maintenance, 482–483
  D&D phase, 477–482
  demolition, 483–484, 483–484f
  S&M, 476–477
Probabilistic risk assessment, 433
Process engineering
  agitation systems, 170–174
  closed cells, 139–143, 139–142f
  end product, 149
  feedstock analysis, 146–149
  mass balance, 143–146, 144f
  solids removal, 177–179
  transfer devices, 150–166
  volume reduction, 175–176
Programmable logic controller (PLC),
  328–329, 329f
Programming
  activity links, 380, 380f
  critical path, 381
  hold points, 382–383
  nuclear factors, 377–379
  programming logic, 379–380
  project phasing, 381–382, 381f
Programming factor, 471–472

Progressive cavity pump, 150
Project controls
  CBS, 374
  integrated breakdown structures, 375, 375f
  OBS, 374
  project execution plan, 375
  risk management, 376–377
  WBS, 372–374, 373f
Project execution plan, 375
Project implementation plan. See Project execution plan
Project inception, waste management
  investment justification, 384
  management strategy, 400–401
  national waste management strategy, 401
  on-site waste treatment options, 401
  research and development (R&D), 384–385
  technology identification, 384
Project phase activities
  commissioning, 392–393, 392f
  construction programming, 391
  equipment installation, 392
  inception, 383–385, 383f
  maintaining design intent, 389
  optioneering, 385–387, 385f
  option refinement, 387–388, 387f
  phased release of information, 391–392
  specification, design, and manufacturing, 389, 389f
Project planning
  client specification, 369–370
  management of, 377
  programming, 377–383
  project controls, 370–377
  project phase activities, 383–393
Pulse jet mixer, 172–173, 173f
Pump-round arrangement, 171, 172f
Push-through filters, 255–256, 255f
PWR fuel loading, 13, 15f

R

Radar systems, 338–339, 338f
Radiation and contamination
  confining light rays, 45, 45f
  neutron activation, 46
  transportation, 45, 46f

Radiation exposure
  background radiation, 50–52, 51f
  dose equivalent, 52–53
  nuclear workers, 53–54, 53f
Radiological changerooms
  airflow, 84, 85f
  boot barrier–inward, 82–84, 82–83f
  boot barrier–return, 85–86
  coveralls, 85
  duplicating equipment, 94
  frisking station, 88–89, 89f
  generic types, 78
  hand monitoring, 87–88, 88f
  health physics, 92–93, 93f
  IPM, 89–91, 90–92f
  locker rooms and basics, 78, 79–80f
  monitoring, 86–91, 86–87f
  monitoring and coverall areas, 79–82, 81f
  nonradiological access routes, 78, 79f
  nuclear facilities, 77, 77f
  nuclear worker basics, 78, 80f
  outdoor radiation controlled areas, 94–95
  restrooms, 94
  security turnstile, 79, 81f
  sub changerooms, 95–103
Radiological waste management hierarchy
  HLW, 400
  ILW, 399
  LLW, 399
  primary and secondary wastes, 398
  VLLW, 399
Radiological zoning
  contamination zones, 65–69, 67f
  designations, 69
  dual radiological classification, 73–74, 75f
  exposure limits, 63–64, 63f
  interfaces, 69, 70f
  monitoring–fail, 72, 72f
  monitoring–pass, 72, 72f
  naming conventions, 62–63
  normal operations, 69
  R1 (white), 64
  R2 (green), 64
  R3 (amber), 64–65
  R4 (red), 65
  start sequence, dual classifications, 71, 71f
  surface contamination, 73–74
  trans-zone protocols, 70
  ventilation system, 62

Radiological zoning and ventilation
   contamination, 520
   radiation, 521–522, 521–522*f*
Radiometric instruments
   aptitudes, 360–361, 360*f*
   assay, 365–366, 365*f*
   CFA. *see* Conditions for acceptance
   cost, 360*f*, 361
   efficiency, 362
   environmental, 363–365, 364*f*
   gas-filled detectors, 350–353, 351*f*
   health physics, 363, 363*f*
   instrument technologies, 366, 367*f*
   monitoring requirements, 349–350
   resolution, 361–362, 361*f*
   safeguards, 367–368
   scintillation detectors, 353–355
   semiconductor detectors, 353–355
   through-wall gamma monitor,
      364–365, 364*f*
Radionuclide fingerprint, 143
Rain, extreme environmental
   design basis, 137
   flood planning, 136, 137*f*
   rainfall management, 137–138, 137*f*
Reactor charge floor, 13, 15*f*
Refine waste management plans, 404
Refurbishment, 462–463, 462–463*f*
Remediation, 486
Remodel and reuse, 464–465, 464*f*
Remotely maintainable pump, 154–156, 155*f*
Remote waste handling, 523–525
Reprocessing, 32–34
   constituents, spent fuel, 34, 34*f*
   De-cladding Magnox fuel rod, 32, 33*f*
   uranium-235, 32
Resistance temperature detector (RTD),
   342, 343*f*
Resources factor, decommissioning
   planning, 470
Respiratory protective equipment (RPE), 429
Reverse flow diverter (RFD), 156, 156*f*
Risk classification, 433–436, 434–435*f*
Robot arms. *See* Manipulators

S

Safety
   conventional and radiological, 438, 438*f*
   hazard analysis studies, 418–426

long-term reviews, 445–446
nuclear, 409–412
occupational, 405–409
OSR, 444
PACSR, 442–444, 443*f*
PCSR, 439–440
PICSR, 441–442
PSR, 439
risk, 426–436
safety case, 413–418
safety case integration, 446–447,
   447–453*f*
short-term reviews, 445
Safety case
   category 1, 415
   category 2, 415
   category 3, 416
   category 4, 416
   committees, 413–414
   evolution of, 416–418, 417*f*
Safety case integration
   construction and installation, 451*f*
   detail design, 450*f*
   inactive and active commissioning, 452*f*
   operations, 453*f*
   optioneering, 448*f*
   option refinement, 449*f*
   project inception, 447*f*
Safety functional requirement, 424*f*, 425
Safety integrity level (SIL), 334–335
Safety, nuclear
   committee, 414
   IAEA, 409
   site license conditions (SLCs), 410–412,
      411–412*f*
Safety performance requirements (SPRs), 425
Scintillation detectors
   detection process, 354
   materials, 353–354
   measurement, 355, 355*f*
   PMT, 354–355, 355*f*
Secondary decommissioning planning phase
   advance D&D, 485
   remediation, 486
Segregation, waste, 526–528, 528*f*
Seismic analysis
   category, 127–128, 128*f*
   DBE, 126–127
   displacement, 128–129

Seismic analysis *(Continued)*
  evaluation, 124
  OBE, 127
  phased release, wall penetration
    information, 123–124, 123*f*
  response spectra, 124–125
  steelwork, 129
  subterranean construction, 129
Self-acting pressure regulators, 270
Self-perpetuating cycle, 506–509,
  506–508*f*
Semiconductor detectors
  conductors, 355–356
  detection process, 357–359, 358–359*f*
  germanium detectors, 359–360
  semiconductors, 356
  transistors, 357, 357*f*
Sequence control
  MCC, 329–330
  PLC, 328–329, 329*f*
Services distribution
  access, 168
  hierarchy, 168–169, 168*f*
  joggle boxes, 169–170, 169*f*
  nuclear plant, 167, 167*f*
  service risers, 167–168
Settling process, 177–178, 177–178*f*
Shield doors
  combination of vertical and horizontal,
    206–207, 206–207*f*
  concrete filled, 209
  construction, 203–204, 203*f*
  maintenance access, 205
  personnel access (PA) doors, 204, 204*f*
  pivoting, 208–209
  pressure relief vent, 204–205
  recovery, 207–208, 208*f*
  vertical, 205–206
Shielding
  bulge, 153–154
  indicative equivalents, 54–55, 54*f*
  ionizing radiation, 48, 48*f*
Shielding windows
  change-out, 191–192
  composition, 187–188, 188*f*
  construction, 188–189, 189*f*
  containment, 190–191, 190*f*
  refraction, 189, 190*f*
  tint and degradation, 191

Shipping containers. *See* Flasks
Sievert, 66
Sievert *vs.* Becquerel, 66
Siphoning, 162, 162*f*
Site license conditions (SLCs), 410–412,
  411–412*f*
Site ring main, 313–314, 314*f*
Site safety committee, 414
Site zoning plan, 465–467, 467*f*
Snow, extreme environmental
    design basis, 134
  shifting snow, 134–135, 135*f*
Socioeconomic planning factors,
  467–470, 468*f*
So far as is reasonably practicable (SFAIRP),
  436
Soft start, 330
Soft stop, 330
Solar heat gain, 258
Solids removal process
  centrifuge, 178
  cross-flow filtration, 179
  settling, 177–178, 177–178*f*
Spent fuel removal
  charge floors, 26–27, 27*f*
  cooling ponds, 28–29, 28*f*
Spent fuel routing, 29–32
Spherical dispersion, 55–56, 55*f*
Stack monitoring ductwork, 559–560, 560*f*
Steam ejectors, 161
Steelwork structures
  bracing, 111–114, 111–112*f*
  floor construction, 110, 111*f*
  portal frame, 114–115, 114–115*f*
  structural grid, 106–110, 106–107*f*
Strakes, 557–558, 557*f*
Stratification, 147–148, 147*f*
Structural grid
  adjustments, 109–110, 110*f*
  arrangements, 107–108, 108*f*
  load paths, 108, 109*f*
  steel stanchion, 106–107, 107*f*
  steelwork grid, 106, 106*f*
Structures, systems, and components (SSCs),
  424–425
Sub changerooms
  equipment transfer lobby, 103
  generic types, 95–96
  open, C2/C3 (S), 102–103

Sub changeroom-C2/C3
  clothing and protective equipment,
    96–98, 97f
  containment, 98
  disposable items, 98
  monitoring procedure, 98–99
  personal monitoring devices, 98
Sub changeroom-C3/C4
  clothing and protective equipment,
    100–101, 101f
  monitoring procedure, 101, 102f
  trans-zone protocol, 100
Submersion, decontamination, 212–213
Suction head, 151–152, 152f
Suitably qualified and experienced person
  (SQEP), 470
Supervisory control and data acquisition
  (SCADA), 332–334, 333–334f
Surface contamination
  indoors, 73
  outdoors, 74
Surveillance and maintenance (S&M) phase,
  476–477
Swabbing, decontamination, 211–212

T

Temperature, extreme environmental
  design basis, 136
  embrittlement, 136
Temperature measurement
  RTD, 342, 343f
  thermocouple, 342
Tertiary decommissioning planning phase
  advance demolition, 488–489, 488–489f
  POCO, 487
Thermocouple, 342
Thermoluminescent dosimeter (TLD), 57–58,
  58–59f
Through-roof stack, 556–557, 556f
Through-wall gamma monitor, 364–365, 364f
Through-wall manipulator. See Master slave
  manipulator (MSM)
Transfer devices
  fluidic pumping, 156–160
  mechanical pumping, 150–156
Transformer, 314–316, 315f
Transistors, 357, 357f
Turbine hall, 13, 14f

U

Ultrafiltration, 179
Ultrasonic measurement, 340–341, 341f
Unbraced cube, 111–112, 111f
Uninterruptible power supply (UPS),
  guaranteed, 324–326, 324–325f
Unmitigated hazards, 430, 431f

V

Variable frequency drive (VFD), 330
Vehicle bay
  buffering, 544–545
  changeroom access, 547–548
  consumables inspection, 546
  crane hook approach, 548–549, 549f
  engine fumes, 546–547
  fire compartmentation, 547
  flask lifting, 549–550, 550f
  floor level, 547
  future-proofing, 550–551
  local office, 546
  means of escape, 546
  platforms, 545–546
  rail, 541–543, 542–543f
  road, 544
  vehicle characteristics, 545
  wind turbulence, 547
Ventilation plant rooms, generic arrangement,
  259–262, 260f
Ventilation stack
  aerial dispersal, 553–554, 554f
  chimney, 551
  C2 stub stack, 558
  flue arrangements, 552, 552f
  future-proofing, 561
  link bridge, 558–559
  prevailing winds, 554–555
  stack location, 555–557, 555–556f
  stack monitoring, 559–560, 560f
  strakes, 557–558, 557f
Ventilation system. See Nuclear ventilation
  system
Vertical shield doors, 205–206
Very-low level (VLL) waste, 399, 478
Vessel ventilation, 267–268
Vibrating forks, 337, 337f
VLLW. See Very low level waste
Volume reduction, 175–176

W

Wall penetrations, concrete structures,
    119–121
Waste management
    commissioning, 404, 404*f*
    construction and installation,
        403–404, 403*f*
    conventional hierarchy, 395–398, 395*f*
    detail design, 403, 403*f*
    embed waste management
        strategy, 403
    evolution of, 400–404
    optioneering, 401–402, 401*f*

option refinement, 402, 402*f*
project inception, 400–401, 400*f*
radiological hierarchy, 398–400, 398*f*
trials, 403
Wind, extreme environmental
    building movement, 129–130
    design basis, 130
    protection from airborne debris, 130–132,
        131–132*f*
    simulation modeling, 133–134, 133*f*
    suction forces, 132–133
Work breakdown structure (WBS),
    372–374, 373*f*